Biocultural Restoration in Hawai'i

Biocultural Restoration in Hawai'i

Editors

Kawika B. Winter
Kevin Chang
Noa Kekuewa Lincoln

MDPI • Basel • Beijing • Wuhan • Barcelona • Belgrade • Manchester • Tokyo • Cluj • Tianjin

Editors
Kawika B. Winter
University of Hawai'i at Mānoa
USA

Kevin Chang
Kua'aina Ulu Auamo (KUA)
USA

Noa Kekuewa Lincoln
University of Hawai'i at Mānoa
USA

Editorial Office
MDPI
St. Alban-Anlage 66
4052 Basel, Switzerland

This is a reprint of articles from the Special Issue published online in the open access journal *Sustainability* (ISSN 2071-1050) (available at: http://www.mdpi.com).

For citation purposes, cite each article independently as indicated on the article page online and as indicated below:

LastName, A.A.; LastName, B.B.; LastName, C.C. Article Title. *Journal Name* **Year**, *Volume Number*, Page Range.

ISBN 978-3-0365-2618-8 (Hbk)
ISBN 978-3-0365-2619-5 (PDF)

Cover image courtesy of Kāko'o 'Ōiwi

Contents

About the Editors

Kawika B. Winter (Assistant Professor of Biocultural Ecology) is a biocultural ecologist who is actively involved in and builds bridges between research, resource management, and policy. He is a strong advocate for Indigenous Peoples and Local Communities in the context of environmental justice and social justice.

Kevin Chang (Executive Director) leads a non-profit organization that empowers Indigenous people and local communities in Hawai'i through a collection of networks focused on reviving Indigenous resource management and engaging in collaborative management.

Noa Kekuewa Lincoln (Professor of Indigenous Cropping Systems) is a leading scholar who conducts research into and is involved with the revival of Indigenous cropping systems in Hawai'i, with an emphasis on their ecological form and functions.

Preface to "Biocultural Restoration in Hawai'i"

The period of Euro-American colonization across the globe, commencing some three centuries prior to the industrial revolution, devastated Indigenous countries, societies, and cultures. This period ushered in an era of population collapses and extinctions across the whole social-ecological spectrum that was not limited to biodiversity, but included a loss of cultures, languages, knowledge and practices, and this trend of co-extinction continues to this day. Of paramount importance is the accompanying loss of the social-ecological systems that Indigenous societies developed and managed, along with the biodiversity within them. Functionally, the process of colonization severed relationships between Indigenous People and their ancestors, their ancestral places, their resources, and the biodiversity that shaped their cultural identity. The historic record and most scholarship indicate that few Indigenous cultures avoided the most devastating impacts of colonization. None were left untouched. In the later part of the 20th century, conversations about 'decolonization' began to emerge, fueled by the ideas of philosophers such as Ngũgĩ wa Thiong'o, poets such as John Trudell, and other dynamic thinkers around the world. These conversations identified the negative impacts of imposing foreign world views and languages on Indigenous Peoples, and they called for shedding those in order to pave the way for a reawakening and revival of Indigenous ontologies, epistemologies, and languages. Such conversations created an opportunity to highlight another path, one that illuminates the many positive outcomes of the place-based approaches of Indigenous cultures.

The focus of this book on 'biocultural restoration,' could be viewed as a restorative stage of the decolonization process. Biocultural restoration endeavors to reconnect the relationships of Indigenous people and their environment with the goal of restoring health and function to both People and Place. The restoration of Indigenous ontologies, epistemologies, and languages is an inherent part of the process. These ways of knowing, often embedded in Indigenous environmental stories and oral histories, perpetuate ancestral memory in knowledge systems that convey virtues, morals ideals, and philosophies. Biocultural restoration, therefore, entails a revival of Indigenous practices at various scales from the individual, to the family, to the community; and when carried out on a landscape scale, such efforts have broader impacts across the social-ecological system.

More than four decades into an Indigenous cultural renaissance, Hawai'i has emerged as a globally recognized model for biocultural restoration. This societal movement is a major reason why Hawai'i was chosen as the host of the World Conservation Congress in 2016. The event brought nearly 10,000 international scholars and policy makers who desired tangible examples of the effectiveness of biocultural restoration. A paucity of publications in Hawai'i and elsewhere served as an impetus for a 2019 Special Issue in Sustainability focusing on this topic. The collection of manuscripts reflect conversations among various grassroots sharing networks. The topics range from philosophical to theoretical to empirical, and collectively reflect the current dynamics of Hawaiian social-ecological systems within the context of temporal ecology. Every contribution to this volume involved Kānaka 'Ōiwi (Indigenous Hawaiians), which is cumulatively the largest collection of scientific publications by Kānaka 'Ōiwi. Moreover, more than half the authors are women, and two of the manuscripts had a 100% women authorship. Perhaps it is not a coincidence that a collection of work regarding caring for and restoring our mother earth is created by Indigenous peoples and women in particular. As Indigenous scholars endeavor to translate ancestral wisdom for a contemporary global audience, science is increasingly becoming one of the more effective ways of doing so. This volume is a tangible example of such efforts.

Kawika B. Winter, Kevin Chang, Noa Kekuewa Lincoln

Editors

 *sustainability*

Editorial

Hawai'i in Focus: Navigating Pathways in Global Biocultural Leadership

Kevin Chang [1], Kawika B. Winter [2,3,4,*] and Noa Kekuewa Lincoln [5]

[1]   Kua'āina Ulu 'Auamo, Kāne'ohe, HI 96744, USA; kevin@kuahawaii.org
[2]   Hawai'i Institute of Marine Biology, University of Hawai'i at Mānoa, Kāne'ohe, HI 96744, USA
[3]   Department of Natural Resources and Environmental Management, University of Hawai'i at Mānoa, Honolulu, HI 96822, USA
[4]   National Tropical Botanical Garden, Kalāheo, HI 96741, USA
[5]   Department of Tropical Plant and Soil Sciences, University of Hawai'i at Mānoa, Honolulu, HI 96822, USA; nlincoln@hawaii.edu
*   Correspondence: kawikaw@hawaii.edu

Received: 2 January 2019; Accepted: 4 January 2019; Published: 8 January 2019

Abstract: As an introduction to the special issue on "Biocultural Restoration in Hawai'i," this manuscript provides background for term 'biocultural restoration,' and contextualizes it within the realms of scholarship and conservation. It explores two key themes related to the topic. First, "Earth as Island, Island as Earth," scales up an island-borne concept of sustainability into a global context. Second, "Hawai'i as a Biocultural Leader," examines the reasons behind the global trend of looking to the most isolated landmass on the planet for solutions to global sustainability issues. We conclude by summarizing the content of the special issue and pointing out the historic nature of its publication. It is the largest collection to date of scientific papers authored by Native Hawaiians and *kama'āina* (Hawai'i-grown) scholars, and more than 50% of both lead and total authorship are women. This Special Issue, therefore, represents a big step forward for under-represented demographics in science. It also solidifies, as embodied in many of the papers in this Special Issue, indigenous methodologies that prioritize working relationships and practical applications by directly involving those on the front lines of biocultural conservation and restoration.

Keywords: Native Hawaiian; social-ecological system; agro-ecology; *'āina momona*

---

1. Introduction to the Special Issue

"It is seeing that establishes our place in the surrounding world; we explain that world with words but words can never undo the fact that we are surrounded by it. The relation between what we see and what we know is never settled . . . . The way we see things is affected by what we know or what we believe . . . .. We never just look at one thing; we are always looking at the relation between things and ourselves."

—John Berger, *Ways of Seeing* [1]

The term 'biocultural' recognizes humanity as part and parcel of the environment. We are not from another planet. In an esoteric manner that is true to some words, 'biocultural' is an etymological and epistemic step towards recognizing symbiotic relationships between societies and environment in the real world. As the late art critic John Berger mentions, it is an expansion of our perspective to how we relate. The term recognizes that even as humanity shapes the environment, the environment shapes us. It also helps us recognize that those who have developed a long-term experience of 'relationship with place' may help root us back to our home and guide us in living on this planet in a more just and sustainable way.

Biocultural diversity is the diversity of life in all its manifestations—biological and cultural. Biocultural approaches to conservation have been defined by Gavin et al. [2] (p. 140) as "conservation actions made in the service of sustaining the biophysical and sociocultural components of dynamic, interacting, and interdependent social-ecological systems". It is a dynamic, integrative approach to understanding the links between nature and culture and the interrelationships between humans and the environment [3].

Biocultural heritage, as discussed in this Special Issue, encompasses indigenous and local community knowledge innovations, and practices that developed within their social–ecological context [4]. Biocultural approaches emphasize co-evolution of people with their biophysical environment [5], and the importance of language in symbolizing and cementing that relationship [6,7]. To speak of the biocultural is to recognize the existence of multiple worldviews as the foundation for different ways of seeing and different ways of knowing [8]. Such diversity can provide society with a greater adaptive capacity to deal with current and future changes [4,9,10].

The related nature of environment and society is captured in a variety of conceptual frameworks which assert that humans—and their behaviors—are integral elements of all environments and ecosystems. For example, Ingold's "dwelling perspective" elaborates on the concept of humans-in-nature, as involving the "skills, sensitivities, and orientations that have developed through long experience of conducting one's life in a particular environment" [11] (p. 25). Berkes and Folke [12] used the term "social-ecological system" to emphasize the integrative concept of humans-in-nature, and to stress that the delineation between social and ecological systems is artificial and arbitrary. Social-ecological systems are integrated complex systems that include social (human) and ecological (biophysical) subsystems in a two-way feedback relationship. The term emphasizes that the two parts (social subsystem and ecological subsystem) are equally important, and they are coupled, interdependent, and co-evolutionary.

The outcome of these frameworks ultimately is that all ecosystems are what Barton et al. [13] calls socioecosystems or contingent landscapes. These terms refer to "the intertwined social and natural landscapes that are the context of human societies and are contingent on the socioecological history as well as the physical conditions under which that history took place" [14]; they are landscapes that emphasize the interrelatedness of the social and the biophysical elements of the environment. The recognition of the dynamics and importance of the human role in ecosystems goes under a variety of names depending on the academic discipline—biocomplexity, new ecology, historical ecology, environmental history, human ecology, and as used in this Special Issue, biocultural relationships. Under any of these names, investigators are essentially concerned with how contingent landscapes interact with societies.

While the dominant academic perspective in the United States during the 20th century viewed humanity as not only separate from, but also inherently destructive to nature, the philosophy of viewing humanity and nature as intrinsically interconnected is not new. In the 5th century BC, the Greek philosopher Herodotus voiced his observation that events shape both people and nature, and that people and nature interact and evolve together through these events. Natural and social scientists re-discovered this unity of people and nature well known to indigenous societies through such concepts as *aski* of the Cree people in northeast Canada (the integrated concept of land, consisting of living landscape, humans, and spiritual beings), *vanua* in Fiji (a named area of land and sea, considered an integrated whole with its human occupants) [8], and *ahupua'a* in Hawai'i as discussed in several papers in this Special Issue.

In recent decades—out of the dialogue around environmental determinism, in which environmental constraints were thought to shape the evolution of human societies; and cultural ecology, which emphasized the influence of humans over their environments—the idea of the unity of people and nature, redeveloped and theorized over several iterations. A merger of these viewpoints led first to a recognition of two-way interactions between humans and the environment, and finally to an acceptance of co-evolution between people and their places. In the 21st century, this notion has revived and is now a part of mainstream conversations embodied in indigenous and local community movements, scholarship, academia, and the professional spheres.

This co-evolutionary perspective addresses the temporal aspects of this two-way influence, in the words of Barton et al. [13], accepting that:

> " ... *humans cannot be viewed either as passive consumers or rapacious exploiters of ecosystems; conversely ecosystems are more than a backdrop for human agency or a larder to fuel human economies. 'Pristine' ecosystems have not existed anywhere for millennia, and humans and cultural systems have played an integral role in the development and maintenance of ecosystems worldwide. Yet humans—even in the context of complex society—are still subject to a wide variety of ecological constraints. This means that human society is constantly reshaping the intertwined cultural and natural components of the socioecological landscape on which its members and their descendants must operate.*"

This perspective is a common thread and a key part of the existential foundation of the people of Hawai'i. Indeed, today our constitution and legal system recognize and have begun to re-invigorate the common law of our land, the indigenous public trust doctrine of *mālama 'āina* (to care for that which feeds). Human–nature systems, indigenous knowledge, and biocultural heritage gain significance in this context.

2. Earth as Island, Island as Earth

As the earth moves into a new age—the Anthropocene—islands, like those in Hawai'i, have much to offer our global society. When the first astronaut to orbit Earth returned home, it was not the wonders of space that struck him the most, but rather the view of the earth as an isolated sphere within that space. Subsequent photos of the earth, such as "Earthrise" and "Blue Marble", arguably had a great impact on environmentalism. They hit home the indisputable fact that our earthly home is indeed a tiny island in the ocean of space, and that that our planet has clear boundaries and limits to its resources.

This is not a new concept for island cultures. Since time immemorial, island communities recognized the constraints of unfettered growth on natural resources. They not only adapted to live within what the environment could reasonably offer, but also discovered innovative and ingenious ways to manage their land and seascapes. Our Hawaiian ancestors sought to stabilize and expand key biomes and ecotones to enhance the provision of ecosystem services from natural systems. This strategy recognizes that the environment has its own *mana*—its own authority to make decisions, its own power to provide outputs, and its own spirit that enhances the world around it. To work with the environment, rather than in imposition to it, is a more efficient means to multiple ends. However, it requires that we adapt ourselves to the land as much as we adapt the land to ourselves. Health of the land and the health of community is inextricably linked.

Our Hawaiian ancestors developed a variety of ways to obtain higher returns from ecosystem services. This Special Issue documents the use of certain agro-ecological concepts that maximize returns. In particular, these systems tend to enhance ecosystems on the margins, the so-called edge effects of ecology that are often highly productive. Examples include:

- Flooded-field agro-ecosystems that demonstrate features of riparian areas and wetlands, and that were extensively developed. The expanded agro-ecological zone retains much of the ecosystem services of the natural areas—flood control, erosion mitigation, habitat for freshwater fauna and birds, groundwater recharge—while at the same time providing for greater cultural necessities such as preferred flora and fauna species for food, medicine, and ceremonies.
- Nearshore aquaculture that utilize walled fishponds to more efficiently provide marine resources and simultaneously maintain ecosystem services of estuaries, such as habitat protection and water filtration. These ponds also enhance nutrient efficiency that ultimately produces marine resources that can be obtained efficiently at a high catch per unit of effort.

- Agroforestry systems that maintain much of the ecosystem services of a natural forest such as nutrient cycling, biodiversity, soil creation, and erosion control, while increasing societal necessities such as resilience and abundance.

The island perspective is borne out of the necessity of limits. Here in Hawai'i, we can see the horizon just out beyond our shoreline. We are the most isolated land mass on the planet, more than 2,000 miles from anywhere else. Island communities are worlds in themselves, bounded by the vast ocean to a finite area of land and resource base. On a remarkably short time scale, many islands in Oceania reached their carrying capacity, forcing their societies to adapt. Survival within a paradigm of perceived limits requires humans to think in an entirely different framework, to make difficult decisions about what is truly needed and desirable. How many people are too many? What standard of living is appropriate? Every island has dealt with its given limits in different ways—some by choice and foresight, others by harsh social restrictions, and some by extreme environmental regulation. As our global island rushes towards its own limits there is much to learn from the Hawaiian experience.

One illustration of an island perspective is in the Hawaiian words for water—*wai;* and wealth, worth, or presiding—*waiwai*. The relationship between these two concepts, within a Hawaiian world view, implies that natural resources—particularly water—was of great importance to prosperity. To protect water resources, prohibitions on private ownership of land and water existed, and resources held in common as a public trust. Large-scale land divisions were generally based on watersheds. Spatial division and prioritization applied to the use of water for drinking, bathing, and irrigation. Agricultural water diversion designs directed flow back into the river to preserve the water flow downstream. As fresh water flowed into the ocean, managed areas inland helped maximize the productivity built on precious nutrients. The allocation of stream resources was very important to each *ahupua'a*, or social-ecological community [15,16]. The practicality of Hawaiians' relationships to nature was, and to a large extent still is, culturally reinforced. Concepts of conservation woven into religion, politics, the economy, and social structures, communicated a kinship with the natural environment. Deities assumed plant and animal forms; every aspect of the world was infused with *mana*—spiritual power to be appropriately respected.

Another illustration of the island perspective is the incorporation of the sacred into conservation practice [17]. In Hawai'i, this includes various kinds of sacred sites such as *wao akua* (sacred forests), *wahi pana* (storied places), and *wahi kapu* (holy places). An example of an area encompassing all three is Mauna Kea, the tallest mountain in the world as measured from its volcanic base on the seafloor. It is one of the holiest places in the Hawaiian archipelago with slopes covered in sacred forest, and its summit home to revered deities, a sacred lake, ancient shrines, the highest burial site in Polynesia, and a foundation for Native Hawaiians' creation story. Biocultural restoration encompasses the restoration of such sacred sites, as they combine both ecological and cultural values. The process of restoration in Hawai'i includes a re-examination of ourselves, our identity, our knowledge systems and our relationship to our place on the path toward re-invigorating a sense of community and righting the canoe.

Despite the onslaught on Hawaiian culture and state of the environment, the sentiment of nature as the provider, and humans as the protector, is still strong. Hawai'i is an island embedded in island earth. The importance of studying, understanding, and unpacking biocultural restoration here is important for conservation in the present, and for the evolving and climatically changing future.

3. Hawai'i as a Biocultural Leader

Hawai'i is an emerging leader and global touchstone in biocultural restoration, knowledge generation, development of both theory and philosophy, and action—partly as a consequence of a revitalization movement that started in the 1970s [6]. Hawai'i is home to many projects to restore the health and function of systems that exist in the confluence of nature and humanity. In this endeavor, multifaceted approaches to facilitate the return to a state of resource abundance—known in the Hawaiian language as *'āina momona*—emerge. Several of the more successful attempts in this movement merged the ancestral and contemporary in the realms of science, technology, and philosophy to inform adaptive practices in multiple fields and initiatives that aim to restore biocultural resource abundance. More broadly, this is a movement to restore Native Hawaiian epistemologies, language, cultural practices, and connection to place. Part of this includes the restoration of cultural landscapes that encompass sacred sites, biocultural resources, and traditional practices—as documented in several papers in this Special Issue.

However, getting to this point has been an uphill battle, and there is a historical and political context for this struggle. The dominant view of nature and ecosystems for the past century and more—imposed on indigenous people in indigenous places—have often been through the eyes of Western European and North American travelers, settlers, nature enthusiasts, observers, activists, or conservation professionals. Towards the last half of the 20th century, 'conservation' had become the modern incarnation of colonization—providing a kinder and gentler, but no less condescending, vision of a foreign worldview imposed on indigenous people and ancestral places. Some contemporary reflections on the history of conservation show that the 'fathers' of the modern U.S. conservation movement might have articulated a very different, and perhaps better, vision of ecosystem management if they had consulted with the native people and advocated for their right to exist in their homelands. Instead, in many cases, they contributed to their displacement.

However, the past two or three decades ushered in a new paradigm. Many conservationists have moved closer to indigenous visions and world views, and some have developed strong collaborative partnerships with indigenous groups. At the forefront of biocultural restoration in Hawai'i—and across the globe—are consultations with, and the participation and perspectives of, the native and local people who inhabit and/or have a deep and long-term relationship with their places. Hawai'i's revitalization movement also owes a great deal to Native Hawaiian scholars who documented indigenous knowledge before it was lost e.g., [18–20]; and to researchers who drew from multiple sources to unravel the cultural complexities of Hawaiian society e.g., [21–23]. Moving forward it will increasingly rely on multi-disciplinary approaches that reveal the landscape complexities of Hawai'i, e.g., [5,24–26]. We have reached a maturation point in Hawai'i where both *kūpuna* (respected elders) and senior scholars have mentored in a new generation of researchers and practitioners who work with both the people and the place. Instead of being expansive, this approach starts in situ with measurable units of place, and incorporates native and local perspectives and relationships.

As an example of the far-reaching depth of these perspectives and efforts, a friend, and young indigenous scientist, Dr. Kiana Frank recently informed us how her study and understanding of her Hawaiian culture and the community *mo'olelo* (stories) helps her to see a nuanced Hawaiian eco-understanding all the way down to the microbial level. She can tie her community and her students directly to this knowledge as a part of their heritage and source of well-being.

On a macro-level, examples of environmental justice and social transformation grow and compound. Grassroots networks caring for Hawai'i's environment have burgeoned and pushed collective interest of a long tradition of community-based natural resource management. Growing and impactful networks include E Alu Pū (a statewide, community-based stewardship network), Hui Mālama Loko I'a (a statewide, fishpond restoration network), Maui Nui Makai Network (focused on community nearshore management efforts in Maui County), Kai Kuleana (a community-based nearshore management network in Kona, Hawa'i), and Hui Loko (a fishpond and anchialine-pond restoration network on the island of Hawai'i). These networks, along with their membership organizations—both individually and collectively—have stoked long-glowing embers of Hawai'i's biocultural heritage, and the fire now burns on every island in the archipelago.

These networks create alliances, and the impacts of their collaboration and coordination culminated in 2015 when their collective support was a critical component in the State's adoption of the first community-based subsistence fishery area (CBSFA) rules in Hā'ena (Kaua'i). This ground-breaking initiative—tantamount to a 'first ever' achievement in a global context—created the governance structure for an approach to fishery co-management that is inclusive and sustainability-oriented [7,27].

This effort was guided by teachings carried in the stories of elders, which helped a community understand the way their ancestors related to the environment, while contextualizing the role of researchers in community-based efforts. Support for this accomplishment was inspired by community-based natural resource management efforts and shared lessons learned that were borne out of efforts in Moʻomomi (Molokaʻi). Hāʻena's CBSFA success was a touchstone event that has opened the door for a statewide movement to better care for nearshore resources in rural Hawaiian communities—such as those like Moʻomomi (Molokaʻi), Kīpahulu (Maui), and Miloliʻi (Hawaiʻi Island), among others—who have been navigating bureaucracy and politics towards this same goal for more than two decades.

Themes of community-based stewardship are being incorporated into the education system at the primary and secondary level inspired, in part, by the efforts described above. The notion that Hawaiʻi needs education programs that produce future elders as much as they produce future professionals has taken root. As such, curriculum development and community efforts are beginning to merge. In charter schools, Hawaiian language education and grassroots environmental stewardship efforts—to care for and nurture the relationship between people and their places—have been embedded in the curriculum. Educational networks are budding forth as well, such as Koʻolau ʻĀina Aloha (a region education and environmental stewardship network in the Koʻolaupoko region of Oʻahu) and others. These networks feed each other and continue to grow in size, strength, and influence as they stand on a firm foundation of biocultural heritage.

The networks described above are composed of people who use their culture, ecological knowledge, and ancestral practice to inform their relationships with, and care for, the environment. This is the approach that will ultimately transform the way Hawaiʻi is cared for in the future. Communities are stepping up to remind us that we have a *kuleana* (right and responsibility) to care for Hawaiʻi. This includes not just a right to benefit from Hawaiʻi's environment (food, recreation, tourism or otherwise), but a duty to both *mālama* (care for) Hawaiʻi, and to ensure the people's concomitant right to do as much to assure abundance and well-being for unborn generations.

4. This Special Issue

Our goal with this Special Issue was to produce a well-rounded collection of papers documenting the state of biocultural restoration in Hawaiʻi from a scholarly perspective. We very much view this as an opportunity to professionally raise up some of Hawaiʻi's thought leaders. The Issue highlights viable models of biocultural conservation in the larger effort to restore *ʻāina momona*, with some focus on the management of forests, streams, nearshore fisheries, traditional crop diversity, and traditional food systems. Although none of the papers directly address health and wellness, and issues related to legal and policy matters, restoring *ʻāina momona* builds a foundation that can facilitate change in these areas as well. We want to emphasize the biocultural foundation of both ecological and cultural restoration. Conserving biocultural diversity and restoring the health of social-ecological systems can, as illustrated herein, be founded on cultural values and aligned with community priorities.

A common theme amongst the efforts examined in this Special Issue is the re-creation of landscape mosaics that included agro-ecological systems designed and managed for cultural and social benefit in such a way that did not irreparably compromise the integrity of native ecosystems. This overarching management approach was at the foundation of indigenous adaptation to island resource scarcity and long-term sustainability. These lessons were not learned lightly—as several island failures can be seen across the Pacific; nor were these systems maintained lightly—many successful islands had strict systems of enforcement put in place to ensure that land use was prioritized in order to provide for the needs of the community above the individual. In our own humble way, we hope this issue helps to translate some of the knowledge accumulated by our island ancestors in a way to not only contribute to the growing momentum in Hawaiʻi, but also to provide viable solutions to global issues for our greater island earth as well.

This Special Issue is the largest collection to date of scientific papers authored by Native Hawaiian and *kamaʻāina* (Hawaiʻi-grown) scholars. Each of the more than one dozen papers is co-authored by at least one Native Hawaiian scholar, with collective contributions from nearly one hundred *kamaʻāina* scholars, which represents an ancestral multi-ethnic mosaic and experience of living in these islands. As expressed in the 'Acknowledgements' sections of each of the papers, this Special Issue embody the perspectives and teachings of several dozen elders, cultural practitioners, and community leaders. It is no surprise these papers have all coalesced around the common theme of biocultural restoration in Hawaiʻi. In the legacy of *Papa-hānau-moku* (Earth mother), more than 50% of lead authors and co-authors are *wāhine* (female), a reflection of this Special Issue which is ultimately focused on how we better nurture and care for our island home.

Also significant is that many of the papers in this Issue employ indigenous methodologies that prioritize working relationships and practical applications by directly involving those on the front lines of biocultural conservation and restoration. Under a different research paradigm, many of these individuals may have served as 'informants' or human 'subjects' for information extraction, rather than as co-authors of the papers. This engaging approach allows those who are 'living' biocultural restoration to tell their own stories, coupled with scientific research that provides both experiential and experimental evidence. This approach to writing and documenting the biocultural restoration efforts in Hawaiʻi parallels efforts on the ground, efforts that critically rely on strong, multifaceted relationships between communities, organizations, scientists, and policy-makers to create successful collaborative partnerships.

You will find in this Issue an exploration of various themes of biocultural restoration in Hawaiʻi. Some touch on philosophical aspects, such as the value system at the foundation of Hawaiian biocultural resource management [28], as well as on theoretical aspects, such as examinations of the structure and function of the Hawaiian social-ecological system [5]. It also includes a comprehensive overview of the systems-based approach to Hawaiian biocultural resource management [15], a multi-faceted approach to rain-fed agro-ecological systems [29], and a case study on monitoring biocultural resources [30]. Historical ecology is utilized in two papers to provide insights into how the Hawaiian archipelago was transformed from an ecosystem into a social-ecological system with the first arrival of Polynesians, and how these social-ecological systems, in turn, underwent a regime shift once Europeans colonized these islands [5,31]. Traditional approaches to biocultural resource management in the 21st century are explored from two angles. One [27] looks at it from a community-based natural resource management perspective, whereas the other [32] looks at it from the perspective of an *aliʻi* (royal) trust organization that is Hawaiʻi's largest private land owner and benefactor of the Native Hawaiian community. An important contribution by Kealiikanakaoleohaililani et al. [17] highlights the spiritual foundations and the role of ritual in biocultural restoration in indigenous places. The issue also includes papers that quantify ecosystem services and cultural services that are the products of biocultural restoration, including flooded field systems [33], agroforestry systems [34], and aquaculture systems [35].

In closing, it is important to ground this Special Issue in its historic and political context. Our effort is just one product of the long-term culmination of collective energies concerning biocultural survival and justice following the overthrow of the Hawaiian Kingdom by business interests and the United States in 1893. It is fueled by the spirit that gave birth to a cultural and civic revival, which ultimately ushered in the Hawaiian renaissance that began in the 1970s. It was then when the call to *mālama ʻāina* began to gather strength. It is the same spirit that subsequently inspired occupations that led to repatriations of history, language, land, of *iwi* (bones), and cultural artifacts among others, as well as the development and growth in influence of semi-sovereign entities like the Office of Hawaiian Affairs, and the maturation of *aliʻi* trusts that were originally developed by Hawaiian royal families.

Each of these progressive threads (language, culture, history, education, law and policy, etc.) became part of an ever-thickening and sturdy rope, which brought our island community into a new era. Our focus on biocultural restoration is just one common thread. On our shores, our home, our community, and our biocultural approach took the center stage at the 2016 World Conservation

Congress in Honolulu where the 'culture-nature/nature-culture' journey in conservation was launched, and then carried forward around the world. The context in which many of the authors in this Special Issue have grown up in the last two generations represent an era of increasing empowerment for a thoughtful, deliberate, and grounded restoration mindset. This mindset not only helps steward our home, but also creates new institutional approaches and practical rewards in the form of jobs and opportunities to allow for the culture and people of Hawai'i to thrive.

Biocultural restoration in the long run will have to be in situ, at the confluence of people and place, and in reality where theory does not always stick and laboratory controls are not available. The process of scientifically documenting the state of biocultural restoration in Hawai'i has provided valuable insight into the past as much as into the present. Restoration is an active term. In Hawai'i it is about reviving the virtues of *aloha 'āina* and the practice of *mālama 'āina*, to love and care, respectively, for the *'āina*, that which feeds. This is the foundation upon which this Special Issue on "biocultural restoration in Hawai'i" is built.

Author Contributions: Conceptualization, K.C., K.B.W. and N.K.L.; Methodology, K.C., K.B.W. and N.K.L.; Validation, K.C., K.B.W. and N.K.L.; Investigation, K.C., K.B.W. and N.K.L.; Resources, K.B.W.; Writing—Original Draft Preparation, K.B.W.; Writing—Review & Editing, K.C., K.B.W. and N.K.L.

Acknowledgments: We acknowledge and appreciate the guidance of our *kūpuna* (esteemed elders) who have laid the path and taught us how to walk it. We express our gratitude to Fikret Berkes, who has inspired, guided, and supported us as scholars, leaders, and guest editors on this Special Issue. We also thank Hawai'i Community Foundation, who, in their continued support of *mālama 'āina* efforts across Hawai'i, has funded the APC for the majority of the articles in this Special Issue.

Conflicts of Interest: The authors declare no competing interests as defined by *Sustainability*, or other interests that might be perceived to influence the results and discussion reported in this paper. The founding sponsors had no role in the design of the study; in the collection, analyses, or interpretation of data; in the writing of the manuscript and in the decision to publish the results.

References

1. Berger, J. *Ways of Seeing*; British Broadcasting Corporation and Penguin Books: London, UK, 1973.
2. Gavin, M.C.; McCarter, J.; Mead, A.T.P.; Berkes, F.; Stepp, J.R.; Peterson, D.; Tang, R. Defining biocultural approaches to conservation. *Trends Ecol. Evol.* **2015**, *30*, 140–145. [CrossRef]
3. Maffi, L.; Woodley, E. *Biocultural Diversity Conservation: A Global Sourcebook*; Routledge: New York, NY, USA; London, UK, 2012.
4. Davidson-Hunt, I.J.; Turner, K.L.; Mead, A.T.P.; Cabrera-Lopez, J.; Bolton, R.; Idrobo, C.J.; Miretski, I.; Morrison, A.; Robson, J.P. Biocultural design: A new conceptual framework for sustainable development in rural indigenous and local communities. *Sapiens* **2012**, *5*, 33–45.
5. Winter, K.B.; Lincoln, N.K.; Berkes, F. The Social-Ecological Keystone Concept: A metaphor for understanding the structure and function of a biocultural system. *Sustainability* **2018**, *10*, 3294. [CrossRef]
6. McGregor, D.P. *Na Kua'Aina. Living Hawaiian Culture*; University of Hawaii Press: Honolulu, HI, USA, 2007.
7. Vaughan, M.B. *Kaiaulu: Gathering Tides*; Oregon State University Press: Corvallis, OR, USA, 2018.
8. Berkes, F. *Sacred Ecology*, 4th ed.; Routledge: New York, NY, USA; London, UK, 2018.
9. Maffi, L. Linguistic, cultural and biological diversity. *Annu. Rev. Anthropol.* **2005**, *29*, 599–617. [CrossRef]
10. Gavin, M.C.; McCarter, J.; Berkes, F.; Mead, A.T.P.; Sterling, E.J.; Tang, R.; Turner, N.J. Effective biodiversity conservation requires dynamic, pluralistic, partnership-based approaches. *Sustainability* **2018**, *10*, 1846. [CrossRef]
11. Ingold, T. *The Perception of the Environment: Essays in Livelihood, Dwelling and Skill*; Routledge: New York, NY, USA; London, UK, 2000.
12. Berkes, F.; Folke, C. (Eds.) *Linking Social and Ecological Systems. Management Practices and Social Mechanisms for Building Resilience*; Cambridge University Press: Cambridge, UK, 1998.
13. Barton, C.M.; Bernabeu, J.; Aura, J.E.; Garcia, O.; Schmich, S.; Molina, L. Long-term socioecology and contingent landscapes. *J. Archaeol. Method Theory* **2004**, *11*, 253–295. [CrossRef]
14. Kirch, P.V. Hawaii as a model for human ecodynamics. *Am. Anthropol.* **2007**, *109*, 8–26. [CrossRef]

15. Winter, K.B.; Lucas, M. Spatial modeling of social-ecological management zones of the *ali'i* era on the island of Kaua'i with implications for large-scale biocultural conservation and forest restoration efforts in Hawai'i. *Pac. Sci.* **2017**, *71*, 457–478. [CrossRef]

16. Winter, K.B.; Beamer, K.; Vaughan, M.; Friedlander, A.M.; Kido, M.H.; Akutagawa, M.K.H.; Kurashima, N.; Lucas, M.P.; Nyberg, B. The Moku System: Managing biocultural resources for abundance within social-ecological regions in Hawai'i. *Sustainability* **2018**, *10*, 3554. [CrossRef]

17. Kealiikanakaoleohaililani, K.; Kurashima, N.; Francisco, K.; Giardina, C.; Louis, R.; McMillen, H.; Asing, C.; Asing, K.; Block, T.; Browning, M.; et al. Ritual + Sustainability Science? A Portal into the Science of Aloha. *Sustainability* **2018**, *10*, 3478. [CrossRef]

18. Kamakau, S.M. *The Works of the People of Old*; Bishop Museum Press: Honolulu, HI, USA, 1976.

19. Malo, D. *Moolelo Hawaii-Hawaiian Antiquities*; Bishop Museum Press: Honolulu, HI, USA, 1951.

20. Pukui, M.K. *'Olelo No'eau: Hawaiian Proverbs & Poetical Sayings*; Bishop Museum Press: Honolulu, HI, USA, 1983.

21. Abbott, I.A. *La'au Hawai'i: Traditional Hawaiian Uses of Plants*; Bishop Museum Press: Honolulu, HI, USA, 1992.

22. Beamer, K. *No Mākou ka Mana: Liberating the Nation*; Kamehameha Publishing: Honolulu, HI, USA, 2014.

23. Kame'eleihiwa, L. *Native Land and Foreign Desires*; Bishop Museum Press: Honolulu, HI, USA, 1992.

24. Vitousek, P.M.; Ladefoged, T.N.; Kirch, P.V.; Hartshorn, A.S.; Graves, M.W.; Hotchkiss, S.C.; Tuljapurkar, S.; Chadwick, O.A. Soils, agriculture, and society in precontact Hawaii. *Science* **2004**, *304*, 1665–1669. [CrossRef] [PubMed]

25. Lincoln, N.K.; Ladefoged, T.N. Agroecology of pre-contact Hawaiian dryland farming: The spatial extent, yield and social impact of Hawaiian breadfruit groves in Kona, Hawai'i. *J. Archaeol. Sci.* **2014**, *49*, 192–202. [CrossRef]

26. Lincoln, N.K.; Vitousek, P.M. Indigenous Polynesian agriculture in Hawai'i. *Environ. Sci.* **2017**. [CrossRef]

27. Delevaux, J.M.; Winter, K.B.; Jupiter, S.D.; Blaich-Vaughan, M.; Stamoulis, K.A.; Bremer, L.L.; Burnett, K.; Garrod, P.; Troller, J.L.; Ticktin, T. Linking Land and Sea through Collaborative Research to Inform Contemporary applications of Traditional Resource Management in Hawai'i. *Sustainability* **2018**, *10*, 3147. [CrossRef]

28. Montgomery, M.; Vaughan, M. Ma Kahana ka 'Ike: Lessons for Community-Based Fisheries Management. *Sustainability* **2018**, *10*, 3799. [CrossRef]

29. Lincoln, N.; Rossen, J.; Vitousek, P.; Kahoonei, J.; Shapiro, D.; Kalawe, K.; Pai, M.; Marshall, K.; Meheula, K. Restoration of 'Āina Malo 'o on Hawai'i Island: Expanding Biocultural Relationships. *Sustainability* **2018**, *10*, 3985. [CrossRef]

30. Morishige, K.; Andrade, P.; Pascua, P.; Steward, K.; Cadiz, E.; Kapono, L.; Chong, U. Nā Kilo 'Āina: Visions of Biocultural Restoration through Indigenous Relationships between People and Place. *Sustainability* **2018**, *10*, 3368. [CrossRef]

31. Gon, S.M., III; Tom, S.L.; Woodside, U. 'Āina Momona, Honua Au Loli—Productive Lands, Changing World: Using the Hawaiian Footprint to Inform Biocultural Restoration and Future Sustainability in Hawai'i. *Sustainability* **2018**, *10*, 3420. [CrossRef]

32. Kurashima, N.; Jeremiah, J.; Whitehead, A.; Tulchin, J.; Browning, M.; Duarte, T. 'Āina Kaumaha: The Maintenance of Ancestral Principles for 21st Century Indigenous Resource Management. *Sustainability* **2018**, *10*, 3975. [CrossRef]

33. Bremer, L.L.; Falinski, K.; Ching, C.; Wada, C.A.; Burnett, K.M.; Kukea-Shultz, K.; Reppun, N.; Chun, G.; Oleson, K.L.; Ticktin, T. Biocultural Restoration of Traditional Agriculture: Cultural, Environmental, and Economic Outcomes of Lo'i Kalo Restoration in He'eia, O'ahu. *Sustainability* **2018**, *10*, 4502. [CrossRef]

34. Langston, B.; Lincoln, N. The Role of Breadfruit in Biocultural Restoration and Sustainability in Hawai'i. *Sustainability* **2018**, *10*, 3965. [CrossRef]

35. Moehlenkamp, P.; Beebe, C.; McManus, M.; Kawelo, A.; Kotubetey, K.; Lopez-Guzman, M.; Nelson, C.; Alegado, R. Kū Hou Kuapā: Cultural restoration improves water budget and water quality dynamics in He'eia Fishpond. *Sustainability* **2019**, *11*, 161. [CrossRef]

 *sustainability*

Article

The *Moku* System: Managing Biocultural Resources for Abundance within Social-Ecological Regions in Hawai'i

Kawika B. Winter [1,2,3,*], **Kamanamaikalani Beamer** [4,5,6], **Mehana Blaich Vaughan** [3,4,6,7], **Alan M. Friedlander** [8,9], **Mike H. Kido** [10], **A. Nāmaka Whitehead** [11], **Malia K.H. Akutagawa** [5,6], **Natalie Kurashima** [11], **Matthew Paul Lucas** [12] and **Ben Nyberg** [2]

[1] Hawai'i Institute of Marine Biology, University of Hawai'i at Mānoa, Kāne'ohe, HI 96744, USA
[2] National Tropical Botanical Garden, Kalāheo, HI 96741, USA; bnyberg@ntbg.org
[3] Natural Resources and Environmental Management, University of Hawai'i at Mānoa, Honolulu, HI 96822, USA; mehana@hawaii.edu
[4] Hawai'inuiākea School of Hawaiian Knowledge—Kamakakūokalani Center for Hawaiian Studies University of Hawai'i at Mānoa, Honolulu, HI 96822, USA; beamer@hawaii.edu
[5] William S. Richardson School of Law—Ka Huli Ao Center for Excellence in Native Hawaiian Law, University of Hawai'i at Mānoa, HI 96822, USA; maliaaku@hawaii.edu
[6] Hui 'Āina Momona Program, University of Hawai'i at Mānoa, Honolulu, HI 96822, USA
[7] University of Hawai'i Sea Grant College Program, University of Hawai'i at Mānoa, Honolulu, HI 96822, USA
[8] Fisheries Ecology Research Lab, University of Hawai'i at Mānoa, Honolulu, HI 96822, USA; friedlan@hawaii.edu
[9] Pristine Seas, National Geographic Society, Washington, DC 20036, USA
[10] Pacific Biosciences Research Center, University of Hawai'i at Mānoa, Honolulu, HI 96822, USA; mkido@hawaii.edu
[11] Natural and Cultural Resources, Kamehameha Schools, Honolulu, HI 96813, USA; nawhiteh@ksbe.edu (A.N.W.); nakurash@ksbe.edu (N.K.)
[12] Department of Geography and Environment, University of Hawai'i at Mānoa, Honolulu, HI 96822, USA; mplucas@hawaii.edu
* Correspondence: kawikaw@hawaii.edu

Received: 31 July 2018; Accepted: 1 October 2018; Published: 4 October 2018

Abstract: Through research, restoration of agro-ecological sites, and a renaissance of cultural awareness in Hawai'i, there has been a growing recognition of the ingenuity of the Hawaiian biocultural resource management system. The contemporary term for this system, "the *ahupua'a* system", does not accurately convey the nuances of system function, and it inhibits an understanding about the complexity of the system's management. We examined six aspects of the Hawaiian biocultural resource management system to understand its framework for systematic management. Based on a more holistic understanding of this system's structure and function, we introduce the term, "the *moku* system", to describe the Hawaiian biocultural resource management system, which divided large islands into social-ecological regions and further into interrelated social-ecological communities. This system had several social-ecological zones running horizontally across each region, which divided individual communities vertically while connecting them to adjacent communities horizontally; and, thus, created a mosaic that contained forested landscapes, cultural landscapes, and seascapes, which synergistically harnessed a diversity of ecosystem services to facilitate an abundance of biocultural resources. "The *moku* system", is a term that is more conducive to large-scale biocultural restoration in the contemporary period, while being inclusive of the smaller-scale divisions that allowed for a highly functional system.

Keywords: Hawaii; biocultural resource management (BRM); ahupuaa; social-ecological community; social-ecological zone

1. Introduction

The small size of many Pacific Islands, coupled with the frequency of catastrophic natural events (i.e., hurricanes, tsunami, drought, flooding, lava flows, etc.) resulted in the development of social-ecological systems around the anticipation of and rapid recovery from environmental change. For this reason, Pacific Islands have been a focus of research into social-ecological system resilience, especially in light of global climate change [1–3]. Understanding traditional approaches to resource management has been a key component of such research. It is apparent that some Pacific Island cultures exceeded resource limits and exhausted their island's carrying capacity early on, while others adapted to resource limitations by adopting conservation measures and, therefore, persisted [4,5]. The Hawaiian archipelago in the era prior to European contact in 1778 (pre-contact era) is a prime example of the latter, making Hawaiian resource management in that era a particular topic of interest with global ramifications.

The "biocultural resource management" (BRM) approaches developed and employed by Hawaiians to manage an archipelago-scale social-ecological system—in the pre-contact era—sustained an abundance of resources for more than a millennium [6]. This state of biocultural resource abundance is known in the Hawaiian language as, "*'āina momona*", and is a term that was particularly attributed to lands that employed aquaculture technologies to increase fish biomass [7]. The word, "*'āina*", is a derivation from the word, "*'ai*",which means, "food, or to eat", with the nominalizer "*na*" added to literally mean, "that which feeds" [8], but is generally used as a noun meaning, "Land, earth" [9]. The word, "*momona*", is an adjective meaning, "Fat; fertile, rich, as soil; fruitful...", [9]. Thus, the term *'āina momona* is commonly translated in the contemporary period as, "fat land", or, "abundant land", in the context of food production. *'Āina momona* was achieved and maintained through careful management on a landscape scale, which extended from the mountains to the sea [6,10].

Through research, restoration of agro-ecological systems, and a renaissance of cultural awareness in Hawai'i, there has been a growing recognition of the ingenuity of Hawaiian biocultural resource management systems. These systems effectively adapted to local conditions, while accumulating a body of knowledge in response to observed effects of management—both successes and failures—in order to sustain resource abundance over time. Researchers [11–16], policy makers, K-12 educators, and others, frequently refer to the Hawaiian system of biocultural resource management as, "the *ahupua'a* system." In this vein, *ahupua'a* are frequently described as self-sustaining units, and put forth as models for sustainability in Hawai'i today [17,18]. *Ahupua'a* have been equated with watersheds, and described as being in alignment with Western scientific management approaches such as "ridge to reef", and ecosystem-based management [19,20]. Our research indicates that while some of the notions aligning Western scientific approaches to resource management with Hawaiian approaches to biocultural resource management may be valid, attributing them to the *ahupua'a* scale does not stand up to scrutiny. For example, some key resources (e.g., adze for felling trees and carving canoes) did not naturally exist within each *ahupua'a*, and the population dynamics of key species managed for the survival of human populations were not confined to *ahupua'a* boundaries. In fact, there are many examples of biocultural resources that were often managed at the scale of larger land divisions. These nuances, discussed in more detail below, refute the notion that *ahupua'a* were self-sustaining. Furthermore, only 5% of *ahupua'a* have boundaries that actually corresponded with watershed boundaries [15], whereas other land-division scales more closely align with this concept (discussion below). There are also land-locked *ahupua'a*, which do not have boundaries that touch the ocean, and coastal *ahupua'a*, which do not have boundaries that extend to the mountains [15]. Therefore, the notion that *ahupua'a* were watershed-based, self-sustaining units is not supported. As such, limiting the contemporary application of Hawaiian biocultural resource management to the *ahupua'a* scale is not conducive to effective, large-scale restoration.

In recognition of knowledge gaps in the understanding of how Hawaiian biocultural resource management strategies functioned and adapted on a system level, this research aims to fill those gaps by synthesizing 21st century research on the topic and coupling that with contemporary understandings about population dynamics of key biocultural resource species. We aim to build a more nuanced understanding of the inner workings of the Hawaiian biocultural resource management system in the pre-contact era, and how it was able to foster long-term biocultural resource abundance. We do this through an examination of six aspects of biocultural resource management. We also aspire to use a more complete understanding to determine a more accurate term to describe this complex system as to be applicable in the contemporary period for large-scale (i.e., system level) biocultural restoration.

2. Methods

The authors of this paper operate in the realms of both biophysical and social science, and have a combined study of various aspects of the social-ecological system in Hawai'i that adds up to well over a century of work. The group includes multi-disciplinary ecologists, botanists, aquatic biologists, and geographers, along with scholars of Hawaiian resource management and governing policy. In this paper we draw upon our collective research that has employed various methods such as archival resource analysis (including maps, governing documents of the Hawaiian Kingdom, Hawaiian language newspapers, etc.), elder interviews, spatial modeling, remote sensing, and biological mapping/monitoring from the mountains to sea. Recent advances in our collective work include several inter-disciplinary projects in the biocultural realm, which have allowed us to synergistically engage with one another's research in the pursuit of better understanding the depth and the breadth of the Hawaiian biocultural resource management system. These collaborations have been key to the development of this article.

3. Results

Our research yielded information that can be grouped into six aspects of biocultural resource management that are relevant to what Winter et al. [21] referred to as the "Hawaiian social-ecological system".

Aspect 1: Nested land divisions provided the framework for systematic management of biocultural resources.

The genesis of landscape-scale biocultural resource management, within the social-ecological system of the Hawaiian archipelago, was born out of necessity when human-population growth began to put a strain on natural resources. Hawaiian historians of the 19th century, such as Kamakau [7] and Malo [22], recounted that at the height of human population in the *ali'i* era, the land was divided into various scales—such as *moku, 'okana, kalana, ahupua'a, 'ili, mo'o, pauku,* and further into various types of agricultural plots (Table 1). Of these land divisions, the *moku* and the *ahupua'a* were key political boundaries in the pre-contact system of governance, managed by positions in the ruling class known as *ali'i 'ai moku* and *ali'i 'ai ahupua'a* respectively. Land divisions below the *ahupua'a* (social-ecological community) level were primarily derived through kinship and cared for by specific extended families [23]. While biocultural resources were managed within the context of those scaled boundaries, there is insufficient understanding of the interplay between the nested land divisions within the biocultural resource management system.

Table 1. Categories of land divisions within an island documented in the 19th century by Kamakau [7] and Malo [22], with contemporary descriptions of the units they represented as interpreted by the authors.

Land Division Term	Unit within the System
moku	A social-ecological region
'okana / kalana	Intermediate category being either a group of *ahupua'a* within a *moku* that collectively compose a larger watershed; or a smaller watershed within a single, large *ahupua'a*
ahupua'a	A social-ecological community

Table 1. *Cont.*

Land Division Term	Unit within the System
'ili	A division within an *ahupua'a*, often associated with an extended family
mo'o	A section of land within an *'ili*
pauku	A strip of land within an *mo'o*
kīhāpai and others	Various types of cultivated plots

The first land division made to manage biocultural resources under the strain of a growing human population was that of *moku* (district or region), and continued population growth later necessitated the subdivision of *moku* into *ahupua'a* (a community-level division) for more localized resource management [6,7,24]. This approach to biocultural resource management was not standardized in a cookie cutter approach, but rather depended on biophysical aspects of the land- and sea-scape [16]. Historical maps and Hawaiian language records detail the proper names and boundaries of some units below the *ahupua'a* level, such as *'ili*. While these place names have been mapped for some individual *ahupua'a* [14], comprehensive mapping of these land divisions for all the islands in the archipelago has yet to be completed.

Aside from the biophysical differences across islands, as well as the regions within them, land divisions varied over time, being shaped by the dynamic and varied needs of each island's human population, as well as the political structure needed to govern people and manage biocultural resources. It is not clear how many times *moku* were re-subdivided into *ahupua'a* in order to manage the needs of a growing human population. The names of some *ahupua'a* seem to indicate that they were at one time larger *ahupua'a* that were later subdivided into two. This is evident by the occasional occurrence of adjacent *ahupua'a* having binomial names that are differentiated only by the epithet, being descriptors of opposing characteristics; whereas all other *ahupua'a* names are monomials. For example, on Kaua'i, Kalihi-wai (Kalihi of fresh water) is adjacent to Kalihi-kai (Kalihi of salt water), and Nu'alolo-kai (Nu'alolo of the sea) is adjacent to Nu'alolo-'āina (Nu'alolo of the land) [16]; and on Hawai'i Island, Pakini-nui (Pakini major) is adjacent to Pakini-iki (Pakini minor) [15]. This may be evidence that *ahupua'a* were subdivided to adjust to the needs of the people. A similar trend is observed in adjacent *moku* of similar aspect—such as Kona 'Ākau and Kona Hema on Hawai'i Island, and Ko'olau Loa and Ko'olau Poko on O'ahu—although it is unknown whether or not these are the result of a historical subdivision process for which records have been lost to time.

All of the Hawaiian terms for land divisions (Table 1), with the exception of two—*'okana and kalana*—were primarily political boundaries associated with governance and systematic biocultural resource management as discussed above [7,22]. Both of these terms are somewhat cryptic, intermediate level social-ecological divisions. Each has a unique definition, but both seem to be applied to the same situation in different places in the archipelago; and, therefore, we suspect that these two terms are synonyms. Synonymy has been documented between the varying classification systems utilized in the pre-contact era [25], including for terms used to classify land designations within the Hawaiian biocultural resource management system [16]. Such synonymy can lead to confusion, which is particularly true for terms that have fallen out of common usage in the contemporary period, and especially for classifications that—by their cryptic nature—do not fit well into tables developed by scholars.

Both *'okana* and *kalana* were units smaller than a *moku* that could have either contained several small *ahupua'a* [9,26], or were distinct areas within large *ahupua'a* [27]. The intermediary nature of this land division has led to confusion about what this unit was, exactly, and how this concept fits into contemporary restoration efforts. It is seemingly more related to biophysical realities and regional identity of the community rather than governance and resource management. "*'Okana*", is a contraction of, "'*oki*," and "'*ana*", meaning, "cutting off", [26] in reference to the partition of a larger land division into smaller units. While its synonym, "*kalana*", can be broken down into, "*kala*", and its nominalizing suffix, "*na*", to literally mean, "that which loosens, frees, releases, removes, unburdens" (translation by authors), in reference, perhaps, to a watershed. These divisions were based upon the biophysical characteristics of the area, rather than the political needs for governance.

There are a few known examples that can inform our contemporary understanding of these terms. The term *kalana* has been applied to the Hanalei region of northern Kaua'i, which includes the *ahupua'a* of Hanalei, Waioli, Waipā, and Waikoko [28]. This appears to reference lands that collectively release *wai* (fresh water) into Hanalei Bay. Other examples are observed on the dry leeward side Hawai'i Island, where the *moku* of Kona is divided into the *kalana* of Kekaha, Kona Kai'ōpua, and Kapalilua [29]. The *kalana* of Kekaha (a contraction of the term, "ke-kahawai-'ole", meaning, "land without streams") in the northern area of Kona is characterized by arid lands with neither streams nor abundant rainfall, but instead has subterranean freshwater flow. Kona Kai'ōpua (Kona of the puffy clouds above the ocean), in the middle section of Kona, is where the *'ōpua* (cumulus) clouds commonly rest in the field of vision in region just off shore. Kapalilua (the double cliff), in the southern region of Kona, is composed of several *ahupua'a* which encompass a region in Kona with a unique topography that is dominated by large sections of sea cliffs.

Uncertainties remain relating to the boundaries of various land divisions, as described above. This arises from several factors: (1) While Hawaiians quickly adopted paper-based mapping, after contact with Europeans, as a crucial means of documenting and asserting knowledge and rule over lands, they did not make such maps in the pre-contact era [30]; (2) several volcanic eruptions have modified or destroyed *ahupua'a* and/or *moku* boundaries; (3) boundaries were well established at the shoreline, but were more ambiguous offshore; (4) the conquest and unification of the islands destroyed sovereign boundaries established by prior dynasties; and (5) current boundaries set by various indigenous and historical authorities are sometimes in conflict [15]. More research into historic land divisions and how their boundaries shifted over time is needed.

Aspect 2: Designation of social-ecological zones (*wao/kai*) allowed for the management of population dynamics for key resource species across social-ecological regions (*moku*).

Terrestrial social-ecological zones (*wao*) within a social-ecological region (*moku*) were designated by a two-word term beginning with "*wao*" and followed by an epithet that described their primary purpose and indicated appropriate activities within each zone [16] (Table 2, Figures 1 and 2). Social-ecological zones in the marine environment (*kai*) have been historically documented within this system [7,29] (Table 3), but these have yet to be comprehensively examined or explored with spatial modeling. Both *wao* and *kai* spanned across the *moku*, which effectively divided each individual social-ecological community (*ahupua'a*) vertically, while connecting it horizontally to adjacent *ahupua'a* within a *moku* (Figure 1). The vertical divisions allowed for system-based management within each *ahupua'a*, while the horizontal connections between *ahupua'a* allowed for coordinated management of the population dynamics of key resource species between *ahupua'a* within each zone spanning a *moku*. This was achieved, in part, by a rotating system of harvest restrictions (described below), which ultimately facilitated management for maximum cumulative abundance and benefit of the entire system—a point that is elaborated below (Aspect 3).

Table 2. The five terrestrial social-ecological zones (*wao*) that appear to have been recognized on the island of Kaua'i. Management implications for each zone are provided (based on Table 3 in Winter and Lucas [16]).

Social-Ecological Zone	Translation	Management Implications
wao akua	Sacred forest	Primary function: Perpetual source population for endemic biodiversity. Designated as "sacred forest", making it a restricted forest zone for a native-only plant community, accessed only under strict protocols. Associated with montane cloud forest, elfin forest.

Table 2. *Cont.*

Social-Ecological Zone	Translation	Management Implications
wao kele	Wet forest	Primary function: Maximize aquifer recharge. An untended forest zone associated with core watershed areas (remote upland, wet forest below the clouds) which was left as a native-dominant plant community. Impractical for access except for transit-through via trails.
wao nāhele	Remote Forest	Primary function: Maximize habitat for native birds. A forest zone that was minimally-tended (generally remote upland, mesic forest) and left as a native-dominant plant community. Impractical for access except by bird catchers and feather gatherers.
wao lāʻau	Agro-forest	Primary function: Maximize the availability of timber and non-timber forest products. A zone allowing for the management of a highly-tended forest via an integrated agroforestry (native and introduced plants) regime: • Native and introduced hardwood timber • Introduced food trees • Native and introduced biofuel sources • Maximization of native biodiversity for non-timber forest products • Cordage and weaving material • Medicine and dyes • Ceremonial and adornment plants
wao kānaka	Habitation zone	Primary function: landscape-scale augmentation to maximize the availability of food, medicine, and housing. A zone allowing for (but not mandating) the conversion of forest to field agriculture, aquaculture, habitation, recreation, and/or temple worship. Native and introduced trees tended, individually or in groves, for regular and specific cultural services.

Table 3. An abridged list of select social-ecological zones (*kai*) within the marine environment as documented by Maly and Maly [29]. Translations of the meaning of these zones are provided by the authors.

Marine Social-Ecological Zone	Translation by Authors
ka poʻina nalu	Fringing reef with breaking waves (representing the seaward boundary of *ahupuaʻa*)
kai lūheʻe	Sea for fishing with octopus lures (outer reefs)
kai koholā	Sea frequented by humpback whales (*Megaptera novaeangliae*) (submerged volcanic shelves)
kai ʻele	Black sea (deep-sea area, possibly between volcanic shelves)
kai uli	Dark sea (deep-sea area, possibly beyond the islands' volcanic foundations)
kai pualena	Sea along the horizon that gets the first touch of the sun's light (deep-sea area)
kai pōpolohua-a-Kāne-i-kahiki	Distant, dark sea associated with the travels of Kāne (deep-sea area beyond sight of land)

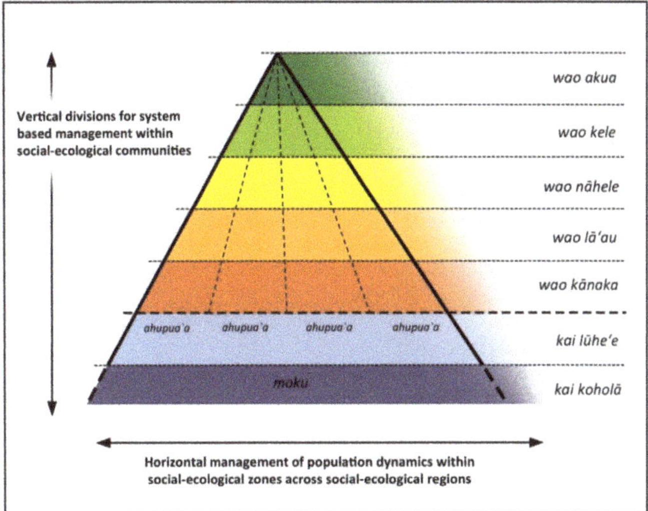

Figure 1. A schematic model depicting the layout of a single social-ecological region (*moku*) including the structure of both social-ecological zones (*wao* and *kai*, designated horizontally) and of social-ecological community boundaries (*ahupuaʻa*, designated vertically) to convey the framework for the biocultural resource management of the *moku* system in the Hawaiian archipelago in the pre-contact period. This framework provided for management in both the horizontal and vertical dimensions. Social-ecological zones are based on those identified from the island of Kauaʻi [16].

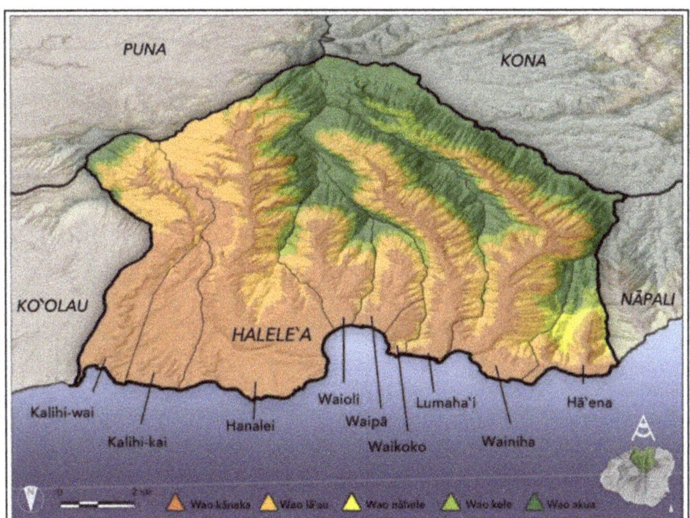

Figure 2. A spatial model depicting the layout for the social-ecological region (*moku*) of Haleleʻa on the island of Kauaʻi, including the social-ecological zones (*wao*) that dictated resource management in each social-ecological community (*ahupuaʻa*), as determined by Winter and Lucas [16]. Each *wao* is represented by a different color as indicated in the key. This *moku* contains nine *ahupuaʻa*, each of which are labeled here by name. Not all *ahupuaʻa* modeled here have all five *wao* documented from the island of Kauaʻi, which indicates that each *ahuapuaʻa* had varying levels of access to and amounts of biocultural resources.

Aspect 3: Population management of key biocultural resources operated on an ecoregion scale.

Moku provide ideal units for examining management systems for key resources [31]. While they are often understood as political boundaries, their alignment facilitated decentralized resource management under *ali'i 'ai moku*, the royal title for those who administered resources in a *moku*. *Moku* boundaries encompass land- and sea-scapes and are aligned with biophysical attributes of island ecosystems—such as landscape aspect, topography, climate regime, wave exposure, watershed classification, forest distribution, substrate type, and aquifer boundaries (Figures 3 and 4). In this regard, *moku* boundaries are more closely aligned with the scientific understanding of an archipelago-scale ecoregion than any other unit of land division recognized in pre-contact Hawai'i. Ecoregions are relatively large units of land containing a distinct assemblage of natural communities and species, with boundaries that approximate the original extent of natural communities prior to major land-use change [32]. While usually referred to on a global scale, we use this term on an archipelago scale. This concept is explored in more detail below.

Owing to Hawai'i's orthographically driven climate patterns across the landscape and shoreline, bio-physical resources—such as sunlight, rainfall, temperature and wave energy [33,34]—ultimately drive natural resource abundance and the potential for cultivating biocultural resources via agro-ecological and aquaculture systems. While there are climatic similarities across *moku*, there are also key differences between *moku*. These differences can be seen with an RGB visualization of equalized temperature ($^{\circ}$C), solar radiation (W/m^2), and rainfall (mm) [35,36] respectively (Figure 3). This can also be visualized in data distributions in histograms of climatic and landscape variables island wide, across *moku*, and within social-ecological zones (Figure 4). The overlay of *moku* boundaries in Figures 3 and 4 reveal clear patterns of climatic similarity within each *moku*. This suggests these divisions optimized land uses and had the potential to contain specialized biocultural resources. In particular, *wao kanaka* zones (including coastal areas) are primarily differentiated between *moku* by solar radiation, rainfall, temperature, and wave energy. This suggests that human interaction with the environment in these areas helped to further distinguish the *moku* from one another and inform appropriate uses. This is evident in the varying forms and intensification of agriculture associated with each *moku* [8], as well as coastal resource development or extraction [29]. This research does not assume that only these physical variables strictly dictated *moku* or *wao* boundaries while disregarding social and cultural drivers; however, an examination of the patterns of both similarities and differences across these spaces does suggest a logical grouping of resource uses as dictated or limited by some bio-physical constraints. *Moku* boundaries also correspond well with the population dynamics of key biocultural resources—such as fish, birds, invertebrates, and plants—that could be more effectively managed in the context of their natural ranges, and in their respective gene pools within ecoregions. Specific examples of key species in these life-form categories are given below.

Fresh-, brackish-, and salt-water vertebrate and invertebrate species were important components of traditional food systems in pre-contact Hawai'i [29]. At the local (*ahupua'a*) and district (*moku*) levels, fishing activities and catch distribution were strictly disciplined by a system of rules and regulations—born out of an understanding about the life cycles of various aquatic species—that were embedded in socio-political structures and religious systems (discussed below). Harvest management was not based on a specific amount of fish, but on identifying the specific times and places that fishing could occur so as not to disrupt basic life-cycle processes and habitats of important food resources [37]. Many of these laws provided protection for important species and allowed Hawaiians to derive sustenance from the ocean for centuries [38]. Knowledge about fish habitat needs, behaviors, and life cycles paved the way for the development of various aquaculture technologies that both increased and stabilized the production of fish biomass [29,39] in the social-ecological system.

Watersheds that contained perennial streams flowing from the mountains to the sea were provided with important vertical dimensions of instream food resources in the form of various species of native fish (*'O'opu*) and macroinvertebrates (*'Ōpae* and *Hīhīwai*) (Table 4). *'O'opu* were the most commonly-referenced fish listed as a traditional food source by native Hawaiians on islands with perennial streams in the middle of the 19th century, which alludes to the importance of these freshwater protein sources in that era [29]. This was particularly true for families living inland from the coast. Hawai'i's native stream species are all amphidromous [40] in that they move out to sea as larvae and return to freshwater as sub-adults to complete their juvenile and adult phases [41,42]. For *'O'opu*, eggs are laid and fertilized in nests, often close to stream mouths. Newly hatched larvae passively drift with stream currents into nearshore areas as marine plankton [43], then metamorphose and recruit into streams as juveniles [44]. The recruiting *'O'opu* are known in Hawaiian as *hinana*, which is the first size class recognized as edible [39]. Adults of each species predictably distribute themselves into high densities along elevational zones in the stream continuum [45], where they may be reliably collected seasonally. Given their amphidromous life histories, sustaining native *'O'opu*, *'Opae* (an ethnogenus comprising *Atyoida* and *Macrobrachium*), and *Hīhīwai* (*Neritina granosa*) larval production from streams within and among watersheds is important to replenish oceanic planktonic populations as cohorts mature to enter streams as juveniles. An ecoregional-scale of resource management, consisting of multiple adjacent streams combined into an ecoregion management unit (*moku*) would, therefore, serve to optimize larval production regionally and be beneficial in sustaining native food resources in streams on all islands.

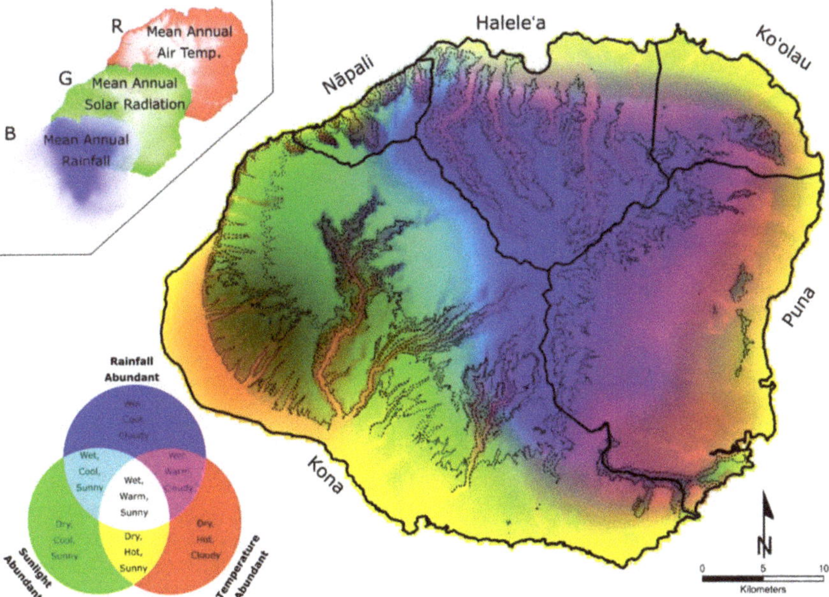

Figure 3. A visual interpretation of climate as delineated by histogram-normalized color combinations of red, green, and blue to simultaneously visualize gradients and combinations of temperature, solar radiation, and rainfall (red: mean annual temperature (°C); green: mean annual solar radiation (W/m^2); blue: mean annual rainfall (mm)). Social-ecological region (*moku*) boundaries (thick black lines), and social-ecological zone (*wao*) boundaries (thin dashed lines) representing the data produced by Winter and Lucas [16] are overlaid atop the island of Kaua'i. All climate data are from Giambelluca [35,36]. Areas with blue dominance represent relative rainfall abundance, areas of green dominance represent relative solar radiation abundance, and areas of red dominance represent relative warmer temperatures. This results in color mixes that demonstrate these climatic variables, with the Venn diagram providing a color key for visual interpretation of the mean annual climatic variability.

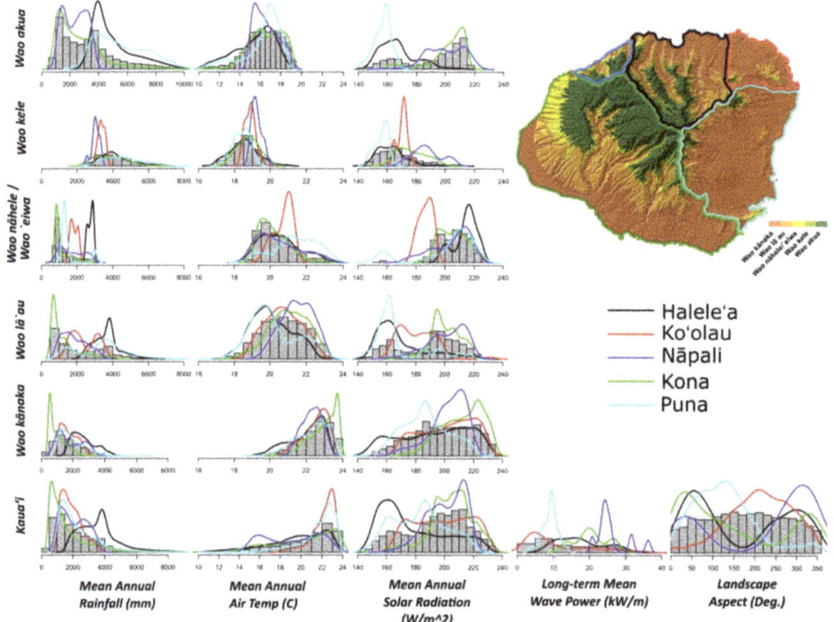

Figure 4. Histograms of climate and landscape variables (columns) for the example island of Kaua'i. From left to right: mean annual rainfall (mm), mean annual temperature (°C), mean annual solar radiation (W/m^2), long-term wave power (Kw/m), and landscape aspect. Rows display island-wide data distribution (bottom) and subsets of socio-ecological zones. Grey histograms represent all data in the zone or island with color coordinated distribution lines display distribution of each according to *moku*. Base-layer image of Kaua'i indicating social-ecological zones is from Winter and Lucas [16]. The boundaries of each of the five *moku* (Halele'a, Ko'olau, Nāpali, Kona, and Puna) for Kaua'i are indicated in separate colors.

Nearshore fish species were also important as a protein source, particularly for people living along the coast, and were managed on an archipelago-based ecoregion scale for abundance [29]. Management tools included the use of temporal and seasonal closures, a practice widely used in traditional Pacific marine tenure systems. Such closures most often applied to reduce intensive harvest of spawning fish or aggregations that occurred during lunar, seasonal, or annual cycles [4,46]. A number of pelagic and migratory species were heavily relied on as food sources, and effective management of their populations was more appropriately addressed at the *moku* level. An example of such management is evident in the ancient fishing regulation of 'Ōpelu (mackerel scad, *Decapterus* spp.)—in the *moku* of Kona Hema, Hawai'i Island, which happened beyond the seaward boundary of the *ahupua'a* in that ecoregion. This regulation mandated that 'Ōpelu be actively fed (*hānai 'ia*) in their natural aggregation areas (*ko'a*) during the restricted (*kapu*) season, which was associated with their spawning period. Each fishing family had a designated *ko'a* to *hānai* during the *kapu* season. If they fulfilled that responsibility they were allowed to fish within any of the *ko'a* during the unrestricted (*noa*) season, after first harvesting from the one they tended. If, however, a family did not fulfill their responsibility to *hānai* their designated *ko'a* in the *kapu* season, they then lost their privilege of fishing for 'Ōpelu in the following *noa* season. This is recalled in the proverb, "*Hānai a 'ai*", [29] that roughly translates to, "Feed [the fish], and [you may] eat", (translation by authors). Regulations that restricted the fishing of key species during their spawning season and calling for the active feeding of them during this period likely increased the fecundity of key resource fish species for the entire *moku*. The six-month *kapu* season for *ʻŌpelu* was the *noa* season for *Aku* (skipjack tuna, *Katsuwonus pelamis*), a predator of juvenile *ʻŌpelu* [29], therefore this restriction/feeding season for *ʻŌpelu* corresponded with a shifted dietary reliance of Hawaiians to top-predator species as a protein source. As such, in addition to limiting pressure on key lower trophic level fish species, harvesting their predators reduced their natural mortality. When the *kapu* was lifted for *ʻŌpelu* fishing, the six-month *kapu* for *Aku* fishing commenced [7,29,39], thus allowing for population recovery of that species. The rotating *kapu/noa, noa/kapu* seasons alternated between these two species on an annual basis. Another important nearshore fish, *ʻAnae holo* (striped mullet, *Mugil cephalus*), was a prized species that migrates along coastal areas and into estuaries within an archipelago-scale ecoregion, and was a focal species in aquaculture systems. Not only were *ʻAnae holo* fished for as they passed through the coastline of an *ahupuaʻa*, they were also attracted into aquaculture systems, which were designed to create or enhance habitat for key resource species in a contained area. This included six classes of fishponds [29,39]. The replenishment of fishponds was dependent on the spawning success of this and other species, which happens on a scale that is more closely aligned with *moku* boundaries than any other scale of land division in ancient Hawaiʻi.

Birds—including forest birds, waterfowl, seabirds, and other migratory species—were another key biocultural resource group as a source of both food for sustenance, and feathers for adornment. As with pelagic and migratory fish, the population dynamics of native birds extended beyond *ahupuaʻa* boundaries. Hawaiian honeycreepers (Fringillidae: Drepanidinae), a highly diverse passerine group relied upon for their feathers, can have home ranges of up to 12 ha [47]. In the context of inland forest at or near the apex of *ahupuaʻa* home ranges of native honeycreepers could most certainly go beyond *ahupuaʻa* boundaries, while staying well within the social-ecological zones (Figure 2) that spanned multiple *ahupuaʻa*—such as the *wao akua* and the *wao nāhele* in the case of forest birds. The *Koloa* (Hawaiian duck, *Anas wyvilliana*), once an important source of food associated with the *wao kānaka* zone [8], has been documented to fly between wetland systems in the same *moku* [48]. Ground-nesting seabird colonies—such as those of the *ʻUaʻu* (Hawaiian petrel, *Pterodroma sandwichensis*), which was another food source when abundant—encompass the upland forest of entire *moku*. An example of this is the colony at Honoonāpali [49]—the region of montane cloud forest encompassing the entire *wao akua* zone in the *moku* of Nāpali on the island of Kauaʻi. Therefore, given that key resource birds have home ranges and population dynamics, which existed in social-ecological zones that spanned across many *ahupuaʻa* yet remained within *moku* boundaries, managing their populations for abundance would have been more effective if done at the *moku* scale.

Species ranges and population dynamics of native plants—as opposed to cultivated crops—were also not limited to *ahupuaʻa* boundaries. Native plants co-evolved with three natural vectors of dispersal—wind, birds (either internally or externally), and ocean currents. Coastal plants tend to be distributed by ocean currents, whereas inland species tend to be distributed by wind or wing [50]. *ʻŌhiʻa lehua* (*Metrosideros polymorpha*), the native tree with the highest biocultural value [51], has wind-born seeds that can be dispersed great distances. As for culturally-important trees with fleshy fruits—such as *Māmaki* (*Pipturus* spp.), *ʻAlaheʻe* (*Psydrax odorata*), and many others—avian dispersers are critically important, and such birds are responsible for the structure and diversity of forests in Hawaiʻi [52]. Therefore, diversity of culturally-important native plants, as well as the structure of forests depended on physical and ecological factors that existed on a scale more closely aligned with those of the *moku* than any other scale of land division in ancient Hawaiʻi.

The abundance of biocultural resources, needed by stewards of the *ahupuaʻa* for their sustenance and well-being, depended on ecological factors, including life cycles of key resource species, that operated on scales larger than that associated with *ahupuaʻa* boundaries. This makes the larger *moku* a more practical unit for management.

Aspect 4: Ensuring high levels of biodiversity resulted in resilient food systems.

Hawaiians in the pre-contact era used taxonomy to attribute names to specific units of biodiversity in their social-ecological system [25], which provided a means to manage the components at the foundation of a diverse range of sociocultural traditions. The management of biocultural diversity has been identified as an important aspect of maintaining—and potentially restoring—the structure, function, and resilience of social-ecological systems [21]. The same concept can be applied to food systems. There is a word in the Hawaiian language for famine—*wī* [9]—which indicates that food was not perpetually abundant in all areas. Periods of famine are noted to have followed natural disasters, such as hurricanes, or climatic shifts which resulted in extended periods of drought [53]. This evidence suggests occasional short-term declines in food abundance, yet points to the importance of biodiversity for resilience of the food system. Some species of plants are referred to as "famine foods" [8,54], and the same is true for some species of marine life [39]. Resource managers had to maintain high levels of biodiversity (Table 4) throughout the social-ecological system as a means to facilitate resilience in the food system. Resource managers had tools to maintain abundance and biodiversity in the food system. These tools included various types of *kapu*, or harvest/access restrictions, to allow for the recovery of populations of key species [29]. When certain species had *kapu* placed upon them, many others in the system could be relied upon as substitutes—as indicated in the alternating *kapu* between 'Ōpelu and *Aku* (discussed above). The high levels of redundancy in wild food sources is indicative of a resilient food system, one that identified food sources that were relied on primarily in periods of scarcity.

Table 4. The amount of native biodiversity functionally relied upon as food sources in the pre-contact era Hawai'i.

Life Form	Edible Species	Source
Freshwater vertebrates	5	Maly and Maly 2003 [27]
Freshwater invertebrates	4	Maly and Maly 2003 [27]
Ocean vertebrates	231	Maly and Maly 2003 [27]
Ocean invertebrates	57	Maly and Maly 2003 [27]
Macro-algae	29	Abbott 1996 [55]
Birds	38	Keauokalani 1859–1860 [56]

Aspect 5: Rotations of harvest restrictions were tools to manage for abundance of biocultural resources.

Maly and Maly [29] comprehensively documented Hawaiian fishing traditions from the pre-contact era, through the Kingdom period, and into the territorial period—based on a compilation of historical records and oral histories. They documented rotating harvest restrictions (*kapu*) that were placed and lifted (making an area *noa* or free from restriction) on either a regular or intermittent basis. The Hawaiian biocultural resource management system employed various kinds of harvest and access restrictions (*kapu*). The punishment for breaking a *kapu* was swift and severe [7,22]. A summary of the types of *kapu* employed in Hawaiian biocultural resource management strategies is described below (Table 5). These various kinds of *kapu* were employed in concert with each other—on both a temporal and spatial scale—to manage for the long-term abundance of key biocultural resources, while at the same time ensuring that local communities could access resources for their daily survival and well-being. The process for deciding which kind of *kapu* to employ and when, with the goal of managing population dynamics within a *moku*, was done by implementing a multi-criteria decision-making process—such as that which is described below (Aspect 6).

Table 5. A list of various types of *kapu* (restriction) along with associated descriptions compiled from Maly and Maly [29] and examples for each.

Kapu Type	Description of *kapu*	Examples
Seasonal harvest restriction associated with spawning periods	Placed an annual ban on the harvest of key fish species during their spawning season, which helped to ensure healthy populations for future fishing seasons.	Annual six-month *kapu* on ʻŌpelu (*Decapterus* spp.)
Monthly harvest restriction associated with particular moon phases	Regulated either specific harvest practices or harvest of particular species on named moon phases, which effectively staggered harvesting pressure throughout the month and protected spawning events occurring on certain moons.	No fishing allowed on the 27th phase of the moon (*Kāne*).
Occasional access restriction, associated with particular areas	Intermittently imposed to restrict human access into areas that needed immediate recovery, or in areas being saved for a planned large harvest in the foreseeable future.	Lāwaʻi (an *ahupuaʻa* in Kona, Kauaʻi) is a place-name commemorating the lifting of a *kapu* over the entire bay fronting that *ahupuaʻa*.
Occasional harvest restriction, associated with a particular taxa	Intermittently imposed to temporarily rest harvest of specific taxa observed to be in decline as a means to facilitate population recovery.	*Kapu* placed on ʻUla (lobster, *Panulirus marginatus*) when population observed to be in decline.
Occasional harvest restriction, associated with a particular life-stage of a specific taxa	Prevented harvest of particular species at key stages in their life cycles, as a means to manage population demographics of that species and enhance reproduction. These restrictions only protected certain life stages while other life stages of that same species could be harvested.	*Kapu* placed on *Moi liʻi* (juvenile threadfin, *Polydactylus sexfilis*) only, while allowing for the harvest of other life stages of the same species.

Aspect 6: Systematic approaches towards holistic evaluation of solutions to biocultural resource problems.

In resource management, solutions born out of a narrow view of a problem have the potential to unintentionally create new problems in other areas of a system. Multi-criteria decision-making processes can be used as a tool to determine the best possible solution to a complex problem [57]. Hawaiians employed such tools in the approach of managing biocultural resources to attain abundance (*ʻāina momona*) in their social-ecological system.

Knowledge of an evaluation process relating to the system-level management of biocultural resources has been documented from the island of Molokaʻi—as developed in the pre-*aliʻi* era prior to the voyage of Pāʻao to Hawaiʻi (approximately 800 years BCE). This evaluation process operated on both the temporal and spatial planes, and in the spiritual realm. It was utilized as a tool by decision-making councils that were composed of recognized experts who were valued for their unique skills and experience—whether that be in agro-ecology, aquaculture, hydrology, meteorology, phenology, etc. The councils operated along certain guiding principles, and themselves guided resource management to ensure the health and integrity of eight resource realms [6,58]. The council's decision-making process entailed consideration of the impact of a proposed solution on each of the eight realms (i.e., the spatial scale, Table 6) as to arrive at solutions that addressed the problems of a specific realm without causing harm to any of the other realms. Once a decision was arrived at, it was implemented by the people in a manner that honored the ancestral past while addressing present needs, and establishing more abundance for future generations (i.e., the temporal scale) [6,58].

Table 6. The eight main components of the systematic evaluation process that was developed on the island of Moloka'i to ensure abundance in all resource realms of the social-ecological system [6,58], with descriptions and contextual interpretations provided by the authors.

Component of Decision Matrix	Component Description and Contextual Interpretation
moana-nui-ākea	The sea from the shoreline to the horizon, as seen from the highest vantage point in the area; and all associated biota.
kahakaipepeiao	The area extending from the place where the ocean meets the land to the place where soil exists. This includes the splash zone where algae, crabs, and other shellfish may be located; sands where turtles nest; dunes where seabirds nest and coastal strand vegetation exists; sea cliffs; and all associated biota.
mauka	The area from where soil begins, extending all the way to the mountaintops; and all associated biota.
nāmuliwai	All the sources of fresh water—artesian springs, streams (including coastal springs that create brackish-water and contribute to healthy and productive estuarine environments); and all associated biota.
kalewalani	The realm inclusive of everything above the land—the air, winds, sky, clouds, rain, rainbows, birds, atmosphere, sun, moon, planets, and stars. This encompasses all the elements and celestial bodies that influence the tides and ocean currents, which directed traditional navigation and guided fishing and planting seasons.
kānakahōnua	The needs of the people. This included the *kānāwai* (laws) that governed behaviors and ensured a functioning society which contributed to the people's health and well-being.
papahelōlona	The intellect and cumulative knowledge built up over generations. This is the knowledge of *kahuna* (keepers of priestly knowledge), knowledge about the connections across the social-ecological system and the correlations between the cycles of nature, and knowledge of expert practitioners in astronomy, healing, and other schools of knowledge.
ke'ihi'ihi	The spiritual realm and the ceremonies needed to maintain *pono* (balance) in the *'āina*. These included elements of nature, ancestral deities, and religious protocols needed to maintain sanctity in the landscape.

The implementation of biocultural resource management tools, such as the coordination of various types of *kapu* (harvest restrictions) across the *moku* (as discussed above), were the kind of issues decided upon by systematic evaluations of both problems and potential solutions. The unilateral placement of *kapu* on the scale of a single *ahupua'a* would not be as effective as collaborative and coordinated efforts between multiple adjacent *ahupua'a*. Various types of rotating *kapu* were employed in concert—between *ahupua'a* within the context of the *moku*—to synergistically yield long-term abundance of key biocultural resources. For example, when a key species was closed in one *ahupua'a*, it might be open in the adjacent *ahupua'a*, with shared harvest rights across both, so that residents could continue to access that resource even while it was rested and rejuvenating in their own home area. The designation of social-ecological zones, which maintained horizontal connections between *ahupua'a* facilitated this management approach, and allowed for the continual replenishment of key species in the archipelago-scale ecoregion without compromising the ability of *ahupua'a* tenants to feed themselves. This was true for key biocultural resources in oceans, estuaries, streams, wetlands, and forested areas. Similar evaluation processes were likely employed in the *ali'i* era—between the arrival of Pā'ao from Tahiti and the arrival of Europeans in 1778—although records of this are not known to exist.

4. Discussion

An analysis of various aspects of managing biocultural resources on a system level has provided some insight into the pathways that pre-contact Hawaiians followed to attain the state of abundance known in the Hawaiian language as *'āina momona*. However, an abandonment of traditional resource management practices in the post-contact era led to a decline in biocultural resources. A good example of this can be seen by the loss of *kapu* (restrictions) as resource management tools.

Kapu were born out of and engrained in the ancient Hawaiian religion in the pre-contact era. These restrictions regulated many aspects of society and human behavior, not just use and management of biocultural resources [7,22,59]. When the ancient Hawaiian religion was abolished in 1819—forty years after Western contact—the *kapu* system was dissolved. With it went a system of regulations for resource extraction, and the authority to enforce violations [29]. Regulations and enforcement were key tools used to manage for long-term abundance of biocultural resources. Loss of the *kapu* system left valuable species unprotected as Hawai'i, an important stop on shipping routes across the Pacific, entered the global trade economy of the 19th century. The massive over-harvesting of *'Iliahi* (Sandalwood, *Santalum* spp.) for export to China contributed to the near extinction of these trees [60]. The example of *'Iliahi* shows how not only key species, but entire ecosystems, were vulnerable to the pressures of capitalism without the *kapu* system in place to protect biocultural resources. After the word *"kapu"* took on a negative connotation in the Christian era—due to its association with the ancient religion—some forms of resource extraction regulations continued under a different term, *"ho'omalu"*, which means, "to rest;" and were codified into law during the Kingdom Era. This was applied locally within *ahupua'a* to particular species or areas, as needed and identified by the designated *konohiki* [29].

The abolishment of the *kapu* system was just one of many changes that undermined the Hawaiian system of biocultural resource management during the 19th century. Depopulation from introduced diseases in the century following European contact was a major contributing factor to the abandonment of agro-ecological systems [61]. Changes in land tenure from the 1840s through the overthrow of the monarchy in 1893 created private ownership in place of communal land holdings [14,30,62,63]. Nearshore fisheries, and local rights to harvest and manage them, were gradually condemned, starting with the Act that annexed Hawai'i as a territory in 1900. This opened fisheries to public access and shifted resource management authority from the *ahupua'a* level to centralized bureaucracies under the territorial and then state governments, and decoupled nearshore resource management from land-based resource management [64,65]. However, in spite of all the change, some *ahupua'a* tenants continued modified forms of biocultural resource management tools into the 20th century, such as the continued practice of designating species and areas for protection (*ho'omalu*). These informal "rests" were designated by respected elders, but were not codified or enforceable except by social pressures [37]. Andrade [14] documents some specific examples of informal community agreements to rest certain areas, or to rotate harvest in the *ahupua'a* of Hā'ena (Halele'a, Kaua'i). Hā'ena is just one of many Hawaiian communities that found novel ways of adapting to continue traditional resource management practices well into the 20th century.

5. Conclusions

Of all the scales of land division in ancient Hawai'i, the *moku* unit is the scale most closely aligned with archipelago-scale ecoregions that encompass population dynamics of key biocultural resources—such as fish, birds, and plants. Biocultural resource management on this scale involved spatial management in both the horizontal and the vertical planes via the designation of social-ecological zones, as well as the concentric scaling of nested land divisions. All of this was done in concert with knowledge about temporal patterns associated with the cycles of lunar months and solar years, which were correlated with life cycles and population dynamics of key resource species. Given the success of this traditional resource management system in ancient Hawai'i, a return to this approach would be an essential component of large-scale biocultural restoration in the 21st century.

We introduce the term "the *moku* system" to describe the Hawaiian biocultural resource management system, practiced in the pre-contact era, which divided large islands into social-ecological regions (*moku*) and further into interrelated social-ecological communities (*ahupua'a*)—each of which contained a network of scaled kinship-derived sections (*'ili, mo'o,* etc.) nested within them. Each *moku* had several social-ecological zones (e.g., *wao* and *kai*) running horizontally as belts across the region. These *wao* divided individual *ahupua'a* vertically while connecting them to adjacent *ahupua'a* horizontally, allowing for holistic management of biocultural resources across human communities. These delineated social-ecological zones created a mosaic that contained forested landscapes, cultural landscapes [66], and seascapes which synergistically harnessed a diversity of ecosystem services to facilitate an abundance of biocultural resources. The richest (*waiwai*) *ahupua'a* cycled enough fresh water (*wai*) through them to allow for aquaculture via various classes of fresh- and/or brackish-water fishponds. Such *ahupua'a* were labeled with the term "*āina momona'* (abundant lands) due to the amount of food and other biocultural resources they were able to sustainably produce over successive generations.

The contemporary trend of framing biocultural conservation efforts around the scale of *ahupua'a* can be effective in some localized instances, such as the creation of Indigenous and Community Conserved Areas (ICCAs). Successful examples of these in the contemporary period include the Hā'ena Community-based Subsistence Fishing Area (CBSFA) on the island of Kaua'i, and the Ka'ūpūlehu Fish Replenishment Area on Hawai'i Island, which employs marine management rules and regulations (e.g., closed areas, closed seasons, size restrictions, restricted entry), within single *ahupua'a*, that have been used for thousands of years by Pacific Islanders [67]. However, limiting discussions of biocultural resource management to the *ahupua'a* scale may not be conducive for the success of large-scale efforts to restore and maintain biocultural resource abundance. While the scale of *ahupua'a* is key, there are multiple additional scales of divisions within *moku* boundaries ('*okana*/*kalana*, *ahupua'a*, '*ili*, *mo'o*, *pauku*) that need to be considered. More research is needed to understand the interplay between these divisions, the organization of human communities in ancient Hawai'i, and to allow for further insight into the historic management of biocultural resources as a means to inform contemporary restoration efforts.

Author Contributions: Conceptualization, K.B.W. and K.B.; Methodology, K.B.W., K.B., M.B.V., and M.P.L.; Validation, A.M.F., M.H.K., A.N.W., N.K. and M.K.H.A.; Formal Analysis, K.B.W and M.P.L.; Investigation, K.B.W., and K.B.; Data Curation, B.N., and M.P.L.; Writing—Original Draft Preparation, K.B.W.; Writing—Review & Editing, K.B.W., K.B., M.B.V., M.H.K., A.M.F., N.K., A.N.W, and M.P.L.; Visualization, M.P.L, and B.N.; Funding Acquisition, K.B.W.

Funding: The APC was funded through the generous support of Hawai'i Community Foundation.

Acknowledgments: We would like to acknowledge those whose teachings and endeavors have laid the foundation for this research. In particular we express our gratitude to Edward Kaanaana and John Ka'imikaua (both now passed), who were lineal keepers of knowledge, wisdom, and practice that descended from ancient times. We also have deep appreciation for Carlos Andrade, a native Hawaiian practitioner, scholar, and philosopher who has provoked our thoughts and challenged our thinking for decades. A special thanks go out to S. Kekuewa Kikiloi, Kā'eo Duarte, Kānekoa Kūkea-Schultz, and others who have shaped our thoughts on the inner workings and management of the Hawaiian social-ecological system. Hawai'iFinally, we thank the communities of the E Alu Pū Network whose continued work in the traditional and customary practice of *mālama 'āina* inspires us all and gives us hope for the future. *Mahalo.*

Conflicts of Interest: The authors declare no conflicts of interest.

References

1.   McMillen, H.; Ticktin, T.; Friedlander, A.; Jupiter, S.; Thaman, R.; Vietayaki, J.; Giambelluca, T.; Campbell, J.; Rupeni, E.; Apis-Overhoff, L.; et al. Small islands, valuable insights: systems of customary resource management and resilience in the Pacific. *Ecol. Soc.* **2014**, *19*, 44. [CrossRef]

2.   Delevaux, J.M.S.; Whittier, R.; Stamoulis, K.A.; Bremer, L.L.; Jupiter, S.; Friedlander, A.M.; Poti, M.; Guannel, G.; Kurashima, N.; Winter, K.B.; et al. A linked land-sea model framework to inform ridge-to-reef management in high oceanic islands. *PLoS ONE* **2018**, *13*, e0193230. [CrossRef] [PubMed]

3.   Ticktin, T.; Dacks, R.; Quazi, S.; Tora, M.; McGuigan, A.; Hastings, Z.; Naikatini, A. Significant linkages between measures of biodiversity and community resilience in Pacific Island agroforests. *Conserv. Biol.* **2018**, in press. [CrossRef] [PubMed]

4.   Johannes, R.E. The renaissance of community-based marine resource management in Oceania. *Annu. Rev. Ecol. Syst.* **2002**, *33*, 317–340. [CrossRef]

5.   Tainter, J.A. Archaeology of overshoot and collapse. *Annu. Rev. Anthropol.* **2006**, *35*, 59–74. [CrossRef]

6.   Minton, N.; Ka'imikaua, J.K. *A Mau A Mau: To Continue Forever*; Oshita, R., Minton, N., Eds.; Nā Maka o ka 'Āina (Vid): Nā'ālehu, HI, USA, 2000.

7.   Kamakau, S.M. *Ka Hana a Ka Poe Kahiko*; Bishop Museum Press: Honolulu, HI, USA, 1976.

8. Handy, E.S.C.; Handy, E.G.; Pukui, M.K. *Native Planters in Old Hawaii: Their Life, Lore, and Environment*; Bishop Museum Press: Honolulu, HI, USA, 1972.
9. Elbert, S.; Pukui, M.K. *Hawaiian Dictionary*; University of Hawaii Press: Honolulu, HI, USA, 1986.
10. Kelly, M. *Ahupua'a Fishponds and Lo'i: A Film for Our Time*; Nā Maka o ka 'Āina: Nā'ālehu, HI, USA, 1992.
11. Kelly, M. Changes in Land Tenure in Hawaii, 1778–1850. Master's Thesis, University of Hawaii, Honolulu, HI, USA, 1956.
12. Minerbi, L. Indigenous management models and protection of the ahupua'a. *Soc. Process Hawaii* **1999**, *39*, 208–225.
13. Mueller-Dombois, D. The Hawaiian Ahupua'a Land Use System: Its Biological Resource Zones and the Challenge for Silvicultural Restoration. *Bishop Mus. Bull. Cult. Environ. Stud.* **2007**, *3*, 23–33.
14. Andrade, C. *Hā'ena: Through the Eyes of Ancestors*; University of Hawaii Press: Honolulu, HI, USA, 2008.
15. Gonschor, L.; Beamer, K. Towards an inventory of ahupua'a in the Hawaiian Kingdom: A survey of nineteenth- and early twentieth-century cartographic and archival records of the island of Hawai'i. *Hawaii. J. Hist.* **2014**, *48*, 53–67.
16. Winter, K.B.; Lucas, M. Spatial modeling of social-ecological management zones of the *ali'i* era on the island of Kaua'i with implications for large-scale biocultural conservation and forest restoration efforts in Hawai'i. *Pac. Sci.* **2017**, *71*, 457–477. [CrossRef]
17. Kaneshiro, K.Y.; Chinn, P.; Duin, K.N.; Hood, A.P.; Maly, K.; Wilcox, B.A. Hawaii's mountain-to-sea ecosystems: Social–ecological microcosms for sustainability science and practice. *EcoHealth* **2005**, *2*, 349–360. [CrossRef]
18. Jokiel, P.L.; Rodgers, K.S.; Walsh, W.J.; Polhemus, D.A.; Wilhelm, T.A. Marine resource management in the Hawaiian archipelago: the traditional Hawaiian system in relation to the western approach. *J. Mar. Biol.* **2011**, *2011*, 151682. [CrossRef]
19. Bridge, T.C.; Hughes, T.P.; Guinotte, J.M.; Bongaerts, P. Call to protect all coral reefs. *Nat. Clim. Chang.* **2013**, *3*, 528. [CrossRef]
20. Oleson, K.; Falinski, K.; Audas, D.M.; Coccia-Schillo, S.; Groves, P.; Teneva, L.; Pittman, S. Chapter 11: Linking Landscape and Seascape Conditions: Science, Tools and Management. In *Seascape Ecology*; Wiley: Hoboken, NJ, USA, 2017; pp. 319–364.
21. Winter, K.B.; Lincoln, N.K.; Berkes, F. The Social-Ecological Keystone Concept: A quantifiable metaphor for understanding the structure, function, and resilience of a biocultural system. *Sustainability* **2018**, *10*, 3294. [CrossRef]
22. Malo, D. *Ka Mo'olelo Hawai'i: Hawaiian Traditions*; Translation by Malcolm Chun; First People's Productions: Honolulu, HI, USA, 2006; p. 274.
23. Handy, E.C.S.; Pukui, M.K. *The Polynesian Family System in Ka'u, Hawai'i*; Bishop Museum Press: Honolulu, HI, USA, 1958.
24. Beamer, K. Huli Ka Palena. Master's Thesis, University of Hawaii, Honolulu, HI, USA, 2005.
25. Winter, K.B. Kalo [Hawaiian Taro, *Colocasia esculenta* (L.) Schott] Varieties: An assessment of nomenclatural synonymy and biodiversity. *Ethnobot. Res. Appl.* **2012**, *10*, 423–447.
26. Andrews, L. *A Dictionary of the Hawaiian Language: To Which is Appended an English-Hawaiian Vocabulary and a Chronological Table of Remarkable Events*; HM Whitney: Honolulu, HI, Hawai'i, 1865.
27. Maly, K.; Maly, O. *He Wahi Mo'olelo No Na Lawai'a Ma Kapalilua, Kona Hema, Hawai'i: A Collection of Historical Interviews with Elder Kama'āina Fisher-People from the Kapalilua Region of South Kona, Island of Hawai'i*; A Kumu Pono Associates report for The Nature Conservancy of Hawai'i; The Nature Conservancy of Hawai'i: Honolulu, HI, USA, 2003.
28. Kimura, L.K.; Mahuiki, R.N. *Ka Leo Hawai'i*; A Hawaiian language program on KCCN 1420AM, archived at University of Hawai'i at Mānoa under HV24.14; University of Hawai'i at Mānoa: Honolulu, HI, USA, 1972.
29. Maly, K.; Maly, O. *Ka Hana Lawai'a a me nā Ko'a o nā Kai 'Ewalu: Summary of Detailed Findings from Research on the History of Fishing Practices and Marine Fisheries on the Hawaiian Islands*; A Kumu Pono Associates report for The Nature Conservancy of Hawai'i; The Nature Conservancy of Hawaii: Honolulu, HI, USA, 2003. Available online: http://www.kumupono.com/Ocean%20Resources/HiPae74_Vol-I_b_reduced.pdf (accessed on February 2, 2018).
30. Beamer, K. *No Mākou ka Mana: Liberating the Nation*; Kamehameha Publishing: Honolulu, HI, USA, 2014.

31. Friedlander, A.M.; Donovan, M.K.; Stamoulis, K.A.; Williams, I.; Brown, E.; Conklin, E.J.; DeMartini, E.E.; Rodgers, K.S.; Sparks, R.T.; Walsh, W.J. Human-induced gradients of reef fish declines in the Hawaiian Archipelago viewed through the lens of traditional management boundaries. *Aquat. Conserv. Mar. Freshw. Ecosyst.* **2018**, *28*, 146–157. [CrossRef]

32. Olson, D.M.; Dinerstein, E.; Wikramanayake, E.D.; Burgess, N.D.; Powell, G.V.N.; Underwood, E.C.; D'amico, J.A.; Itoua, I.; Strand, H.E.; Morrison, J.C.; et al. Terrestrial Ecoregions of the World: A New Map of Life on Earth: A new global map of terrestrial ecoregions provides an innovative tool for conserving biodiversity. *BioScience* **2001**, *51*, 933–938. [CrossRef]

33. Wedding, L.M.; Lecky, J.; Gove, J.M.; Walecka, H.R.; Donovan, M.K.; Williams, G.J.; Jouffray, J.B.; Crowder, L.B.; Erickson, A.; Falinski, K.; et al. Advancing the integration of spatial data to map human and natural drivers on coral reefs. *PLoS ONE* **2018**, *13*, e0189792. [CrossRef] [PubMed]

34. Li, N.; Cheung, K.F.; Stopa, J.E.; Hsiao, F.; Chen, Y.-L.; Vega, L.; Cross, P. Thirty-four years of Hawai'i wave hindcast from downscaling of climate forecast system reanalysis. *Ocean Model.* **2016**, *100*, 78–95. [CrossRef]

35. Giambelluca, T.W.; Chen, Q.; Frazier, A.G.; Price, J.P.; Chen, Y.-L.; Chu, P.-S.; Eischeid, J.K.; Delparte, D.M. Online Rainfall Atlas of Hawai'i. *Bull. Am. Meteorol. Soc.* **2013**, *94*, 313–316. [CrossRef]

36. Giambelluca, T.W.; Shuai, X.; Barnes, M.L.; Alliss, R.J.; Longman, R.J.; Miura, T.; Chen, Q.; Frazier, A.G.; Mudd, R.G.; Cuo, L.; et al. *Evapotranspiration of Hawai'i*; Final report; U.S. Army Corps of Engineers—Honolulu District, and the Commission on Water Resource Management, State of Hawai'i: Honolulu, HI, USA 2014.

37. Poepoe, K.; Bartram, P.; Friedlander, A. The use of traditional Hawaiian knowledge in the contemporary management of marine resources. In *Fishers' Knowledge in Fisheries Science and Management*; Haggan, N., Neis, B., Baird, I., Eds.; UNESCO: Paris, France, 2007; pp. 117–141.

38. McClenachan, L.; Kittinger, J.N. Multicentury trends and the sustainability of coral reef fisheries in Hawai'i and Florida. *Fish Fish.* **2013**, *14*, 239–255. [CrossRef]

39. Titcomb, M.; Pukui, M.K. *Native Use of Fish in Hawaii*, 2nd ed.; University of Hawaii Press: Honolulu, HI, USA, 1972.

40. McDowall, R.M. Hawaiian stream fishes: The role of amphidromy in history, ecology and conservation biology. *Bishop Mus. Bull. Cult. Environ. Stud.* **2007**, *3*, 3–9.

41. Kinzie, R.A. Habitat utilization by Hawaiian stream fishes with reference to community structure in oceanic island streams. *Environ. Biol. Fishes* **1988**, *22*, 179–192. [CrossRef]

42. Fitzsimons, J.M.; Nishimoto, R.T.; Devick, W.S. Maintaning biodiversity in freshwater ecosystems on oceanic islands of the tropical Pacific. *Chin. Biodivers.* **1996**, *4*, 23–27.

43. Kido, M.H.; Heacock, D.E. The spawning ecology of 'o'opu nakea (*Awaous stamineus*) in Wainiha River and other selected north shore Kauai rivers. In *New Directions in Research, Management and Conservation of Hawaiian Freshwater Stream Ecosystems: Proceedings of the 1990 Sympo*; Technical Report 96-01; Department of Land and Natural Resources: Honolulu, HI, USA, 1991; pp. 142–157.

44. Radtke, R.L.; Kinzie, R.A., III; Folsom, S.D. Age at recruitment of Hawaiian freshwater gobies. *Environ. Biol. Fishes* **1988**, *23*, 205–213. [CrossRef]

45. Kido, M.H. A persistent species assemblage structure along a Hawaiian stream from catchment-to-sea. *Environ. Biol. Fishes* **2008**, *82*, 223–225. [CrossRef]

46. Johannes, R.E. Traditional marine conservation methods in Oceania and their demise. *Annu. Rev. Ecol. Syst.* **1978**, *9*, 349–364. [CrossRef]

47. VanderWerf, E.A. Breeding biology and territoriality of the Hawaii Creeper. *Condor* **1998**, *100*, 541–545. [CrossRef]

48. Engilis, A., Jr.; Pratt, T.K. Status and population trends of Hawaii's native waterbirds, 1977–1987. *Wilson Bull.* **1993**, *105*, 142–158.

49. Troy, J.R.; Holmes, N.D.; Veech, J.A.; Raine, A.F.; Green, M.C. Habitat suitability modeling for the endangered Hawaiian petrel on Kauai and analysis of predicted habitat overlap with the Newell's shearwater. *Glob. Ecol. Conserv.* **2017**, *12*, 131–143. [CrossRef]

50. Wagner, W.L.; Herbst, D.R.; Sohmer, S.H. *Manual of the Flowering Plants of Hawai'i, Vols. 1 and 2 (No. Edn 2)*; University of Hawai'i and Bishop Museum Press: Honolulu, HI, USA, 1999.

51. Burnett, K.; Ticktin, T.; Bremer, L.; Quazi, S.; Geslani, C.; Wada, C.; Kurashima, N.; Mandle, L.; Pascua, P.; Depraetere, T.; et al. Restoring to the Future: Environmental, Cultural, and Management Tradeoffs in Historical versus Hybrid Restoration of a Highly Modified Ecosystem. *Conserv. Lett.* **2018**, *2018*, e12606. [CrossRef]

52. Chimera, C.G.; Drake, D.R. Patterns of seed dispersal and dispersal failure in a Hawaiian dry forest having only introduced birds. *Biotropica* **2010**, *42*, 493–502. [CrossRef]

53. Businger, S.; Nogelmeier, M.P.; Chinn, P.W.; Schroeder, T. Hurricane with a history: Hawaiian newspapers illuminate an 1871 storm. *Bull. Am. Meteorol. Soc.* **2018**, *99*, 137–147. [CrossRef]

54. Abbott, I.A. *La'au Hawai'i: Traditional Hawaiian Uses of Plants*; Bishop Museum Press: Honolulu, HI, USA, 1992.

55. Abbot, I.A. *Limu: An Ethnobotanical Study of Some Hawaiian Seaweeds*; Pacific Tropical Botanical Garden: Lawai, HI, USA, 1996.

56. Keauokalani, Z. *Birds, by Kepelino*; Hawaiian Ethnographic Notes; Bishop Museum Archive: Honolulu, HI, USA, 1859–1860; Volume 1, pp. 1127–1155.

57. Kiker, G.A.; Bridges, T.S.; Varghese, A.; Seager, T.P.; Linkov, I. Application of multicriteria decision analysis in environmental decision making. *Integr. Environ. Assess Manag.* **2005**, *1*, 95–108. [CrossRef] [PubMed]

58. Akutagawa, M.K.H. The 'Aha Moku Rules of Practice and Procedure: Weaving 'Ōiwi Governance and Expertise in Mālama 'Āina. *Hūlili* **2019**. accepted.

59. Kamakau, S.M. *Ka Po'e Kahiko*; Bishop Museum Press: Honolulu, HI, USA, 1991.

60. Morgan, T. *Hawaii: A Century of Change (1778–1876)*; Harvard University Press: Cambridge, UK, 1948.

61. Kurashima, N.; Jeremiah, J.; Ticktin, A.T. I Ka Wā Ma Mua: The Value of a Historical Ecology Approach to Ecological Restoration in Hawai'i. *Pac. Sci.* **2017**, *71*, 437–456. [CrossRef]

62. Kame'eleihiwa, L. *Native Land and Foreign Desires*; Bishop Museum Press: Honolulu, HI, USA, 1992.

63. Beamer, K.; Tong, W. The Mahele Did What? Native Interest Remains. In *Hulili: Multidisciplinary Research on Hawaiian Well-Being*; Kamehameha Publishing: Honolulu, HI, USA, 2016; Volume 10.

64. Kosaki, R.H. *Konohiki Fishing Rights*; Report No. 1, June 1954 (Request No. 3642); Legislative Reference Bureau, University of Hawai'i: Honolulu, HI, USA, 1954.

65. Vaughan, M.B.; Ayers, A.L. Customary Access: Sustaining Local Control of Fishing and Food on Kaua'i's North Shore. *Food Cult. Soc.* **2016**, *19*, 517–538. [CrossRef]

66. Molnar, Z.; Berkes, F. Role of traditional ecological knowledge in linking cultural and natural capital in cultural landscapes. In *Reconnecting Natural and Cultural Capital: Contributions from Science and Policy*; Paracchini, M.L., Zingari, P.C., Blasi, C., Eds.; European Union: Luxembourg, 2018; pp. 183–193.

67. Delevaux, J.; Winter, K.; Jupiter, S.; Blaich-Vaughan, M.; Stamoulis, K.; Bremer, L.; Burnett, K.; Garrod, P.; Troller, J.; Ticktin, T. Linking Land and Sea through Collaborative Research to Inform Contemporary applications of Traditional Resource Management in Hawai'i. *Sustainability* **2018**, *10*, 3147. [CrossRef]

 *sustainability*

Article

'Āina Momona, Honua Au Loli—Productive Lands, Changing World: Using the Hawaiian Footprint to Inform Biocultural Restoration and Future Sustainability in Hawai'i

Samuel M. Gon III *, Stephanie L. Tom and Ulalia Woodside

The Nature Conservancy of Hawai'i, Honolulu, HI 96817, USA; stom@tnc.org (S.L.T.);
ulalia.woodside@tnc.org (U.W.)
* Correspondence: sgon@tnc.org or hawaii@tnc.org

Received: 1 August 2018; Accepted: 21 September 2018; Published: 25 September 2018

Abstract: Pre-Western-contact Hawai'i stands as a quintessential example of a large human population that practiced intensive agriculture, yet minimally affected native habitats that comprised the foundation of its vitality. An explicit geospatial footprint of human-transformed areas across the pre-contact Hawaiian archipelago comprised less than 15% of total land area, yet provided 100% of human needs, supporting a thriving Polynesian society. A post-contact history of disruption of traditional land use and its supplanting by Western land tenure and agriculture culminated in a landscape less than 250 years later in which over 50% of native habitats have been lost, while self-sufficiency has plummeted to 15% or less. Recapturing the *'āina momona* (productive lands) of ancient times through biocultural restoration can be accomplished through study of pre-contact agriculture, assessment of biological and ecological changes on Hawaiian social-ecological systems, and conscious planned efforts to increase self-sufficiency and reduce importation. Impediments include the current tourism-based economy, competition from habitat-modifying introduced species, a suite of agricultural pests severely limiting traditional agriculture, and climate changes rendering some pre-contact agricultural centers suboptimal. Modified methods will be required to counteract these limitations, enhance biosecurity, and diversify agriculture, without further degrading native habitats, and recapture a reciprocal Hawaiian human-nature relationship.

Keywords: human land use footprint; traditional ecological knowledge; biocultural restoration; social-ecological system; Hawaiian Islands; biocapacity; sustainability

---

1. Introduction

> *E Kāne-au-loli-ka-honua*
>
> *Honu ne'e pū ka 'āina*
>
> O Kāne-who-transforms-the-world
>
> Like a sea-turtle crawling, so the land (changes)

The opening lines out of a traditional *pule* (prayer) for cultivation evokes a Hawaiian god who transforms the world, an acknowledgement of the dynamic nature of ecosystems. The second line is evocative of the nature of changes; occurring slowly over the course of generations, but, as a sea turtle's surges of movement upward from the shore towards her nesting site, sometimes more abrupt, noticeable, dramatic. The wisdom incorporated within oral traditions in Hawai'i (and elsewhere in the world) may be, at first blush, obscure and incomprehensible, but ultimately a huge wealth of information pertinent to today's challenges can be found within them. This paper describes how an effort to combine biological monitoring, archeological databases, and oral traditions created the first geospatially explicit rendering of the human land use footprint in the pre-contact Hawaiian archipelago.

While this geospatial footprint allowed for a variety of very useful extrapolations, including better estimates of the pre-contact human population in Hawai'i, not only for the entire archipelago, but per island, it also offered a milestone in the story of landscape changes in Hawai'i from those times to present, and can inform future strategies for biocultural restoration and sustainability.

Hawaiian biological diversity has seen losses and changes as a result of the presence of people and their biological introductions. So too has Hawaiian culture seen losses, in language, knowledge, and sovereignty; yet traditional knowledge provides some of our best sources directly describing the pre-contact world. Our efforts to understand the magnitude of changes to natural systems in Hawai'i led us, at about the turn of the millennium, to model the patterns of major ecosystems in Hawai'i, so that we have a fair idea of the pre-human ecological settings to contrast with the sometimes startling and staggering losses of our natural heritage in today's world.

1.1. The Rich Ecological Setting in the Hawaiian Islands

A variety of sources have documented the biotic richness of the Hawaiian Archipelago, recognizing it as a unique Biogeographic Ecoregion whose isolation has generated extremely high levels of endemism in both terrestrial and marine realms (e.g., ~90% endemism of native flowering plants; >98% endemism of native terrestrial invertebrates; 25% endemism of native reef fishes) [1,2]. An estimated 15,000 species are found nowhere else [3]. When a Holdridge Lifezone analysis [4] was conducted for the Hawaiian Islands by the U.S. Forest Service [5] it revealed that of the 38 lifezones defined in a system designed to cover the full range of terrestrial ecosystems on Earth, 27 could be found in the 17,400 sq km land area of the Hawaiian archipelago, making the archipelago the single most ecosystem-rich known on the planet [6]. This explains the many natural communities endemic to Hawai'i that comprise its broad native habitat zones.

Biocapacity, defined as the ability of an area of land or sea to provide for natural resources [7], is acknowledged as varying site by site according to a number of factors, including ecological richness. The extremely high diversity of biophysical conditions in Hawai'i suggests strongly that its biocapacity, although never formally determined numerically, is higher than the global average. This has probably facilitated both the prominent adaptive radiations of endemic Hawaiian species into a broad range of ecological niches, as well as the remarkably large pre-contact Hawaiian population supported by the archipelago. As a social-ecological territory, it was as close as possible to being an independent unit—relying on no external trade for survival.

1.2. The Current Loss of Major Terrestrial Native Habitats in Hawai'i

Recent mappings of the remaining native-dominated vegetation in Hawai'i have been conducted (e.g., Figure 1), and largely agree on the areal extent of remaining native-dominated habitats [8–11]. They point to major losses of certain broad categories of natural communities, such as the Lowland Dry Communities, which have been almost entirely lost on smaller islands, and have been reduced to 31% of their original extent on the largest island of Hawai'i. In contrast, certain zones, in large part much less suitable for human occupation or uses, have retained much larger percentages of their original cover, as seen in Table 1. Geospatial documentation of the remaining native-dominated areas have guided conservation efforts of both public (Federal and State) as well as private agencies and organizations, focusing efforts on the maintenance of intact areas, augmented by restoration of damaged or destroyed ecosystems [12].

It is apparent that the elevation and moisture zones most compatible with human residence and uses, such as agriculture, have resulted in a bias toward loss of lowland native ecosystems in Hawai'i. With few exceptions, areas below 600 meters elevation have been almost entirely displaced by a growing human footprint of land use, and by that of the non-native plant and animal introductions that have naturalized and spread, further displacing native species habitats [13].

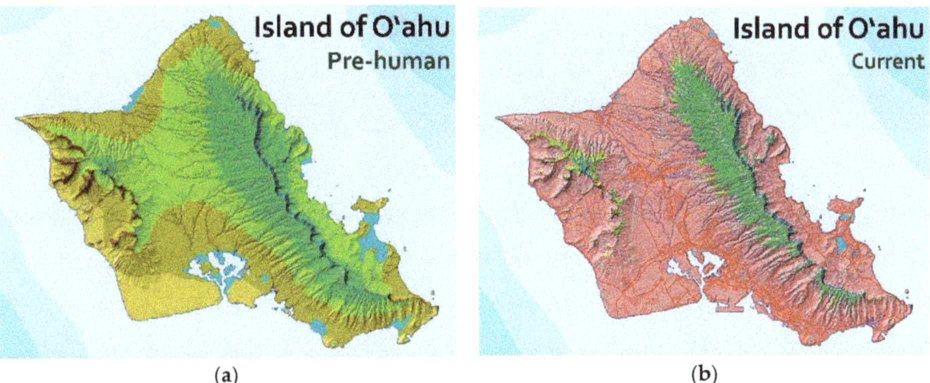

Figure 1. (a) Native habitats on Oʻahu before humans. (b) Current extent of native habitats (via [10]). Pink = human footprint. Over 80% of native habitat has been lost.

Table 1. Example remaining native habitat zones on Hawaiʻi Island, the largest island in Hawaiian archipelago.

Native Habitat	Remaining Extent as of 2015
Montane Mesic	73%
Montane Dry	59%
Lowland Wet	45%
Lowland Mesic	28%
Lowland Dry	30%

The history of social-ecological landscape change in Hawaiʻi occurred over the course of about 1000 years, beginning with the initial migration of Polynesians from the nearest archipelagoes of Oceania, those of the Marquesas and Tahiti. For centuries, the human population grew and spread across the Hawaiian archipelago and developed a unique indigenous Hawaiian culture, marked by an epistemology that regarded the surrounding biotic community as familial and ancestral, thereby establishing a strongly biocultural society [14–18]. The rich ecosystems of the Hawaiian Islands generated an equally rich cultural system in the pre-contact society that developed within it.

Another major milestone occurred in 1778 when the Hawaiian Islands were encountered by Captain James Cook and this initial contact with the Western World resulted in increasing presence and influence of Western culture and land uses in the islands, establishing a different social-ecological context based on commodification of land and natural resources, culminating in the footprint of the early 21st Century. Although there have been discussions of the pre-contact and post-contact impacts of humans on the native biota and ecosystems of Hawaiʻi [19], there had been no geospatially explicit reconstructions of landscape change offered specifically focusing on native habitat loss. Many of those early observations by Westerners were made from the ocean with limited geographic view plane and often by those with no familiarity with Hawaiian vegetation. Instead, we had only the reconstructions of the pre-human extent of terrestrial native-dominated vegetation zones in Hawaiʻi [20] to compare against the current extent (see example for Island of Oʻahu below). Oʻahu offers one of the more dramatic examples of the impacts that our human presence has wrought on native habitats.

However, for every "before and after" situation that spans centuries of time, it is instructive to provide intermediate stages that speak to the human factors, such as population growth, changes in religion, economic systems, and land tenure, and key introductions of both species and activities that influenced the trajectory, rate, and intensity of social-ecological change.

2. Materials and Methods

Mapping of the Human Land Use Footprint in Pre-Contact Hawai'i

Models of pre-Western contact agriculture in Hawai'i were combined with archeological and oral tradition to create an explicit geospatial footprint of human-occupied and transformed areas across the pre-contact Hawaiian archipelago. The goal was to determine the explicit geospatial areas that, by 1770 (the decade of Western contact), had been chronically occupied, directly manipulated, and significantly changed from pre-existing native ecosystem types into traditional Hawaiian uses: house sites, agricultural fields, fishponds, religious sites, major roads, and trails.

At the onset, we point out that this is not to be confused with the ecological footprint used in modern assessments of human sustainability [7,21,22], but is related to it because it describes explicitly the geospatial extent of human land uses related to elements of ecological assessments: agricultural use, resource areas utilized for shelter, energy, medicines, material resources, and other needs of a human population. It is also the inverse of the presence of pre-human ecosystems, and allows for assessments of impacts on ecosystem services and their historic decline in the course of increasing human modification and displacement of those native ecosystems. The Hawaiian social-ecological system of land management has been described as the *ahupua'a* system [23,24], and in this issue, as the *moku* system [18], based on units of land and sea that typically included a cross section of ecosystems from the summit of an island to the coast, and outward to include nearshore marine habitats. Nested within the *ahupua'a* were smaller units, while both clusters of *ahupua'a* and larger-scale units called *moku* comprised the major basis for Hawaiian social-ecological regions and management communities. Integration of human society and its processes with the endemic biota and a small set of transported Polynesian plant and animal introductions, shown in Appendix A, created a system in which biological resources were deeply woven via explicit genealogical ties, rendering them as biocultural relationships [25].

We recognized that Hawaiian management of *ahupua'a* and *moku* in the pre-contact era tended to minimize the human footprint by delineating portions of the landscape as *wao kanaka* (realm of human influence, typically in coastal and lowland areas) and designating sacred (typically upland) habitats such as the *wao akua* (realm of deities) [18,26].

Pertinent to the impacts of intensive agriculture on this social-ecological system, we incorporated the work of Ladefoged et al. [27] who created a geospatial model expressing the optimal conditions for the cultivation of the two major staple crops in Hawai'i: *kalo* (taro, *Colocasia esculenta*) and *'uala* (sweet potato, *Ipomoea batatas*). It was tested and refined via comparison to known archeological complexes associated with agriculture [27,28]. Because practically all of the lands of greatest potential for agriculture had been developed for agriculture (as seen by high congruence of agricultural archeology with the agriculture models), applying formulae for deriving human population estimates from agricultural area for Pacific Island nations yielded a pre-contact Hawaiian population of 400,000 to 800,000, with the largest populations on the islands of Hawai'i, Maui, O'ahu and Kaua'i [27]. For this paper we explicitly derived population estimates for the eight main islands by applying the island footprint percentages to a total population of 500,000. The uneven populations of the islands were further discussed in Kirch 2011 [29] in terms of the population basis of the great Hawaiian chiefdoms of the four most populous islands, supported by their exceptional agricultural and biocultural potential. Such highly productive agricultural lands, the basis for not only political power but cultural proliferation, were called *'āina momona*, sweet/productive lands [30,31]—the most important lands for maintaining biocultural vitality and biocapacity in those times, and an important focus for restoration of social-ecological systems and biocultural revitalization today.

From 2009 to 2012, working in cooperation with the research staff of the Office of Hawaiian Affairs (OHA), we expanded on the agricultural model by mapping known *loko i'a* (estuarine walled fishponds, a major source of protein foods), and continuing the reviews of archeological geospatial databases compiled by the State of Hawai'i Historic Preservation Division (SHPD) [28] of the Department of Land and Natural Resources (DLNR), as well as historical maps from the Department of Accounting and General Services (DAGS).

Major compilations of oral history out of a variety of sources in both English and Hawaiian were gleaned for further information on *wahi pana* (storied localities), terrestrial trail systems, religious sites, including *heiau* (temples) and *ko'a* (shrines), to set against the emerging geospatial depiction of areas of habitation, agriculture, or other traditional uses listed in Appendix B. Because the oral traditional accounts were extremely place-specific, and because current land boundaries retained the *ahupua'a* designations largely intact from pre-contact times [32], descriptions of places in oral accounts were readily placed geospatially, to corroborate models and archeological mappings. It is becoming apparent that in terms of indigenous knowledge archives in written form, the millions of pages of Hawaiian language newspapers represent the single largest of such first-peoples archives known in the world [33]. Appendix B offers an overview of some of the major sources that were consulted.

We applied the agricultural model, augmented by documentation of the historical trails and fishponds, locations of *heiau* and other archeological geospatial data, and corroborated this with traditional accounts of the chiefly centers of governance and famous population centers. We unified all the layers, buffered them, and created the Hawaiian footprint.

3. Results

What emerged from this multidisciplinary combination of sources was the first geospatially explicit footprint of pre-contact human activity that modified or displaced the original native terrestrial habitats in the Hawaiian Islands, as seen in Figures 2–7. It was coined "the Hawaiian Footprint Project". This process was applied to all of the eight main Hawaiian Islands, and an example is available for public scrutiny online [34], with GIS layers provided by request via The Nature Conservancy of Hawai'i.

We demonstrated that the footprint affected pre-existing native ecosystems in an uneven manner, with the largest impacts in wetlands that were converted into *lo'i kalo* (flooded field system) agriculture and *loko i'a* (estuarine walled fishponds), in lowland dry and mesic areas, where wood was collected for houses, cooking fires, tools, and other needs, and land was cleared for habitation, with regular fires set to promote pili grass fields for thatching. Other native ecosystems at higher elevations were negligibly affected.

A similar analysis of land uses one century later, applied to the Island of Hawai'i, documented greatly increased disruption of native vegetation [35]. Table 2 lists selected extents of habitats displaced by the 1870 human footprint and their current status. The geospatial depiction comparing these same pre- and post-contact situations, seen in Figure 7, clearly demonstrates the greatly accelerated rate of social-ecological disruption and loss of the original biocultural landscape.

Table 2. Extensive native habitat loss on the Island of Hawai'i in the first 100 years after Western contact by 1870 was driven primarily by large-scale ranching and the advent of sugarcane monoculture.

Native Habitat Loss	1770 (ha)	% Lost	1870 (ha)	% Lost	2015 (ha)	% Lost
Lowland Mesic	14,400	21%	29,900	44%	48,800	72%
Lowland Dry	42,200	19%	93,100	43%	151,300	70%
Lowland Wet	20,700	9%	27,500	12%	124,600	55%
Montane Dry	2100	1.4%	55,400	37%	61,000	41%
Montane Mesic	800	1%	12,900	17%	19,400	26%
Alpine/Subalpine	1300	<1%	13,600	6%	18,700	9%

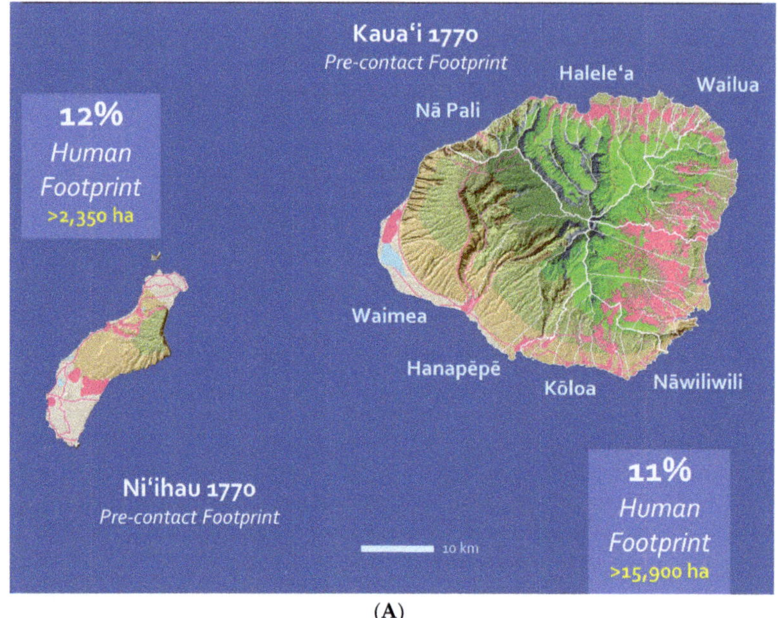

(**A**)

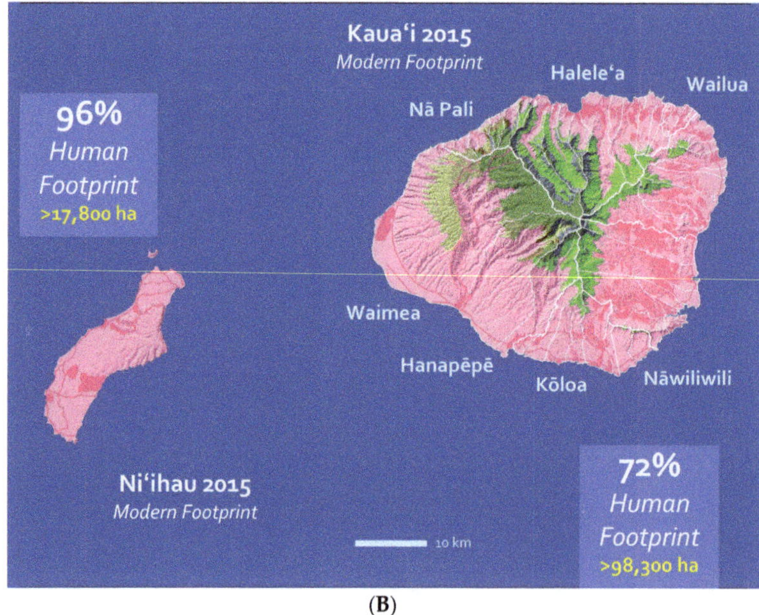

(**B**)

Figure 2. (**A**) Hawaiian footprint, prior to Western contact, resulted in <12% native habitat loss on the islands of Kaua'i and Ni'ihau. (**B**) Modern footprint resulted in 72% and 96% native habitat loss, respectively. Key: dark pink = pre-contact footprint; light pink (for comparison) = modern footprint; white line = *moku*, districts and *ahupua'a*; colored basemap = major native vegetation zones, after the Hawai'i Ecoregion Plan [36].

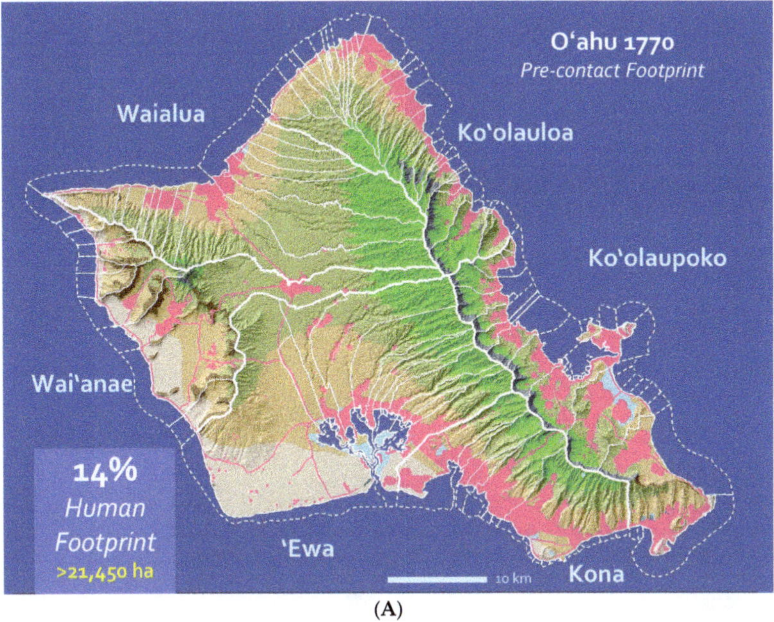

(A)

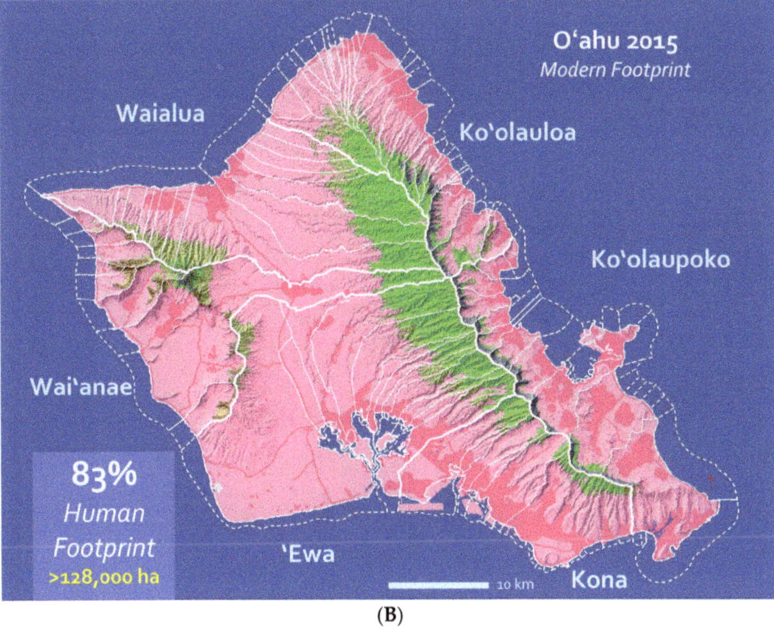

(B)

Figure 3. (**A**) Hawaiian footprint, prior to Western contact, resulted in 14% native habitat loss on the island of O'ahu. (**B**) Modern footprint resulted in 83% native habitat loss. Key: dark pink = pre-contact footprint; light pink (for comparison) = modern footprint; white line = *moku* (districts) and *ahupua'a*; dotted white line = historical nearshore fisheries, makai part of *ahupua'a*; colored basemap = major native vegetation zones, after the Hawai'i Ecoregion Plan [36].

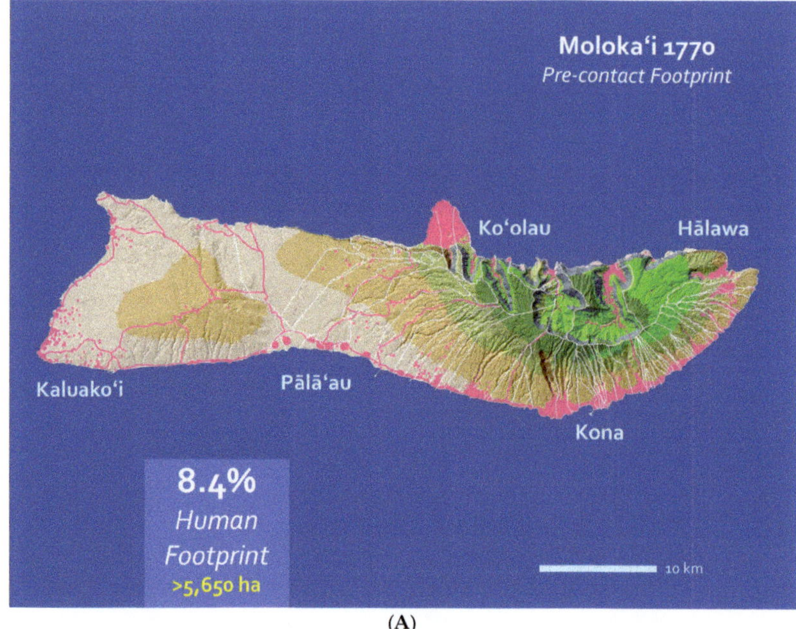

(A)

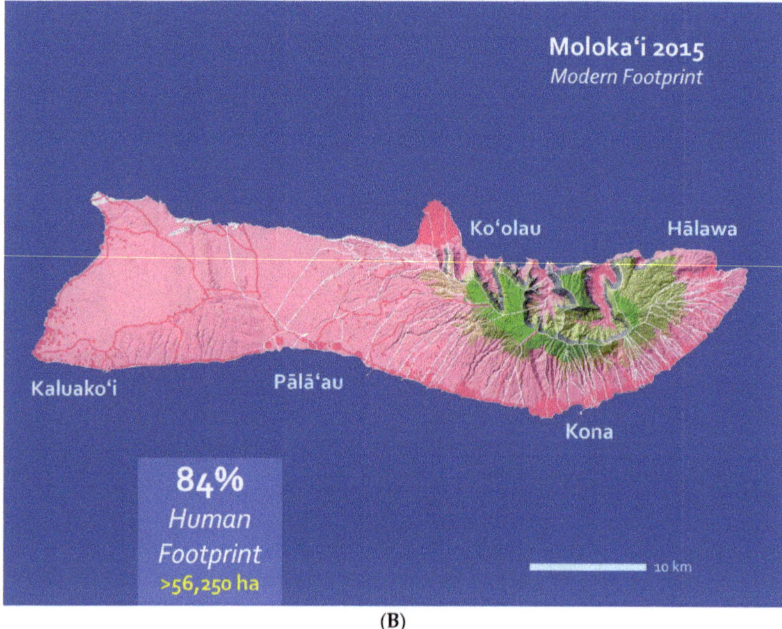

(B)

Figure 4. (**A**) Hawaiian footprint, prior to Western contact, resulted in <9% native habitat loss on the island of Moloka'i. (**B**) Modern footprint resulted in 84% native habitat loss. Key: dark pink = pre-contact footprint; light pink (for comparison) = modern footprint; white line = *ahupua'a*; colored basemap = major native vegetation zones, after the Hawai'i Ecoregion Plan [36].

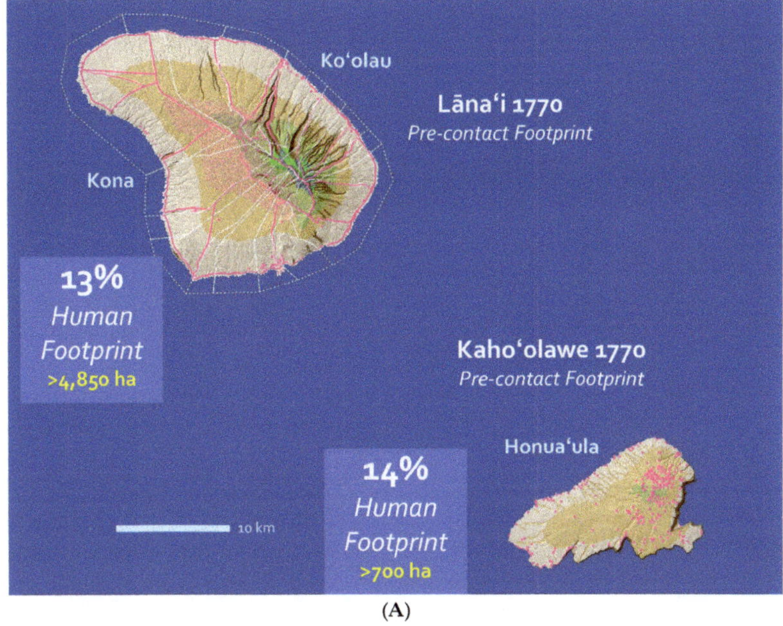

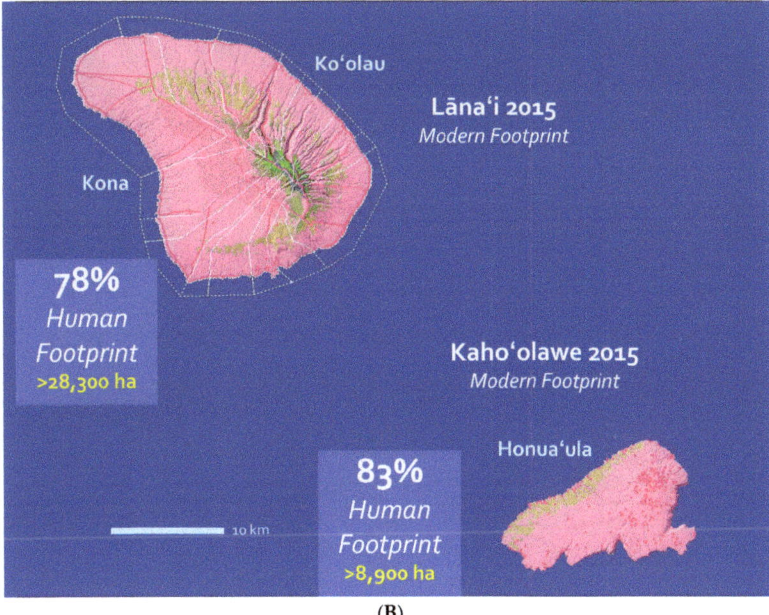

Figure 5. (**A**) Hawaiian footprint, prior to Western contact, resulted in <14% native habitat loss on the islands of Lāna'i and Kaho'olawe. (**B**) Modern footprint resulted in >78% native habitat loss. Key: dark pink = pre-contact footprint; light pink (for comparison) = modern footprint; white line = *ahupua'a*; dotted white line = historical nearshore fisheries, makai part of *ahupua'a*; colored basemap = major native vegetation zones, after the Hawai'i Ecoregion Plan [36].

(A)

(B)

Figure 6. (**A**) Hawaiian footprint, prior to Western contact, resulted in 11% native habitat loss on the island of Maui. (**B**) Modern footprint resulted in 70% native habitat loss. Key: dark pink = pre-contact footprint; light pink (for comparison) = modern footprint; white line = *moku* (districts) and *ahupua'a*; colored basemap = major native vegetation zones, after the Hawai'i Ecoregion Plan [36].

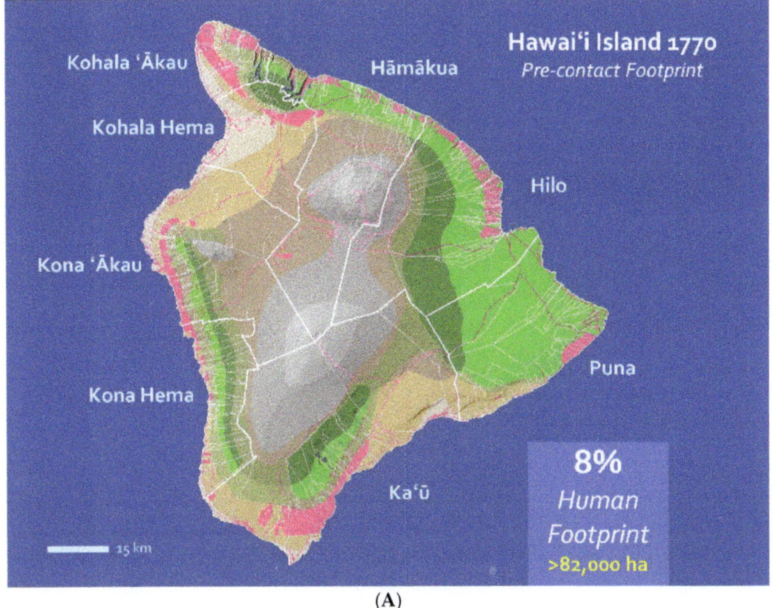

(**A**)

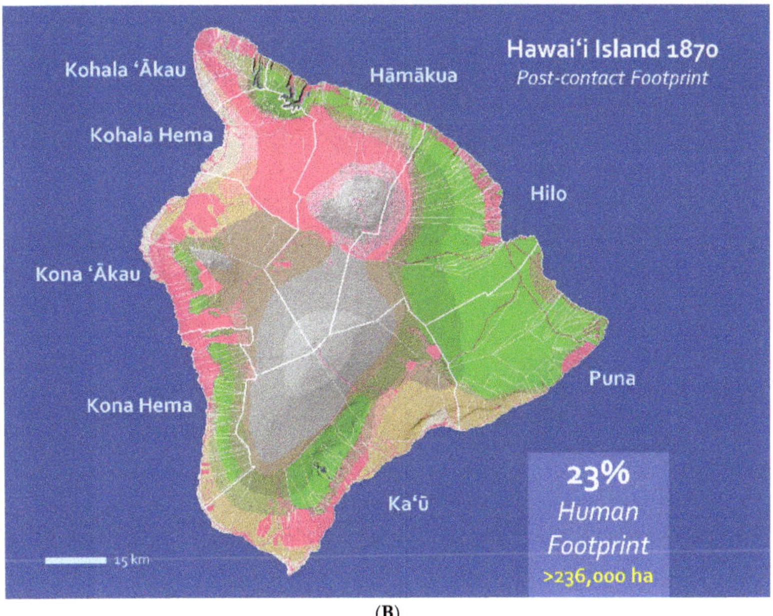

(**B**)

Figure 7. *Cont.*

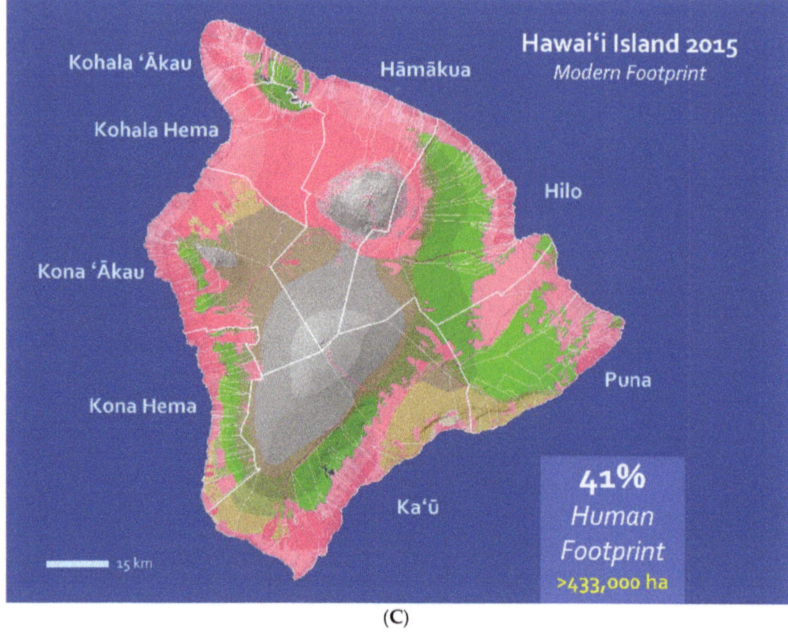

(C)

Figure 7. (**A**) Hawaiian footprint, prior to Western contact, resulted in 8% native habitat loss on the island of Hawaiʻi. (**B**) The human footprint tripled 100 years after Western contact. (**C**) Modern footprint resulted in 41% native habitat loss. Social-ecological change over two centuries reflects the effects of commodification of land and resources, and loss of pre-contact biocultural relationships. Key: dark pink = pre-contact human footprint; medium pink (for comparison) = 1870 footprint, light pink (for comparison) = modern footprint; white line = *moku* (districts) and *ahupuaʻa*; colored basemap = major native vegetation zones, after the Hawaiʻi Ecoregional Plan [36].

The pattern of wet valley occupation and working of large seasonal fields applies across the archipelago. Using the population estimate methods described in Ladefoged et al. [27] and Kirch [29] which yielded population estimates of 400,000 to 800,000, we derived, using the pre-contact footprint for individual islands, population estimates for each of those islands. Table 3 depicts the distribution of the population among islands when using a total population of 500,000. As might be expected, the majority of the population was on the large Island of Hawaiʻi. It is remarkable to look on these results in terms of the human geography of ancient Hawaiʻi; when used as a backdrop for traditional stories and accounts, every prominent place name and every celebrated place was included, as shown in Appendix B.

Table 3. Pre-contact Hawaiian population estimates for the main Hawaiian Islands. A total population of 500,000 was selected for this table as it falls within the 400–800k range and simplifies presentation.

Island	Footprint (Ha)	%	Est. Population
Hawaiʻi	81,800	53.0	265,000
Oʻahu	21,600	14.0	70,000
Maui	21,100	13.6	68,000
Kauaʻi	16,000	10.3	51,500
Molokaʻi	5600	3.7	18,500
Lānaʻi	5200	3.4	17,000
Niʻihau	2300	1.5	7500
Kahoʻolawe	700	0.5	2500

4. Discussion

4.1. The Hawaiian Social-Ecological System As a Model of Sustainability and Self-Sufficiency

Based on the best available data at the time, the two major conclusions of the Hawaiian Footprint Project were that prior to Western contact in 1778, a substantial human population in the Hawaiian archipelago (estimated at 400,000–800,000 people) had affected less than 15% of the original area of native terrestrial ecosystems, and was necessarily 100% self-sufficient, that is, did not rely on any significant external inputs from the rest of global humanity. Thus pre-contact Hawai'i stands as a quintessential sustainability example of a large human population that practiced intensive agriculture, yet minimally displaced the native habitat that was the foundation of its vitality and development. This example of human sustainability in a finite (but extremely rich) high island setting was achieved because of a Hawaiian worldview that regarded nature as familial and ancestral, sacred and of immense value [17,18].

4.2. Using Pre-Contact Models of Sustainability in Transformed Landscapes

When the models for pre-contact agriculture were published and made publicly available [32], it generated many inquiries regarding the use of the mapped extent of pre-contact agriculture as guidance for revitalization of current biocultural restorations. To the extent that areas of pre-contact agriculture remain available for agricultural use in our times, it stands to reason that the model could indicate areas of greatest potential for successful social-ecological revitalization of Hawaiian traditional agriculture.

4.3. Post-Contact Changes to the Social-Ecological Landscape of Hawai'i

In the 240 years that followed initial contact with the Western world, much has changed in both the social-ecological setting and the biocultural setting of Hawai'i. The acceleration of native ecosystem loss since Western contact has been dramatic with the smaller, drier islands such as Ni'ihau losing essentially everything. Hawai'i Island, by virtue of relatively vast and remote interiors, too high and cold for cultivation, retains the highest percentage in modern times, the only island with less than a 50% footprint today. Several different reviews of these changes point to the imposition of Western worldviews that viewed land and natural resources as commodities to be exploited to feed capitalist economies, leading to practices such as large-scale ranching and mono-crop agriculture of sugarcane and pineapple that supplanted multi-crop and semi-wild systems of the pre-contact Polynesian social-ecological system and induced wholesale erasure of native biodiversity across hundreds of thousands of hectares. [37–39]. Our recent geospatially explicit review of land use changes on Hawai'i island between 1770 (pre-contact) and 1870 (one century after contact), demonstrated that the human footprint had more than tripled in size. These changes entirely transformed lowland social-ecological landscapes, and extended high into the montane zones on the highest islands of Maui and Hawai'i, displacing biocultural resources there and reducing inherent biocapacity. This is a trend that has continued into the 21st Century, resulting in the modern human footprint that is more than five times larger than the pre-contact Hawaiian Footprint on the Island of Hawai'i. Self-sufficiency, expressed as a lack of importation of goods, has plummeted from 100% in pre-contact times to 15% or less in the 21st century [40,41].

The same phenomenon noted when assessing the biocapacity of urban areas, such as large cities, can be applied to Hawai'i. Any given city's biocapacity is largely appropriated from areas outside of the city limits [21], and treats the metropolitan core as a social-ecological island that has low inherent biocapacity and extremely high population density, compensated for via importation of resources from other areas both within immediately adjacent regions and increasingly more broadly. In like manner, the economy of Hawai'i, currently driven by tourism, sees both an increased effective population size made up of a varying stream of transient visitors (1.4 million permanent residents, +7–10 million additional visitors per year in Hawai'i) whose demands far exceed local biocapacity, and has created a growing urbanization in many areas that were once prime agricultural lands, further limiting efforts to increase self-sufficiency and sustainability [40]. This is compensated for by a high importation rate, and contributes to our low self-sufficiency.

4.4. Non-Native Species

During the 1000 years of the pre-contact period, perhaps 50–60 species of plants had been introduced into the highly endemic Hawaiian Islands terrestrial flora, summarized in Appendix A [41]. The majority of these Polynesian introductions were agricultural crops, plants used in cordage and plaiting, other ethnobotanical species, and a handful of agricultural weeds inadvertently introduced. Nearly all of these were largely confined to agricultural settings, and did not naturalize readily into surrounding native vegetation. *Kukui* (*Aleurites moluccana*), and possibly *hau* (*Hibiscus tiliaceus*), are exceptions and have naturalized readily, frequently as canopy dominants in lowland riparian situations on all of the larger islands [37,42]. The otherwise non-invasive nature of the majority of the Polynesian plant introductions meant that even in areas completely converted to croplands, any fallow areas would have converted back into native successional communities. Even in those areas dominated by *kukui*, native subcanopy and groundcover diversity would have remained, and a mixed forest with strong native composition would still be present. The greater impact of Polynesian introduced animals, in particular *'iole*, the Polynesian rat (*Rattus exulans*), is undeniable see: [37,43,44], but the same patterns of native vegetation recovery and dominance in response to this disturbance would hold true.

Two-hundred and forty years of plant introductions without adequate biosecurity measures since Western contact have completely changed that picture and disrupted the process of vegetation succession in Hawai'i. Perhaps 15,000 or more taxa of vascular plants had been introduced to Hawai'i [45]. Among these are hundreds of habitat-modifying species that not only degrade native vegetation composition and structure, but can disrupt traditional agriculture and greatly increase the labor required to remove aggressive weeds and successfully grow desired crops. Introduced animals, including wetland invertebrates such as Apple snails (*Pomacea canaliculata*) and crayfish (*Procambarus clarkii*) damage both the plants and the traditional infrastructure of *lo'i kalo* (flooded field system), adding further impediments to biocultural restoration of traditional agriculture. Introduced bacterial and fungal diseases are another major challenge to *kalo* and other traditional crops [46]. The post-contact introduction and spread of non-native ungulates, such as cattle, goats, and sheep, and their wholesale denudation of the forested watershed on all islands created the watershed crisis of the turn of the 20th century [47].

It becomes more and more clear that inadequate biosecurity stands as one of the greatest current and future impediments to biocultural restoration and sustainability in Hawai'i [48]. Invasive non-native species have already caused significant harm to natural and cultural resources, economy, and way of life; for example, they affect critical native ecosystem services, such as long-term reliability of freshwater resources, as well as agricultural productivity, human health and community well-being. We must support, implement and augment efforts to establish stronger biosecurity in Hawai'i as the current context of highly appropriated biocapacity to support a tourism economy continues into the future. Moreover we must develop more effective tools for dealing with a long history of intentional and unintentional introductions of habitat modifying non-native species that greatly impair the potential for biocultural restoration.

4.5. Climate Change

In an era of increasing climate change affecting both marine and terrestrial systems, predicted effects on precipitation and temperature could affect the potential for biocultural restoration. The high islands of Hawai'i exhibit elevation zonation in both temperature and moisture, as seen in Figure 7, and it is anticipated that zones will shift in their placement, and that novel zones currently not present will come into being [49]. Because the models for both *kalo* and *'uala* are sensitive to precipitation (*'uala* particularly so), the archeology of sweet potato agriculture on Maui already demonstrates a mismatch: archeological complexes associated with seasonal *'uala* agriculture on the western slope of Haleakalā extend into areas with annual precipitation that is currently insufficient for the crop, as shown in Figure 8. It is a clear indication that over 240 years ago, slopes that are currently too hot and dry for growing sweet potatoes were seasonally worked for that crop. This means that in the decades to come, with warming and drying trends predicted for the lowlands of the Hawaiian Islands, the model generated for the pre-contact Hawaiian footprint will have to be adjusted in various ways to track the optimal rainfall conditions of the future.

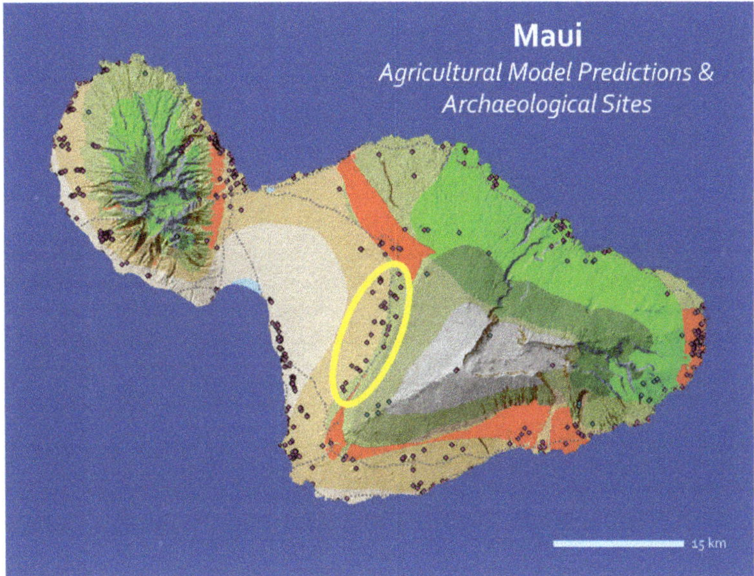

Figure 8. Map showing mismatch of *'uala* agricultural model prediction (red) with archeological complexes (dots) on the west flank of East Maui (yellow oval).

In a somewhat less direct manner, drying trends may convert streams that are currently continuous and perennial (and therefore suitable for *kalo*) into intermittent streams that may provide insufficient water for the crop. If the predicted trends are for warming and drying, this likely means an overall reduction in the potential area for wet *lo'i kalo* production.

In similar manner, each of the traditional crops of Hawai'i, and indeed all future potential crops, should be assessed for their optimal climate envelopes, and plans made to shift the areas designated for those crops according to shifting climate patterns in the decades to come. A similar analysis was already conducted for every native flowering plant in Hawai'i [50], and this tool is already being promulgated and applied in conservation efforts involving assisted migration of rare plants out of habitat that is becoming climatically suboptimal. This has broad relevance to biocultural restoration planning, adding another complex factor to consider in the geographic placement and selection of species involved, anticipating future optimal climate envelopes.

4.6. Diversification

One of the major advantages of the broad range of life zones in Hawai'i is the great potential for diversification of agriculture, enhancing biocapacity. While the models for the pre-contact footprint were based on the optimal range of the two major staple crops of those times, modern agriculture in Hawai'i has already seen an expansion to include a wide variety of agricultural products, including coffee, macadamia nuts, tropical fruit, ornamentals, and vegetable crops that were not available in pre-contact times. While we should likely never again consider a large-scale monoculture approach that was the signature of the sugarcane and pineapple eras of agriculture in Hawai'i, the future offers a broad range of possibilities. It may be feasible to develop agroforestry models such as those used traditionally and successfully in other island nations (e.g., Pohnpei) and gain both agricultural diversity as well as the benefits of ecosystem functions that derive from maintaining forest cover and diverse understory structure. These have the potential to minimize erosion and sedimentation of our streams and nearshore marine habitats, increasing both terrestrial and marine habitat viability and the potential for food production and biocultural restoration.

5. Conclusions

Reconstructions of pre-contact agricultural hotspots are instructive in demonstrating the potential for a closed island social-ecological system to sustainably support a large human population in an entirely self-sufficient manner while creating a relative small land-use footprint that allows for maintenance of strong native biological diversity and vital ecosystem processes and services. While it might be desirable to recapture that ancient situation, several factors have imposed themselves over the last 240 years of post-contact history and greatly complicate any simple schemes to restore that pre-contact state. One is the presence of thousands of non-native plants and animals that impose their own ecological influences that impede agricultural success via competition, predation, and pathologies that did not exist in pre-contact times. Another is the irreversible land developments that have displaced many areas of formerly rich agricultural production. A third is the effect of sheer numbers of people present in the islands, far exceeding the estimated 400,000–800,000 Hawaiians that comprised the archipelagic human population prior to contact. Finally, the anticipated changes in climate, including temperature and precipitation, will require adjustments of the models of optimal agricultural output, and may render some of the original areas unusable, while other areas may emerge as optimal in the future. Knowing these limitations is a vital step toward addressing and surmounting them. While we may not be able to turn the clock back, we are more able than ever to take intelligent action to frame our future.

More importantly however, is the lesson of the thousand years of pre-contact Hawaiian presence, and the social-ecological system that developed as a result of a worldview with a strong foundation of biocultural relationships. These regarded the natural world as family in a reciprocal and caring relationship wherein human health and welfare was viewed as one with the health and welfare of the surrounding living community. In such a context, humans stand not intrinsically apart from nature, and not solely as a threat to nature, but acknowledge that we are a force of nature with potential to damage or to repair. The consequences of shifting from this social-ecological system into one of land and resources as economic commodities has clearly resulted in a post-contact history of loss of native habitats, sustainability, and self-sufficiency. Recapturing and reestablishing those traditional island values in a modern context is a core underpinning in biocultural restoration.

In our analyses of pre-contact Hawai'i we see that it is possible to support a thriving human population, practice intensive sustainable agriculture, and establish a social-ecological system that maintained the native habitat that was the foundation of *'āina momona*. It becomes clear that a future shift that strives to recapture the best of the pre-contact social-ecological system is sorely needed in Hawai'i and by extension, Planet Earth. Achieving this biocultural restoration will take the best of indigenous values combined with the best of 21st Century knowledge to realize.

Author Contributions: Conceptualization, S.M.G., III; Methodology, S.M.G., III, S.L.T.; Formal Analysis, S.M.G., III, S.L.T.; Investigation, S.M.G., III, S.L.T., U.W.; Resources, U.W.; Data Curation, S.L.T.; Writing—Original Draft Preparation, S.M.G., III; Writing—Review & Editing, S.M.G., III, S.L.T., U.W.; Visualization, S.L.T., S.M.G., III; Funding Acquisition, S.M.G., III, S.L.T., U.W.

Funding: This research was funded by The Nature Conservancy of Hawai'i and the Office of Hawaiian Affairs. The APC was funded by Hawai'i Community Foundation.

Acknowledgments: We acknowledge the support of the Office of Hawaiian Affairs, particularly Kale Hannahs, Kamana'opono Crabbe, Kamoa Quitevis, Andrew Choy, and Zack Smith, as well as other partners, data sources, and reviewers of the Hawaiian Footprint project: Pua Aiu, Meghan Brasier, Kawika Burgess, Erik Burton, Patricia Cockett, Ross Cordy, Dorothe Curtis, Theresa Donham, Ka'eo Duarte, Rowan Gard, Nakoa Goo, Bob Hobdy, Jim Jacobi, Jason Jeremiah, Nahaku Kalei, Russell Kallstrom, Keala Kanaka'ole, Sabra Kauka, Jody Kaulukukui, Eric Komori, Natalie Kurashima, Kepa Maly, Dwight Matsuwaki, Nancy McMahon, Theresa Menard, Richard Stevens, Kanoe Steward, Jason Sumiye, Bert Weeks, Chipper Wichman, and Brad Wong.

Conflicts of Interest: The authors declare no conflicts of interest.

Appendix A

Known and Potential Polynesian Introductions of Plant Species to the Hawaiian Archipelago [42] including well-known species of Polynesian biocultural significance as well as others potentially introduced (questionably indigenous or early post-contact introductions). Status and scientific names as listed in Wagner et al.

Hawaiian Name	Biocultural Relationship	Status	Scientific Name
kou	wood, *lei*	Polynesian introduction	*Cordia subcordata*
kamani	*lei*, wood	Polynesian introduction	*Calophyllum inophyllum*
'uala	staple crop	Polynesian introduction	*Ipomoea batatas*
ipu	containers, music	Polynesian introduction	*Lagenaria siceraria*
kukui	oil, medicinal, wood, *lei*, relish, dye	Polynesian introduction	*Aleurites moluccana*
'auhuhu	fish poison	Polynesian introduction	*Tephrosia purpurea*
'ulu	staple food crop, medicinal, sap, wood	Polynesian introduction	*Artocarpus altilis*
wauke	fiber, clothing	Polynesian introduction	*Broussonetia papyrifera*
'ōhi'a 'ai	fruit	Polynesian introduction	*Syzygium malaccense*
'awa	ritual drink, medicinal	Polynesian introduction	*Piper methysticum*
noni	medicinal, dye	Polynesian introduction	*Morinda citrifolia*
kī	food, medicinal, ritual	Polynesian introduction	*Cordyline fruticosa*
'ape	famine food	Polynesian introduction	*Alocasia macrorrhiza*
kalo	mainstay food crop	Polynesian introduction	*Colocasia esculenta*
niu	food, wood, fiber	Polynesian introduction	*Cocos nucifera*
uhi	secondary crop, not naturalized	Polynesian introduction	*Dioscorea alata*
hoi	famine food, naturalized	Polynesian introduction	*Dioscorea bulbifera*
pi'a	famine food, naturalized	Polynesian introduction	*Dioscorea pentaphylla*
mai'a hē'ī	wild food source	Polynesian introduction	*Musa troglodytarum*
mai'a (varieties)	staple crop	Polynesian introduction	*Musa x paradisiaca*
kō	food	Polynesian introduction	*Saccharum officinarum*
pia	food	Polynesian introduction	*Tacca leontopetaloides*
'ōlena	dye, medicinal, ritual	Polynesian introduction	*Curcuma longa*
'awapuhi	medicinal	Polynesian introduction	*Zingiber zerumbet*
pā'ihi'ihi	uncommon medicinal, accidental?	Polynesian introduction?	*Rorippa sarmentosa*
kāmole	wetland, accidental w/*kalo*?	Polynesian introduction?	*Ludwigia octovalvis*
'ihi	medicinal; indig? seeds in pre-contact sites	Polynesian introduction?	*Oxalis corniculata*
—	cultiv. central Pac, 3 records from HI	Polynesian introduction?	*Solanum viride*
'ohe Kahiki	tools, wood, music, container; indig?	Polynesian introduction?	*Schizostachyum glaucifolium*
—	seeds in pre-contact sites; indig NA, SA	naturalized?	*Daucus pusillus*
pohe	indig NA; pre 1871 HI records	naturalized?	*Hydrocotyle verticillata*
koali 'ai	famine food, poss indig?	naturalized?	*Ipomoea cairica*
koali kuahulu	pantropical, indig?	naturalized?	*Merremia aegyptia*
kākalaioa	indig/early intro; also *hihikolo*	naturalized?	*Caesalpinia major*
maunaloa	indig Honduras; 1st record HI 1825	naturalized?	*Dioclea wilsonii*
pāpapa	native to tropical Asia? edible	naturalized?	*Lablab purpureus*
—	pantropical weed	naturalized?	*Sida rhombifolia*
kāmole	accidental w/*kalo*?	naturalized?	*Polygonum glabrum*
pōniu	also *haleakai'a*; medicinal	naturalized?	*Cardiospermum halicacabum*
'aka'akai	also *kaluhā*, indigenous to NA & SA	naturalized?	*Schoenoplectus californicus*
—	cosmop., accidental on *kalo*?	naturalized?	*Lemna aequinoctialis*
—	cosmop., accidental on *kalo*?	naturalized?	*Spirodela polyrrhiza*
—	indig Asia, Malesia; 1st HI coll pre 1871	naturalized?	*Garnotia acutigluma*
'ili'ohu	once noted near *kalo* fields; extinct?	indigenous?	*Cleome spinosa*
—	widespread in the S. Pacific	indigenous?	*Ipomoea littoralis*
kākalaioa	indig/early intro; *lei*, medicinal	indigenous?	*Caesalpinea bonduc*
—	widesp trop Indo-Pac, but 1st HI rec 1920	indigenous?	*Entada phaseoloides*
pakaha	indig NA, pretty flowers, no descr uses	indigenous?	*Lepechinia hastata*
pūkāmole	1st HI record 1794; medicinal	indigenous?	*Lythrum maritimum*
ma'o	indigenous NA	indigenous?	*Abutilon incanum*
hau	wood, fiber, medicinal	indigenous?	*Hibiscus tiliaceus*
milo	wood	indigenous?	*Thespesia populnea*
pōpolo	medicinal, dye, food	indigenous?	*Solanum americanum*
uhaloa	medicinal	indigenous?	*Waltheria indica*
'ahu'awa	fiber, plaiting	indigenous?	*Cyperus javanicus*
—	prob indigenous	indigenous?	*Carex thunbergii*
kohekohe	low elev marshes	indigenous?	*Eleocharis calva*
hala	brought, but also indig; plaiting, food	indigenous?	*Pandanus tectorius*

Hawaiian Name	Biocultural Relationship	Status	Scientific Name
mānienie ʻula	1st HI record 1819, widespread	indigenous?	*Chrysopogon aciculatus*
pili	thatch	indigenous?	*Heteropogon contortus*
mauʻu laiki	no rec uses; post-contact Hawn name	indigenous?	*Paspalum scrobiculatum*
—	accidental w/*kalo*? leafy pondweed	indigenous?	*Potamogeton foliosus*
—	accidental w/*kalo*? long-leaved pondweed	indigenous?	*Potamogeton nodosus*

Key: **Hawaiian name** (— = no known Hawaiian name); **Biocultural relationship** (Hawaiian uses, other salient info); **Status** (indigenous? = possibly indigenous; naturalized? = possibly early naturalized post-contact introduction).

Appendix B

Some major sources of Hawaiian oral tradition, place names, and agricultural areas consulted and incorporated into the Hawaiian Footprint Project:

a. Beckwith, M. 1951. *The Kumulipo*; University of Chicago Press: Chicago, IL. (The Hawaiian chant of creation.)

b. Cordy, R. 2002. An Ancient History of Waiʻanae—Ka Moku o Waiʻanae: He Moʻolelo o ka Wā Kahiko; Mutual Publishing: Honolulu.

c. Dibble, S. 1838 (1984). *Ka Moʻolelo Hawaiʻi*; University of Hawaiʻi Press: Honolulu. (In original Hawaiian language, with English translations.)

d. Ellis, W. 1825 (2004). A Narrative of an 1823 Tour Through Hawaiʻi or Owhyhee, with remarks on the History, Traditions, Manners, Customs and Language of the inhabitants of the Sandwich Islands; Mutual Publishing: Honolulu.

e. Emerson, N. 1906 (2013). *Unwritten Literature of Hawaii: The Sacred Songs of the Hula*; Charles Tuttle: Tokyo. (Includes text of hundreds of chants in original Hawaiian, with translations and discussion in English.)

f. Emerson, N. 1915. *Pele and Hiʻiaka: A Myth from Hawaii*. Charles Tuttle, Tokyo.

g. Fornander, A. 1880. An Account of the Polynesian Race: Its Origin and Migrations and the Ancient History of the Hawaiian People to the Times of Kamehameha I. Vol. II. Trubner, London.

h. Fornander, A. 1916-1920. Fornander Collection of Hawaiian Antiquities and Folklore. Memoirs of the Bernice Pauahi Bishop Museum, Vols. 4–6. Honolulu. (In original Hawaiian and English translation.)

i. Handy, E.S. 1940. The Hawaiian Planter—Volume 1. *Bishop Mus. Bull. 161*. Bishop Museum Press: Honolulu.

j. Handy, E.S., Handy, E.G. and Pukui, M.K. 1972. Native Planters in Old Hawaii: Their Life, Lore, and Environment. *Bishop Mus. Bull. 233*. Bishop Museum Press: Honolulu.

k. Hawaiian Historical Society. 2001. Nā Mele ʻAimoku, Nā Mele Kūpuna, a me Nā Mele Ponoʻi o ka Mōʻī Kalākaua I. Dynastic Chants, Ancestral Chants, and Personal Chants of King Kalākaua I. Hawaiian Language Reprint Series, Honolulu.

l. Hawaiian Studies Institute. 1987. Map: Oʻahu Pre-Māhele Moku and Ahupuaʻa. Kamehameha Schools, Honolulu.

m. Iʻi, J.P. 1959. *Fragments of Hawaiian History*. Bishop Museum Press, Honolulu.

n. Kamakau, S.M. The People of Old. Bishop Museum Press, Honolulu. (A series of compilations of Hawaiian traditional knowledge contributed to the Hawaiian language newspapers, published in three volumes: *Ka Poʻe Kahiko: The People of Old* 1964; *Ka Hana a Ka Poʻe Kahiko: The Works of the People of Old* 1976; *Nā Moʻolelo a Ka Poʻe Kahiko: Tales and Traditions of the People of Old.* 1991.)

o. Kamakau, S.M. 1961. *Ruling Chiefs of Hawaiʻi*. Kamehameha Schools Press, Honolulu. (A series of articles focused on the exploits of the aliʻi (ruling chiefs) of Hawaiʻi from ancient times to the time of the establishment of the Hawaiian Kingdom.)

p. Kanahele, G. 1995. Waikīkī 100 BC to 1900 AD: An Untold Story. privately published, Queen Emma Land Co.: Honolulu.

q. Kepelino. 1932 (2007). Traditions of Hawaiʻi. Bishop Museum Press, Honolulu. (In original Hawaiian language and English translation.)

r. Luomala, K. Voices on the Wind. Polynesian Myths and Chants. Bishop Museum Press.

s. Malo, D. Ka Moʻoʻolelo Hawaiʻi. Bishop Museum Press, Honolulu. (Original Hawaiian language volume, companion volume in English entitled Hawaiian Antiquities.)

t. Manu, M. 1884-85. *He Moolelo Kaao no Keaomelemele*. Bishop Museum Press, Honolulu. (Extracted from the Hawaiian language newspaper *Ka Nupepa Kuokoa* and published as a compilation, in original Hawaiian language with English translation.)

u. Multiple authors. 1856-1885. *He Lei No Emmalani: Chants for Queen Emma Kaleleonālani.* (200+ Hawaiian chants and songs composed by as many authors, compiled, translated and edited by Mary Kawena Pukui, Theodore Kelsey, and M. Puakea Nogelmeier, in original Hawaiian language with English translations.)

v. Nakuina, M.K. 1902. Moʻolelo Hawaiʻi o Pākaʻa a me Kū-a-Pākaʻa, Nā Kahu Iwikuamoʻo o Keawenuiaʻumi, ke aliʻi o Hawaiʻi, a ʻo nā moʻopuna a Laʻamaomao! Compiled and reprinted by Kalamaku Press, Honolulu. (In original Hawaiian language, with English translation in a separate partner volume.)

w. Office of Hawaiian Affairs. 2011-1018. Papakilo Database: Kūkulu ka ʻike i ka ʻŌpua. (A database of land grant data, Hawaiian language Newspapers, Place Name databases, and other sources of information on Hawaiian lands.) https://www.papakilodatabase.com/main/about.php

x. Pukui, M.K. & A. Korn. 1973. *The Echo of Our Song: Chants and Poems of the Hawaiians.* University of Hawaiʻi Press, Honolulu. (In Hawaiian language with English translations.)

y.   Pukui, M.K., S. Elbert, & E. Mookini. 1976. *Place Names of Hawaii*. University of Hawai'i Press, Honolulu.

z.   Pukui. M.K. 1983. *'Ōlelo No'eau*. Bishop Museum Press, Honolulu. (A compilation of 2942 Hawaiian proverbs and poetical sayings, in Hawaiian language, with English translations.)

aa.   Pukui, M.K., & L. Green. 1995. *He Mau Ka'ao Hawai'i: Folktales of Hawai'i*. Bishop Museum Press, Honolulu. (In Hawaiian language and English translations.)

bb.   Stokes, J. 1991. Heiau of the Island of Hawai'i: A Historic Survey of Native Hawaiian Temple Sites. Bishop Museum Press, Honolulu.

cc.   Sterling, E. & C. Summers. 1978. *Sites of O'ahu*. Bishop Museum Press, Honolulu.

dd.   Summers C. 1971. Moloka'i: A Site Survey. Pac. *Anthropological Records 14*. Bishop Museum, Honolulu.

ee.   Sterling, E. P. *Sites of Maui*. Bishop Museum, Honolulu.

ff.   Ulukau.org (Hawaiian Electronic Library, a Hawaiian language compilation site, searched for numerous Hawaiian language newspaper articles too numerous to list (nupepa.org), providing descriptions of the biocultural geography of pre-contact and early post-contact Hawai'i). Maintained by Alu Like, Inc., Hale Kuamo'o, and the Bishop Museum, Honolulu.

References

1.   Loope, L.L. Hawai'i and the Pacific Islands. In *Status and Trends of the Nation's Biological Resources*; Mac, M.J., Opler, P.A., Puckett Haecker, C.E., Eds.; U.S. Geological Survey: Washington, DC, USA, 1999; pp. 747–774, ISBN 016053285X.

2.   DeMartini, E.E.; Friedlander, A.M. Spatial patterns of endemism in shallow-water reef fish populations of the northwestern Hawaiian Islands. *Mar. Ecol. Prog. Ser.* **2004**, *271*, 281–296. [CrossRef]

3.   Eldredge, L.G.; Evenhuis, N.L. *Hawai'i's Biodiversity: A Detailed Assessment of the Numbers of Species in the Hawaiian Islands*; Bishop Museum Press: Honolulu, HI, USA, 2003; Volume 76, pp. 1–28.

4.   Holdridge, L.R. *Life Zone Ecology*; Tropical Science Center: San Jose, Costa Rica, 1967; ISBN BWB13752780.

5.   Tosi, J.A.J.; Watson, V.; Bolaños, R. *Life Zone Maps: Hawaii, Guam, American Samoa, Northern Mariana Islands, Palau, and the Federated States of Micronesia*; Tropical Science Center and the Institute of Pacific Islands Forestry, USDA Forest Service: San Jose, Costa Rica; Hilo, HI, USA, 2002.

6.   Asner, G.P.; Elmore, A.J.; Hughes, F.R.; Warner, A.S.; Vitousek, P.M. Ecosystem structure along bioclimatic gradients in Hawai'i from imaging spectroscopy. *Rem. Sens. Environ.* **2005**, *96*, 497–508. [CrossRef]

7.   Hopton, M.E.; White, D. A simplified ecological footprint at a regional scale. *J. Environ. Manag.* **2012**, *111*, 279–286. [CrossRef] [PubMed]

8.   Jacobi, J.D.; Price, J.P.; Fortini, L.B.; Gon III, S.M.; Berkowitz, P. *Carbon Assessment of Hawaii: U.S. Geological Survey Data Release*; U.S. Geological Survey: Reston, VA, USA, 2017.

9.   Gon, S.M., III; Allison, A.; Cannarella, A.J.; Jacobi, J.D.; Kaneshiro, J.Y.; Kido, M.H.; Lane-Kamahele, M.; Miller, S.E. *A GAP Analysis of Hawaii—Final Report*; U.S. Geological Survey, Research Corporation of the University of Hawai'i: Honolulu, HI, USA, 2006.

10.   U.S. Geological Survey. LANDFIRE.HI_100EVT—Hawai'i Existing Vegetation Type Layer: U.S. Geological Survey. 2009. Available online: http://landfire.cr.usgs.gov/viewer/viewer.html?bbox=-164,15.79,-151.67,25.48 (accessed on 25 June 2013).

11.   Gon, S.M., III; Matsuwaki, D. *Hawaiian Ecosystems: Before and after 1500 Years of Human Presence. Geographic Information System Shape Files Depicting Major Ecosystem Extent in the Hawaiian Archipelago in Pre-Human and Current Context*; The Nature Conservancy of Hawai'i: Honolulu, HI, USA, 1998.

12.   Hawaii Conservation Alliance. *HCA Strategic Action Plan 2017–2022*; East-West Center: Honolulu, HI, USA, 2017.

13.   Armstrong, R.W.; Bier, A.J. *Atlas of Hawaii*, 3rd ed.; University of Hawai'i Press: Honolulu, HI, USA, 1983; ISBN 9780824821258.

14.   Kirch, P.V. *Feathered Gods and Fishhooks: An Introduction to Hawaiian Archaeology and Prehistory*, 1st ed.; University of Hawai'i Press: Honolulu, HI, USA, 1985; ISBN 9780824819385.

15.   Kirch, P.V. Rethinking east Polynesian prehistory. *J. Polyn. Soc.* **1986**, *95*, 9–40.

16.   Kirch, P.V.; Green, R.C. History, phytogeny and evolution in Polynesia. *Curr. Anthropol.* **1987**, *28*, 431–456. [CrossRef]

17.   Dudley, M.K. *A Hawaiian Nation I: Man, Gods, and Nature*; Nā Kāne o ka Malo Press: Honolulu, HI, USA, 1990; ISBN 1878751158.

18. Winter, K.B.; Beamer, K.; Vaughan, M.B.; Friedlander, A.M.; Kido, M.; Whitehead, A.N.; Akutagawa, M.K.H.; Kurashima, N.; Lucas, M.P.; Nyberg, B. The *Moku* System: Managing biocultural resources for abundance within social-ecological regions. *Sustainability* **2018**, in press.

19. Kirch, P.V. The impact of the prehistoric Polynesians on the Hawaiian ecosystem. *Pac. Sci.* **1982**, *36*, 1–14.

20. Price, J.P.; Gon, S.M., III; Jacobi, J.D.; Matsuwaki, D. *Mapping Plant Species Ranges in the Hawaiian Islands: Developing a Methodology and Associated GIS Layers*; Hawai'i Cooperative Studies Unit Technical Rep. HCSU-008; University of Hawai'i at Hilo: Hilo, HI, USA, 2007.

21. Rees, W.E. Ecological footprints and appropriated carrying capacity: What urban economics leaves out. *Environ. Urban.* **1992**, *4*, 121–130. [CrossRef]

22. Wackernagel, M. Ecological Footprint and Appropriated Carrying Capacity: A Tool for Planning Toward Sustainability. Ph.D. Thesis, The University of British Columbia, School of Community and Regional Planning, Vancouver, BC, Canada, 1994.

23. Kamehameha Schools Bishop Estate. *Life in Early Hawai'i: The Ahupua'a*; Kamehameha Schools Press: Honolulu, HI, USA; 1994; ISBN 9780873360388.

24. Mitchell, D.D.K. *Resource Units in Hawaiian Culture*; Kamehameha Schools Press: Honolulu, HI, USA, 1982; ISBN 9780873360166.

25. Johnson, R.K. *The Kumulipo Mind: A Global Heritage*; Privately Published: Honolulu, HI, USA, 2000.

26. Winter, K.B.; Lucas, M. Spatial modeling of social-ecological management zones of the ali'i era on the island of Kaua'i, with implications for large-scale biocultural conservation and forest restoration efforts in Hawai'i. *Pac. Sci.* **2017**, *71*, 457–477. [CrossRef]

27. Ladefoged, T.N.; Kirch, P.V.; Gon, S.M., III; Chadwick, O.A.; Hartshorn, A.S.; Vitousek, P.M. Opportunities and constraints for intensive agriculture in the Hawaiian archipelago prior to European contact. *J. Archaeol. Sci.* **2009**, *36*, 2374–2383. [CrossRef]

28. Hawai'i State Historic Preservation Division (SHPD). *State Inventory of Historic Places Database (SIHP) as of 2013*; SHPD: Kapolei, HI, USA, 2013.

29. Kirch, P.V. *Roots of Conflict: Soils, Agriculture, and Sociopolitical Complexity in Ancient Hawai'i*; School for Advanced Research Press: Santa Fe, NM, USA, 2011; ISBN 9781934691267.

30. Anonymous Author. Aloha 'Āina. *Ka Nūpepa Puka Aloha 'Āina* **1893**, *1*, 18.

31. Poepoe, K. *Systems and the Ahupua'a: Nā Mea Ola o Nā Wao (Teacher Resources in Mālama 'Āina)*; Hui Mālama o Mo'omomi and The Pacific American Foundation: Kualapuu, HI, USA, 2009.

32. Office of Hawaiian Affairs. Kipuka Database. Available online: http://kipukadatabase.com/kipuka/ (accessed on 7 May 2013).

33. Arista, N. *The Kingdom and the Republic: Sovereign Hawai'i and the Early United States*; University of Pennsylvania Press: Philadelphia, PA, USA, 2018; ISBN 9780812250732.

34. Office of Hawaiian Affairs. Webmap: Moloka'i Pre-Contact vs Current Land Use Footprint. Available online: http://kipukadatabase.com/apps/footprint%20molokai/ (accessed on 7 May 2013).

35. Gon, S.M., III; Tom, S.L.; Jacobi, J.D. Human Footprint & Native Ecosystem Extent in 1870 on the Island of Hawai'i: A pivotal period in the history of ecosystem change. In Proceedings of the Hawai'i Conservation Conference, Honolulu, HI, USA, 18–20 July 2017; Hawai'i Conservation Alliance: Honolulu, HI, USA, 2017.

36. The Nature Conservancy (TNC). Hawaiian High Islands Ecoregional Plan. Web-based biodiversity conservation planning document by the Nature Conservancy of Hawai'i and other partner organizations. Available online: http://hawaiiecoregionplan.info (accessed on 6 June 2018).

37. Cuddihy, L.W.; Stone, C.P. *Alteration Native Hawaiian Vegetation: Effects of Humans, their Activities and Introductions*; University of Hawai'i Press: Honolulu, HI, USA 1990; ISBN 9780824813086.

38. Kirch, P.V. Archaeology and global change: The Holocene record. *Annu. Rev. Environ. Resour.* **2005**, *30*, 409–440. [CrossRef]

39. Ziegler, A.C. *Hawaiian Natural History, Ecology, and Evolution*; University of Hawai'i Press: Honolulu, HI, USA; 2002; ISBN 9780824842437.

40. Leung, P.S.; Loke, M. *Economic Impacts of Increasing Hawaii's Food Self-Sufficiency. Economic Issues 6*; College of Tropical Agriculture and Human Resources, University of Hawai'i: Honolulu, HI, USA, 2008.

41. Kent, G. *Food Security in Hawai'I*; Department of Political Science, University of Hawai'i: Honolulu, HI, USA, 2016.

42. Wagner, W.L.; Herbst, D.R.; Sohmer, S.H. *Manual of the Flowering Plants of Hawai'i*; Bishop Museum Special Publication 83; University of Hawai'i and Bishop Museum Press: Honolulu, HI, USA, 1999; Volume 2, ISBN 0824821661.

43. Loope, L.L.; Hamann, O.; Stone, C.P. Comparative conservation biology of oceanic archipelagoes. *BioScience* **1988**, *38*, 272–282. [CrossRef]

44. Athens, J.S.; Tuggle, H.D.; Ward, J.V.; Welch, D.J. Avifaunal extinctions, vegetation change, and Polynesian impacts in prehistoric Hawai'i. *Archaeol. Ocean.* **2002**, *37*, 57–78. [CrossRef]

45. Neal, M. *In Gardens of Hawaii*; Honolulu, Hawaii; Bishop Museum Press: Honolulu, HI, USA, 1965; ISBN 9780910240338.

46. Ooka, J.J. *Taro Diseases: A Guide for Field Identification*; College of Tropical Agriculture and Human Resources, University of Hawai'i: Honolulu, HI, USA, 1994.

47. Conry, P.J.; Cannarella, R. *Hawaii Statewide Assessment of Forest Conditions and Trends*; Department of Land and Natural Resources, Division of Forestry and Wildlife: Honolulu, HI, USA, 2010; pp. 154–178.

48. State of Hawai'i. *Hawai'i Interagency Biosecurity Plan. 2017-2027*; Hawaii Department of Agriculture and the Department of Land and Natural Resources: Honolulu, HI, USA, 2016. Available online: https://hdoa.hawaii.gov/wp-content/uploads/2016/09/Hawaii-Interagency-Biosecurity-Plan.pdf (accessed on 26 September 2016).

49. Fortini, L.B.; Jacobi, J.D.; Price, J.P. Projecting end-of-century shifts in the spatial pattern of moisture zones across Hawai'i to assess implications to vegetation shifts. In *Baseline and Projected Future Carbon Storage and Carbon Fluxes in Ecosystems of Hawai'i*; Zhu, Z.L., Reed, B.C., Eds.; U.S. Geological Survey Professional Paper 1834; Reston, VA, USA, 2017; pp. 21–42. [CrossRef]

50. Fortini, L.B.; Price, J.P.; Jacobi, J.D.; Vorsino, A.E.; Burgett, J.M.; Brinck, K.W.; Amidon, F.; Miller, S.; Koob, G.; Paxton, E.H. *A Landscape-Based Assessment of Climate Change Vulnerability for All Native Hawaiian Plants*; Hawai'i Cooperative Studies Unit Report 044; University of Hawai'i at Hilo: Hilo, HI, USA, 2013.

 *sustainability*

Article

'Āina Kaumaha: The Maintenance of Ancestral Principles for 21st Century Indigenous Resource Management [†]

Natalie Kurashima *, Jason Jeremiah, A. Nāmaka Whitehead, Jon Tulchin, Mililani Browning and Trever Duarte

Kamehamea Schools, Natural and Cultural Resources, Honolulu, HI 96813, USA; jajeremi@ksbe.edu (J.J.); nawhiteh@ksbe.edu (A.N.W.); jotulchi@ksbe.edu (J.T.); rebrowni@ksbe.edu (M.B.); trduarte@ksbe.edu (T.D.)
* Correspondence: nakurash@ksbe.edu; Tel.: +1-808-322-5348
† We choose not to italicize Hawaiian words, in recognition of the indigenous language and space that we write from within Hawai'i. We refer to Hawaiian organisms by their Hawaiian names.

Received: 1 August 2018; Accepted: 25 October 2018; Published: 31 October 2018

Abstract: Globally, there is growing recongition of the essential role indigneous people have in biocultural conservation. However, there are few cases of applied indigenous resource management today, especially from the indigenous standpoint. In this paper, we provide an example of the maintenance and adaptation of an indigenous resource management system in Hawai'i from the perspective of an instrumental 'Ōiwi (Indigenous Hawaiian) social institution, Kamehameha Schools. Kamehameha Schools is not only the largest private landowner in Hawai'i, but is uniquely tied to a lineage of traditional ali'i (chiefs) resulting in present-day influence, decision-making authority, and wealth to fund a perpetual vision for its ancestral lands and communities. Notably, we share our journey from the perspective of indigenous resource managers, using the 'Ōiwi methodology of mo'okū'auhau (genealogy and continuity) to guide our (re)discovery of what it means to steward in an indigenous way. First, we ground ourselves in 'Ōiwi worldviews, recognizing our genealogical and reciprocal connections to 'āina (land and sea). Then, we examine the functions of the traditional institution of the ali'i and the chiefly principle of 'āina kaumaha—a heavy obligation to steward the biocultural health of lands and seas in perpetuity. We detail how 'āina kaumaha has manifested and transferred over generations, from traditional ali'i to the royal Kamehameha line, to Kamehameha Schools as an ali'i institution. Finally, we discuss how we endeavor to meet inherited obligations through Kamehameha Schools' resource management approach today, which includes active stewardship of vast tracts of native ecosystems and Hawai'i's most important cultural sites, influencing biocultural well-being through representing 'Ōiwi perspectives in diverse industries, and developing the next generation of 'Ōiwi stewards. We provide a guide for indigenous organizations (re)defining their ancestral ways of stewardship, as well as for the many non-indigenous agencies with obligations to native lands and people today working to incorporate indigenous systems into their current management. Given that much of the world's lands are indigenous spaces, we argue that the restoration of effective biocultural resource management systems worldwide requires the maintenance, and in some cases reestablishment, of indigenous institutions at multiple levels.

Keywords: indigenous resource management; Hawai'i; biocultural conservation

---

1. Introduction

As indigenous people are increasingly recognized as critical to biocultural resource conservation globally [1–4], some indigenous communities and organizations are regaining management of their ancestral lands and resources [5–9]. Yet, the question remains, how can the world manage biocultural land and seascapes in an indigenous way in the 21st century, given the immense environmental and social changes that indigenous systems have endured over the last several hundred years. Hawai'i exemplifies intense socio-ecological change. The archipelago has some of the highest rates of endemism on land [10] and in the sea [11] in the world, but is challenged with extreme threats of invasive species and habitat loss, leading to high degrees of extinction [12]. Kānaka 'Ōiwi (Indigenous Hawaiian people) who have stewarded the lands and seas of Hawai'i for a millennium, have been systematically dispossessed of sovereign governance, including loss of stewardship and access to ancestral lands as well as loss of traditional management and tenure of shoreline and ocean resources [13], and face disproportionately high rates of poverty, homelessness, health issues, abuse, and incarceration [14,15]. As the health and well-being of indigenous people are inextricably linked to the health of their ancestral places [16–18], it is critical to understand past and current systems of indigenous stewardship not only for potential resource management benefits, but for the plethora of interrelated social values which drive collective well-being.

We know that indigenous management systems are by their nature responsive, adaptive to social and ecological change, and transform over time [19], and therefore could be applicable to address today's sustainability challenges that fundamentally bridge disciplines, such as climate change adaptation [13]. Moreover, many indigenous resource management institutions and their governance have been shown to be especially sustainable and resilient over long periods of time [20]. Yet, there is a scarcity of examples of applied indigenous resource management [21–23]. Furthermore, existing models of application and integration of indigenous knowledge in resource management often focus on data-oriented knowledge that can easily fit into western science frameworks such as knowledge of weather and climate [24,25], animal ecology [26], phenology [27] or management practices like marine prohibitions [28], forest patch protection [29], agroecological practices [30], and watershed-based management [31]. Focusing on knowledge and management practices alone ignores the fact that indigenous knowledge is inextricably nested within systems of practice and belief [19], and successful application of indigenous knowledge systems for resource management depends on the conservation of these through the system's social institutions and worldviews [29,32,33].

Additionally, the socio-ecological literature championing indigenous systems of stewardship lacks cases from the perspective of indigenous authors, although there are a few examples [34–36]. When working to understand and bridge indigenous systems of resource management, it is critical to learn directly from the indigenous point of view. Beyond that, indigenous-led examples allow for a self-determined approach and expression grounded in ancestral ways of knowing [36–38]. In this paper, we provide an example of the maintenance and adaptation of an indigenous resource management system in Hawai'i from the perspective of an influential 'Ōiwi (Indigenous Hawaiian) social institution, Kamehameha Schools. Kamehameha Schools is not only the largest private landowner in Hawai'i, but as we explain below, is tied to a lineage of traditional chiefs resulting in unique influence and wealth today.

1.1. Brief History of Kamehameha Schools

Shortly after Western Contact in 1778, the famous ali'i (chief) Kamehameha I, united the Hawaiian Islands under his rule for the first time in Hawai'i history, forming the Hawaiian Kingdom [39]. This consolidation of power allowed for the hereditary passing of lands to his descendants, who would remain the dominant chiefs through the Hawai'i Kingdom era. Even throughout the process of Western-based land privatization in the mid-1840's, Kamehameha's descendants retained control of vast tracts of lands. These lands were passed down within the family until the last direct descendant of the Kamehameha line, Bernice Pauahi Bishop. During her lifetime (1831–1884), Pauahi witnessed a substantial 35% decline in the Native Hawaiian population mainly due to Western disease. With that decline came a loss of 'Ōiwi lifeways and tradition. Because Pauahi believed that education would offer her people the best future, she left her entire estate, nine percent of the Hawaiian Islands, to establish the Kamehameha Schools. At her passing, Pauahi's estate totaled 375,500 acres of land assessed at about $474,000. Today, Kamehameha Schools' estate includes nearly 365,800 acres or ten percent of Hawai'i's land (Figure 1), and combined with other assets, is valued at $11.5 billion [40]. As a result of Pauahi's vision, Kamehameha Schools' mission focuses on the creation of educational opportunities in perpetuity to improve the capability and well-being of people of Kānaka 'Ōiwi ancestry.

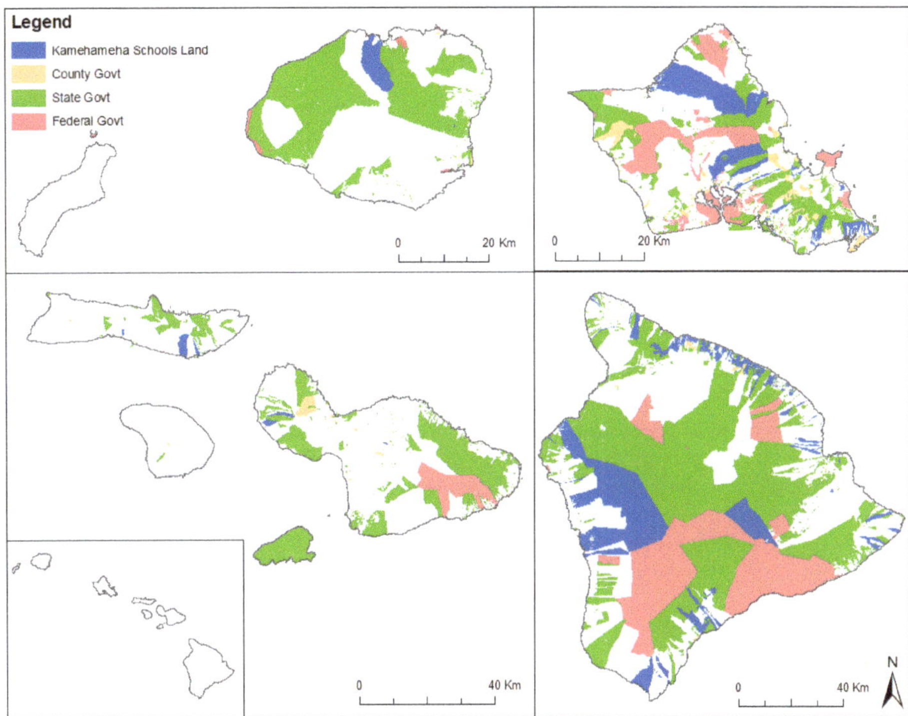

Figure 1. Map showing distribution of Kamehameha Schools lands (blue) along with lands managed by the Counties (orange), the State (green), and the Federal (pink) governments.

1.2. Development of Kamehameha Schools' Natural and Cultural Resource Management

In 2002, the department responsbile for Natural and Cultural Resources stewardship was created. We, the authors of this paper, currently make up this team at Kamehameha Schools, which is responsible for the stewardship of all of the native ecosystems and cultural sites and landscapes within the institution's 365,800 acre landholdings. At its inception, our team was charged with the task of "mālama i ka 'āina", an 'Ōiwi concept literally translated as "care for the land", which the institution defines as "ethical, prudent, and culturally-appropriate stewardship of lands and resources" [41]. Those of us who were there at the time, along with our predecessors and former leaders (Neil Hannahs, Ulalia Woodside, and Kekuewa Kikiloi), set out to understand and (re)discover what it means to steward Kamehameha Schools' land in an 'Ōiwi way, something that was customary in our history, but had been largely lost within the institution in the recent century. Since 2002, we have taken deep dives into resources, including within historical documents, 'Ōiwi scholarship, and our own institutional archives, so that we could understand why and how to do our work of appropriate 'Ōiwi stewardship of Kamehameha Schools land.

Of the authors, five of us are Kānaka 'Ōiwi and all of us are kama'āina (born and raised in Hawai'i; literally "land child" [42]) with collectively over 40 years of experience managing resources at Kamehameha Schools from an indigenous institutional perspective. This figure is even more extraordinary when considering the fact that our team's average age is currently just 35 years old. Our backgrounds include training in both 'Ōiwi and Western knowledge systems in the fields of: Ritual, Oli (the practice of chant), 'Ōlelo Hawai'i (Hawaiian Language), Traditional Hawaiian Land Use, Biocultural Resource Management, Ethnoecology, Botany, Archaeology, Anthropology, Hawaiian Studies, Urban and Regional Planning, Agroecology. We consistently bridge these complementary systems and disciplines in our work directing the maintenance and restoration of biocultural integrity on Kamehameha Schools' lands. This paper is a result of our personal experience of years developing an indigenous way of stewardship through on-the-ground natural and cultural resource management.

As an indigenous organization, Kamehameha Schools is unique, in Hawai'i and within the broader global context, because it is (1) a large-scale owner and steward of ancestral lands with (2) substantial decision-making authority and (3) financial wealth to fund a perpetual vision for its lands and communities. Albeit, within a larger colonial hierarchy (U.S. state, federal, and international regulations and law). So, what does it look like, when an indigenous organization has the power to be in full control of resources for the betterment of their native communities in perpetuity? There are far-reaching implications of sharing our story, for other indigenous organizations around the world that are (re)discovering their own ways of stewardship. Moreover, there are state, federal, and other non-governmental organizations that do not have indigenous-focused missions in Hawai'i and around the world, but are now responsible for the management of former indigenously-stewarded lands and seas. In Hawai'i, those include some of the most bioculturally important lands that were originally bestowed to the Hawaiian Kingdom Government or to the Crown, including the summits of sacred mountains (e.g., Mauna Kea (see [43]), Mauna Loa, Kīlauea, Haleakalā, Ka'ala) as well as the majority of the watersheds, and all of the nearshore and marine seascapes across the archipelago (Figure 1). Because of the growing recognition of the value of biocultural resource management [44], many of these agencies are working towards incorporating 'Ōiwi ways of knowing into their current management [45,46]. Our case study can provide a potential roadmap for these non-indigenous agencies working to bridge multiple knowledge systems to manage or restore biocultural abundance in the lands and communities they are responsible for today.

In this case study of applied indigenous resource management, we have used 'Ōiwi methodologies to define for ourselves what it means to steward 'āina (land and seascapes) in an 'Ōiwi way. Therefore, this paper does not follow the standard Introduction, Methods, Results, and Discussion format of a scientific journal article, but rather is framed by the 'Ōiwi concept of mo'okūauhau (genealogy). First, we work to understand our mo'okūauhau, our continuity in our positions as 'āina stewards of an ali'i institution. We look to indigenous worldviews and traditional institutions particular to us as Kamehameha Schools land stewards through examining primary 'Ōiwi sources, such as mele (chants) and ka'ao (cosmologies), mo'olelo (life stories) and 'ōlelo no'eau (wise sayings of biocultural significance), as well as secondary source interpretations of traditional concepts from early Kānaka 'Ōiwi scholars. Second, we examine the transference of obligation genealogically and through landscapes. We look to chiefs of the Kamehameha lineage and how they related to their land holdings during their lifetime through chants written for them, mo'olelo in Hawaiian language newspaper articles of their time, as well as documentation of their land stewardship in their letters and land reports. The task is to figure out who we are, not in the sense of our name, or title, but *what is our function*? We do this by determining where we sit in the genealogical framework of the institution. Third, we discuss the ways in which our resource management predecessors took on the responsibility, to both our 'āina and our Kānaka 'Ōiwi communities, again examining internal land documents and letters of our institutional predecessors. Fourth, we consider our research and resource management experiences to discuss how we as an indigenous institution endeavor to meet these genealogical obligations through our resource management program today. We seek to answer the question, how do we serve to continue the function of those before us our institutional genealogy in today's socio-ecologic context?

2. 'Ōiwi Worldview: Mo'okū'auhau & Aloha (Genealogy & Reciprocity)

First we had to understand how and where we as humans sit in relation to the rest of our Hawaiian universe, including the natural and cultural resources we steward, through an 'Ōiwi worldview. From this perspective, kānaka (people) have a shared ancestry with 'āina (land and sea), inclusive of the earth, sky, the celestial bodies, all living things in the sea, on land and in the atmosphere, as well as the dynamic processes that sustain these systems [47–51]. In the Kumulipo cosmology, which details the creation of the Hawaiian universe, in the first wā (time) there is only darkness [48]. The chant states that all life originates from the primordial slime. First is born Kumulipo, a male, and Po'ele a female, then the 'ukuko'ako'a (coral polyp), the many creatures and algae of the ocean and their counterpart plants on land. The two thousand line chant recognizes the birth of all our fish and invertebrates and their plant counterparts, the insects, forest and seabirds, turtles, lobsters, those that cling, the pig, the rat, and the dog. In the eighth wā, the female La'ila'i and the male Ki'i were born and the gods Kāne and Kanaloa were born. It is from La'ila'i and Ki'i that generations of ali'i and then commoners were born. The Kumulipo also recognizes the genealogy of Hāloa—the first kalo and elder brother of the Hawaiian people, who came from Wākea (expansive sky), Papa (earth foundation), and Ho'ohōkūkalani (star establisher). In this single chant, one can see the intimate kinship that the people of Hawai'i share with the Hawaiian universe, from the tiny coral polyp to every fish or plant or animal, each is foundational to our genealogy as kānaka.

"I ola 'oe, i ola mākou nei."
When you live, so do we.
—Hi'iakaikapoliopele to an 'Ōhi'a (Metrosideros polymorpha) forest [52]

In genealogies such as the Kumulipo, we recognize that 'āina—inclusive of all the native lifeforms and ecological processes, came before us as kānaka, and have created the foundations upon which we live and thrive. Thus, we all have inherited the responsibility through our genealogy as Kānaka 'Ōiwi to ensure the continuation of such foundations. The health of 'āina is inherently and reciprocally related to the health and well-being of its people [18,53]. For example, our existence depends on the health and existence of the forest, all lifeforms and ecological processes therein, which provides us with fresh water, climate regulation, materials for construction, and medicine, as well as our ancestral plants like the endemic and bioculturally foundational 'ōhi'a (*Metrosideros polymorpha*), which also provides cultural inspiration, reminding us how to adapt and flourish in harsh conditions, while also nourishing those around us. The opposite is true as well—that the forest's health depends on our health. We must ensure that our human activities do not adversely affect the forest, not overharvest any of its elements, protect it from weeds and invasive animals, and continue to chant, sing, and dance to honor its existence. Within this worldview, we participate in a reciprocal relationship with the natural world—to take and give in kind, as a sound and necessary means to ensure our collective well-being inclusive of 'āina.

3. The Ali'i Institution

In order to understand where we sit in Kamehameha Schools' continuity as an ali'i institution stewarding lands, we must first understand the basic traditional systems of how ali'i related to and were responsible for 'āina. It is outside of the scope of this paper to give a comprehensive summary of the ali'i institution. Here, we provide our understanding of the ali'i particularly in regards to stewardship of 'āina in the past, so that we can think about application of these concepts today.

3.1. Reciprocal Relationships between ʻāina, aliʻi, and makaʻāinana

"ʻIliʻili o Hāloa."
Pebbles of Hāloa.
Descendants of chiefs of Hāloa, grandson of Wākea and Papa, or any chiefs descended from the gods. [54] #1227.

In this section we discuss the social institution of the aliʻi class in traditional Hawaiʻi, describing the reciprocal relationships between (1) the ʻāina, (2) the aliʻi, and (3) the makaʻāinana (commoners) (Figure 2). In the ʻŌiwi worldview, the aliʻi were recognized as close descendants of the akua (elemental deities or natural phenomena often translated as gods) because they came before makaʻāinana genealogically [48,55]. The high aliʻi were considered to be direct descendants of akua [56], with some accounts calling highest niʻaupiʻo chiefs "gods among men" [48]. Akua are natural forces and elements that sustain life. For example the major akua Kāne and his many forms include fresh water, the sun, air currents, as well as the associated forces such as the gathering of clouds, the red-hued setting sun, or the dark density within a storm [47] (p. 104–105). Akua are also the lifeforms on earth, including the animals, rocks, and plants, such as the ʻieʻie (*Freycinetia arborea*), the climbing monocot in the forest also associated with Kāne due to the ways in which it intercepts and distributes rain into the forest understory [57]. One can see that akua *are* ʻāina, as they make up the land and seas and all of the processes therein [33,44]. ʻāina, the term we use for land and seascapes, is translated as "that which feeds" physically, spiritually, mentally. Akua, and ʻŌiwi spirituality more generally, cannot be separated from ʻāina.

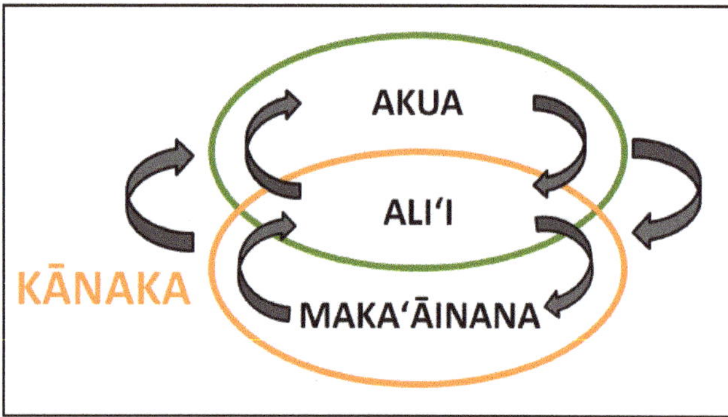

Figure 2. Hierarchical and reciprocal relationships between akua, aliʻi, and makaʻainana.

Akua were described by Pukui, Haertig, and Lee as "the impersonal gods of Hawaii, powerful, distant deities whose origins were lost in dim corridors of time" [56] (p. 23). However, because of their close genealogical relationship, the aliʻi could maintain an intimate connection with higher level akua, that makaʻāinana could not. Although, there were other less significant personal gods that makaʻāinana had access to [56].

"Hānau ka ʻāina, hānau ke aliʻi, hānau ke kanaka."
Born was the land, born were the chiefs, born were the commoners.
The land, the chiefs, and the commoners belong together [54] #466.

Unlike Western feudal relationships between chiefs and commoners, in traditional Hawaiʻi, aliʻi had a close kinship with the makaʻāinana. The makaʻāinana considered the aliʻi, an elder sibling [53]. They needed one another, while the makaʻāinana cultivated abundance on both land and sea with expert skills spanning realms of engineering, botany, medicine, navigation, psychology, sport, fishing, farming, architecture, etc. [58], and the aliʻi provided access to akua, protection from war, enforcement of social norms, as well as the maintenance of land and seascape integrity. Because of the genealogical relationship to akua (who came before) and maka'ainana (who came after), the ali'i had a duty to care for both akua/'āina and their people. There were many types of ali'i situated within a hierarchical structure, correlating to various scales of resource management. The highest ali'i (e.g., Ali'i Nui) was entrusted with the coordination of the largest scale of resource management at the scale of the island. While the ali'i below him were in charge of the stewardship according to various smaller socio-ecological divisions [50] (Figure 3).

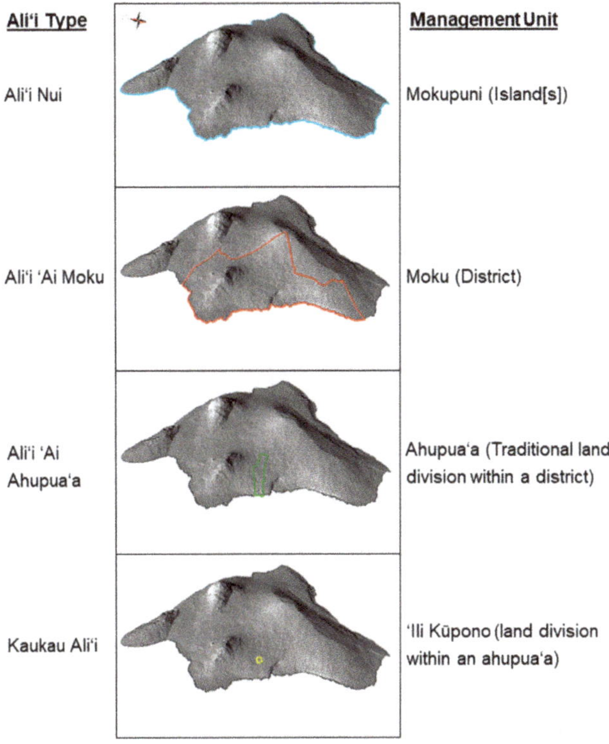

Figure 3. The various scales of socio-ecological management by ali'i in traditional Hawai'i [50].

3.2. Kaumaha-Chiefly Obligation to Steward in Perpetuity

As we (re)understood the relationship ali'i had with 'āina and maka'āinana, we thought more about the vast responsibility that the ali'i shouldered to ensure the functioning of the socio-ecological system. We enlisted Kānaka 'Ōiwi scholars at the indigenous organization, the Edith Kanaka'ole Foundation (EKF) to assist us in understanding these profound concepts. We describe the weighty or heavy burden of ali'i to safeguard and perpetuate resources in perpetuity for his or her people as kaumaha [47,57]. We use kaumaha to describe a deep, imperative responsibility that one cannot easily relinquish, such as that of landscape-level 'āina stewardship.

The word kaumaha is colloquially most often understood as the feeling of sadness, grief and sorrow, and contemporarily, students often learn this word early in 'ōlelo Hawai'i classroom training as an example stative verb in sentences like: "Kaumaha au, I am sad." However, connotations of grief and sadness are not what we wish to convey in this context of 'āina stewardship. Pukui, Haertig and Lee (2002) explain that kaumaha as grief and sorrow is derived from the *"original use of kaumaha meaning weight or heavy weight and from use of the separate syllables, kau (place, put, set) and maha (relief or rest). From the most literal connotation, that holding a physical weight is followed with relief when it is set down, came the abstract idea that grief is a heavy weight followed by relief."* [56] (p. 132). The literal and figurative kaumaha is somewhat like the English word burden, which refers to a heavy load that one carries, but figuratively and colloquially is understood as misfortune or hardship. In our context describing the perpetual duty to maintain and steward 'āina, we refer to the original meanings of kaumaha: heavy, weighty, profound, deep and significant, and not its subsequent often cited figurative definitions.

Furthermore, we contrast the concept of kaumaha—an ali'i obligation of stewardship, with the more familiar term kuleana, meaning right, privilege, responsibility [42]. Traditionally, when a new Ali'i Nui ascended to power, all of the lands of the Island would be redistributed to the various ali'i allies under him or her. The rights and responsibilities (or kuleana) that maka'āinana had on those lands would not change through these conversions. On the other hand, the rights and obligations (or kaumaha) that individual ali'i had to steward lands given to him or her could change drastically [59]. We view the right and responsibility maka'āinana had to live on and care for their lands and seas as a kuleana, and indeed this is the name given to the lands that maka'āinana claimed during the land privatization process of the 1840's [45]. Both kaumaha and kuleana are inherited responsibilities through land and mo'okū'auhau, however, we argue that because of the intimate interdependence that ali'i had with akua and maka'āinana across various landscape scales, they carried a different obligation to maintain critical 'āina functions for all (i.e., water cycling—discussed below).

'O ke akua ke komo, 'a'oe komo kānaka
The gods may enter, man cannot enter
'O ke kāne huawai, he akua kēnā
Man with the water gourd, he is the god
Kumulipo, Lines 111–112 [48]

In the Kumulipo lines above, the man with the kāne huawai (water gourd) is refered to as an akua. Why does this kāne huawai elevate this man to akua status? We understand this passage to be about the maintenance of the water cycle, and that the man who perpetuates the sources and driving forces of water is not a merely a man but is an ali'i and god-like [57]. The ali'i as the principle land steward had the kaumaha or imperative duty to maintain the water cycle that sustains all life—the streams, the groundwater, the forested watershed, evaporation, condensation, precipitation [57]. The maintenance of the function of the watershed ensured that all lifeforms in the Hawaiian realm would be supported, including people. This involved the protection of the geologic formations that drive weather systems, the large trees that interact with the atmosphere to produce rain and cloudfall, the multi-layered forest that slows the force of descending raindrops, the groundcovers that absorb moisture and prevent erosion, as well as the aquifers and conduits that hold and distribute fresh water throughout the landscape.

In order to fulfill the kaumaha to ensure maintenance of 'āina in perpetuity, the ali'i imposed rules for resource interaction, which included kānāwai (laws, rules, or protocols) and kapu (sacred, prohibitions, taboos) at different scales [42,60,61]. Kānāwai and kapu are both essential elements of 'Ōiwi religion more broadly. Kānāwai and kapu indicate the relationship between natural phenomenon and natural phenomenon, Kānaka and natural phenomenon, and Kānaka to Kānaka", [47] (p. 45), meaning these concepts tell us (1) how different elements of the environment will interact, (2) how we as people and the environment should interact, and (3) how we as people should interact with one another. All three relationships are vital for biocultural resource health. The ali'i, as the institutors of kapu and kānāwai in traditional Hawai'i, governed through these relationships. First, they understood environmental element interactions and cycles through careful observation over generations. Second, using this expert scientific understanding of their environments, they understood resources had specific kapu or a sacredness which prohibited kānaka to access those resources at specific times. Ali'i enacted kānāwai of how kānaka should access those kapu resources. For example, seasonal kānawai were put upon many kapu fish according to observations of their spawning times, knowledge of their maturity rates, food abundance and availability, and other information [49,50,62]. And third, ali'i created and enforce rules of how people should interact with one another, an example being Kamehameha I's famous Kānāwai Māmalahoe, or Law of the Splintered paddle, which declared protections for all people, young and old, from violent assault [63].

3.3. Kaumaha Inherited Throughout the Kamehameha Line

As we began to understand kaumaha as a foundational concept of traditional ali'i stewardship, we wanted to determine how the ali'i who came before us demonstrated kaumaha in their 'āina management and how kaumaha was transferred. At the start of this journey to (re)discover our function, the genealogy of Kamehameha Schools' lands was forgotten. We knew that lands originally were consolidated under Kamehameha I and eventually were bequeathed to Bernice Pauahi, yet we did not know which other ali'i had kaumaha for these lands, and thus how they stewarded. One of the first tasks taken on by our predecessors was to determine which ali'i are connected to lands in the over 70 ahupua'a we have tenure for today. This was done by looking through the institutions property title archives, historic maps, as well as Māhele records (documents from land privatization in the 1840's), most of which are now electronically available online (see kipukadatabase.com, papakilodatabase.com, Avakonohiki.org). The results of this research are shown in Figure 4.

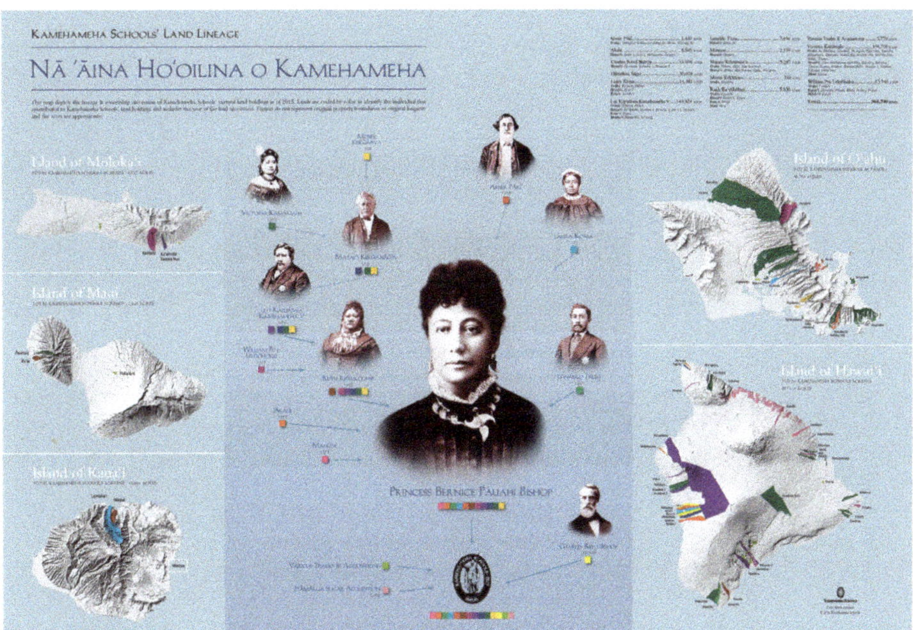

Figure 4. Map depicting 'āina ho'oilina or inherited lands from Bernice Pauahi Bishop's ali'i. ancestors.

More recently, we have looked to specific ali'i predecessors to understand how they carried kaumaha to steward the same lands we manage today. We found that the kaumaha that the ali'i held to ensure the persistence and prosperity of 'āina, including its flora, fauna, and human communities for generations to come are illustrated throughout the Kamehameha line of chiefs, starting with Kamehameha I. We found the principle of kaumaha evident even through Western contact and colonization, and the vast socio-ecological changes in Hawai'i. Stories of Kamehameha's life originally printed in Hawaiian language newspapers in the 19th and 20th centuries, often layout his accomplishments in stewardship and restoration of biocultural integrity [64–67]. One mo'olelo found in multiple accounts showed us how he managed resources during his rule, at a time just after first Western contact. When seeing his people harvesting small sandalwood (*Santalum* spp.; 'iliahi) trees from upland forests in the early 1800's at the behest of other lower-ranking ali'i who were taking advantage of new capitalistic economies, Kamehameha said to them:

> **"No ke aha la oukou i mana'o ai e kua i kēia lā'au'ala li'ili'i, 'oiai e nui aku ana 'o ia ma kēia mua aku? Ua pane mai la nā kākau ali'i me nā kānaka. Ua elemakule 'oe a ua kokoke mai nā lā hope o kou ola 'ana, a na wai la ia mau lā'au'ala, 'a'ole mākou i 'ike i ka mea nāna ia lā'au'ala ma kēia mua aku. Ia wā pane aku 'o Kamehameha iā lākou penei: 'A'ole anei 'oukou i 'ike i ka'u po'e keiki? 'O lākou auane'i ka po'e nāna ia mau lā'au'ala, a o lākou nō auane'i ko'u po'e kaulana 'āina ma kēia hope aku, ke huli a'e nā lā o ko'u ola 'ana ma kēia ao."**

> *Why did you folks think to cut these small sandalwood when it still has yet to grow? The royal scribe and the people replied, You are getting old and the end of your life is getting near, and for who are these sandalwoods? We don't see the ones who these sandalwoods belong to from here on out. Kamehameha responded to them: You folks don't see my future descendants? From this time, they are the ones these sandalwoods belong to, and they are indeed my land stewards hereafter, my living days are turning over in this realm.*

(Rev. J.F. Pokuea in Ke Aloha Aina 1896 [64]; Translated and diacriticals added by author)

This account suggests that Kamehameha may have instituted a kapu on sandalwood harvest during this time, as would be custom for an ali'i of his status. The story alludes to Kamehameha's (1) responsibility to manage of the forest resources at a landscape scale, larger than areas that are the responsibility of lesser ali'i; (2) recognition of the mutual dependence of kānaka and the sandalwood tree, (3) and that he is considering this mutla dependence not only in the current generation but for the numerous of generations of kānaka 'ōiwi, even those still unseen to us today.

Because ali'i, including Kamehameha maintained a close relationship with the akua, they were often called upon by their maka'āinana to serve as an intermediary in times of need [47]. Kamehameha exhibits his relationship during the 1801 lava flow in the Kekaha region of Kona, Hawai'i Island where he is asked by his maka'āinana to intercede with Pele (the Hawaiian deity of volcanism, fire, lava, eruption) as a lava flow was threatening the lands in the region. Even a century after Western contact, the people still looked to the ali'i as the social institution that communicated with natural phenomena, the akua. For example, the kaumaha to communicate with akua continues on to Kamehameha's great-granddaughter Ruth Ke'elikōlani (Figure 5). In 1881, Ke'elikōlani is also called upon by the people of Hilo to intercede with Pele in upper Hilo, where she successfully entreats with Pele to stop a lava flow heading towards Hilo [68].

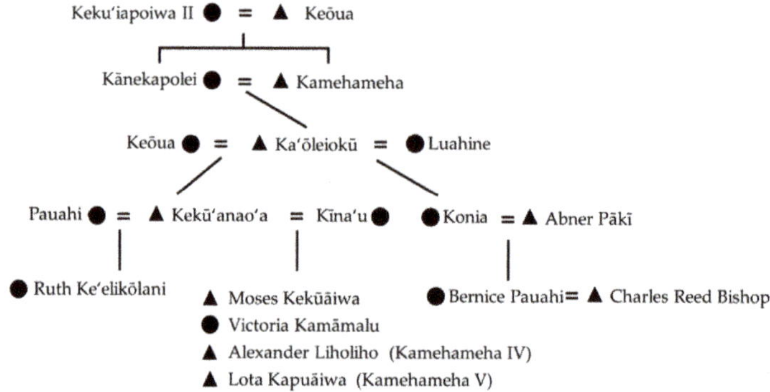

Figure 5. Mo'okū'auhau showing the ancestry of kaumaha from Kamehameha to his great-granddaughters Ruth Ke'elikōlani and Bernice Pauahi.

In this same era of the 1800's, we see the kaumaha of the ali'i being passed down from chief to their heirs not just genealogically, but also through the land tenure. When Ke'elikōlani, the largest landowner in Hawai'i, passes away in 1883, her land lineage from the Kamehameha line as well as the responsibility of the ali'i to be the link to akua is passed down to our organization's founder, Bernice Pauahi Bishop. Pauahi's name, inherited from her aunt (Ke'elikōlani's mother) translated as "destroyed by fire" [69], suggests a relationship between her kaumaha as a Kamehameha ali'i and the fire or Pele [47]. In Hawai'i tradition, mele inoa (name chants) were created to honor a newborn's genealogy from the baby's human family to the ancestral deities that came before them [55,68] Pauahi's name chant (a portion below) provides information codifying her connection with the akua Pele as well as her kaumaha to protect and maintain the water, and her connection as an ali'i to the akua [47].

He kapu he moe wai no ka uka
The taboo is the taboo of "water storage" for the uplands
Kahu ka 'ena, kai mu i kai ko'o, o Nauahi loa
Steward the heat, from silent seas to billowing seas, longevity of all manner of uahi (smoke)
(He Inoa No Pauahi -Birth Chant of Bernice Pauahi Bishop), Lines 12 and 81 [68]; Translated in [47])

3.4. Kamehameha Schools Inherits Kaumaha

Throughout her life, Pauahi inherited lands from her ali'i family members (Figure 4), with the largest set of lands from Ruth Ke'elikōlani at her death in 1883. At that time, Pauahi was the last reamining heir of the Kamehameha line, and her estate of over 370,000 acres of land represented the lands once controlled by Kamehameha. In the same year, Bernice Pauahi completed her will, which bequeathed all of her estate to create and maintain the Kamehameha Schools. She died just one year later [68]. This relationship and responsibility of the ali'i is passed on even after the biological end of the Kamehameha line at Pauahi's death, through the creation of the Kamehameha Schools. In this establishment of the Kamehameha Schools as an "ali'i institution", as the steward and keeper of the Kamehameha line's ali'i lands, Kamehameha Schools serves the ali'i function, with the kaumaha to care for its resources and people in perpetuity.

In the late 19th and early 20th century, extractive land use practices were widespread across Hawai'i, and included largescale clearing of lands for pasture, sugar, and pineapple. Despite this, early Kamehameha Schools Trustees instead made decisions to ensure the protection of resources on the estate's landholdings. The quote below from Charles Reed Bishop, Pauahi's husband and one of the first Trustees of Kamehameha Schools, demonstrates his understanding of the ali'i kaumaha to maintain the natural resources, not only for the short-term utilitarian benefits to the institution, but more importantly also for the maintenance of climate regulation, erosion control, water supply and quality into the future.

> *So much is already known regarding the great value of forests, not only for furnishing fuel, building material and furniture woods, but in preserving the rainfall, restraining the violence of freshets, perpetuating the springs and rivulets of water, tempering the atmosphere and preventing the waste of the soil... It would be well for large landowners to reserve suitable localities—hilltops ... for tree planting. The results, if not directly profitable in a pecuniary point of view, would be advantageous in the local effects upon the climate, and protection against landslides and storms. It has come to be a necessity.*

—Charles Reed Bishop, 1883 [69]

The actions of Kamehameha Schools' early Trustees recognize the ali'i responsibility of maintaining the water cycle, as explained earlier. Agents of the ali'i institution at that time voluntarily set aside vast tracts of native forest lands for the express purpose of watershed and forest resource protection. They chose not to convert lands to pasture by clearcutting, as many other large landowners were doing, and instead chose to maintain these forested uplands for generations to come. Beginning in the 1890's, they allocated resources to fencing, removal of livestock, establishment of nurseries, and replanting of trees, years before the creation of the United States Forest Service (1905) or the territory of Hawai'i's Board of Commissioners of Agriculture and Forestry (1903) [70]. The forethinking by these early Trustees has resulted in many native forested landscapes that serve as primary watersheds for many communities today. For example, in the leeward district of Kona, Hawai'i Island, we currently actively steward much of the remaining mesic and wet native-dominated forest in the region (Figure 6), which continue to serve as vital sources for much of the community's drinking water.

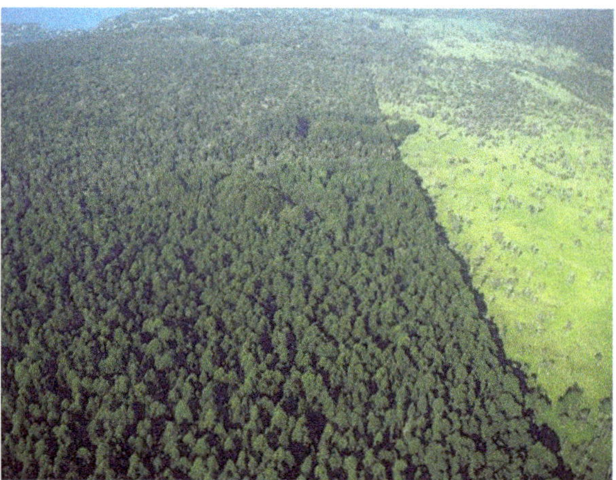

Figure 6. Results of early KS decisions to manage forests. An aerial view of KS forested lands in Hōnaunau, Kona, Hawai'i Island on the left, and neighboring lands that were converted to pasture use.

In the same era, Charles Reed Bishop, as an agent of the ali'i institution, recognized that the lands held by Kamehameha's lineage hold arguably some of the most sacred and spiritual sites in all of Hawai'i. Yet, as culture, language and practices, were being extinguished across Hawai'i, and whole cultural landscapes erased by practices such as plantation agriculture and logging, he identified the ali'i Trust's responsibility to protect cultural resources. He wrote,

There is a matter that should not be lost sight of. I mean the acquisition and control of the Heiaus [religious temples] and Puuhonuas [religious complex; place of refuge], say those of Mookini in Kahala [Kohala], of Puukohola at Kawaihae, of Pakaalana in Waipio, of Honaunau in Kona, and perhaps one of the Islets of Mokuola in Hilo Bay, and any others of interest and worth preserving . . . once in the control of the Museum they should be protected perpetually . . .

—Charles Reed Bishop to Henry Holmes, 1897 [71]

The "museum" referenced here is the Bishop Museum, Hawai'i's largest museum today. Founded by Bishop at the request of his wife Pauahi as a perpetual place to house the many cultural items of her family, and as this quote suggests, maintain the wahi kūpuna (ancestral spaces) that are critical to 'Ōiwi identity. It is important to note is that Charles Reed Bishop was an American as well as a Hawaiian Kingdom citizen, however not Kānaka 'Ōiwi by ancestry. Yet, as a representative of the ali'i institution, he inherits and carries the kaumaha to maintain the biocultural health of Kamehameha Schools' resources in perpetuity. This idea is powerful. We are not biological descendants of the ali'i who originally held and cared for the lands we are responsible for today, but, we have inherited the kaumaha held in the mo'okū'auhau of the 'āina. We carry the kaumaha of our predecessors irrelevant of our biological ancestry or cultural background. Our team has personally come to these realizations through our research and through our physical, mental, and spiritual work within these lands day in and day out, year in and year out.

4. The 21st Century: Returning to Ancestral Principles

4.1. Remembering our Foundations

Though we found references and examples of prudent management of natural and cultural resources and perpetual vision by early leaders at Kamehameha Schools, these concepts were not widespread within the institution for much of the 20th century. In fact, one of the main reasons we undertook this exercise to understand our function through genealogy is because there are even more cases in the recent memory of the institution's damaging decisions [72]. For example, long-term leases to vast sugar plantations (i.e., Kawailoa, O'ahu); transformation of biodiverse forests to pasture and agriculture (i.e., Keauhou-Ka'ū, Hawai'i Island); residential developments in traditionally abundant agroecosystems and aquaculture areas (i.e., Ka'elepulu and He'eia, O'ahu), commercial and resort development in areas with widespread and important cultural sites (i.e., Ka'ūpūlehu, Kahalu'u, and Keauhou, Hawai'i Island); legacy land sold then subsequently bulldozed (i.e., Kapu'a, Hawai'i Island which held one of the most diverse mesic forests in the area). Kamehameha Schools was "land rich, and cash poor", and at that time, and its 'āina was viewed solely as an economic resource to fund the educational mission. Yet, this approach completely ignored the 'Ōiwi worldview that kānaka and 'āina are ancestrally and reciprocally connected and disregarded the mo'okū'auhau of Kamehameha Schools' ali'i lands including the kaumaha transferred within. There were many other Kamehameha Schools proposals to monetize ali'i lands for marinas, resorts, golf courses, gentlemen estates, and even an amusement park in the late 1900's, however, these were met with opposition from kānaka 'ōiwi communities around the Islands. These communities are analogous to maka'āinana, as they continued to care for and live on their ancestral lands. Ultimately, it was the voices of the community that compelled Kamehameha Schools into transformative realignment evident in its 2000 Strategic Plan centered around Kānaka 'Ōiwi community discussions and concerns [41].

Goal 6: Kamehameha Schools will mālama i ka 'āina: practice ethical, prudent and culturally appropriate stewardship of lands and resources.

6.1—Manage lands and other resources to optimize their support of the educational mission.

- *Manage the portfolio of resources to derive an overall balance of economic, educational, cultural, environmental and community returns.*
- *Develop and incorporate educational programs and curricula into resource stewardship programs.*
- *Develop and incorporate resource stewardship into educational programs and curricula.*
- *Inventory and manage Kamehameha Schools' nonfinancial resources (e.g., historic, cultural, human, and intellectual).*

6.2—Manage lands to protect and enhance ecosystems and the wahi kūpuna (ancestral sites inclusive of all cultural resources and iwi [burials]) they contain.

- *Integrate Hawaiian cultural values and knowledge into resource stewardship practices.*
- *Incorporate ahupua'a—land division—management principles which recognize the interdependencies of ecosystems and create a synergy of uses in land use decisions.*
- *Promote a broad understanding of stewardship efforts and, as appropriate, cultural resource management programs. (Kamehameha Schools Strategic Plan 2000–2015)*

In this new era, decisions around lands and resources returned, for the first time in a generation, to the ancestral principle that kānaka are part of a larger "'ohana (family) that includes the rest of the universe: land; sky; fresh water; salt water; plants and animals" [73]. Around this time, the Kamehameha Schools leadership made successful financial investments, resulting in a shift from a land rich and cash poor entity, to a Trust that was "land rich" and growing a diverse endowment of real estate and financial assets. This wealth allowed Kamehameha Schools the opportunity to once again make strategic investments into the stewardship of its landholdings.

Today, Kamehameha Schools is an $11.5 billion trust, and the organization as a whole recognizes that our institutional genealogy is deeply tied to our 'āina, which we can directly attribute our current economic, educational, and biocultural wealth to [41]. We recognize that because of its inheritance from our ali'i (Pauahi, Ke'elikōlani, to Kamehameha I and beyond), our landbase (ten percent of Hawai'i), and now our financial wealth, our kaumaha or profound responsibility to both steward 'āina and reciprocally support community is even greater. Kamehameha Schools continues to be looked at by many beneficiaries to be the ali'i that provides for community in a modern context beyond just education, but for community and social programming, agricultural opportunities, natural and cultural resource management, publishing, political advocacy, and commercial real estate development.

4.2. The Manifestation of 'Ōiwi Principles in Stewardship Today

I Hawai'i nŌ nā Hawai'i i ka 'āina
Hawaiians are Hawaiians because of the land [74]

How do we continue the function of the ali'i and leaders of our institutional genealogy in today's socio-ecological context? In 2008, our department composed the adage above. The saying recognizes (1) the shared ancestry that kānaka have with all of the elements of 'āina and (2) our reciprocal relationship and interdependence. Just as the 'Ōiwi lifeforms evolved over many years to create the unique native ecosystems and landscapes of Hawai'i, so too have those ecosystems shaped the cultural identity, traditions and practices of Kānaka 'Ōiwi as a people. These concepts of mo'okū'auhau and aloha are at the foundation of our stewardship—which includes natural and cultural resource management, advocacy, community engagement, and beyond.

Additionally, we hold kaumaha as a programmatic principle and basis for many of our decisions, as we strive to meet our obligation as an ali'i land steward by contributing to biocultural resource stewardship across Hawai'i. Twelve percent of Hawai'i's remaining native ecosystems are on Kamehameha Schools' lands and over half of all of Hawai'i's threatened and endangered species depend on increasingly rare native habitats within these landholdings. Like our predecessors, we recognize that healthy, functioning native ecosystems provide critical ecosystem services that sustain life and quality of life in our islands, for native plants and animals, and all those who drink our water, breathe our air, and are inspired by our landscapes. Though we are not mandated to do so by any governmental authority, we know our function is to maintain the summits, headwaters, forests, coastal zones, hills, as sources of our ancestral identity for the benefit of our 'Ōiwi communities forever. An entire one fifth of the watershed forests that sustain Hawai'i's aquifers are on Kamehameha Schools lands. We manage these vital watershed resources in association with regional watershed partnerships, voluntary alliances of private landowners and governmental agencies, who agree to work together to steward resources across landownership boundaries. Due to our extensive landholdings, we belong to 7 of Hawai'i's 10 watershed partnerships and are one of the very few private landowners that provide financial resources to support the critical watershed conservation work of these partnerships.

Over the past generation, we've shifted over 100,000 acres of land from pasture back into management for ecosystem health. Our approach is to protect intact native communities; enable natural regeneration, where possible, by suppressing priority threats; and, where necessary, reintroduce biocultural diversity. For example, at Keauhou, Ka'ū, we are restoring native forests and shrublands on 30,000 acres of former cattle pasture. We have installed protective fencing and removed cattle, goats, sheep, and pigs. Native canopy is being restored through replanting. Our communities participate in stewardship through 'āina-based education and community engagement programs that are strengthening and reestablishing ancestral connections to forested landscapes. Most recently, we've formalized our commitment to stewardship in this area through a Safe Harbor Agreement with the State of Hawai'i Department of Land and Natural Resources and the U.S. Fish and Wildlife Service. The agreement is a cooperative effort that supports recovery of threatened and endangered species. It is the first of its kind in Hawai'i, the longest term in the United States, and the first to cover 7 species of native birds, the 'ōpe'ape'a (*Lasiurus cinereus semotus*; Hawaiian hoary bat), and 25 species of native plants. The Safe Harbor Agreement is a commitment to mālama Kamehameha Schools 'āina at Keauhou, Ka'ū, inclusive of rare species recovery, through landscape-scale restoration of native forests and protection of native species, while preserving Kamehameha Schools ability for beneficiaries to interact with these landscapes and resources.

The landscapes we steward are also inclusive of critical cultural resources, comprised of wahi kūpuna which include heiau, agricultural and aquacultural systems, trails, habitations areas, as well as koehana (artifacts), and iwi kūpuna (ancestral burials). Like our predecessors, we recognize our obligation to the perpetuation of our cultural resources for future generations, as they are embodiments of our ancestors' presence, existence, and sustainability. We acknowledge that these wahi kūpuna and koehana are repositories of ancestral knowledge and energy [75]. Thus, we are committed to stewarding our cultural resources in perpetuity. In order to understand the mo'okū'auhau, history, and ancestral knowledge of our lands, we have completed ethnohistoric studies for all of our landholdings. We have also inventoried close to four thousand cultural sites through various archaeological studies have to inform our stewardship of these places into the future.

We are entrusted not only to know where these sites are, but to also protect, maintain, and restore Kānaka 'Ōiwi connections with these significant wahi kūpuna. Though often challenging, we work to maintain the conditions of the many heiau in different conditions across our lands, often in remote areas and heavily impacted by natural and man-made threats. We conduct consistent management of these sacred sites and places to ensure the integrity of the wahi kūpuna, but also to maintain access to these sites for community use. We have pioneered cultural resource community engagement strategies, first providing opportunities to access to 'ike vāina (knowledge of lands) through the development of online resources including a website of 3D models of select cultural sites and artifacts. Community engagement strategies also involve improving physical access to wahi kūpuna with the intention of rekindling pilina or the ancestral bond between kānaka and 'āina. This strategy can take many forms, and includes the development of interpretive trails at sites, hosting community workdays, supporting field schools, and facilitating restoration of traditional agriculture and loko i'a (aquaculture systems). It is through both the knowledge and physical interaction with these cultural resources that our communities access and reestablish their connection to place and identity.

4.3. Influencing Stewardship Beyond Kamehameha Schools 'āina

Like our ali'i predecessors, we work across large landscapes, and are sometimes looked to by our communities as well as government agencies to be the representative voice of Kānaka 'Ōiwi in 'āina stewardship. Consequently, beyond our direct resource management functions, we have taken on the significant duty to drive 'Ōiwi perspectives in statewide consortiums and alliances. We participate in leadership roles and provide support, especially in topics concerning cultural perspectives, community-based management, and biocultural stewardship, in influential groups such as the Hawai'i Conservation Alliance, Pacific Islands Climate Change Cooperative, Environmental Funders Group, International Union for the Conservation of Nature—Indigenous Peoples Organizations and others. Such organizations and their members influence the agenda for research, funding, and policy at state, regional, and international levels. By actively participating in these groups, we uplift 'Ōiwi perspectives across scales of stewardship industries.

Furthermore, we also serve to represent our Kānaka 'Ōiwi communities with state agencies, such as those that regulate cultural sites (State of Historic Preservation), the Division of Forestry and Wildlife, and Department of Aquatic Resources. For example Kamehameha Schools supports an effort which created the first community-based marine 10-year rest area in Hawai'i in the ocean off Ka'ūpūlehu, a Kamehameha Schools landholding. We provided support in the decade of social and ecological research of the fishery, and provided testimony in favor of the initiative which required a change to the Hawai'i State Administrative Rules. Today, with the marine rest area in place, we are actively engaged in the community's planning and management of the natural and cultural resources in the near-shore and coastal areas as a partner. We have served as the lead communicator with regulatory expertise with state agencies which have jurisdiction over some of the community's stewardship activities. The Ka'ūpūlehu marine rest area embodies the duty we feel to support our communities in their capacity to manage their ancestral biocultural abundance. Additionally, Kamehameha Schools directly partners with community individuals and organizations through 'āina-based education programing, stewardship or resource management agreements, as well as community-based management of 'āina across all of the islands we own land.

Finally, in recognition of the overall lack of Kānaka 'Ōiwi and 'Ōiwi perspectives in 'āina stewardship industries across the archipelago, our team has developed internship and other professional training opportunities to foster the development of future Kānaka 'Ōiwi resource managers in Hawai'i. These internships focus on training participants in culturally-grounded approaches to research, science, and 'āina stewardship, while also simultaneously contributing to Kamehameha Schools' on-the-ground resource management in a variety of forms. For example, due to historically little to no 'Ōiwi representation in the cultural resource management field in Hawai'i, including in the field of contractors that Kamehameha Schools looks for its own work, Kamehameha Schools supports the Wahi Kūpuna Internship Program (WKIP), developed by the Kānaka 'Ōiwi organization Huliauapa'a. The WKIP is an immersive summer internship training undergraduate Kānaka 'Ōiwi students in 'Ōiwi cultural resource management, including appropriate cultural protocol, technical skill building in ethnographic and archival research, geographic information systems (GIS), community interviews, archaeological field methods, research writing, and presentations. In addition, each student chooses a management-aligned topic to research throughout the summer tailored to questions from Kamehameha Schools resource managers or community stewards, in order to ground their newly learned skills in applied management issues. Similarly, Kamehameha Schools supports a summer Ecosystem Monitoring Internship Program focused on training undergraduate and senior level high school Kānaka 'Ōiwi students in biological survey and forest management strategies, while also providing a solid grounding in 'Ōiwi perspectives on 'āina. Many alumni of this program have gone on to seek advanced degrees in natural resources fields, and most are now employed in 'āina stewardship careers, working as kahu 'āina (land stewards), foresters, researchers, and educators. Through the half-dozen 'āina stewardship training opportunities we directly manage, and numerous others we support financially, we are working to uplift the next generation of vŌiwi thought leaders in the 'āina space. This is indeed our kaumaha, to not only care for Hawai'i's lands for our communities today, but that we ensure there are capable stewards for every aspect of aloha 'āina (love for the land)—from those who will manage the lands at Kamehameha Schools, or those who will steward from the community, to those who will do the research on our lands and seas, to those who are entrusted to teach 'Ōiwi resource management to future generations of stewards.

5. Discussion

There is increasing consensus that indigenous people worldwide are essential to biocultural conservation, because of their widespread tenure, resilient and persistent management systems, sustainable practices, and innovative conservation techniques [3]. Yet, there is a lack of examples of applied indigenous resource management in the systems context, especially from the indigenous perspective. Our case study provides an example of the maintenance of an indigenous resource management system over hundreds of years fixed within traditional worldviews, social institutions, and principles. We have shown that this resource management system was not, and is not static, but instead like other socio-ecological governance systems, has transformed and adapted overtime [19,76,77]. We provide an important case of how an organization defines itself using ancestral concepts and methodologies to execute leadership in indigenous stewardship today.

Our process is replicable, and could serve as a guide for both indigenous organizations determining for themselves how to steward in a way that is appropriate to their ancestral ways of knowing, as well as for non-indigenous agencies which have obligations to indigenous lands and people. First, we let 'Ōiwi methodologies guide us in the process of (re)discovery. Second, we looked to the foundation, our ʻŌiwi worldviews (moʻokūʻauhau and aloha). It should be noted that genealogical and reciprocal connections to landscapes are inherent in many other indigenous cultures [38], and could potentially also be a foundation for other organizations. Third, we worked to understand the institution we sit, from traditional functions, to the transference of responsibility (in our case kaumaha) through time, actors, and lands. It was through the self-determined indigenous process that we were able to answer the questions: what is our function and how do we continue that function today? It is important to note that the process is continual and iterative, as with ʻŌiwi ways of knowing, adapting, and stewarding. We continue to hold team discussions dedicated to better understanding our position in the continuity of aliʻi function, and consequently, we consistently have new realizations, both personal and collective, about why and how we manage ʻāina.

There are challenges and limitations beyond our individual and institutional control in maintaining an indigenous system of resource management. This continues to be an issue globally, where the structures of governance and community that indigenous institutions sit within have changed quite drastically. Even within our own indigenous institution, we face cycles of changing leadership and vision. Beyond that, as a private landowner we have limited control over what aliʻi traditionally regulated (i.e., fresh water allocation, public access, nearshore and marine tenure, etc.) because these are now under the control of county, state, and federal agencies. Furthermore, traditional relationships between aliʻi and community have been radically altered, with traditional social structures of community centered on ʻāina, reciprocally connected to a hierarchical system of social regulation, greatly damaged through generations of American colonization, though examples of persistence exist [78,79]. Therefore, it is critical for our ʻŌiwi stewardship in this contemporary social context to look beyond our conservation and management practices alone, but to consider our greater kaumaha, our chiefly obligation to steward biocultural systems in perpetuity. We have found that one way we can fulfill our kaumaha in today's governance context, is to collaborate with governance agencies that now also carry a responsibility and heaviness once only granted to aliʻi, as well as to restore our mutually benefitting relationships with community through supporting their own stewardship of their ancestral resources.

Indigenous institutions are among the world's most durable and enduring for governance, offering inspiration for the development of sustainable systems, while others, acting on vested interests, opt for short-term gains at substantial social and environmental cost. [80] (p. 340)

Like in traditional indigenous societies, effective indigenous resource management systems require social institutions that as we have shown, coordinate resource management at different scales, take on roles of governance of resources, and plan for stewardship in perpetuity [19]. For all of us working towards biocultural abundance on indigenous lands (in Hawaiʻi those are all landowners), it is essential to restore the institutional functions and continuity for today's context. We provide one roadmap of an institution refinding our continuity and purpose. The restoration of indigenous knowledge for resource management today necessitates the maintenance, and in some cases reestablishment, of indigenous institutions at the regional, county, state, or national level.

Author Contributions: Conceptualization, N.K., J.J., J.T., N.W., M.B., T.D.; Investigation, N.K., J.J., N.W.; Writing—Original Draft Preparation, N.K., J.J., J.T., N.W., Writing—Review & Editing, N.K., J.J., N.W., M.B.

Acknowledgments: We are grateful to Ulalia Woodside, Neil Hannahs, Kamakani Dancil, Kealakaʻi Kanakaʻole for shaping Kamehameha Schools' thinking around ʻŌiwi ʻāina stewardship; We mahalo the Edith Kanakaʻole Foundation and its scholars including: Pua Kanakaʻole Kanahele, Huihui Kanahele-Mossman, Kekuhi Kealiikanakaoleohaililani, Kuʻulei Kanahele, and Kalei Nuʻuhiwa for their research that has greatly influenced our understanding of contemporary aliʻi stewardship.

References

1. United Nations State of the World's Indigenous Peoples. *UNPFII. State of the World's Indigenous Peoples*; United Nations: New York, NY, USA, 2009.
2. IUCN. *Report Submitted by IUCN—International Union for Conservation of Nature to the Tenth Session of the United Nations Permanent Forum on Indigenous Issues*; International Union for Conservation of Nature: New York, NY, USA, 2011.
3. Garnett, S.T.; Burgess, N.D.; Fa, J.E.; Fernández-Llamazares, Á.; Molnár, Z.; Robinson, C.J.; Watson, J.E.M.; Zander, K.K.; Austin, B.; Brondizio, E.S.; et al. A spatial overview of the global importance of Indigenous lands for conservation. *Nat. Sustain.* **2018**, *1*, 369–374. [CrossRef]
4. Sobrevila, C. *The Role of Indigenous Peoples in Biodiversity Conservation: The Natural but Often Forgotten Partners*; The World Bank: Washington, DC, USA, 2008.
5. Borrini-Feyerabend, G.; Kothari, A.; Oviedo, G. *Indigenous and Local Communities and Protected Areas: Towards Equity and Enhanced Conservation*; International Union for Conservation of Nature: New York, NY, USA, 2004. [CrossRef]
6. Altman, J.; Buchanan, G.; Larsen, L. *The Environmental Significance of the Indigenous Estate: Natural Resource Management as Economic Development in Remote Australia*; Paper No. 286; Centre for Aboriginal Economic Policy Research: Canberra, Australia, 2007.
7. Aho, L.T. Indigenous challenges to enhance freshwater governance and management in Aotearoa New Zealand—The Waikato river settlement. *J. Water Law* **2009**, *20*, 285–292.
8. Magallanes, C.J.I. Maori Cultural Rights in Aotearoa New Zealand: Protecting the Cosmology that Protects the Environment. *Widener Law Rev.* **2015**, *21*, 273–327.
9. Scholtz, C. *Negotiating claims: The Emergence of Indigenous Land Claim Negotiation Policies in Australia, Canada, New Zealand, and the United States*; Routledge: London, UK, 2006; ISBN 0415976901.
10. Eldredge, L.G.; Evenhuis, N.L. *Hawaii's Biodiversity: A Detailed Assessment of the Numbers of Species in the Hawaiian Islands*; Bishop Museum Occasional Papers—Hawaii Biological Survey 2001–2002; Bishop Museum Press: Honolulu, HI, USA, 2003; Volume 76, pp. 1–30.
11. DeMartini, E.; Friedlander, A. Spatial patterns of endemism in shallow-water reef fish populations of the Northwestern Hawaiian Islands. *Mar. Ecol. Prog. Ser.* **2004**, *271*, 281–296. [CrossRef]
12. Loope, L. Hawaii and the Pacific Islands Overview of the Pacific. In *Status and Trends of the Nation's Biological Resources*; Mac, M.J., Opler, P.A., Puckett Haecker, C.E., Eds.; United States Department of the Interior, United States Geological Survey: Reston, VA, USA, 1999; pp. 747–774.
13. Mackenzie, M.K.; Serrano, S.K.; Sproat, D.K. *Native Hawaiian Law: A Treatise*; Kamehameha Publishing: Honolulu, HI, USA, 2015.
14. Kana'iaupuni, S.M.; Malone, N.J.; Ishibashi, K. *Income and Poverty among Native Hawaiians: Summary of Ka Huaka'i Findings*; Kamehameha Schools—PASE: Honolulu: HI, USA, 2005; Volume 10.
15. Moy, K.L.; Sallis, J.F.; David, K.J. Health indicators of native Hawaiian and pacific islanders in the United States. *J. Commun. Health* **2010**, *35*, 81–92. [CrossRef] [PubMed]
16. Kimmerer, R. Restoration and reciprocity: The Contributions of Tradtional Ecological Knowledge. In *Human Dimensions of Ecological Restoration: Integrating Science, Nature, and Culture*; Egan, D., Hjerpe, E.E., Abrams, J., Eds.; Island Press: Washington, DC, USA, 2011; pp. 257–276.
17. Berkes, F. Environmental governance for the anthropocene? Social-ecological systems, resilience, and collaborative learning. *Sustainability* **2017**, *9*, 1232. [CrossRef]
18. Pascual, P.; McMillen, H.; Ticktin, T.; Vaughan, M.; Winter, K. Beyond Services: A Process and Framework for Incorporating Cultural, Genealogical, and Place-Based Relationships into Ecosystem Service Assessments. *Ecosyst. Serv.* **2017**, *26*, 465–475. [CrossRef]
19. Berkes, F.; Colding, J.; Folke, C. Rediscovery of Traditional Ecological Knowledge as adaptive management. *Ecol. Appl.* **2000**, *10*, 1251–1262. [CrossRef]
20. Trosper, R.L. Northwest coast indigenous institutions that supported resilience and sustainability. *Ecol. Econ.* **2002**, *41*, 329–344. [CrossRef]
21. Toledo, V.M.; Ortiz-Espejel, B.; Cortés, L.; Moguel, P.; de Jesús Ordoñez, M. The multiple use of tropical forests by indigenous peoples in Mexico: A case of adaptive management. *Conserv. Ecol.* **2003**, *7*, 9. [CrossRef]

22. Johannes, R.E. The Renaissance of Community-Based Marine Resource Management in Oceania. *Annu. Rev. Ecol. Syst.* **2002**, *33*, 317–340. [CrossRef]

23. Clarke, W.C.; Thaman, R.R. *Agroforestry in the Pacific Islands: Systems for Sustainability*; United Nations University Press: Tokyo, Japan, 1993.

24. Green, D.; Billy, J.; Tapim, A.; Green, D.; Billy, J. Indigenous Australians' knowledge of weather and climate. *Clim. Chang.* **2010**, *100*, 337–354. [CrossRef]

25. Lefale, P.F. Ua 'afa le Aso Stormy weather today: Traditional ecological knowledge of weather and climate. The Samoa experience. *Clim. Chang.* **2010**, *100*, 317–335. [CrossRef]

26. Huntington, H.P. Using Traditional Ecological Knowledge in Science: Methods and Applications. *Ecol. Appl.* **2000**, *10*, 1270–1274. [CrossRef]

27. Lantz, T.C.; Turner, N.J. Tradittional Phenological Knowledge of Aboriginal Peoples in British Columbia. *J. Ethnobiol.* **2003**, *23*, 263–286.

28. Friedlander, A.M.; Shackeroff, J.M.; Kittinger, J.N. Customary Marine Resource Knowledge and use in Contemporary Hawai'i. *Pac. Sci.* **2013**, *67*, 441–460. [CrossRef]

29. Gadgil, M.; Berkes, F.; Folke, C. Indigenous Knowledge for Biodiversity Conservation. *AMBIO A J. Hum. Environ.* **1993**, *22*, 151–156.

30. Altieri, M.A. Linking ecologists and traditional farmers in the search for sustainable agriculture. *Front. Ecol. Environ.* **2004**, *2*, 35–42. [CrossRef]

31. Kaneshiro, K.Y.; Chinn, P.; Duin, K.N.; Hood, A.P.; Maly, K.; Wilcox, B.A. Hawai'i's Mountain-to-Sea Ecosystems: Social–Ecological Microcosms for Sustainability Science and Practice. *Ecohealth* **2005**, *2*, 349–360. [CrossRef]

32. Berkes, F. *Sacred Ecology*; Routledge: London, UK, 2012.

33. Reo, N.J. The Importance of Belief Systems in Traditional Ecological Knowledge Initiatives. *Int. Indig. Policy J.* **2011**, *2*, 8. [CrossRef]

34. Kimmerer, R.W.; Lake, F.K. The Role of Indigenous Burning in Land Management. *J. For.* **2001**, *99*, 36–41. [CrossRef]

35. Kimmerer, R.W. Native Knowledge for Native Ecosystems. *J. For.* **2000**, *98*, 4–9. [CrossRef]

36. Morishige, K.; Andrade, P.; Pascua, P.; Steward, K.; Cadiz, E.; Kapono, L.; Chong, U. Nā Kilo 'āina: Visions of Biocultural Restoration through Indigenous Relationships between People and Place. *Sustainability* **2018**, *10*, 3368. [CrossRef]

37. Oliveira, K.-A.R.K.N.; Wright, E.K. *Kanaka 'ōiwi Methodologies: mo'olelo and Metaphor*; University of Hawai'i Press: Honolulu, HI, USA, 2015; ISBN 082485585X.

38. Smith, L.T. *Decolonizing Methodologies: Research and Indigenous Peoples*; Zed Books: London, UK, 2012; ISBN 1848139500.

39. Desha, S.; Frazier, F.N.; Hawaii. *Historic Preservation Division. Kamehameha and his warrior Kekūhaupi'o*; Kamehameha Schools Press: Honolulu, HI, USA, 2000; ISBN 9780873360562.

40. Kamehameha Schools. *Makawalu: 2017 Kamehameha Schools Annual Report*; Kamehameha Schools: Honolulu, HI, USA, 2017.

41. Kamehameha Schools. *Kamehameha Schools Strategic Plan 2000–2015*; Kamehameha Schools: Honolulu, HI, USA, 2000.

42. Pukui, M.K.; Elbert, S.H. *Hawaiian Dictionary*; University of Hawaii Press: Honolulu, HI, USA, 1986.

43. Peralto, L.N. Mauna a Wākea: Hānau ka Mauna, the Piko of our Ea. In *A Nation Rising*; Goodyear-Ka'ōpua, N., Hussey, I., Wright, E.K., Eds., Duke University Press: Durham, UK; London, UK, 2014; pp. 233–243.

44. Hawaii Conservation Alliance. *Making the Case for Community-Based Adaptive Collaborative Management in Hawai'i*; Hawaii Conservation Alliance: Honolulu, HI, USA, 2018.

45. Kealiikanakaoleohaililani, K.; Giardina, C.P. Embracing the sacred: An indigenous framework for tomorrow's sustainability science. *Sustain. Sci.* **2016**, *11*, 57–67. [CrossRef]

46. Kealiikanakaoleohaililani, K.; Kurashima, N.; Francisco, K.; Giardina, C.; Louis, R.; McMillen, H.; Asing, C.; Asing, K.; Block, T.; Browning, M.; et al. Ritual + Sustainability Science? A Portal into the Science of Aloha. *Sustainability* **2018**, *10*, 3478. [CrossRef]

47. Edith Kanakaole Foundation. *Kanawai Honuamea*; Edith Kanakaole Foundation: Hilo, HI, USA, 2013.

48. Beckwith, M.W. *The Kumulipo*; University of Hawaii Press: Honolulu, HI, USA, 1951.

49. Malo, D. *Hawaiian Antiquities (Moolelo Hawaii)*; Bernice Pa.; Bishop Museum Press: Honolulu, HI, USA, 1951.
50. Kamakau, S.M. *Tales and Traditions of the People of Old: Nā Moʻoleo a ka Poʻe Kahiko*; Translated from the newspaper Ke Au ʻOko ʻa by Mary Kawena Pukui, arranged and edited by Dorothy B. Barrère. Bernice P, Bishop Museum Special Publication 51; Bishop Museum: Honolulu, HI, USA, 1991.
51. Oliveira, K.-A.R.K. *Ancestral Places: Understanding Kanaka Geographies*; Oregon State University Press: Corvallis, OR, USA, 2014.
52. Hoʻoulumāhiehie. *Ka Moʻolelo o Hiʻiakaikapoliopele*; Ka Naʻi ʻAupuni: Hilo, HI, USA, 1905.
53. Andrade, C. *Hāʻena: Through the Eyes of the Ancestors*; University of Hawaiʻi Press: Honolulu, HI, USA, 2008; ISBN 0824834100.
54. Pukui, M.K. *ʻŌlelo Noʻeau: Hawaiian Proverbs and Poetical Sayings*; Bishop Museum Press: Honolulu, HI, USA, 1983.
55. Kepelino, Z. *Kepelino's Traditions of Hawaii*; Beckwith, M.W., Ed.; Bishop Museum Press: Honolulu, HI, USA, 2007.
56. Pukui, M.K.; Haertig, E.W.; Lee, C.A. *Nānā i ke kumu (Look to the Source) Volume I*; Hui Hānai, an auxiliary of the Queen Liliʻuokalani Children's Center; Queen Liliʻuokalani Children's Center: Kaneohe, HI, USA, 2002; ISBN 096167380X.
57. Edith Kanakaʻole Foundation. *Kūmokuhāliʻi: ʻaina Kaumaha no Ke Kula o Kamehameha Mahele ʻāina*; Edith Kanakaʻole Foundation: Hilo, HI, USA, 2011.
58. Abbott, I. *Lāʻau Hawaiʻi: Traditional Hawaiian Use of Plants*; Bishop Museum Press: Honolulu, HI, USA, 1992.
59. Beamer, K. *No mākou ka mana: Liberating the Nation*; Kamehameha Publishing: Honolulu, HI, USA, 2014; ISBN 0873363299.
60. Kirch, P.V. *How Chiefs Became Kings: Divine Kingship and the Rise of Archaic States in Ancient Hawaiʻi*; University of California Press: Berkeley, CA, USA, 2010; ISBN 9780520947849.
61. Cachola-Abad, K. *The Evolution of Hawaiian Socio-Political Complexity: An Analysis of Hawaiian Oral Traditions*; University of Hawaiʻi at Mānoa: Hilo, HI, USA, 2000.
62. Kahāʻulelio, D.; Nogelmeier, P. *Ka ʻoihana lawaiʻa: Hawaiian Fishing Traditions*; Bishop Museum Press: Honolulu, HI, USA, 2006; ISBN 9781581780383.
63. Kamakau, S.M. *Ruling Chiefs of Hawaii*; Kamehameha Schools Press: Honolulu, HI, USA, 1992; ISBN 0873360141.
64. Pokuea, R.J.F. *Hookumuia ana o Hawaii*; Ke Aloha Aina: Honolulu, HI, USA, 1896; Volume 7.
65. Desha, S.L. *Moolelo Kaao no Kekuhaupio ke Koa Kaulana o ke Au o Kamehameha ka Nui*; Ka Hoku o Hawaii: Honolulu, HI, USA, 1924.
66. Kamakau, S.M. *Ka Moolelo o Kamehameha I*; Ka Nupepa Kuokoa: Honolulu, HI, USA, 1867.
67. He Moolelo Kaao Hawaii no Kamehameha I: Kana Mau Kaua ame na Poe Kaulana o ia Wa. *Ke Alakai o Hawaii*, 15 November 1928; 4.
68. Kanahele, G.H.S. *Pauahi: The Kamehameha Legacy*; Kamehameha Schools Press: Honolulu, HI, USA, 1986.
69. Bishop, C.R. Report of Committee on Forestry. *The Planter's Monthly*, 16 October 1883.
70. State of Hawaii Department of Agriculture Hawaii Board of Agriculture. Available online: http://hdoa.hawaii.gov/chair/boa/ (accessed on 5 October 2018).
71. Kent, H.W. *Charles Reed Bishop: Letter File*; Royal Mausoleum Honolulu: Honolulu, HI, USA, 1972; Volume 103.
72. Hannahs, N. Indigenizing Management of Kamehameha Schools' Land Legacy. In *I Ulu i ka ʻāina*; Osorio, J., Ed.; University of Hawaii Press: Honolulu, HI, USA, 2014; pp. 62–75.
73. Kamehameha Schools. *Kamehameha Schools Strategic Planning Process Working Group Presentations*; Kamehameha Schools: Honolulu, HI, USA, 2000.
74. Kamehameha Schools. *Natural Resources Management Plan*; Kamehameha Schools: Honolulu, HI, USA, 2011.
75. Kamehameha Schools. *Cultural Resources Management Plan*; Kamehameha Schools: Honolulu, HI, USA, 2011.
76. Folke, C.; Hahn, T.; Olsson, P.; Norberg, J. Adaptive governance of social-ecological systems. *Annu. Rev. Environ. Resour.* **2005**, *30*, 441–473. [CrossRef]
77. Kawharu, M. Kaitiakitanga: A maori anthropological perspective of the maori socio-environmental ethic of resource management. *J. Polym. Soc.* **2000**, *109*, 349–370.
78. McGregor, D.P.; Morelli, T.P.; Matuoka, J.K.; Minerbi, L. An Ecological Model of Wellbeing. *Int. Handb. Soc. Impact Assess. Concept. Methodol. Adv.* **2003**, *10*, 106–128.

79. Vaughan, M.B.; Vitousek, P.M. Mahele: Sustaining Communities through Small-Scale Inshore Fishery Catch and Sharing Networks 1. *Pac. Sci.* **2013**, *67*, 329–344. [CrossRef]
80. Howitt, R. Indigenous rights vital to survival. *Nat. Sustain.* **2018**, *1*, 339–340. [CrossRef]

 *sustainability*

Article

Ritual + Sustainability Science? A Portal into the Science of Aloha

Kekuhi Kealiikanakaoleohaililani [1], Natalie Kurashima [2,*], Kainana S. Francisco [3], Christian P. Giardina [3], Renee Pualani Louis [4], Heather McMillen [5], C. Kalā Asing [6], Kayla Asing [7], Tabetha A. Block [3], Mililani Browning [2], Kualii Camara [8], Lahela Camara [9], Melanie Leilā Dudley [5], Monika Frazier [10], Noah Gomes [11], Amy Elizabeth Gordon [12], Marc Gordon [13], Linnea Heu [14], Aliah Irvine [15], Nohea Kaawa [5], Sean Kirkpatrick [16], Emily Leucht [9], Cheyenne Hiapo Perry [17], John Replogle [18], Lasha-Lynn Salbosa [19], Aimee Sato [20], Linda Schubert [21], Amelie Sterling [9], Amanda L. Uowolo [3], Jermy Uowolo [6], Bridget Walker [22], A. Nāmaka Whitehead [2] and Darcy Yogi [23]

[1] Hālau ʻŌhiʻa—Hawaiʻi Stewardship Training, Hilo, HI 96720, USA; ohaililani@gmail.com
[2] Kamehameha Schools, Natural and Cultural Resources, Honolulu, HI 96813, USA; rebrowni@ksbe.edu (M.B.); nawhiteh@ksbe.edu (A.N.W.)
[3] USDA Forest Service, Institute of Pacific Islands Forestry, Hilo, HI 96720, USA; ksfrancisco@fs.fed.us (K.S.F.); cgiardina@fs.fed.us (C.P.G.); tabethaablock@fs.fed.us (T.A.B.); auowolo@fs.fed.us (A.L.U.)
[4] Hālau ʻŌhiʻa—Aʻa a Mole Cohort, Hilo, HI 96720, USA; reneel@hawaii.edu
[5] Hawaiʻi Department of Land and Natural Resources, Division of Forestry & Wildlife, Honolulu, HI 96813; heather.l.mcmillen@hawaii.gov (H.M.); mldudley@hawaii.edu (M.L.D.); kaawa.nohea@gmail.com (N.K.)
[6] Mauna Kea Forest Restoration Project, Hilo, HI 96720, USA; kakuhihewa@gmail.com (C.K.A.); jermyuowolo@yahoo.com (J.U.)
[7] Pūnanaleo o Hilo, Hilo, HI 96720, USA; kayla.asing@gmail.com
[8] Department of Hawaiian Homelands, Hilo, HI 96720, USA; kualiic@hotmail.com
[9] ʻImi Pono no ka ʻĀina, Three Mountain Alliance, Hawaiʻi Volcanoes National Park, HI 96718, USA; lahelacamara@gmail.com (L.C.); emily.imipono@gmail.com (E.L.); ameliesterling@gmail.com (A.S.)
[10] Aloha Kuamoʻo ʻĀina, Kailua-Kona, HI 96740, USA; monikameilanifrazier@gmail.com
[11] Kamehameha Schools, Kealapono, Honolulu, HI 96813, USA; noahjgomes@gmail.com
[12] Gig Called Life Coaching Services, Kamuela, HI 96743, USA; aegimago@gmail.com
[13] State of Hawaiʻi Department of Human Services, Honolulu, HI 96813, USA; gourdo@gmail.com
[14] Pacific Internship Programs for Exploring Sciences, University of Hawaiʻi at Hilo, Hilo, HI 96720, USA; lheu@hawaii.edu
[15] Oʻahu Army Natural Resources Program, Schofield Barracks, HI 96857, USA; aliah@hawaii.edu
[16] Hawaii Community College, University of Hawaiʻi at Hilo, Hilo, HI 96720, USA; seank808@hawaii.edu
[17] Mauna Kea Watershed Alliance, Hilo, HI 96720, USA; cheyennehiapo@gmail.com
[18] Hālau ʻŌhiʻa—ʻŌhiʻalaka Cohort, Hilo, HI 96720, USA; jrepsr@gmail.com
[19] US Fish and Wildlife Service, Pacific Islands Fish and Wildlife Office, Honolulu, HI 96850, USA; lasha-lynn_salbosa@fws.gov
[20] Department of Botany, University of Hawaiʻi at Mānoa, Honolulu, HI 97822, USA; aimeeysato@gmail.com
[21] Volcano School of Arts and Sciences, Volcano Village, HI 96785, USA; schubiedobert@yahoo.com
[22] Kamuela Hardwoods, Kamuela, HI 96743, USA; walkerb715@gmail.com
[23] Natural Resources and Environmental Management, University of Hawaiʻi at Mānoa, Honolulu, HI 96822, USA; dyogi2@hawaii.edu
*   Correspondence: nakurash@ksbe.edu; Tel.: +1-808-322-5348

Received: 1 August 2018; Accepted: 19 September 2018; Published: 28 September 2018

Abstract: In this paper, we propose that spiritual approaches rooted in the practice of Hawaiʻi ritual provide a powerful portal to revealing, supporting, and enhancing our collective *aloha* (love, fondness, reciprocity, as with a family member) for and dedication to the places and processes that we steward. We provide a case study from Hawaiʻi, where we, a group of conservation professionals known as Hālau ʻŌhiʻa, have begun to foster a collective resurgence of sacred commitment to the places and processes we steward through remembering and manifesting genealogical relationships to our landscapes through Indigenous Hawaiian ritual expression. We discuss how a ritual approach to our lands and seas makes us better stewards of our places, better members of our families and communities, and more fulfilled individuals. We assert that foundations of the spiritual and the sacred are required for effectively advancing the science of sustainability, the management of natural resources, and the conservation of nature.

Keywords: sacred ecology; biocultural conservation; Hawai'i

---

1. *Welina*—Welcome and Orientation

You have come to Hilo to the USDA Forest Service to visit the *hālau* (traditional Hawaiian school of learning). You arrive, park your car, and wait a little bit. If you leave your car now, you will be drenched because Hilo is still raining until we can bid Hurricane Lane *"aloha"* and greet the next storm. What was his name? Anyway, someone runs out wth umbrellas to bring you into the *lānai* (outdoor covered area) where we meet before transforming the facility's conference room into our learning space. As you transition from dry to wet, your attention turns to the voices of men, women, children swelling and pulsing with song in rhythm with the *pakapaka* (pitter patter) of the rain. You do not even notice that your left shoe is soaked through to the sock. As you get closer, your vision glimpses a wonderful eclectic collection of the world in welcoming, chanting you into Hālau 'Ōhi'a:

> *Ua lū kinikini ka hua 'ōhi'a lehua mai 'ō a 'ō o Lononuiākea*
> Two million lives in the seeds of *'ōhi'a* strewn about from near and far in Hawai'i
> *Halihali 'ia e ka 'ēheu hulu makani*
> Carried on the wings of the wind
> *Hi'ipoi 'ia e ka Poli mahana o Kānehoa, o Honuamea*
> Caressed in the warmth of Honuamea, the volcanic earth; nourished by Kānehoa, the sun
> *Ua a'a, ua mole, ua mōhala a'ela*
> We are rooted, tapping the source of water—unfurling and peaking towards full bloom
> *'O ka 'apapane, 'o ka mamo, 'o ka nuku 'i'iwi, 'o ka 'āhihi*
> A diversity of hues, brilliant scarlet, golden, salmon, and the rare white
> *Mai hiki lalo a i hiki luna e waiho nei i hāli'i moku lā*
> We are blankets of *'ōhi'a* forests that extend beyond the horizons of my vision
> *Ua 'ikea! A he leo nō ia.*
> It is done with the simple offering of the voice.

"The real root of these [sustainability] issues, both cause and cure, lies not in our science or technology but in our own spiritual and intellectual poverty or more hopefully, in our own spiritual and intellectual resources". [1] (p. 3)

1.1. Why the Need for Ritual in Conservation?

In Hawai'i, spiritual foundations continue to define relationships among many cultural practitioners, community members, places, and processes [2–4]. We propose that sacred ritual plays a central role in elevating these foundations and enhancing the well-being of all members of the coupled socioecological system. Specifically, this paper makes the case that spiritually oriented ritual is a powerful portal to revealing, supporting, and building up our collective love for and devotion to the places and processes that we steward. It is this path that we believe is required for effectively advancing the science of sustainability, the management of natural resources, and the conservation of nature. In advancing these disciplines, we also believe that spiritual approaches that engage different levels of personal and communal ritual enhance our ability to interact with our landscapes and seascapes and so can best position Hawai'i to achieve biocultural well-being.

Hālau ʻŌhiʻa is both a venue for and a process whereby we can explore the meaning of family life and our connections to a broadly defined genealogy of place. *"Hālau"* translates to traditional Hawaiian school of learning, literally meaning "many breaths," and is often associated with the traditional dancing art of *hula.* *"ʻŌhiʻa"* is the name of Hawaiʻi's most common, widespread, and bioculturally important native tree (*Metrosideros polymorpha* Gaudich, Myrtaceae), and the name literally means "to gather." The spiritual venue and the sacred process are created by engaging native Hawaiian rituals, which include the use of Hawaiian language, the retelling of sacred stories, the performing of traditional chants and dance, and the creation of our own poetic texts and art forms. Through these practices and the resulting deep learning of cultural and physical geographies that surround us, we are able to establish and deepen our kincentric relationships to the world around. The ultimate goal of this learning is no less than to transform the way we view and steward our lands and seas. As in the *hālau* setting, this paper is made up of the many breaths, voices, and ideas from our group. Like the *ʻōhiʻa*, we are a diverse group of resource managers, field technicians, researchers, interns, educators, cultural practitioners, administrators, students, and program leaders representing many organizations, generations, and life experiences. In short, we are people whose functions are foundational to the well-being of our Hawaiʻi landscapes, seascapes, and communities.

1.2. What Is Ritual?

Ritual rooted in spirituality is an ubiquitous feature of the human experience across planet Earth and throughout human history, and takes many forms across and within cultures. Ritual of a spiritual nature has been examined by countless scholars over many centuries, and has been characterized as serving a wide diversity of societal functions, including to name just a few ritual practices: bringing about an altered state, as with healing and shamanistic rituals [5]; expressing or presenting a system of beliefs, for example, about the structure of society or kinship relationships [6]; conserving resources, for example, by defining the taking of resources [7,8] or the imposition of food taboos [9]; managing resources and horticultural practices based on weather, phenology, and astronomical cycles [10]; avoiding contagion [11]; improving social cohesion [12] and protective social bonds that increase survival [13]; making pilgrimages to natural sacred sites [14], including to redefine oneself [15]; and burying family and friends [6,16]. While early theories framed rituals as functioning to protect the status quo, to resist change, and to relieve anxiety over uncertainty about observed or experienced phenomena [17], contemporary perspectives point to rituals as also serving as agents of cultural change, in both historic and contemporary contexts, as rituals are often "created by families, secular and religious celebrants, civil servants, or volunteers" [18] (p. 2). As such, rituals can play subversive, creative, or socially critical roles [19]. Where ritual catalyzes social transformations [18,20] through their performative, structured, and collaborative natures [18], they can be seen as providing "breakthroughs to the knowledge of the 'sacred'" the functions of which are "seen in a future we are not likely to be able to even guess" [19] (p. viii).

For our group, ritual has become a means to:

(1) enter into a sacred space within which members of Hālau ʻŌhiʻa can holistically (mind, body, and spirit) embrace widely ranging topics of existential importance to being human;

(2) deepen our kinship relationships with the world around us; then from this,

(3) catalyze personal and professional transformation and growth;

(4) recognize and embrace the deep linkages binding together *haumāna* (student/students) and *kumu* (master teacher), *haumāna* and *kūpuna* (ancestors broadly defined), and *haumāna* and *ʻāina* (lands and seas; that which sustains); and

(5) identify, engage, and express gratitude to and *aloha* for the diverse linkages that sustain us physically (evolutionarily, nutritionally, biogeochemically), mentally (psychologically, professionally, academically), and spiritually (our relationships and ancestral connections to persons and places).

1.3. Case Study: Hālau 'Ōhi'a and Ritual

The unique Hālau 'Ōhi'a program was developed and is taught by Kekuhi Kealiikanakaoleohaililani, a master teacher, who is trained in and has been practicing for over 40 years the Hawai'i traditions of *hula*, chant, and ritual. She is one of the *kumu hula* (teacher of traditional Hawaiian environmental dance) in the traditional dance school of learning Hālau o Kekuhi, a position previously held by her mother and grandmother. We, Hālau 'Ōhi'a, began our journey in 2016 because of a novel question posed by Kumu Kekuhi's research assistant, who asked: (1) "How can Hawaiian culture help us do our jobs?" and (2) "How can this work place become a community?" From that profound query, the idea of Hālau 'Ōhi'a was born. The last two and a half years have included: 35 sessions; a *pāmaomao* (international exchange among communities) with Maori communities of Aotearoa (New Zealand); many *kīpaepae* (a term created by *Kumu Hula* and faculty member from the University of Hawai'i, Taupōuri Tangarō, for the process of setting the foundation for engaging relationships through traditional ritual); and *huaka'i* (journeys) to Kanaloa (a small, very sacred island off the coast of Maui that was confiscated by the US Military, denuded and defiled over 40 years of intensive bombing practice, and through nonviolent protest returned to the Hawaiian people for restoration and reconciliation), as well as *huaka'i* through all the *moku* (land divisions or sub-county districts) of Hawai'i Island: Kona, Ka'ū, Puna, Hilo, and Hāmākua. During these *huaka'i*, we meet and work with *kama'āina* (children of a place), perform bioculturally structured ritual to enter place and perform the work of culturally-grounded restoration, conservation and resource management. We also engage larger audiences through academic presentations (for example, at the annual Hawai'i Conservation Conference, which attracts 1000+ participants from across Hawai'i and the Pacific region), and we also serve the ritual needs of our conservation community (for example, by helping to lead *kīpaepae* for community, educational, or scientific events). Through these experiences, we understand more clearly now that if we are to succeed in our professions as stewards, then practice of our professions demand nothing less than the *aloha* and conviction of a devoted parent for an adored child.

We also understand that we must foster a collective resurgence of sacred commitment to the places and processes we steward, a change that we believe is required if we are to heal the biogeochemical wounds of unsustainable resource extraction and restore sacred relationships across our evolutionary family that together will ultimately foster socioecological well-being. We have, effectively, reimagined our personal and therefore our professional relationship to the places that we steward: the plants, the animals, the corals and microbes, the elements, the human people, the mountains and the valleys, the rivers and the shorelines, and the bays and the open ocean.

1.4. Why We Need to Tell Our Story This Way

Hālau 'Ōhi'a creates a space for collectively recognizing and celebrating deeply held personal motivations that often drive one's relationship with land, river, and sea. This kind of relational dialogue was either not present in our professional work environments or present in very limited ways; this contemporary reality had many of us thinking (to ourselves or in conversations with like-minded colleagues): how can we do our work better and more aligned with personal beliefs and practices? Kekuhi challenged us to use this writing opportunity to articulate our Hālau 'Ōhi'a learning in article format. To be absolutely honest, we struggled with this task, but through extensive discussions, have decided to share our learning in the form of ritual process manifested in the following journal article. You, the reader, may be surprised to learn that simply by arriving at this point of the paper, you have begun the ritual with us, which in the context of our learning as *haumāna* of Hālau 'Ōhi'a begins with a *Welina* (physical and spiritual welcome).

What follows in each section of this paper is an opportunity, if you choose, to engage your own ritual experience. The format then is quite different from what is encountered in indexed scientific journals, including *Sustainability*. Specifically, drawing from elements of our experience of Hawai'i practice, our ritual follows these five steps: the *Welina* or the welcome and orientation (Section 1 above); this is followed by the *Ho'omākaukau* (To set intentions; Section 2), or personal and collective call to preparation that includes setting personal intentions; after setting intentions, the *Hōʻīnana* (To come to life; Section 3) follows and includes the sacred process of initiating, entering into, and moving through multiple layers of knowing and meaning; Section 4 is the *Pani* (Closing), where individually and collectively, we recognize that the ritual has been performed and it is time to transition to Section 5, the *Hoʻokuʻu* (To release from ritual), which allows the participant to return to the mundane after having engaged, embraced, and absorbed sacred lessons provided by the ritual catalyzed experience.

We have made this decision to go with a ritual-based format because this writing effort is not only focused on transferring information, but is intent on providing *hua ʻōhiʻa lehua* (the seed-laden fruits of *ʻōhiʻa*) that lead each reader and author into an opportunity for transformation—both yours and ours. So, by aligning the structure and intention of this paper with this particular ritual process, by making the writing and reading of this paper a ritual in itself, we feel that we are more able to effectively and authentically convey the transformative power of ritual in the pursuit of sustainable resource management and effective conservation. Finally, we believe that it is remarkably appropriate that this paper should be published in a sustainability-focused journal because sacred connections to self, community, and to place are foundational to maintaining the resilience and sustainability of any system.

2. *Hoʻomākaukau*—Setting Intentions

Ritual and Multiple Layers of Meaning

In setting our intentions for writing this ritual, we felt it important to demonstrate how ritual expression can provide a path forward for sustainability, resource management, or conservation professionals to actively be in sacred and intimate relationship with the places that we serve—much as one would be in relationship with one's family or closest friends. To do this, we build on our growing awareness of and commitment to the sacred relationships that define who we are in relation to self, family, community, as well as the world of organisms and processes that sustain us and that are sustained by us. To be clear, developing these spiritual relationships does not require a dismantling of one's personal/professional belief system, but only to consider the notion that spiritually based relationships promote well-being and support a more sustainable path into the future.

As part of the *Hoʻomākaukau* phase, we take time in our daily lives to practice, study, interpret, and learn from *moʻolelo* (life stories), *kaʻao* (stories of, for example, creation and cosmologies), *mele* (traditional songs and chants), *oli* (vocalizing), *hei* (performed string art linked to *oli*), *hula* (Hawaiʻi's environmental dance), and traditional Hawaiian scientific knowledge, such as that which is captured in *ʻōlelo noʻeau* (wise sayings of biocultural significance). A central part of this practice is being aware of and prepared for embracing multiple sources of knowledge, multiple layers of meaning, and multiple ways of interacting with the world [21].

A central but sometimes overlooked feature of Indigenous knowledge systems is the very formal and structured botanical, ecological, agricultural, hydrological, atmospheric, oceanographic, etc., observations that shape Indigenous knowledge of a place [3]. This celebration of diverse ways of knowing is powerfully exemplified within the multilayered *Kiʻi* (reflections) framework, composed of *Kiʻi ʻIaka* (reflections of self), *Kiʻi Honua* (reflections of community), and *Kiʻi Ākea* (reflections of the universal), upon which we rely heavily to convey our lessons learned to you the reader. So, this article, a physical manifestation of the ritual into which we are asking you to engage, seeks to teach and transform at three different scales, perspectives, or levels including the deeply personal, the collective family or community or even regional, and the universal.

In reading a sacred text, interpreting a chant, or in creating a poem, we are drawn personally and uniquely to the exchange because our being is uniquely engaging the elements of a story or chant in that very moment and in a particular place. For example, you, the reader, in reading a story may connect to the sacrifice of an elder brother for his younger sibling because you are the eldest sibling of your family, perhaps have taken on much of the responsibility of raising younger siblings, and by making this connection to the story, certain sections of text or themes have a specific message for your unique experience as elder sibling. This *Ki'i 'Iaka* reflection might be particularly poignant if you have just experienced a powerful sharing with your younger sibling. As the regional reflection, *Ki'i Honua* evokes for the participant a particular set of shared experiences—experiences that might bind together a family, community, or culture. For example, a chant might evoke the importance of a journey across a body of water for accessing new lands or escaping harmful conditions, and you may see your own family's or even community's immigrant journey reflected in the story. In Hawai'i, engaging this theme might conjure images of the *wa'a* (canoe) and the literal and metaphorical importance of the *wa'a* to the Hawaiian people as a vessel for discovery, for connecting peoples across the Pacific, but also as a vehicle for coordination, elevated cooperation, and in the best cases, collaboration. The *Ki'i Ākea* asks the participant to find that which is universal within the images, themes, or ideas that are being shared. For example, loss and sacrifice in preparation for the birthing of something new can be seen as broadly foundational to the human experience, and in the engaging of this cycle, we become part of and are provided an opportunity to learn from the global human experience of transformation by sacrifice.

In engaging this *Ki'i* framework here, this paper structured as ritual expression seeks to:

(1) identify and share the global importance of being genealogically tied to our places—a fundamental feature of the human experience (*Ki'i Ākea*);

(2) show how we have relied on Hālau 'Ōhi'a to help us transition from a Western, colonial model of sustainability science (resource as commodities to be maximized to support human consumption), natural resource management (resources as objects to be managed through centralized, agency-controlled decision making), and conservation (systems of organisms to be protected from human use), towards a kinship-based model where stewardship is defined by sacred relationship to place and process, with traditional Hawaiian scientific knowledge and ritual fostering this transition/transformation (*Ki'i Honua*); and

(3) demystify what ritual can mean for the individual practitioner in a sustainability, resource management, and conservation context (*Ki'i 'Iaka*) through the sharing of our individual experiences in ritual.

A final critical aspect of *Ho'omākaukau* is the identification of one's genealogical (not necessarily genetic) and biogeographical relationships with places or processes. This is a fundamental concept, as these connections define one's reciprocal stewardship relationship with one's surroundings as much as elucidating one's human family genealogy helps us to understand our connection to parents and grandparents, uncles and aunts, the migrations that brought our families to specific geographies, and the cultural identity and traditions that shape and enrich our lives. Viewed more broadly, genealogy as understood within a Hawai'i perspective pushes us to consider broader connections defined by biogeochemical and evolutionary ties, including to sources of food and water that literally make up a resident's physical and spiritual being, and that person's connections to all members of the evolutionary tree of life. By becoming familiar with, engaging, and then cultivating gratitude for one's familial (*Ki'i 'Iaka*), biogeochemical (*Ki'i Honua*), and evolutionary (*Ki'i Ākea*) relationships, those relationships that make up the broadly defined genealogies that sustain us, we are better prepared to enter into *Hō'īnana*, engage in ritual, learn from ritual and then apply lessons to our daily professional and personal lives.

3. *Hō'īnana*—To Come to Life

3.1. What Does Ancestral Ritual Look Like?

In Hālau 'Ōhi'a, ritual begins with two practices—the first involves formally requesting permission to physically and spiritually enter into a sacred space that for our process is the *hālau*. When we have been welcomed into this space that is the *hālau*, our ritual continues with the building of *kuahu* (altar as portal to the sacred) that is the act of physically and spiritually entering into a sacred space shared by all participants. These are foundational practices that achieve several things. The first practice reminds us of our humble status as stewards: at the level of individual entering into a shared space with other students; at the level of a student collective entering into sacred dialogue with each other and our broadly defined communities, including as professionals in places we steward; and at the largest level as a fleeting presence on earth defined by *aloha* for all members in our genealogy, including honoring and expressing gratitude for that which precedes us, stewarding and expressing gratitude for that which sustains us today, and cultivating and expressing gratitude for that which will sustain our genealogy into the future.

The second practice guides us to leave behind mundane distractions (work schedules, shopping lists, household tasks) and focus on fundamental and sacred aspects of being a human in community to more fully engage what it means to be in community, to love and be loved, to care for and be cared for, to sustain and be sustained. Entering into the practice of *kuahu* demands that we fully engage what it means to embrace our genealogical connections to place and process (who broadly defined is sacred to us and why), and through embracing, how can we contribute to a collective exploration and deepening of relationship with and *aloha* for our genealogy (how do we make good on sacred devotion; what is the quality and motivation for this devotion; what are the physical, psychological, emotional, and spiritual trajectories for our relationships). By humbly asking for permission to enter into a space and then literally creating a physically, psychologically, emotionally, and spiritually safe space for sacred dialogue and connection, we open the portal for ritual practice to manifest for each participant's connection, growth, learning, and ultimately, personal and collective transformation.

3.2. Remembering Genealogical Relationships through Ritual

In this section, we share how our genealogical connections can be elucidated and literally manifested through the vehicle of ritual expression. We also share the multiple layers of transformation that have occurred during our time with Hālau ʻŌhiʻa, with the hope that you will as part of this paper-as-ritual process identify how to engage and cultivate a sacred space with your colleagues to identify, discuss, and reflect upon the substantial topics discussed here. From the perspective of genealogy, ritual helps us to understand, honor, and enhance our relationships to the places and processes that we steward. Through much of our formal disciplinary training (e.g., sustainability science, natural resource management, conservation biology), many of us were taught explicitly with learning reinforced implicitly that we, as people, are separate from the natural world, that we have dominion over this world, and so it is within our rights and responsibilities to manage and control the resources of this world in ways that maximize the goods, services, and benefits provided to a society. Relationships have only recently become part of academic considerations of the management calculus [22], but when discussed in broader contemporary contexts, relationships are still portrayed as being ancillary to achieving management success.

For example, the practitioner is often asked to distill down the how, when, and where of resource management to simple economic metrics of success, with metrics of success fully occupying the decision-sphere. Within this framework, sacred relationships can be viewed as hindrances: when formed or held by professionals, these relationships may obscure objective evaluation of metrics of success and so complicate assessments of management; when formed or held by biocultural practitioners and communities who are connected to place and process, these sacred relationships may interfere with centralized decision making about place and process; when formed or held by professionals, practitioners, and communities, these sacred relationships may drive outright conflict that prevents implementation of agency-driven decisions. Conversely, by not embracing sacred relationship in sustainability science, resource management, and ecological conservation, professionals limit their capacity to communicate with biocultural practitioners and communities, and engage practitioners and communities in reciprocal stewardship—with each other and with the places and processes of interest.

An important feature of this conflict is in our training as resource and conservation professionals. Specifically, we are trained in universities, and this training is reinforced in the work place in a way that engrains the notion that in order to protect the plants, animals, resources, and places that we care about, we need to support the designation of these places as protected areas, pay professionals to exclude threats (including people) from these areas, federally list species as being of concern, all with the goal of preserving these areas in an isolated and as close to human-free condition as possible. These approaches identify the natural world as commodity (acres treated, numbers of individuals of a listed species saved) to be isolated and locked away. While resource management approaches or conservation practices are often reasonable and important for perpetuating species of concern, the ritual practiced in Hālau ʻŌhiʻa has shifted assumptions about our role, specifically the role of kincentric connections, in the care of these places and the sustenance we give to but equally important receive from these places.

Ritual is helping each of us, individually and collectively, to connect to our shared and personal landscapes and seascapes, to the organisms and processes that bring life to these places, and to each other and ourselves as genealogical members of these places. At the foundation of this connection is knowing our places geographically, connecting to our processes that sustain us hydrologically, ecologically, and biogeochemically, and engaging our organisms evolutionarily and taxonomically. However, to attain this depth of understanding, ritual asks us to pause, think, notice, consider, and engage with a readiness to listen, receive, and to express gratitude for that which is living and nonliving in a place. In short, as we might bring many ways of knowing to our relationships with friends and family, so ritual asks us to bring many ways of knowing—intimate, artistic, fun, committed, patient, and sacred ways of knowing—to our places.

Returning to the ritual of presenting yourself to a forest, coastal ecosystem, classroom, or gathering space by first setting your intentions and asking permission to enter (the Hawaiʻi ritual of the *mele komo* and the act of *kāhea*), this practice establishes a tone of humility and respect that helps us to open our minds and hearts so that we can learn from that place on multiple levels. We are driven to know more intimately and patiently and with greater commitment the human, plant, and animal-people of that place. We use art to express this *aloha* for these places and the beings that make these places home. We express gratitude to these places and beings because we know that they literally sustain us, as a parent who provides for us physically, psychological, emotionally, and spiritually. We know that without these places, we are left impoverished, much as a life without friendship or deep family ties is lesser existence.

Finally, it is through the ritual that we physically offer our voice, our sweat, and our intentions as part of a reciprocal exchange with those places that we are genealogically connected to, and this exchange promotes well-being. The fields of psychology, animal (including human) cognition, and epigenetics, among others, all provide conclusive evidence that the quality of our relationships shape our health, our joy, our capacity for thriving—in short, our well-being. Experiments with non-human primates and more contemporary lessons from understaffed orphanages have reminded us of the simplicity, universality, and ancestral nature of this truth. And while early philosophical writings about our relationships to nature are rich with notions of well-being, contemporary agency-based approaches to conservation and resource management uncomfortably cling to a strictly biophysical model of stewardship that in our view disempowers the steward and the stewarded.

Kiʻi Ākea—Why is it important for humans to recognize our genealogical connections to place? The need to belong and form attachments is a universal *kiʻi* among humans. Biophysically, we know that all life on this planet and all forms in this universe come from a single cosmic event—the big bang. The atoms that make up our human bodies, the bodies of our plants, animals, the ocean body, the atmosphere, every form on this planet and beyond, all originate and share an ancestry with stars and the most ancestral of cosmic events. Beyond being physically made up of the same building blocks as our stellar and earth landscapes, environments across the planet all physically nurture us. Our mountains give us life through driving our weather patterns, by being the foundation of our forests, which in turn cover the watersheds that form our water sources, and by providing the alluvial substrates for our farmlands where cherished members of our human community cultivate the food we eat while sustaining enormously complex ecosystems. This water and food from our mountains, plains, and seas physically sustains us, providing the building blocks for our cells—our skin, brain, intestines, hair, and muscles—in short, our beings. Research demonstrates that when our connections to these places, from childhood [23] to adults [3], includes acknowledging this genealogical connection to our place—our mountain, our stream, our ocean, the socio-ecological landscape, its fabric and features—we recognize that we are as connected and reliant upon them as we are on our life-giving parents and grandparents. With this relationship of connection and reliance come the same responsibilities to care for these mountains and streams that we have to care for our elder family members. One does not need to be Indigenous to a particular place to take responsibility for one's relationship with the places that give us life and sustain us.

Ki'i Honua—Yet, we can learn from Indigenous cultures, which often codify kincentric relationships between people and the elements of a regional landscape through legends or tales, poetic texts, dances, or other sources. In Hālau ʻŌhiʻa, the first *mele* (chant) and accompanying *hei* (string art) learned by students is "ʻO Wākea Noho iā Papahānaumoku," which details the genealogy of Hawaiʻi—all of its islands and its people. It begins with the male entity Wākea (the expansive sky) joining the female entities Papahānaumoku (she who births islands) and Hoʻohōkūkalani (she who affirms the stars in the heavens) to give birth to the Hawaiian archipelago. As part of this genealogical chant or *koʻihonua*, the union of Wākea and Hoʻohōkūkalani resulted in the birth of a stillborn child, who is buried in the earth. From his body grows the first *kalo* or taro plant (*Colocasia esculenta* (L.) Schott, Araceae), Hāloanakalaukapalili (Hāloa, literally great breath of the quivering leaf), which becomes the most important staple crop in Hawaiʻi. Through this union, a second child is born, also named Hāloa, but this child lives to become the first man and original ancestor of all Hawaiian people [24].

Ki'i ʻIaka—The *kalo* plant is foundational to Pacific island communities because for millennia, it was the main focus of one of the most remarkable traditional Indigenous breeding programs known to science as well as being a source of sustenance for Pacific peoples including settlers of Hawaiʻi. Today, *kalo* continues to be culturally vital despite massive social, agricultural, and ecological changes to Hawaiʻi's food system [25,26]. In understanding the shared genealogy of the Hawaiian people, the *kalo* plant, the islands, the earth, the sky, the stars, through this chant, we are charged with cultivating, caring for, and protecting those plant, land, ocean, and element siblings and ancestors as if they were family. At a personal level, when we plant, maintain, harvest, and prepare the next generation of *kalo*, we do so with the utmost thought and love. We make sure to never step near the roots of the plant, we diligently weed the patch, we learn the names of the dozens of varieties, and when it is time to harvest, we spend hours cleaning its corms and cuttings, always with an eye to replant and where ever possible share the *huli* (pruned stalk) and the *ʻohā* (intact stalk with leaf and some corm), from which forms the next generation of planting material (Figure 1). This is done so that Hāloa is sustained into the future, and in turn, we as people of Hawaiʻi are sustained for generations to come. Manu Meyer [4] (p. 15) quotes a legendary *kalo* farmer from Waipiʻo Valley, who describes the literal and metaphorical importance of planting the elder sibling *kalo* with integrity and sacred devotion because to do otherwise would hamper the growth and integrity of the harvest and genealogical perpetuation of this foundational agricultural resource. More metaphorically, our relationship with the physical crop is a reflection of how we speak, cultivate, and harvest the fruits of our ideas and actions. Do "we speak powerfully, truthfully, and with purpose or do we think ill, speak ill, and act ill" [4]?

Another example of kinship manifested in action can be found in our marine realm. As a descendent of all of the lifeforms starting from the sky, earth, and stars, we are kin to the *ʻopelu* fish (*Decapterus macarellus* Cuvier; mackerel scad), a staple of the Kaʻū region of Hawaiʻi Island and coastal communities across the archipelago. For some of us, when we are harvesting *ʻopelu*, we look at the fish eye to eye, and we tell it, "I'm going to take your life to sustain me and my family;" we recognize the physical and spiritual reciprocity between us as people and the *ʻopelu* as an ancestor. After we have eaten him, we return its body to the ocean. Akin to the relationship between Hawaiians and *kalo*, we honor the *'opelu* for sustaining and nourishing us now and into the future through the entire process from recognition, harvest, ingestion, and returning the ancestor who has fed us back to the source. We, in return, work to sustain the *'opelu* through proper care and management of its surroundings—the coral, the algae, the reef and pelagic fish community, the shellfish—basically, all the features of the *'opelu*'s genealogy that are required to support this species. These features have been gleaned over countless generations of keenly observing trial and error responses of this fish to natural variation in the environment and to traditional management.

Figure 1. Hālau 'Ōhi'a (inclusive of our children) carefully planting *kalo* in Pu'ueo, Waipi'o, Hawai'i Island.

3.3. Manifesting Genealogical Relationships through Ritual

Ki'i Ākea—As we have shown, identification of the genealogical relationships you have with your place is fundamental to recognizing, conceptualizing, and ultimately participating in the reciprocal relationship you have with your surroundings. This approach can be seen as a tool that is applicable and accessible globally. In this section, we discuss how these genealogies are physically manifested through the vehicle of ritual expression and the emotional transformation that occurs in this process. Ritual creates a space and establishes a context for understanding and honoring our relationships to the places we steward.

Ki'i Honua—Returning to the *mele* "'O Wākea Noho iā Papahānaumoku," the recitation of this chant, the application of our breath to these words and names, and the recreating of the images of Wākea, Papahānauamoku, and their island children with the *hei* (Figure 2) allows us to experience the deep, raw, and universal emotions that solidify the genealogical (familial, biogeochemical, evolutionary) connections we have with our surroundings. For us, this transformation can come through a body motion in *hula*, a *hei* figure, or speaking the name and replicating the actions of the volcano deity Pelehonuamea. In this recreation, we allow ourselves to be overtaken by gratitude, as manifested by the *mele* "Lei o Hilo"; by heightened awareness and respect for elemental forces in the *hula* "Kūkulu ka Pahu"; or by the perpetuation of our species in the *hula ma'i* (procreation dances). Once we are touched by these images and emotions, they are a part of us, with each ritual serving as a pathway to making seen and available for learning these vital connections.

Ki'i 'Iaka—As a *hālau*, our learning gained practical expression when we were asked to participate in a *Ho'ola'a 'Āina* ritual led by Kumu Kekuhi. This ritual took place in a healthy native forest ecosystem where low impact construction was to take place for establishing an ecological monitoring tower that is

1 of 20 sites (the Pacific Domain) that make up the National Ecological Observatory Network (NEON). The goal of ritual was to communicate with the site—the soil, trees, birds, and sky—through *oli*, *hula*, and *hei* that construction was going to occur but that healing and regeneration would follow the disturbance. Through our offerings, we became the ritualized exchange for the sacrifices absorbed by the site for the production of scientific information. In the moment of the ritual, during our performance, divisions between performer and the forest people—the trees, the birds, the mist, the wind—dissolved away. For some of us, this moment was the first time ritual expression served as the becoming of the object of the ritual. The dissolving boundaries and resulting connections helped us to match our movements with those of the forest swaying in the wind. The mind was released from what was happening, and footstep instinctively followed footstep in the performing of our hula, motion after motion, until the ritual was complete. Similarly, sounds of chanting flowed as we lost ourselves and became the forest, syllable after syllable, line after line. This ritual allowed us to create and enter into a sacred relationship with this forest; to do this, we left behind mundane considerations and expressed our gratitude for a sacrifice that would provide long-term monitoring data on changes to the health of the forest. In many ways, this event marked an important step in the integration of biophysically defined Western science (concerned with measurable phenomena external to or independent of the measurer) with relationally defined Indigenous science (concerned with observable and sensed phenomena including role of observer in the observed or sensed network of relationships). It reframed impact as redeemable through exchange while clearly elevating the importance of making every effort to honor the sacrifice of a place to science.

Figure 2. Final image created during "'O Wākea Noho iā Papahānaumoku", illustrating the Hawaiian Islands birthed by Papahānaumoku.

3.4. Applying Genealogical Relationships and Ritual in Conservation

We have discussed the transformative nature of ritual expression and the elevating of relationship from ideas and concepts to the realm of physically manifested sacred reality. This transformative becoming is powerful because embracing one's genealogical connection to the world greatly enriches our work as sustainability, natural resources, and conservation professionals. We realize that our effectiveness is influenced by the sacredness with which we engage the places and processes that we steward. We see our ties to a place as akin to our ties with beloved family members, such as a grandparent. In this section, we discuss how we apply and integrate these genealogy and ritual practices in our lives and work as sustainability, resource, or conservation professionals.

Ki'i Ākea—As we have shown, the portal of ritual and its imagery goes beyond human-to-human connections and allows for environmental and elemental beings to become more accessible and relatable to our human experience. The ritual identifies and helps us develop the linkages we have with the other organisms present, the place itself, and the challenges our places face. When we feel connected to a conservation issue of a place, we are able to persist and push through obstacles we encounter because our commitment is not to a job, programmatic theme, or achieving annual statistics, but rather to a family member under threat. This concept of a personal and familial bond is a powerful counter-example to how we humans all too often treat places and processes: enter; take what is needed (be it timber, water, or data); and then when degraded, no longer useful, or no longer funded, abandon.

While there are many examples of agency-based approaches to stewarding place that are positively increasing the well-being of person and place, we suggest that our approach to sustainability, resource management, and conservation could enhance already effective practices. Where approaches are not effective, we suggest that our approach could transform ineffective practices if all practitioners were supported throughout their organization to acknowledge, honor, and engage with their places—the plant people, the animal people, the forest people and the water people, as they would with cherished family members. Through practicing these rituals, recognizing these genealogies, and engaging with places as living, thinking, feeling people, we prepare the internal space and cultivate the awareness for feeling accepted by that place—in short to prepare for being *hānai'd* (raised, reared, fed, nourished, sustained, adopted) by place [27]. Engaging in, practicing, and performing these rituals helps us to embody the idea that we are not separate from (as humans) or in control of (as managers) these places, but that we are enmeshed, or in the words of socioecological systems thinking, that we are part of feedback loops woven into the systems of which we are a part.

Ki'i Honua—

'O Hualālai me Mauna Loa ku'u mau mauna
Hualālai and Mauna Loa are my beloved mountains

'O ke kai mālino ku'u kai
The calm sea is my beloved ocean

'O ka 'eka, ke kai 'ōpua, ke kēhau ku'u mau makani a me ku'u mau ua
The 'Eka (onshore), Kai 'Ōpua (distant horizon clouds), and Kēhau (gentle off shore breeze and dew) are my beloved winds and my beloved rains

Ola!
Life!

In our Hālau 'Ōhi'a journey, at all of the different places we have engaged with throughout our islands and across oceans, we have introduced ourselves to lands and people through our biocultural genealogies and ritual. We present ourselves, not with our name, title, and agency position, but by calling out the name of the mountain of our home: Mauna Kea! Mauna Loa! Hualālai! Kīlauea! Kohala! By dancing in honor of the waterfalls that feed our ancestral food systems. By singing and chanting to the hill, the tree, the birds, and the people that we visit in these special places. In doing this, we are saying, "This is who I am, these are the lands and waters from which I was born, or which now feeds and nourishes me." We are saying, "My extended genealogy honors you."

Ki'i 'Iaka—Some of us take this process to our offices and field sites, teaching our workmates the ritual-based process of engaging new work sites, for example. Letting a place know a visitor's intentions is important to place and to self to create the highest quality work possible for the healing of a landscape or seascape. This sharing of intention through our voices via traditional or new *oli*, through spoken words in the language with which you are comfortable, or even silent thoughts of communication with the place allow us to become more strongly tied to the place, which becomes enhanced through planting, sweating, and working to steward an area. Throughout, we are also learning the patterns of the wind, the path of the animals, and timing of the rain. Over time, we share details about the place with others. We share its genealogy, the mountain that birthed this place, the rains that feed this landscape, members of its ecological community, and past actions that have left these scars on its substrate. We become responsible for that place through this sharing of genealogies; before we realize it, we have become *hānai*'d by that place.

For those of us who work in environmental education or community engagement to promote biocultural conservation, these same processes can be applied to educational groups. Having the visitors to any site first ask permission to enter begins to open the door to other layers of learning and then understanding. We tell the stories of the place, and for the visitors, that begins to reveal more layers. We have them stop and see what the winds, clouds, and birds are doing—exposing more layers. We ask them to smell, taste, and feel the place—more layers still. Why is this important? Because when you come to know a place on these levels—physically, emotionally, spiritually, intellectually, historically—the place becomes part of your genealogical story and you begin to treat the place differently. This may seem strange to those committed to objective purity, but most conservation educators know it is a personal connection between child and forest plant or animal, the awe of a volunteer in the power of planting a tree knowing that the tree will live 300+ years and support countless generations of forest birds, and the love of place that most often drives people to sacrifice so much for the protection of a place dear and near. It could be argued that connection to and reverence for place is a force like no other in the management of lands, waters and seas, and so for thriving stewardship [22,28].

Thus, ritual can be applied in biocultural conservation and resource management to initiate and develop a person's intimate relationship with a place in order to be better stewards. At another collective level, ritual can be a tool that conservation practitioners use to introduce a place to educational groups as if introducing a family member. Critically, the process of ritual can serve to transform human to human relationships in ways that differ from the ad hoc relationship building or formal team building that happens in most organizations serving lands and seas. We are humbled and inspired by singing to the mountains, forest, ocean, and rivers; by sharing our intentions and offering, giving, and honoring the reciprocal relationship with the places and processes that sustain us, and deepening those reciprocal, kincentric relationships between human and plant, plant and mountain, human and human [29]. These rituals act as accelerators or catalysts for relationship building—in our backyards, our stewardship lands, or when traveling as a group to distant lands.

On this last point, engaging, introducing ourselves to, and humbly thanking our Maori hosts in Aotearoa helped them to know that we were paying attention, that we respected them and their mountains and waters, and that we were humbled by their work. Importantly, we also were able to show that we, too love our places and the many and diverse members of our communities, and in this love of our places, we were able to quickly form intimate relationships with our hosts, their families, and their storied places. For some of us who are used to the professional exchanges and encounters (annual society meetings or agency workshops), this radically different approach with radically different outcomes was profound, intense, soul lifting and a powerful lesson of how ritual can manifest transformation.

We hope that we have shown that ritual creates the space for relating to our environments, to each other, and with visitors and hosts on an intimate level, and that ritual operates at various scales. Applying this learning to the work environment has helped us build these relationships in our work, allowing for conversations and actions that were not possible in Hawai'i just a few years ago.

4. *Pani*—Closing

All of us authors have genealogical connections to lands, rivers, and oceans far from Hawai'i, while some of us are also tied by deep ancestry to places in Hawai'i. What we have learned is that regardless of our origins, we must steward our places as family. We must acknowledge that while we have other ancestral homes, ritual supports our continued understanding of who we are in THIS place. The use of *ki'i* helps us understand perspectives from multiple ways of learning. *Mo'okū'auhau* (genealogies) and *ko'ihonua* (cosmologies) biogeochemically and evolutionarily connect us to the water that we drink, the food we eat, and the *'āina* we live on as nothing less than our most beloved family member. So, we leave you, the reader, with this. We encourage you to know your mountain, your water source, your socioecological district, and the stories of these places. How did your stream get its name? What have your people and the people who came before called your significant places? How can you better honor your relationship to these places? Can creating art, new stories, *mele*, and *hula* for them provide an avenue for this furthering of connections? Sing to them. *Hula* for them. Be with these places as you would be with a beloved grandparent. Honor these places by knowing their intricacies while working to enhance their well-being with the commitment and love that you might have for the raising of a child or caring for a loved one. By singing to your places, dancing to your mountains, telling stories to your children about your waters, you build community with your children, your places, with your families, neighbors, colleagues, and with yourself. We sing you our final offering:

A Pō Ē
(*Hei* & *Mele* by Taupōuri Tangarō)

A pō ē, a pō ē
It is night, transitioning to dream time

Kau mai nā hui hōkū
Stars appearing, we are them

A ao a'ela
Day appearing, we are consciousness

Helele'i wale iho nō
Stars fall from the sky, time to awaken

5. *Ho'oku'u* (Release)—What's Next?

Before we depart from a place, we ask permission to leave. So, we ask your permission for release from this ritual expression of engaging you with this article. It is always appropriate to leave *makana* (gifts) of thanks with a host. So, we leave you with the very tools that aided us in our own journey to re-establish our relationship with the genealogies of the places and people of Hawai'i and beyond. First is the *mele* "'O Wākea Noho iā Papahānaumoku," which serves to orient you as a human being to your global, regional, and personal genealogical relationships to Hawai'i and beyond the horizon. To learn this *mele* is to engage one of many of the genealogies of the Hawai'i landscape. To engage the *mele* is to become a part of it.

'O Wākea Noho iā Papahānaumoku [24]

'O Wākea noho iā Papahānaumoku
Wākea resides with Papahānaumoku

Hānau 'o Hawai'i, he moku
Hawai'i is the first-born island child

Hānau 'o Maui, he moku
Maui is born, an island child

Ho'i a'e 'o Wākea noho iā Ho'ohōkūkalani
Diurnal space turns to nocturnal space, the Dome-of-Space intercourses with She-who-populates-the-night-sky

Hānau 'o Moloka'i, he moku
Moloka'i is the first to be born of the stars

Hānau 'o Lāna'ikaula, he moku

Lāna'ikaula an island child is born

Lili'ōpū pūnālua 'o Papa iā Ho'ohōkūkalani
Chaos abounds between earth and stars

Ho'i hou 'o Papa noho iā Wākea
Papa reclaims Sky-father

Hānau 'o O'ahu, he moku
O'ahu is born, an island

Hānau 'o Kaua'i, he moku
Kaua'i is born, an island

Hānau 'o Ni'ihau, he moku
Ni'ihau is born, an island

He 'ula a'o Kaho'olawe
Kaho'olawe is born, the royal one

Second, is a template one can use to learn, know, and call out your human and landscape genealogies or *mo'okūauhau* in a Hawai'i format. The *mo'okū'auhau* is your personal continuum, or genealogical chant. Using the format of the below *ko'ihonua*, or cosmology, you will be able to create your own *mele mo'okū'auhau* or genealogical chant. Though this template provides a Hawai'i context example, the process it illustrates can be applied in landscapes outside of Hawai'i.

Mele Mo'okū'auhau **Template**

'O _______________________ no _______________________
    (name of ancestor A i.e., grandmother) (place that ancestor A is from)
Noho iā _______________________ *no* _______________________
    (name of ancestor B, i.e., grandfather) (place that ancestor B is from)
Hānau 'o _______________________, *he* ___________
    (child of ancestor A & B = ancestor C, i.e., mother) (gender of ancestor C—"*kāne*" if male, "*wahine*" if female)
'O _______________________ no _______________________
    (name of ancestor C i.e., grandmother) (place that ancestor C is from)
Noho iā _______________________ *no* _______________________
    (name of ancestor D i.e., grandfather) (place that ancestor D is from)
Hānau 'o _______________________, *he* ___________
    (child of ancestor C & D = ancestor E, i.e., father) (gender of ancestor E—"*kāne*" if male, "*wahine*" if female)
'O _______________________ no _______________________
    (name of ancestor C i.e., mother) (place that ancestor C is from)
Noho iā _______________________ *no* _______________________
    (name of ancestor E, i.e., father) (place that ancestor E is from)
Hānau 'o _______________________, *he* ___________ (*you*)
    (your name) (your gender)
'O _______________________ *ko'u ahupua'a ma ka moku 'o* _______________________
    (traditional land division where you reside) (district where you reside)
'O _______________________ *ko'u pu'u/mauna*
    (mountain or hill where you reside)
'O _______________________ *ka wai/ke kai*
    (fresh water source or ocean where you reside)
'O *ka wao* _______________________ *ku'u 'āina e noho nei. OLA!*
    (socioecological zone where you reside)

Author Contributions: This work was supported by a team of contributors who participated in the work in the following areas: Conceptualization, K.K.; Methodology, K.K., K.S.F., C.P.G. N.K., R.P.L. H.M.; Investigation, K.K., C.P.G., N.K., H.M., R.P.L., C.K.A., K.A., T.A.B., M.B., K.C., L.C., M.L.D., K.S.F., M.F., N.G., A.E.G., M.G., L.H., A.I.,

S.K., N.K., E.L., C.H.P., J.R., L.-L.S., A.S. (Aimee Sato), L.S., A.S. (Amelie Sterling), A.L.U., J.U., B.W., A.N.W and D.Y.; Resources, C.K.A., K.A., T.A.B., M.B., K.C., L.C., M.L.D., K.S.F., M.F., C.P.G., N.G., A.E.G., M.G., L.H., A.I., N.K., S.K., N.K., E.L., R.P.L., H.M., C.H.P., J.R., L.-L.S., A.S. (Aimee Sato), L.S., A.S. (Amelie Sterling), A.L.U., J.U., B.W., A.N.W. and D.Y.; Writing-Original Draft Preparation, C.P.G., N.K., R.P.L., K.K., and H.M.; Writing-Review & Editing, K.K., C.P.G., N.K., R.P.L., H.M., K.S.F., N.G., C.H.P., A.S. (Aimee Sato); Funding Acquisition, K.K., K.S.F., M.B., C.P.G., N.K., A.N.W.

Funding: The APC was funded by Hawai'i Community Foundation.

Conflicts of Interest: The authors declare no conflict of interest

References

1. Holthaus, G.H. *Learning Native Wisdom: What Traditional Cultures Teach Us about Subsistence, Sustainability, and Spirituality*; University Press of Kentucky: Lexington, KY, USA, 2008; ISBN 0813124875.

2. Kealiikanakaoleohaililani, K.; Giardina, C.P. Embracing the sacred: An indigenous framework for tomorrow's sustainability science. *Sustain. Sci.* **2016**, *11*, 57–67. [CrossRef]

3. Pascua, P.; McMillen, H.; Ticktin, T.; Vaughan, M.; Winter, K.B. Beyond services: A process and framework to incorporate cultural, genealogical, place-based, and indigenous relationships in ecosystem service assessments. *Ecosyst. Serv.* **2017**, *26*, 465–475. [CrossRef]

4. Meyer, M.A. *Ho'oulu: Our Time of Becoming: Collected Early Writings of Manulani Meyer*; Ai Pōhaku Press: Honolulu, HI, USA, 2003; ISBN 1883528240.

5. Balzer, M.M. Flights of the Sacred Symbolism and Theory in Siberian Shamanism. *Am. Anthropol.* **1996**. [CrossRef]

6. Wilson, M. *Rituals of Kinship among the Nyakyusa*; Oxford University Press: London, UK, 1957.

7. Bhagwat, S.A.; Rutte, C. Sacred groves: Potential for biodiversity management. *Front. Ecol. Environ.* **2006**, *4*, 519–524. [CrossRef]

8. Rutte, C. The sacred commons: Conflicts and solutions of resource management in sacred natural sites. *Biol. Conserv.* **2011**, *144*, 2387–2394. [CrossRef]

9. Harris, M.; Bose, N.K.; Klass, M.; Mencher, J.P.; Oberg, K.; Opler, M.K.; Suttles, W.; Vayda, A.P. The Cultural Ecology of India's Sacred Cattle [and Comments and Replies]. *Curr. Anthropol.* **1966**, *7*, 51–66. [CrossRef]

10. Mondragón, C. Of winds, worms and mana: The traditional calendar of the Torres Islands, Vanuatu. *Oceania* **2004**. [CrossRef]

11. Douglas, M. *Purity and Danger: An Analysis of the Concepts of Pollution and Taboo*; Routledge: London, UK; New York, NY, USA, 2003; ISBN 978-0-203-37861-8.

12. Durkheim, E. *The Elementary Forms of Religious Life*; Swain, J.W., Ed.; Dover Publications, Inc.: Mineola, NY, USA, 2008; ISBN 0-19-954012-8.

13. Atran, S. Religion's social and cognitive landscape: An evolutionary perspective. In *Handbook of Cultural Psychology*; Kitayama, S., Cohen, D., Eds.; The Guilford Press: New York, NY, USA, 2007; pp. 417–453.

14. Mazumdar, S.; Mazumdar, S. Religion and place attachment: A study of sacred places. *J. Environ. Psychol.* **2004**. [CrossRef]

15. Rountree, K. Performing the Divine: Neo-Pagan Pilgrimages and Embodiment at Sacred Sites. *Body Soc.* **2006**. [CrossRef]

16. Jamieson, R.W. Material Culture and Social Death: African-American Burial Practices. *Hist. Archaeol.* **1995**, *29*, 39–58. [CrossRef]

17. Malinowski, B. *A Scientific Theory of Culture, and Other Essays*; University of North Carolina Press: Chapel Hill, NC, USA, 1944.

18. Wojtkowiak, J. Towards a psychology of ritual: A theoretical framework of ritual transformation in a globalising world. *Cult. Psychol.* **2018**. [CrossRef]

19. Turner, E. Preface. In *The Nature and Function of Rituals*; Heinze, R.-I., Ed.; Bergin & Garvey: Westport, CT, USA; London, UK, 2000; pp. vii–xiii.

20. Turner, V.W. *The Forest of Symbols: Aspects of Ndembu Ritual*; Cornell University Press: Ithaca, NY, USA, 1967; ISBN 0801491010.

21. McMillen, H.; Ticktin, T.; Springer, H.K. The future is behind us: Traditional ecological knowledge and resilience over time on Hawai'i Island. *Reg. Environ. Chang.* **2016**, 1–14. [CrossRef]

22. Chan, K.M.A.; Balvanera, P.; Benessaiah, K.; Chapman, M.; Díaz, S.; Gómez-Baggethun, E.; Gould, R.; Hannahs, N.; Jax, K.; Klain, S.; et al. Opinion: Why protect nature? Rethinking values and the environment. *Proc. Natl. Acad. Sci. USA* **2016**, *113*, 1462–1465. [CrossRef] [PubMed]
23. Louv, R. *Last Child in the Woods: Saving our Children from Nature-Deficit Disorder*; Algonquin Books of Chapel Hill: Chapel Hill, NC, USA, 2008; ISBN 156512605X.
24. Malo, D. *Hawaiian Antiquities (Moolelo Hawaii)*; Bernice Pa.; Bishop Museum Press: Honolulu, HI, USA, 1951.
25. Handy, E.S.C.; Handy, E.G.; Pukui, M.K. *Native Planters in Old Hawai'i: Their Life, Lore, and Environment*; Bishop Museum Bulletin: Honolulu, HI, USA, 1972; p. 233.
26. Kurashima, N.; Jeremiah, J.; Ticktin, A.T. I Ka Wa Ma Mua: The Value of a Historical Ecology Approach to Ecological Restoration in Hawai'i. *Pac. Sci.* **2017**, *71*. [CrossRef]
27. Pukui, M.K.; Elbert, S.H. *Hawaiian Dictionary*; University of Hawaii Press: Honolulu, HI, USA, 1986.
28. Nash, R.F. *Wilderness and the American Mind*; Yale University Press: New Haven, CT, USA, 2014; ISBN 0300091222.
29. Kimmerer, R.W. *Braiding Sweetgrass: Indigenous Wisdom, Scientific Knowledge and the Teachings of Plants*; Milkweed Editions: Minneapolis, MN, USA, 2013; ISBN 1571313567.

 *sustainability*

Article

Nā Kilo ʻĀina: Visions of Biocultural Restoration through Indigenous Relationships between People and Place

Kanoeʻulalani Morishige [1,2,*], Pelika Andrade [1,3], Puaʻala Pascua [1,4], Kanoelani Steward [1], Emily Cadiz [1], Lauren Kapono [1] and Uakoko Chong [1,5]

[1] Nā Maka o Papahānaumokuākea, Kamuela, HI 96743, USA; pelikaok@hawaii.edu (P.A.); ppascua@amnh.org (P.P.); ksteward@hawaii.edu (K.S.); emilyac@hawaii.edu (E.C.); lmkapono@hawaii.edu (L.K.); autumn77@hawaii.edu (U.C.)
[2] Department of Biology, University of Hawaiʻi at Mānoa, Honolulu, HI 96822, USA
[3] University of Hawaiʻi Sea Grant College Program, Honolulu, HI 96822, USA
[4] Center for Biodiversity and Conservation, American Museum of Natural History, New York, NY 10024-5192, USA
[5] University of Hawaiʻi at Hilo Keaholoa STEM Scholars Program, Hilo, HI 96720, USA
* Correspondence: kimmhkm@hawaii.edu; Tel.: +1-808-722-1366

Received: 1 August 2018; Accepted: 5 September 2018; Published: 20 September 2018

Abstract: Within the realm of multifaceted biocultural approaches to restoring resource abundance, it is increasingly clear that resource-management strategies must account for equitable outcomes rooted in an understanding that biological and social-ecological systems are one. Here, we present a case study of the Nā Kilo ʻĀina Program (NKA)—one approach to confront today's complex social, cultural, and biological management challenges through the lens of biocultural monitoring, community engagement, and capacity building. Through a series of initiatives, including Huli ʻIa, Pilinakai, Annual Nohona Camps, and Kūkaʻi Laulaha International Exchange Program, NKA aims to empower communities to strengthen reciprocal pilina (relationships) between people and place, and to better understand the realistic social, cultural, and ecological needs to support ʻāina momona, a state of thriving, abundant and productive people and places. After 10 years of implementation, NKA has established partnerships with communities, state/federal agencies, and local schools across the Hawaiian Islands to address broader social and cultural behavior changes needed to improve resource management. Ultimately, NKA creates a platform to innovate local management strategies and provides key contributions to guiding broader indigenous-driven approaches to conservation that restore and support resilient social-ecological systems.

Keywords: biocultural monitoring; community engagement; community-based management; indigenous knowledge; indigenous science; Hawaiʻi

---

1. Introduction

The term biocultural continues to gain momentum in research and conservation circles around the world, but the underlying concept of linked biological and cultural systems is something place-based and indigenous communities have known for generations. Broadly described in the literature as work at the intersection of biological, cultural, and linguistic diversity [1], research that examines the relationship between diverse cultures and their varied ecological contexts [2], and approaches that start with and are based upon cultural, place-based perspectives [3], a number of interdisciplinary and multifaceted efforts have attempted to characterize biocultural-oriented research. These studies, supported by social–ecological research exploring the feedbacks between humans and natural systems [4], highlight a broad-sweeping need to develop biocultural approaches to understand the linkages and feedback between human well-being and ecological systems [5,6]. Yet there remains a need for case studies and programmatic examples sharing cultural approaches to building biocultural frameworks that are applicable at multiple scales.

A growing body of literature across academic disciplines asserts the importance of using a biocultural approach that recognizes the connections between people and place in order to inform adaptive management strategies [6,7], community-based management initiatives [7], and environmental literacy projects [8]. For example, Kimmerer (2011) uses the term "reciprocal restoration" to describe "the mutually reinforcing restoration of land and culture such that repair of ecosystem services contributes to cultural revitalization, and renewal of culture promotes restoration of ecological integrity" [9]. Winthrop (2014) uses the terminology "culturally reflexive stewardship" to describe stewardship practices grounded in cultural foundations, affirming social identity, and sharing cultural knowledge and motivations [10]. Pascua et al. (2017) use the concept of cultural ecosystem services as a mechanism to characterize "the ways place-based and indigenous groups interact with their surroundings to derive all forms of sustenance and maintain connection to place [11]".

Understanding sociocultural and ecological systems requires a holistic understanding of the relationships and feedbacks that encompass intangible cultural-ecosystem services [11–13]. Recognizing that humans and the environment are one system is integral to improving adaptive management and governance [5]. Indigenous approaches have been an important means to enhance this understanding and recognition by highlighting the importance of relationships, values, and principles in guiding equitable and effective long-term outcomes [14]. The health of the environment is inextricably and reciprocally linked to the spiritual, emotional, physical, and overall cultural health and well-being of indigenous people [6,15]. In Alaska's Inuit communities, climate change is threatening sea-ice ecosystems, a culturally and spiritually significant landscape, and subsequently contributing to the physical and emotional displacement of these groups to the landscapes that support their elements of social and cultural well-being [16]. In Hawai'i, these indigenous approaches have been applied on a larger scale managing biocultural seascapes, such as Papahānaumokuākea Marine National Monument and other large-scale marine protected areas [17].

Research that explores the restoration of social–ecological systems, and complementary efforts to better understand coupled human and environmental systems, require interdisciplinary tools and techniques as well as holistic perspectives that acknowledge reciprocal feedback between people and place [18,19]. In particular, social-ecological systems that encompass place-based and communities provide time-tested and context-specific insight into biocultural restoration in present day [15].

Putting aside preconceived notions of how science is defined in the modern-day context, in this paper we use the words "indigenous knowledge" and "indigenous science" interchangeably as a purposeful and meaningful way to respect the value of traditional knowledge. Traditional knowledge is a knowledge–practice–belief system that forms unmatched repositories of lived and experienced knowledge of natural resource management, acquired over generations, and often millennia, of interactions between people and place [20,21]. These repositories of long-term observations are born from indigenous inquiry and life experiences that shaped adaptive practices and allowed that culture to survive. It is crucial to be aware of how integration of traditional ecological knowledge into resource management can force indigenous people to fit into non-indigenous interpretations of what traditional and customary practices are and try to conform their knowledge systems into existing management systems [21]. Avoiding predetermined roles within collaborative research partnerships, it is critical to consider a mental shift from declaring the modern scientist as the principal investigator to declaring both indigenous peoples and academic scientists as co-researchers [22]. Indigenous science is a form of indigenous knowledge that "relates to both the science knowledge of long-resident, usually oral-culture peoples, as well as the science knowledge of all peoples who as participants in the culture are affected by the worldview and relativist interests of their home communities" [23]. We use "indigenous science" to honor the biocultural knowledge encompassed in indigenous knowledge–practice–belief systems perpetuated through cultural values and practices.

Indigenous knowledge has been widely recognized for its value in providing alternative approaches to create adaptive ecosystem-based management, providing mechanisms for cultural institutions, leadership capacity, and perpetuating values and practice through intergenerational knowledge transmission [5]. Biocultural approaches present an opportunity for indigenous communities to build adaptive collaborative resource management built on indigenous values, worldview, and knowledge while accounting for social, cultural, and ecological factors [24]. There is substantial potential to support the development of equitable two-way research partnerships to bridge knowledge systems and create solutions based on a local-level understanding of the cultural and social factors that support resilient communities. Co-management approaches should be considered a 'knowledge partnership' that has far-reaching impact into supporting resilience through social learning networks, trust building, knowledge exchange, and collaborative problem solving [6,25]. Some inherent challenges in merging two knowledge systems surround the nature of the process ensuring that both systems are valued and equally respected, and indigenous people are not further marginalized from the partnership process and subsequent management decisions [26,27]. Adaptive governance of social-ecological systems will only be successful through first recognizing that humans and the environment are interconnected in one coevolving system [6].

Bridging the gap between the local and global scale, indigenous communities are integral to the development of biocultural approaches that are relevant to the local social, economic, and political environments that communities live in [28–30]. At a local-level, place-based approach is essential to assess aspects of resilience in social-ecological systems and to identify how specific environments and geographies affect holistic health of people and place [31]. Socio-ecological frameworks should also expand to integrate measures of community health and development helping communities find their strengths, strengthening their social systems and sense of place, among other important aspects of resilience [32]. However, transparent communication about the trade-offs between biodiversity conservation and human well-being is necessary to develop realistic solutions [33].

1.1. Weaving Indigenous Research, Community Engagement, and Capacity Building into our Biocultural Approach to Restoration

In building biocultural frameworks from the community level with applications for broader social-ecological systems, this paper presents a novel approach to weaving indigenous research, community engagement, and capacity building into biocultural restoration stemming from an indigenous worldview. As definitions of biocultural conservation continue to expand, and scholars and practitioners alike continue to weave ancestral and contemporary knowledge, technology, and philosophy, there is a critical need to demonstrate what these tools and approaches might look like in action, and from the perspective of indigenous communities. The Nā Kilo ʻĀina (NKA) Program represents one initiative to provide guidance for building measures and frameworks based on indigenous worldviews, perspectives, and values.

In this paper, we present a case study of the NKA Program, a programmatic approach to biocultural restoration of social–ecological systems that aims to address today's complex social, cultural, and biological management challenges through weaving biocultural research, community engagement, and capacity building to impact local resource management and influence national and global management and policies. First, we focus on how NKA addresses biocultural restoration through an indigenous-based framework that creates a platform to collectively address cultural and social behavior changes needed to improve the holistic health and well-being of ʻāina, Hawaiʻi's biocultural landscapes and seascapes. Second, we provide an overview of the biocultural monitoring tools and community engagement strategies of the NKA Program. The NKA biocultural approach is explained through a programmatic framework that operates through Native Hawaiian community networks and partnerships from local to statewide resource management. NKA community networks provide guidance for developing holistic measures of culture-based, social-ecological resilience based on local-community needs. NKA's work contributes to a recent movement to develop and implement culturally grounded indicators of social–ecological resilience [34]. This highlights a novel contribution towards developing biocultural indicators of linked cultural and ecological health to develop effective place-based management and contribute to the creation of culturally grounded frameworks for social-ecological resilience on a broader scale. Lastly, we highlight the role of NKA in building capacity within Native Hawaiian communities to respond to the challenges of research and management partnerships. This approach and the tools addressing social, cultural, and ecological health are applicable to other programs that aim to utilize a place-based or culture-based approach to biocultural restoration of social–ecological systems.

1.2. Strengthening Indigenous-Driven Initiatives to Support Resilient Social-ecological Systems

Globally, there are many examples of how indigenous people can guide and improve adaptive ecosystem-based approaches, supporting social-ecological resilience [16]. Yet, in many instances, conservation has marginalized indigenous people through management strategies that displace and subsequently negatively impact the well-being of indigenous communities [28,35]. As a result of colonization, numerous indigenous peoples have been disconnected from their ancestral lands and stripped of the power to control decisions that affect the well-being of their indigenous culture and the environment to which they are connected [36]. This paper provides an additional case study of an indigenous group engaging in collaborations to confront the systemic disconnect between people and nature.

A growing body of recent work aims to develop culturally grounded sustainability well-being indicators to better understand how ecological and sociocultural factors and feedbacks operate on multiple scales [3,34]. Indigenous-driven biocultural frameworks from Australia [37] and Aotearoa (New Zealand) [24,38,39] provide important contributions to resource management that include cultural well-being. Māori from the Ngāi Tahu Iwi in South Island, Aotearoa (New Zealand) are at the forefront of working in partnership with the University of Waikato developing a Māori-based framework for management of freshwater systems [24,38,39]. These frameworks are based on understanding cultural well-being through intimate knowledge of the relationships between people and the environment [31].

Several case studies aim to provide guidance for empirical research that can address paired human and environmental health in the context of social-ecological resilience, yet more are needed [32,40]. Ens et al. (2016) showed that indigenous biocultural knowledge plays a key role in joint efforts in protecting cultural and biological-diversity hotspots in Australia's terrestrial systems [41]. Additionally, partnerships with small-island communities in Indonesia, the Philippines, and Timor-Leste demonstrated the potential for indigenous knowledge to inform biodiversity conservation, disaster risk reduction, and climate change adaptation strategies [42]. Local communities can increase the relevance of scientific information to a broader group of stakeholders, produce communication materials that depend on the sociocultural environment, while also revitalizing traditional knowledge systems and strengthening intergenerational knowledge transmission [42].

In the midst of innovating biocultural approaches to collaborative co-management of social-ecological systems on multiple scales, it is essential to share the perspective of indigenous-driven efforts highlighting the importance of relationships, values and principles in guiding equitable and effective long-term outcomes in mainstream conservation [14]. These types of partnerships also serve as a mechanism for social justice and require engagement around issues of community capacity building, differential power dynamics, and the lessons from research and management partnerships [13,16,22,26,40,41,43]. It is important to acknowledge the need to decolonize Western discourse in research and create space for indigenous people to represent themselves [36]. In Aotearoa (New Zealand), Māori are at the forefront of building and implementing culturally grounded frameworks of holistic freshwater stream system health through Māori worldview and practice [38,39]. These efforts, supporting a collective voice advocating for indigenous communities to gain more control over their management decision making, fill a larger role than just consultation in cooperative, community-based, and collaborative management [21,27,36,40,44].

2. Nā Kilo ʻĀina: A Biocultural Programmatic Approach

Established by Nā Maka o Papahānaumokuākea (NMP) ten years ago, the NKA Program utilizes a multifaceted, culturally grounded approach to address the complex resource management issues of today through biocultural monitoring tools and programs for community engagement that support positive cultural and social behavior shifts (Table 1). The systemic disconnect between people and nature underscores the need to develop measures of holistic health through a cultural understanding of social–ecological systems. The NKA Program works towards healing disconnects between people and the environment through honoring the importance of pilina. Pilina (defined as relationships) are threads that bind people to the places to which they connect, and to each other, to encourage a return to indigenous knowledge systems. Ultimately, NKA emphasizes the need to develop, build, and nurture pilina within the community to become more aware of dominant patterns of both the environment and people.

In the face of environmental, social, and political change, NKA operates on the shared understanding that, in order to improve place-based resource management, it is vital to strengthen pilina to ʻāina through building communities of kilo, defined as both the practice and role of keen observer. Native Hawaiian knowledge systems of kilo support multigenerational communities to build a collective and intimate understanding of biocultural landscapes and seascapes. NKA initiatives are designed to empower communities through biocultural monitoring and community engagement to gather and build relationships between people who are committed to deepening place-based knowledge and expanding culturally grounded research. The program also builds the capacity of Native Hawaiians and local students, conservation professionals, and educators to serve communities and increase their voice and participation in management. This initiative weaves ecological and sociocultural information together to explore the holistic interconnectedness of the paired human and natural environment and assesses intertidal ecology (marine invertebrates and algae), algal diversity, population densities, and reproductive seasons and size of resource invertebrates such as ʻopihi (*Cellana* spp.) and hāʻukeʻuke (*Colobocentrotus atratus*).

Building networks through a biocultural approach to community-based resource management increases knowledge sharing and empowers communities to navigate through highly complex social and cultural systems. NKA creates a safe space for critical discussion within community-based resource management for communities to co-develop management solutions that ensure the continuation of a productive and resilient ʻāina. Ultimately, it is important to consider the sociocultural impacts of management decisions [45].

2.1. Overarching Vision: Restoring ʻĀina Momona, the Holistic Health of People and Place

NKA is made up of several initiatives that strive to strengthen indigenous visions of healthy and productive social-ecological systems, or ʻāina, a community of people and place. Feeding from the places that feed you continues a lifelong pilina that binds your commitment to care for these places and share this deep pilina and understanding into the next generations. In this special issue, ʻāina momona (lit. fat, sweet, or fertile lands) is described as a state of perpetual resource abundance. Based on the foundation of NKA, we expand that definition to include abundant and productive communities that are inclusive of people and places. Our approach views supporting social-ecological resilience as a mechanism to return to to ʻāina momona, thriving and productive communities of people and places. ʻĀina momona is the ultimate long-term goal for biocultural restoration in Hawaiʻi that speaks to the productive, healthy, and resilient lands and oceans, including the intimate reciprocal relationships our ancestors had with ʻāina, which we are re-remembering today. Though ʻāina is commonly used to reference land and resources, it is important to clarify that a deeper meaning of the term centers around the reciprocal relationships between the lands, oceans, and people which feed and sustain well-being.

Beyond the physical and/or material aspect of provisioning sustenance, this concept also includes feeding and sustaining the emotional, mental, and spiritual dimensions of well-being. Through this expanded definition of ʻāina and its broader implications of the meaning of ʻāina momona, we identify a greater collective movement to adjust our behaviors to support health and productivity together with lands, watersheds, and oceans with which we all share space. Seeing conservation as healing our people and for collective conversation about shifting behavior based on the holistic needs of a place and the practices those landscapes and seascapes can sustain [46]. Many Native Hawaiian scholars share related insights through their research on the value of intimate relationships to places and how inseparable, continual connections to places allow place-based and indigenous peoples to thrive [47–49]. These relationships are at the core of well-being as Native peoples acknowledge cultural relationships through genealogies and traditional cultural expressions and archival documents that connect Native Hawaiians to the lands and oceans across the Hawaiian Archipelago, including the Northwestern Hawaiian Islands [50]. These connections broaden the perspective of participants to honor and respect themselves/their individual self, family, community, surroundings, and places [51].

2.2. Creating a Foundation Based on Native Hawaiian Place-Based Values and Perspectives

NKA is a biocultural monitoring and community capacity-building program, established by the nonprofit group, Nā Maka o Papahānaumokuākea (NMP), and implemented in partnership with the University of Hawaiʻi Sea Grant Program and a network of community partner organizations, and state and federal agencies. In 2009, a small group of Native Hawaiian undergraduates and graduate students at the University of Hawaiʻi at Hilo recognized the importance of indigenous science in developing meaningful guidance to support community-based biocultural approaches in research and resource management. To address this gap, these scholars drew from their strong cultural backgrounds and formal training in ecology to develop the NKA program.

Established by NMP ten years ago, NKA initiatives are centered around biocultural monitoring tools, community engagement, and capacity building. Community-engagement strategies focus on the understanding of pilina as an important component to ʻāina momona, the holistic vision of Native Hawaiian communities. The focal point of NKA is indigenous inquiry and multidisciplinary research applied in locally-relevant, experientially-driven programs, activities, and tools designed for multigenerational communities. Through focusing on intertidal ecosystems, NKA builds capacity within communities to collect quantitative and qualitative data from intertidal ecosystems and extending across terrestrial and marine ecosystems.

2.3. Addressing Complex Resource Management in Hawaiʻi

Consistent with other approaches in community-based conservation that use a systematic approach, recognize coupled systems inclusive of humans, and utilize participatory methods in resource management [7], customary marine-resource management in Hawaiʻi is characterized by traditional and local practices grounded in a sophisticated understanding of and familiarity with an area, resulting from generations of interaction with the natural resources of that place [52,53]. Developed by necessity as a means for the native tenants to not only survive on one of the world's most remote island chains, but to thrive, these place-based interactions have come to represent the deep-seated connections between people and the places they descend from, relate to, and identify with [11]. Traditional knowledge is based on a traditional system of knowing, founded on fundamental observations, relationships, and practice. This knowledge lives on through Hawaiian communities that function as both physical places and social groups that are regarded as "cultural kīpuka", where knowledge is passed on through active transmission of generational and ancestral knowledge through cultural practices [54].

Traditionally, Native Hawaiians possessed a sophisticated land- and ocean-resource management system built on a strict religious and social norms [52,53,55,56]. Traditional management systems were self-sufficient for more than 1500 years, providing for estimated populations of 400,000 to

800,000 people [57]. Yet, the current health of Hawai'i's coastal fisheries is extremely threatened by major anthropogenic stressors [58–60]. As of 2013, Hawai'i's estimated population is approximately 1.4 million people. The effects of this growing population are reverberating through the political, social, cultural, and environmental communities as Hawai'i, and the world, prepares for a future dealing with overpopulation, urban development, and the deteriorating health of fisheries [60].

Drawing from traditional knowledge to support community-based marine resource management provides a promising path to respond to these issues and can facilitate the creation of collaborative, innovative approaches to conserve marine resources [46]. Recently, managers and practitioners in communities across Hawai'i have begun to explore formal co-management agreements, in particular those grounded in place-based cultural norms, values, and practices, between community groups and resource managers, like the State of Hawai'i [61–63]. These efforts are oriented around uplifting both people and place towards a vision of 'āina momona. However, it is important to note that compromises on both sides are necessary across both parties if co-management is truly the desired goal [33].

Though communities are involved in participatory co-management approaches, there remains additional room to empower communities through building a community's ability to trust in their knowledge systems and advocate for their priorities and vision of health and balance to ultimately restore biocultural landscapes/seascapes on their own terms. Pacific Island scholar, 'Epeli Hau'ofa (2000), explains, "We cannot do away with the global system, but we can control aspects of its encroachment and take opportunities when we see them in order to create space for ourselves [64]". This underscores the self-determination of Pacific Islanders to create equitable engagement in management to develop solutions that will guide the future health and well-being of their biocultural environment and future generations.

3. Programmatic Initiatives

Community-based resource monitoring depends on the trust, reciprocity, and inclusivity of indigenous peoples in decision-making and management [65,66]. Examining the patterns of indigenous knowledge and relationships to freshwater systems across Aotearoa (New Zealand), Australia, and North America, scholars use the term "cultural keystone species" as a focal point to better understand holistic freshwater-ecosystem processes through the interconnectedness of people to these ecosystems [67]. Māori developed a cultural health index focused on indicators of human–environment relationships through indigenous worldviews for a variety of river types that can grow national datasets of holistic health of people and ecosystems [38].

While there is indeed ecological research conducted under the NKA Program, our programacknowledges community data-sharing protocols regarding the research component of this work. This is part of a long-term partnership with local communities in Hawai'i to build local capacity of culturally grounded research and community engagement and to ultimately improve community-based resource management. The research is protected for the community to approve its use. The quantitative data NKA has collected is community-owned and part of collective discussions and co-management efforts to improve local, place-based resource management in Hawai'i. Due to the sensitive nature of the information, in particular target species populations and locations, the findings are protected as a principle of respect to the communities with which we partner and can only be shared in a more generalized format, pending community approval.

Table 1. Summary overview of Nā Kilo ʻĀina Program (NKA) initiatives.

Initiative	Outputs/Deliverables	Scope/Scale of Engagement (*n* = No. of Communities Engaged)	Key Outcomes
Huli ʻIa Qualitative Biocultural Monitoring	Over 20 Huli ʻIa posters distributed for educational use	Since 2008, NKA has facilitated over 84 monthly Huli ʻIa discussions engaging ~138 participants ranging from 9–83 years old (*n* = 7)	Using this information, place-based Mauka-Makai (landscape and seascape) seasonal calendars were developed in three communities
Pilinakai Integrated Approach to Community-Based Monitoring including Quantitative Biocultural Monitoring	A Master's thesis developed in 2011 through the Hawaiʻinuiākea program at UHM	Over 72 monitoring and data-collection weekends (2010–present) (*n* = 5) 20 University of Hawaiʻi at Hilo and University of Hawaiʻi at Mānoa undergraduate interns Three Graduate in Hawaiian Studies, Natural Resources and Environmental Management degrees built upon the Pilinakai and Nā Kilo ʻĀina Program	Participants engaging in nature (both lands and ocean) to encourage personal connection, re-connecting to nature to discover responsibilities as native peoples
NKA Annual Nohona Community Engagement Camps and Programs	Over the past six years, NKA has hosted 12 camps, including annual camps in Kawaihae, Hawaiʻi (since 2011) and Hāʻena, Kauaʻi (since 2016), and contributed to more than 10 other leadership and Lawaiʻa ʻOhana camps throughout Hawaiʻi	To date, NKA has worked with communities on Hawaiʻi Island, Maui, Molokaʻi, Lānaʻi, Oʻahu, and Kauaʻi (*n* = 9) Cumulative total of 550 participants ranging from 2–85 years of age. Over 33 interns trained in NKA monitoring tools	Developed biological inventory and monitoring methodology together with communities
Kūkaʻi Laulaha Leadership Building through Growing Indigenous Networks	50 photo books gifted to international partners	Annual cultural exchange in Mangaia, Cook Islands, and Aotearoa (New Zealand) established in 2014 (*n* = 5) Over 50 participants ranging from 16–70 years of age	Participants share cultural values, social issues, and resource-management strategies

4. Biocultural Community-Based Research and Monitoring Tools

4.1. Huli 'Ia

The concept of adaptation is embodied by indigenous peoples whose ancestors survived through adapting to a multitude of environmental changes. Traditional knowledge is part of a knowledge–practice–belief system [20] and in the Pacific Islands, indigenous communities are entering into community-based approaches rooted in contemporary extensions of traditional knowledge systems that provide valuable high-resolution insight into merging monitoring, customary management, and social mechanisms to support resilient holistic systems [68]. To perpetuate oral transmission of this contemporary knowledge base, we engage in training our memories to identify the changes in the environment and what how that can inform human behavior. This encourages local communities to engage in the knowledge system of kilo, keen place-based observations, that enabled our indigenous ancestors to understand their surroundings well enough to know when and where to gather food in order to sustain themselves for generations.

Huli 'Ia is an NMP tool that supports the NKA Program and engages participants in a process of conducting recurring biocultural monitoring activities to quantitatively assess coastal ecosystems while also qualitatively documenting observations—for example, storm systems, cloud patterns, flowering/fruiting plants, reproductive events of land and ocean organisms, fish schooling/aggregating, and size classes (see [34] for additional details). In a facilitated process, participants document seasonal changes and shifts across entire landscapes over time in an effort to identify correlations between and across species and zones including the ocean, land, and sky. In a facilitated discussion with community biocultural-monitoring participants, the group discusses observations of dominant patterns at a particular time-bound scale (usually one month).

Through a discussion of individual observations and group comparisons, participants collectively learn how to reawaken their senses to pay attention to detailed occurrences from the changing of wind direction, wind speed, dominant cloud formations, and rain patterns, and begin to recognize connections between those observations across time and space. This internalization is ultimately intended to inform how people interact with the environment through Native Hawaiian knowledge systems. For example, this might include avoiding harvesting species in a particular location when it is known to be spawning. Huli 'Ia illuminates a dominant seasonal shift in shoreline communities that can inform future monitoring in these highly variable ecosystems. Built through long-term observation tested by environmental challenges throughout time [52], traditional knowledge can guide ecological monitoring and climate-change resilience frameworks [68,69]. For example, it was important to be down at the shoreline throughout both the rough surf in Ho'oilo (wet season) and the calm conditions of Kauwela (dry season) to record collective observations. Over time, the number of recorded observations grew as people enhanced their ability to observe the environment at the shoreline and in the respective areas where they reside. Driven by patterns of rain, storms, and high surf, the lands and ocean became a teacher. We learned how to empower our knowledge systems to increase the capacity of communities to adapt to conditions under climate change and increasing anthropogenic pressures.

Huli 'Ia is a platform to record place-based cycles of productivity in relation to seasons and lunar cycles to guide and inform management practices. Huli 'Ia aims to awaken the ancestral mindset of paying attention to our environment and our impact on it, and encourages participants to ingrain observations into memory. Community participation in this type of research generates greater social awareness and systemic change [70]. After engaging in participatory methods (as described in [46]) and discussing observations through Huli 'Ia for two years in one of our study sites, our NKA team reviewed the data in an attempt to identify cultural and ecological indicators of ecosystem health. We looked for dominant patterns of occurrences and the relationships between space and time of each traditional Hawaiian month and season. Native Hawaiian knowledge systems are intimately attentive to environmental changes related to the seasons—Kauwela (dry season) and Ho'oilo (wet season)—moon phases, and periods of growth that guide Native Hawaiian approaches to co-management [53]. Native Hawaiian knowledge systems are extremely holistic in nature. Through the use of poetic imagery embedded in traditional knowledge systems and physical pictures, we developed a seasonal calendar showcasing these dominant natural cycles and their correlations. These cycles provide a place-based timeline of social–ecological cycles to guide discussions and implementation of best practices in support of these cycles and, ultimately, their productivity. The seasonal calendar also includes 'ōlelo no'eau, or traditional Hawaiian proverbs. 'Ōlelo no'eau is a traditional process of composing easily remembered wise sayings in order to document and transmit information through poetry [71]. Based on monitoring activities, participants compose contemporary 'ōlelo no'eau to document new knowledge, perpetuating a traditional-knowledge transmission mechanism passing on information to the next generation. To look at patterns from terrestrial and marine systems, we worked in collaboration with the team who manages one of the last remaining remnants of Hawaiian dryland-forest ecosystems. We collaborated to develop a mauka to makai (lit., the mountains to the ocean) seasonal calendar identifying patterns informed through combining their place-based knowledge of the dominant drivers of ecosystem and landscape level changes. The NKA team continues to share the Huli 'Ia methodology through partnerships ranging from local community organizations to state and federal agencies in Papahānaumokuākea Marine National Monument (PMNM). Monthly observations were distilled and compiled into seasonal calendars for North Kona, Hawai'i Island and Hōlanikū (Kure Atoll) in the PMNM.

4.2. Pilinakai

Pilinakai led to building integrating tools to understand intertidal-ecosystem health and build collective place-based knowledge of how to guide behaviors that support the holistic health and productivity of these ecosystems. Biocultural approaches rooted in intertidal ecosystems lead to understanding how resource management needs to reflect the social, cultural, and biological needs of the place. Developed through major partners, the Pilinakai team pulled the strands of integration of knowledge systems as biocultural approaches to empower communities in creating management decisions that support productive and resilient ecosystems inclusive of people. Additional contributions of this research approach include increasing opportunities for two-way mentorship between local undergraduate and graduate students including UHH faculty.

NKA's Pilinakai initiative is coordinated and implemented by indigenous and place-based professionals who are committed to helping communities identify management tools that support productive ecosystems in a biocultural framework. The genealogy of Pilinakai is extensive with important foundational stepping stones that provided safe places and support systems for Native Hawaiians to integrate knowledge systems into practice from the community, state, and federal levels. Focused on intertidal ecosystems, Pilinakai was initially developed from a master's thesis in Hawaiian Studies at the Kamakakūokalani Center for Hawaiian Studies at the University of Hawai'i at Mānoa [72] utilizing standard biological survey protocols introduced by Dr. Chris Bird and implemented and evolved by project mentors as tools for community-based monitoring throughout the Hawaiian Islands. Our study examines temporal patterns in spawning behavior, invertebrate population densities and size structure, and community composition. Huli 'Ia, a monitoring tool, was developed and refined through the Pilinakai project in Ka'ūpūlehu and extended into other communities in West Hawai'i, Kaua'i, and into PMNM. Pilinakai blended both Huli 'Ia and the biological intertidal surveys into a biocultural approach applying these tools in different capacities with different communities. Through growing the vision of integrated intertidal monitoring during the Holo-I-Moana Cruise, the Pilinakai leadership advocated to establish the Annual PMNM Intertidal Research Cruise and the Pilinakai team helped develop and implement the cruise with major partners, the National Oceanic and Atmospheric Administration (NOAA) PMNM office, Office of Hawaiian Affairs, The Nature Conservancy, Nā Maka o Papahānaumokuākea, University of Hawai'i, Conservation International, and Dr. Chris Bird at Texas A&M University Corpus Christi. Members of Pilinakai joined this multidisciplinary group of community members from the ʻOpihi Monitoring Partnership, managers, and academic researchers, to better understand the ecology of ʻopihi populations and intertidal communities through a biocultural lens.

The Keaholoa STEM Scholars Program (KSSP) and the Kūʻula Traditional Marine-Resource Management course at UHH inspired and provided a platform to grow the application of these approaches into practice. It was through the KSSP framework that Pilinakai was able to establish itself into the programs implemented within community-based marine resource management today. The KSSP Pilinakai initiative innovated a way for culturally-grounded research and monitoring pillars to develop Huli ʻIa and quantitative biological tools to assess local populations and intertidal-ecosystem health. The central question driving Pilinakai intertidal monitoring is, "How do we know our environment in such a way that, when we interact with it, it's in a healthy, sustainable way?" This foundational question drove the subsequent implementation of the Huli ʻIa initiative along with evolving quantitative biological surveys with a focus on understanding intertidal ecosystems and overharvested limpets known as ʻopihi (*Cellana* spp.).

In response to concerns about overharvesting of ʻopihi, the Kaʻūpūlehu community was interested in monitoring intertidal resources at Kalaemanō. From 2010 to 2012, a group of Native Hawaiian undergraduate student scholars from the UHH KSSP started an intensive research effort with a community in West Hawaiʻi. Starting from the central questions of Pilinakai, this project used Huli ʻIa and conducted quantitative survey methods adapted from the standardized intertidal monitoring tools. The Pilinakai team entered into this community partnership on a shared commitment of dedicating at least five years of monitoring in Kaʻūpūlehu. As Native Hawaiian and local students and mentors with backgrounds in Hawaiian Studies and Marine Science, the Pilinakai team conducted monthly monitoring using transects to assess intertidal invertebrate diversity, population densities and size structures. Findings included information on peak ʻopihi recruitment and population-size structure and abundance. Understanding the most abundant size classes of ʻopihi and sizes of highest reproductive potential are essential to creating rules that protect present and future ʻopihi abundance.

One important output of the Pilinakai initiative has been the development and refinement of intertidal monitoring methods resulting in a suite of ecological data. In order to examine spawning seasons and the effect of ʻopihi size on reproductive output, we collected ten individuals of the three ʻopihi species, measured body size, dissected out the gonads, and calculated the gonad index (gonad weight/total weight × 100). The gonad indices revealed two spawning seasons within a year, and larger sizes were more fecund than smaller sizes during their peak spawning season. Currently, the State of Hawaiʻi Division of Aquatic Resources enacts minimum-size limits for ʻopihi in one blanket rule for all three species and there are no mandated rules that protect spawning season. This is the first study in Hawaiʻi to investigate spawning timing for the three endemic ʻopihi species and provides more detailed place-based and species-specific information needed for effectively managing local populations.

The third survey method integrated into the Pilinakai biocultural monitoring is the ʻopihi and hāʻukeʻuke rapid assessments being implemented in multiple communities on Hawaiʻi Island, Oʻahu, and Kauaʻi, and into PMNM. NMP is a partner organization in the statewide ʻOpihi Monitoring Partnership that also conducts intertidal chain transects and rapid assessments on Maui and PMNM. The objective of this assessment was to collect information of the distribution of ʻopihi by size and location on shoreline. The dataset provides critical information of ʻopihi abundance by species, size, and location to ultimately develop an additional monitoring protocol. The long-term objective is to use this information to implement biannual monitoring of ʻopihi populations. Another application has been to create maps of hotspots where ʻopihi are most abundant. Examining the spatial and temporal distribution of these populations can help to develop management strategies that account for points of human access, harvesting, and further insight into how to investigate environmental factors linked to productive areas.

From 2010, this research effort established a baseline of intertidal communities including natural seasonal fluctuations in population sizes and size structure of culturally prized intertidal limpet, including the most abundant sizes. This is critical information to understand what level of harvesting these intertidal ecosystems can sustain. Coupled with knowledge of spawning seasons and productive larger sizes of 'opihi, this information has become a platform for creating place-based sustainable harvesting practices. In West Hawai'i, the Ka'ūpūlehu community supported the official designation of the Ka'ūpūlehu 10 year Try Wait Rest Area designed to replenish historical abundance to coastal fish populations and also included the protection of intertidal resources. Relationship building is integral to our long-term commitment to Ka'ūpūlehu and we continue to monitor these areas with students and community members participating in NKA programs throughout the year. Creating this framework and implementing it into practice in Ka'ūpūlehu has helped our team understand seasonal changes in intertidal communities through quantitative and qualitative methodologies. In Ka'ūpūlehu, this is critical information to inform the long-term sustainable fisheries management plan developed through the Ka'ūpūlehu Marine Life Advisory Committee. This is part of a growing dataset of intertidal community diversity, 'opihi densities, and algal composition across Hawai'i. Through a dedicated long-term commitment to local communities on Hawai'i, Kaua'i, and extended partnerships with communities throughout the Main Hawaiian Islands, Pilinakai has extended these tools from undergraduate- and graduate-level research, and then extends these tools into community engagement strategies implemented under NKA.

5. Community Engagement to Support Positive Cultural- and Social-Behavior Shifts

5.1. NKA Annual Nohona (Community Engagement Camps and Programs)

To date, NKA programs have hosted approximately 12 Annual Nohona community engagement camps and contributed to more than 10 other cultural and community-based camps throughout Hawai'i over the past six years. Culture-based education is one that stems from the foundation of a culture and is a framework of teaching and learning that is grounded in "the values, norms, knowledge, beliefs, practices, experiences, places, and language" of a culture [36]. NKA Annual Nohona are rooted in a culture-based and place-based educational framework to honor community resources of people and place and build capacity for youth to become future leaders in their community. Participants of these nohona build reciprocal relationships with place to understand its capability to feed the community, and the community's capability to feed and intimately tend to that place. NKA has worked with nine communities on Hawai'i Island, one community on Maui, one community on Moloka'i, one community on Lāna'i, four communities on O'ahu, and six communities on Kaua'i. NKA has hosted thousands of school-aged children from one-day-only field trips to recurring workshops throughout the years of implementing NKA programs. Through the implementation of trainings and programs, we have trained over 20 UHH undergraduate interns in biocultural monitoring tools and supported the successful completion of Master's of Science thesis drawing heavily from these methods (described in detail in [46]). NKA creates more opportunities for graduate and undergraduate research to expand on applying research towards indigenous-based approaches to research and resource management in intertidal, freshwater, and terrestrial ecosystems.

The students, educators, academic researchers, and conservation professionals involved in the administration of NKA initiatives collaborate across projects and disciplines to continually advance NKA's initiatives towards indigenous approaches to community engagement based on indigenous values honoring relationships. Family and community are key components of culture-based education, and involving families and members of the community supports the growth and success of learners [73]. NKA programs encourage parents and elders of the community to become educators through sharing their stories and knowledge of place to contribute to the cultural identity and sense of belonging of the next generation.

Reflective processes are important and necessary to build confidence, be accountable, and responsive to learning and to ultimately internally strengthen participants, as an individual and a collective [74]. NKA implements strategies focused on reflection in the program curriculum where NKA leadership facilitates safe spaces for participants to critically reflect on their learning and how the NKA activities are supporting their growth as a leader in the community. The NKA workshops serve as an opportunity to engage the next generation in this ancestral mindset to challenge behaviors in our community that hinder our relationships to people and place.

5.2. Kūka'i Laulaha: Broadening Vision of Biocultural Restoration and Building Leadership through Larger Indigenous Networks

Kūka'i Laulaha (KL) is an international cultural exchange built on the framework of pilina that extend between indigenous communities across the Pacific. KL has established and maintained connections with indigenous communities that face similar challenges of addressing biocultural resource abundance in their communities. It is an initiative to grow perspectives on the social, cultural, and biological management challenges that disrupt indigenous relationships to place, and it addresses this issue through strengthening foundations in indigenous language, history, and genealogy. This exchange creates opportunities to grow local leadership in Hawai'i and enables participants to critically think and evoke discussion about the impacts of management strategies and conservation models that are inclusive of human dimensions. It is an exchange for aspiring leaders, active community members, conservation managers and professionals as it is vital that our leadership and work ethic is grounded in honoring our places, our many cultures, and the reciprocal relationships that have become a shared responsibility.

Over the duration of this initiative, over 50 students and community members have participated in this exchange to Aotearoa and the Cook Islands and in turn, these participants have hosted numerous groups comprised of approximately 150 individuals from these communities. Participants are invited to the exchange program after contributing to our local NKA initiatives through discussions, research, monitoring, and/or community service. Past participants include high school students, UH Hilo and UH Mānoa undergraduate and graduate students, community members of Waimea, Kailapa, Molokai, and Hā'ena, and organizations including the Queen Lili'uokalani Children's Center, Hui Maka'āinana o Makana, and Kailapa Community Association. KL immerses participants in indigenous communities to gain understanding of cultural traditions, beliefs, and practices and how cultural values influence the way they manage their natural resources. Then, upon returning home, participants are encouraged to think about their role in the communities they serve, and create effective, multifaceted strategies guided by indigenous relationships to place.

More importantly, KL introduces participants to the realities and sometimes overwhelming sociocultural conflicts experienced by Pacific indigenous peoples, and ways to begin to heal to shift normalized behaviors. For example, for the past five years in Aotearoa, KL has been working with Te Taitimu Trust (TTT), a nonprofit organization whose goal is to motivate youth to become leaders in their communities. Each year, participants have been involved with TTT's annual camp that focuses on whanaungatanga, building familial relationships through shared experiences. It brings together rangatahi (youth) from various backgrounds who each deal with different realities at home. Some of these realities include suicide, gang involvement, and drug and alcohol abuse. It is through connections and conversations from communities such as TTT that KL participants have realized a different source of disconnect that affects the families of indigenous communities. In an effort to empower communities in natural-resource management, it important to consider these social environments and the potential consequences of resource-management decisions on long-term social and cultural health. KL provides an important opportunity to experience first-hand the struggles and strategies of environmental and community issues across the Pacific.

Within community engagement, NKA work on strategies to provide cultural foundations for cultivating a generation of young leadership within communities. Having older youth learn through

NKA and start to teach these lessons to younger children is part of a cord of succession to support the healing of 'āina, our community of people and place. Through long-term relationship building with young leaders, we support their role and contributions to community decision-making that benefit the overall health and wellness of 'āina.

6. Capacity Building within Indigenous Communities

Capacity building is a major challenge in creating long-term partnerships with indigenous communities [24]. Other major challenges in management partnerships include the short-term nature of grant-funded research projects and the potential for mismatch in research project and community-valued timelines [65]. In order to confront the recurring issue of inconsistent funding and subsequent high turnover of short-term partnerships with communities, the NKA team dedicates at least five years to communities with whom we work and operates through supporting community partnerships from a local, state, and federal level. As an extension of the value of pilina in managing for 'āina momona, healthy, productive, and resilient systems of people and place, NKA focuses on leadership and succession building that holds individuals and partnerships accountable to honoring pilina and reciprocity.

The NKA leadership team confronts the problem through a developing local leadership based on indigenous values of pilina and commitment to contributing to healthy communities. The team actively participates in NKA Annual Nohona and other activities throughout the year on the shared commitment to the collective work. This work is important for NKA leadership, which is a group of Native Hawaiian and local women from undergraduate and graduate levels spanning marine science, Hawaiian Studies, natural-resource management, and culture-based education who work across the Hawai'i to empower indigenous science and community engagement. Throughout the year, many of the same middle-school and high-school youth continue to participate in NKA activities and NKA leadership grow into the commitment to mentorship of the next generation of community members. Seeing examples of Native Hawaiian and local leadership provides a source of support for their individual paths. This addresses the need for succession building where youth can recognize their value to their community and positively contribute to foster critical thinking of the social, cultural, and ecological needs for their community to be healthy and thriving.

In the long-term, these youth will contribute positively to their community and stay actively involved in contributing to their community in some capacity. NKA creates opportunities to grow local and Native Hawaiian mentors and role models for the next generation of community leaders who understand the realities of their communities and how to support a path to heal these relationships that support a resilient social-ecological system. Through a shared long-term commitment to the communities we serve, NKA leadership facilitates discussions of social and cultural shifts needed to improve resource management and the holistic well-being of our communities. Collectively, these solutions enable the adaptive governance of social–ecological systems in the face of today's global environmental pressures and changes by prioritizing knowledge coproduction, collaboration, and social and institutional learning [6].

This case study highlights views and practices on cultivating reciprocal pilina with communities and within broader conservation partnerships and indigenous networks, provided through examples such as KL. The deeper the relationships grow within Hawai'i's communities and indigenous communities in Aotearoa and Mangaia in the Cook Islands, the more future generations can gain the experience and insight to lead NKA back home and how to apply it to their communities.

7. Discussion: Healing Communities to Support Healthy and Resilient Communities of People and Place

This paper offers a case study that defines biocultural restoration through indigenous relationships to people and place, sharing biocultural monitoring tools and community engagement and capacity building strategies that address holistic social-ecological systems. NKA is an example of how a biocultural approach can inform community-based marine resource management ranging from a local to Archipelago-wide scale. It is part of a biocultural approach to developing culturally grounded indicators of well-being from a local to regional scale [3,30,34].

Huli ʻIa and Pilinakai are integral to the growth of NKA that contributed to how we as Native Hawaiians perpetuate indigenous-knowledge systems, biocultural monitoring, and community engagement to identify social and cultural factors to support resource management and holistic restoration of indigenous relationships to place. Huli ʻIa provides a culturally based methodology for understanding cultural and ecological connections based on a Native Hawaiian worldview. Learning from the ʻāina and community through Huli ʻIa and Pilinakai and sharing NKA in other communities throughout Hawaiʻi provides one example of how the value of reciprocal pilina builds trust allowing for NKA to grow a community network and a larger indigenous network through KL.

As biocultural restoration of social–ecological systems and the sustainable use of natural resources continues to be a priority in discussions at regional, national, and global scales, many have realized that a first step is to acknowledge and aim to better understand the intrinsic connection between people and place [6]. The relationship between humans and the environment are widely acknowledged in indigenous epistemology, passed down through creation stories and other traditional forms of information dissemination for generations. ʻĀina momona is the ultimate long-term goal that speaks to the productive, healthy, and resilient lands and oceans including the intimate reciprocal relationships our ancestors had with ʻāina which we are re-remembering today. This reciprocal relationship is ingrained in the cultural memory of place-based and indigenous communities around the world. A growing number of disciplines including sustainability science and ecological restoration, as evidenced by this special issue, provides critical pathways to explore the multifaceted biocultural approaches to addressing resource abundance through strengthening the intimate connections between people and place.

Pilina and Reciprocity within Conservation and Research Partnerships

Conservation goals do not always align between the indigenous people and partner organizations, and new approaches are needed to respect and value indigenous knowledge and worldviews [35]. One area attempting to bridge this gap on multiple scales lies in developing culturally grounded indicators of natural, cultural, and socioeconomic well-being with application and relevance on a local scale [3,34]. For example, in Melanesia, agreeing on a shared vision and clear expectations is essential to create transparent communication and equitable outcomes [13,75].

NKA offers another dimension to community-driven research specifically focused on indigenous self-empowerment through capacity building in restoring ʻāina momona, thriving, productive, and healthy biocultural communities. Empowering cultural perspectives and values provides invaluable insight into the feedbacks in a social-ecological system [3,13]. Indigenous approaches have woven cultural, social, and ecological into many cords of knowledge that have the power to address social justice and equity of costs and benefits, and the impact of conservation actions on cultural identity [13]. This is part of a rise in broadening the definition and advocating for self-determination of indigenous communities within conservation where communities define well-being [43]. However, because conservation goals do not always align with a collective solution developed from indigenous communities, it takes long-term commitment and personal investment to building relationships, trust, and reciprocal partnerships with indigenous communities in conservation. As previously noted, the potential mismatch in time frames between communities and partners is a challenge for grant-based work [65].

In the long-term, the goal of NKA is to empower community voices and decision-making as an ʻohana (family), gathering around building pilina to place and perpetuating ancestral-knowledge systems. Only the community itself can identify the best ways to reach out to their peers and to initiate the hard conversations about behavior changes. Through community self-empowerment, we support perpetuating traditional knowledge systems and building collective contemporary knowledge of biocultural systems. In the long-term, we are supporting a movement to create place-based management and behavior shifts based on the collective and equitable needs of people and place. Ultimately, this approach provides thought-provoking insight into resource management and decision-making process and empowers community members to move collectively and to critically assess how management decisions may affect the environment and community into future generations.

8. Closing

Community-based resource management in the Pacific Islands is well-positioned to move forward within partnerships where indigenous people are at the core and not just the periphery. Pacific Islanders descend from ancestors who survived through harsh conditions and high degrees of environmental variability from which they possessed intimate traditional knowledge and values of reciprocity and respect for the environment [68]. Thus, those who possess intimate knowledge of place should be considered the most capable of making decisions about that place. As the tide continues to turn towards empowering indigenous communities in natural resource management, it is essential to share indigenous-driven initiatives that can guide future direction in addressing social, cultural, and ecological factors to address within resource management and in the broader restoration of social-ecological systems. This case study lays a foundation for empowering indigenous initiatives built on a collective vision of healthy thriving social–ecological systems and is part of a growing effort to clear a path forward for indigenous communities to bring their priorities to the forefront.

In closing, NKA is one approach honoring the importance of pilina as the important threads that bind our communities closer to each other and the places that feed our well-being. NKA gathers communities around pilina, in particular how maintaining healthy pilina to place and one another is an essential element of 'āina momona, thriving and productive communities of people and place. Ultimately, restoring biocultural health means healing indigenous relationships to place and each other.

As Native Hawaiians return to our core values of honoring reciprocity in pilina to the 'āina and to one another, we can improve the way we can rely on each other for research, community engagement, education, resource management, and policy. By coming together and trusting in ancestral knowledge systems, we are able to take steps forward together to build resilient and adaptive communities. As the community-based marine-resource-management movement grows in Hawai'i and the Pacific Islands, NKA strives to be present on all fronts of the social, biological, and cultural needs to create culturally grounded resource management designed to restore abundance and productivity to our biocultural lands and oceans.

Author Contributions: Conceptualization, P.A., K.M., E.C., P.P., K.S., L.A., U.C.; Methodology, P.A., K.M., P.P., E.C., K.S., L.K., U.C.; Formal Analysis: P.A., K.M., P.P., E.C., K.S.; Invesigation, P.A., K.M., E.C., P.P., K.S., L.A., U.C.; Resources, N/A; Data Curation, N/A; Writing-Original Draft Preparation, K.M., P.P., K.S., E.C.; Writing-Reviewing & Editing, K.M., P.P., K.S., E.C., L.K.; Supervision, P.A., E.C., K.M.; Project Administration, P.A., E.C., K.M.; Funding Acquisition, P.A.

Funding: The APC was funded by Hawai'i Community Foundation.

Acknowledgments: This work was funded in part by Kamehameha Schools, Office of Hawaiian Affairs, the University of Hawai'i Hilo (PACRC), the University of Hawai'i Sea Grant College Program, National Oceanic and Atmospheric Administration (NOAA) Papahānaumokuākea Marine National Monument ('Aulani Wilhelm, Moani Pai, Hōkū Johnson), NOAA Pacific Islands Climate Change Cooperative, NOAA B-WET Program, Hawai'i Tourism Authority, and conducted with support from community partnerships in North and South Kohala with UH Sea Grant, Kalaemanō through Kalaemanō Cultural Interpretive Center (Ku'ulei Keakealani and Aunty Lei Lightner), the University of Hawai'i at Hilo (Misaki Takabayashi), Keaholoa STEM Scholars Program (Kēhau Springer, Makani Gregg, Pat Springer, Kahoane Aiona, Tyson Fukuyama, Nākoa Goo, and support from Kamalani Pua, Brawn Albino, 'Iolani Kauhane, Rodney Balais, Moani Pai, Rob Paige, Trisann Bambico, Niegel Rozet, and many more), Kanu o Ka 'Āina New Century Public Charter School, Kohala Middle School, Kailapa Community Association, Ka'ūpūlehu Marine Life Advisory Committee, The Nature Conservancy, and Conservation International and on Kaua'i at Nā Pali and Hā'ena with Hui Maka'āinana o Makana, Kawaikini, Hawai'i Department of Land and Natural Resources-Natural Area Reserve System, Ke Kula o Ni'ihau, Waimea High School, Nōhili Pacific Missile Range Facility, and Kōke'e Resource Conservation Program. Mahalo nui to the Hawai'i Community Foundation (Dana Okano) for your sponsorship making it possible to share our work in this special issue! We wish to also acknowledge the many communities and partnerships that contribute to this collective journey to listen to ʻāina and build a collective, diverse understanding of place and our pilina to it. In listening and engaging, we open our senses to internalizing the needs of our communities and places and re-remember the path paved by our ancestors to guide the solutions and behavior shifts we need for ʻāina to be productive and thriving as our communities will too. To the international communities (Mangaia and Rarotonga in the Cook Islands, Ruatoria, Whanganui, and Hastings in New Zealand) we exchange with through Kūkaʻi Laulaha, we mahalo you for being a beautiful reminder of all the threads interconnected in our ʻupena of Moananuiākea. You allow us to be able to pass these pilina to another generation reinforcing these threads that stretch far into our histories and, if tended well, far into our futures! Mahalo maoli nō to our communities in Hawaiʻi and across Moananuiākea for joining us in cultivating the soil so one day these seeds will grow and thrive and continue the pilina for many generations to come.

Conflicts of Interest: The authors declare no conflict of interest. The funders had no role in the design of the study; in the collection, analyses, or interpretation of data; in the writing of the manuscript, and in the decision to publish the results.

References

1. Maffi, L. *On Biocultural Diversity: Linking Language, Knowledge, and the Environment*; Smithsonian Institution Press: Washington, DC, USA, 2001.

2. Kassam, K.A.S. *Biocultural Diversity and Indigenous Ways of Knowing: Human Ecology in the Arctic*; University of Calgary Press: Calgary, AB, Canada, 2009.

3. Sterling, E.J.; Filardi, C.; Newell, J.; Albert, S.; Alvira, D.; Bergamini, N.; Betley, E.; Blair, M.E.; Boseto, D.; Burrows, K.; et al. Biocultural approaches to well-being and sustainability indicators across scales. *Nat. Ecol. Evol.* **2017**, *1*, 1798. [CrossRef] [PubMed]

4. Binder, C.R.; Hinkel, J.; Bots, P.W.N.; Pahl-Wostl, C. Comparison of frameworks for analyzing social-ecological systems conservation. *Ecol. Soc.* **2013**, *18*. [CrossRef]

5. Berkes, F.; Colding, J.; Folke, C. Rediscovery of traditional ecological knowledge as adaptive management. *Ecol. Appl.* **2000**, *10*, 1251–1262. [CrossRef]

6. Berkes, F. Environmental governance for the Anthropocene? Social-ecological systems, resilience, and collaborative learning. *Sustainability* **2017**, *9*, 1232. [CrossRef]

7. Berkes, F. Rethinking community-based conservation. *Conserv. Biol.* **2004**, *18*, 621–630. [CrossRef]

8. Kulnieks, A.; Longboat, D.R.; Young, Y. *Contemporary Studies in Environmental and Indigenous Pedagogies: A Curricula of Stories and Place*; Springer Science & Business Media: Berlin, Germany, 2013.

9. Kimmerer, R. Restoration and reciprocity: The contributions of traditional ecological knowledge. In *Human Dimensions of Ecological Restoration*; Island Press: Washington, DC, USA, 2011; pp. 257–276.

10. Winthrop, R.H. The strange case of cultural services: Limits of the ecosystem services paradigm. *Ecol. Econ.* **2014**, *108*, 208–214. [CrossRef]

11. Pascua, P.A.; McMillen, H.; Ticktin, T.; Vaughan, M.; Winter, K.B. Beyond services: A process and framework to incorporate cultural, genealogical, place-based, and indigenous relationships in ecosystem service assessments. *Ecosyst. Serv.* **2017**, *26*, 465–475. [CrossRef]

12. Chan, K.M.; Goldstein, J.; Satterfield, T.; Hannahs, N.; Kikiloi, K.; Naidoo, R.; Vadeboncoeur, N. Cultural services and non-use values. In *Natural Capital: Theory and Practice of Mapping Ecosystem Services*; Oxford University Press: Oxford, UK, 2011; p. 206.

13. Zafra-Calvo, N.; Pascual, U.; Brockington, D.; Coolsaet, B.; Cortes-Vasquez, J.A.; Gross-Camp, N.; Palomo, I.; Burgess, N.D. Towards an indicator system to assess equitable management in protected areas. *Biol. Conserv.* **2017**, *211*, 134–141.

14. Louis, R.P. Can you hear us now? Voices from the margin: Using indigenous methodologies in geographic research. *Geo Res.* **2007**, *45*, 130–139. [CrossRef]

15. Berkes, F.; Folke, C.; Colding, J. *Linking Social and Ecological Systems: Management Practices and Social Mechanisms for Building Resilience*; Cambridge University Press: Cambridge, UK, 2000; pp. 1–27.

16. Durkalec, A.; Furgal, C.; Skinner, M.W.; Sheldon, T. Climate change influences on environment as a determinant of Indigenous health: Relationships to place, sea ice, and health in an Inuit community. *Soc. Sci. Med.* **2015**, *136*, 17–26. [CrossRef] [PubMed]

17. Kikiloi, K.; Friedlander, A.M.; Wilhelm, A.; Lewis, N.A.; Quiocho, K.; Āila, W., Jr.; Kahoʻohalahala, S. Papahānaumokuākea: Integrating culture in the design and management of one of the world's largest marine protected areas. *Coast. Manag.* **2017**, *45*, 436–451. [CrossRef]

18. Hicks, C.C.; Levine, A.; Agrawal, A.; Basurto, X.; Breslow, S.J.; Carothers, C.; Charnley, S.; Coulthard, S.; Dolsak, N.; Donatuto, J.; et al. Engage key social concepts for sustainability. *Science* **2016**, *352*, 38–40. [CrossRef] [PubMed]

19. Liu, J.; Dietz, T.; Carpenter, S.R.; Alberti, M.; Folke, C.; Moran, E.; Pell, A.N.; Deadman, P.; Kratz, T.; Lubchenco, J.; et al. Complexity of coupled human and natural systems. *Science* **2007**, *317*, 1513–1516. [CrossRef] [PubMed]

20. Berkes, F. *Sacred Ecology: Traditional Ecological Knowledge and Management Systems*; Taylor & Francis: Abingdon, VA, USA, 1999; p. 203.

21. Nadasdy, P. The politics of TEK: Power and the "integration" of knowledge. In *Arctic Anthropology*; University of Wisconsin Press: Madison, WI, USA, 1999; pp. 1–18.

22. Ober, R.; Bat, M. Self-empowerment: Researching in a both-ways framework. *Ngoonjook* **2008**, *33*, 43.

23. Snively, G.; Corsiglia, J. Discovering indigenous science: Implications for science education. *Sci. Educ.* **2001**, *85*, 6–34. [CrossRef]

24. Tipa, G.; Welch, R. Co-management of natural resources: Issues of definition from an indigenous community perspective. *J. Appl. Behav. Sci.* **2006**, *42*, 373–391. [CrossRef]

25. Berkes, F. Evolution of co-management: Role of knowledge generation, bridging organizations and social learning. *J. Environ. Manag.* **2009**, *90*, 1692–1702. [CrossRef] [PubMed]

26. Muller, S. Two ways': Bringing indigenous and nonindigenous knowledges together. In *Country, Native Title and Ecology*; Australian National University E-Press and Aboriginal History Incorporated (Monograph 24): Canberra, Australia, 2012; pp. 59–79.

27. Natcher, D.C.; Davis, S.; Hickey, C.G. Co-management: Managing relationships, not resources. In *Human Organization*; Society for Applied Anthropology: Oklahoma City, OK, USA, 2005; pp. 240–250.

28. Popova, U. Conservation, traditional knowledge, and indigenous peoples. *Am. Behav. Sci.* **2014**, *58*, 197–214. [CrossRef]

29. Tanguay, J. Alternative indicators of well-being for Melanesia: Cultural values driving public policy. In *Making Culture Count*; Palgrave Macmillan: London, UK, 2015; pp. 162–172.

30. McCarter, J.; Sterling, E.; Jupiter, S.; Cullman, G.; Albert, S.; Basi, M.; Betley, E.; Boseto, D.; Bulehite, E.; Harron, R.; et al. Biocultural approaches to developing well-being indicators in Solomon Islands. *Ecol. Soc.* **2018**, *23*, 32. [CrossRef]

31. Panelli, R.; Tipa, G. Placing well-being: A Maori case study of cultural and environmental specificity. *EcoHealth* **2007**, *4*, 445–460. [CrossRef]

32. Berkes, F.; Ross, H. Community resilience: Toward an integrated approach. *Soc. Natl. Resour.* **2013**, *26*, 5–20. [CrossRef]

33. McShane, T.O.; Hirsch, P.D.; Trung, T.C.; Songorwa, A.N.; Kinzig, A.; Monteferri, B.; Mutekanga, D.; Van Thang, H.; Dammert, J.L.; Pulgar-Vidal, M.; et al. Hard choices: Making trade-offs between biodiversity conservation and human well-being. *Biol. Conserv.* **2011**, *144*, 966–972. [CrossRef]

34. Sterling, E.J.; Ticktin, T.; Morgan, K.; Cullman, G.; Alvira, D.; Andrade, P.; Bergamini, N.; Betley, E.; Burrows, K.; Caillon, S.; et al. Culturally grounded indicators of resilience in socio-ecological systems. *Environ. Soc.* **2017**, *8*, 63–95. [CrossRef]

35. Howitt, R. Indigenous rights vital to survival. *Nat. Sustain.* **2018**, *1*, 339. [CrossRef]

36. Smith, L.T. *Decolonizing Methodologies: Research and Indigenous Peoples*; Zed Books Ltd.: London, UK, 2013.

37. Sangha, K.K.; Le Brocque, A.; Costanza, R.; Cadet-James, Y. Ecosystems and indigenous well-being: An integrated framework. *Glob. Ecol. Conserv.* **2015**, *4*, 197–206. [CrossRef]

38. Tipa, G.; Teirney, L.D. *A Cultural Health Index for Streams and Waterways: A Tool for Nationwide Use*; Ministry for the Environment: Wellington, New Zealand, 2006; pp. 1–58.

39. Morgan, T.K.K.B. Waiora and cultural identity: Water quality assessment using the Mauri Model. *AlterNative Int. J. Indig. Peoples* **2006**, *3*, 42–67. [CrossRef]

40. Bohensky, E.L.; Maru, Y. Indigenous knowledge, science, and resilience: What have we learned from a decade of international literature on "integration"? *Ecol. Soc.* **2011**, *16*, 6. [CrossRef]

41. Ens, E. Conducting two-way ecological research. In *People on Country: Vital Landscapes, Indigenous Futures*; Altman, J., Kerins, S., Eds.; The Federation Press: Annandale, Australia, 2012; pp. 45–64.

42. Hiwasaki, L.; Luna, E.; Shaw, R. Process for integrating local and indigenous knowledge with science for hydro-meteorological disaster risk reduction and climate change adaptation in coastal and small island communities. *Int. J. Dis. Risk Reduc.* **2014**, *10*, 15–27. [CrossRef]

43. Biedenweg, K.; Gross-Camp, N.D. A brave new world: Integrating well-being and conservation. *Ecol. Soc.* **2018**, *23*, 32. [CrossRef]

44. Thaman, K.H. Decolonizing Pacific studies: Indigenous perspectives, knowledge, and wisdom in higher education. *Contemp. Pac.* **2003**, *15*, 1–17. [CrossRef]

45. Tsuji, L.J.; Ho, E. Traditional environmental knowledge and western science: In search of common ground. *Can. J. Nat. Stud.* **2002**, *22*, 327–360.

46. Cadiz, E. *Pilina-Mālama-'Āina Momona: A Community Driven Monitoring Program to Understand Health and Well-Being of People and Place in Hā'ena, Kaua'i.* M.S.; University of Hawai'i at Mānoa, ProQuest Dissertations & Theses Global: Honolulu, HI, USA, 2017.

47. Kame'eleihiwa, L. *Native Land and Foreign Desires: Pehea Lā E. Pono Ai?* Bishop Museum Press: Honolulu, HI, USA, 1992; p. 25.

48. Handy, E.S. *Native Planters in Old Hawai'i*; Bishop Museum Press: Honolulu, HI, USA, 1972.

49. Nāone, C.K. 'O Ka 'Āina, Ka 'ōlelo, A Me Ke Kaiāulu. In *Kamehameha Schools Hūlili: Multidisciplinary Research on Hawaiian Well-Being*; Kamehameha Schools Publishing: Honolulu, HI, USA, 2008; Volume 5, pp. 315–339.

50. Kikiloi, K.; Graves, M. Rebirth of an archipelago: Sustaining a Hawaiian cultural identity for people and homeland. In *Hūlili: Multidisciplinary Research on Hawaiian Well-Being*; Kamehameha Schools Publishing: Honolulu, HI, USA, 2010; Volume 6, pp. 73–114.

51. Kawai'ae'a, K.K.C. Ho'i hou i ke kumu! Teachers as Nation Builders. Indigenous Educational Models for Contemporary Practice. In *Our Mother's Voice*, 1st ed.; Ah Nee-Benham, M.K.P., Ed.; Taylor & Francis: Abingdon, VA, USA, 2008; Volume 2, pp. 41–45.

52. Titcomb, M. *Native Use of Fish in Hawai'i*; University of Hawai'i Press: Honolulu, HI, USA, 1972; Volume 29.

53. Poepoe, K.K.; Bartram, P.K.; Friedlander, A.M. The use of traditional knowledge in the contemporary management of a Hawaiian community's marine resources. In *Fishers' Knowledge in Fisheries Science and Management*; Haggan, N., Neis, B., Baird, I.G., Eds.; UNESCO: Paris, Fance, 2005; pp. 90–111.

54. McGregor, D. *Nā Kua'āina: Living Hawaiian Culture*; University of Hawai'i Press: Honolulu, HI, USA, 2007.

55. Maly, K.; Maly, O. *Ka Hana Lawai'a a Me Nā Ko'a O Na Kai 'ewalu: A History of Fishing Practices and Marine Fisheries of the Hawaiian Islands*; Kumu Pono Associates LLC: Lanai City, HI, USA, 2003.

56. Kahā'ulelio, D.; Nogelmeier, P. *Ka 'Oihana Lawai 'a: Hawaiian Fishing Traditions*; Bishop Museum Press: Honolulu, HI, USA, 2006.

57. Kittinger, J.N.; Pandolfi, J.M.; Blodgett, J.H.; Hunt, T.L.; Jiang, H.; Maly, K.; McClenachan, L.E.; Schultz, J.K.; Wilcox, B.A. Historical reconstruction reveals recovery in Hawaiian coral reefs. *PLoS ONE* **2011**, *6*, e25460. [CrossRef] [PubMed]

58. Friedlander, A.M.; Shackeroff, J.M.; Kittinger, J.N. Customary marine resource knowledge and use in contemporary Hawai'i. *Pac. Sci.* **2013**, *67*, 441–460. [CrossRef]

59. Friedlander, A.M.; DeMartini, E.E. Contrasts in density, size, and biomass of reef fishes between the northwestern and the main Hawaiian islands: The effects of fishing down apex predators. *Mar. Ecol. Prog. Ser.* **2002**, *230*, 253–264. [CrossRef]

60. Friedlander, A.M. Status of Hawai'i's coastal fisheries in the new millennium. In Proceedings of the 2001 American Fisheries Society Hawaii Chapter, Honolulu, HI, USA, 14 October 2004.

61. Ayers, A.L.; Kittinger, J.N. Emergence of co-management governance for Hawai'i coral reef fisheries. *Glob. Environ. Chang.* **2014**, *28*, 251–262. [CrossRef]

62. Schemmel, E.; Friedlander, A.; Andrade, P.; Keakealani, K.; Castro, L.; Wiggins, C.; Wilcox, B.; Yasutake, Y.; Kittinger, J. The codevelopment of coastal fisheries monitoring methods to support local management. *Ecol. Soc.* **2016**, *21*, 34. [CrossRef]

63. Vaughan, M.B.; Thompson, B.; Ayers, A.L. Pāwehe Ke Kai a 'o Hā'ena: Creating state law based on customary indigenous norms of coastal management. *Soc. Natl. Resour.* **2016**, *30*, 31–46. [CrossRef]

64. Hau'ofa, E. Epilogue: Pasts to remember. In *Remembrance of Pacific Pasts: An Invitation to Remake History*; University of Hawai'i Press: Honolulu, HI, USA, 2000; pp. 453–471.

65. Adams, M.S.; Carpenter, J.; Housty, J.A.; Neasloss, D.; Paquet, P.C.; Service, C.; Walkus, J.; Darimont, C.T. Toward increased engagement between academic and indigenous community partners in ecological research. *Ecol. Soc.* **2014**, *19*, 5. [CrossRef]

66. Wilson, N.J.; Mutter, E.; Inkster, J.; Satterfield, T. Community-Based Monitoring as the practice of Indigenous governance: A case study of Indigenous-led water quality monitoring in the Yukon River Basin. *J. Environ. Manag.* **2018**, *210*, 290–298. [CrossRef] [PubMed]

67. Noble, M.; Duncan, P.; Perry, D.; Prosper, K.; Rose, D.; Schnierer, S.; Tipa, G.; Williams, E.; Woods, R.; Pittock, J. Culturally significant fisheries: Keystones for management of freshwater social-ecological systems. *Ecol. Soc.* **2016**, *21*, 22. [CrossRef]

68. McMillen, H.; Ticktin, T.; Friedlander, A.; Jupiter, S.; Thaman, R.; Campbell, J.; Veitayaki, J.; Giambelluca, T.; Nihmei, S.; Rupeni, E.; et al. Small islands, valuable insights: Systems of customary resource use and resilience to climate change in the Pacific. *Ecol. Soc.* **2014**, *19*, 44. [CrossRef]

69. McMillen, H.; Ticktin, T.; Springer, H.K. The future is behind us: Traditional ecological knowledge and resilience over time on Hawai'i Island. *Reg. Environ. Chang.* **2017**, *17*, 579–592. [CrossRef]

70. Jason, L.A.; Keys, C.B.; Suarez-Balcazar, Y.; Taylor, R.R.; Davis, M.I. *Participatory Community Research: Theories and Methods in Action*; American Psychological Association: Washington, DC, USA, 2004.

71. Pukui, M.K. *'Olelo No'eau: Hawaiian Proverbs & Poetical Sayings*; Bishop Museum Press: Honolulu, HI, USA, 1983; Volume 71.

72. Andrade, P. Ho'i i ka Pilina Kai: Re-Establishing a Relationship with Our Ancestors. Available online: http://www2.hawaii.edu/~pelikaok/aboutproject.html (accessed on 10 February 2018).

73. Kana'iaupuni, S.M.; Kawai'ae'a, K.K.C. E Lauhoe Mai Nā Wa'a: Toward a Hawaiian Indigenous Education Teaching Framework. In *Hūlili: Multidisciplinary Research on Hawaiian Well-Being*; Kamehameha Publishing: Honolulu, HI, USA, 2008; Volume 5, pp. 67–90.

74. Goethals, M.S.; Howard, R.A.; Sanders, M.M. *Student Teaching: A Process Approach to Reflective Practice*, 2nd ed.; Pearson Education Inc.: London, UK, 2004.

75. Jupiter, S. Culture, kastom and conservation in Melanesia: What happens when worldviews collide? *Pac. Conserv. Biol.* **2017**, *23*, 139–145. [CrossRef]

 *sustainability*

Article

The Social-Ecological Keystone Concept: A Quantifiable Metaphor for Understanding the Structure, Function, and Resilience of a Biocultural System

Kawika B. Winter [1,2,3,*], Noa Kekuewa Lincoln [4] and Fikret Berkes [5]

[1] Hawai'i Institute of Marine Biology, University of Hawai'i at Mānoa, Kāne'ohe, HI 96744, USA

[2] Department of Natural Resources and Environmental Management, University of Hawai'i at Mānoa, Honolulu, HI 96822, USA

[3] National Tropical Botanical Garden, Kalāheo, HI 96741, USA

[4] Department of Tropical Plant and Soil Sciences, University of Hawai'i at Mānoa, Honolulu, HI 96822, USA; nlincoln@hawaii.edu

[5] Natural Resources Institute, University of Manitoba, Winnipeg, MB R3T 2N2, Canada; Fikret.Berkes@umanitoba.ca

* Correspondence: kawikaw@hawaii.edu

Received: 31 July 2018; Accepted: 11 September 2018; Published: 14 September 2018

Abstract: Social-ecological system theory draws upon concepts established within the discipline of ecology, and applies them to a more holistic view of a human-in-nature system. We incorporated the keystone concept into social-ecological system theory, and used the quantum co-evolution unit (QCU) to quantify biocultural elements as either keystone components or redundant components of social-ecological systems. This is done by identifying specific elements of biocultural diversity, and then determining dominance within biocultural functional groups. The "Hawaiian social-ecological system" was selected as the model of study to test this concept because it has been recognized as a model of human biocomplexity and social-ecological systems. Based on both quantified and qualified assessments, the conclusions of this research support the notion that taro cultivation is a keystone component of the Hawaiian social-ecological system. It further indicates that sweet potato cultivation was a successional social-ecological keystone in regions too arid to sustain large-scale taro cultivation, and thus facilitated the existence of an "alternative regime state" in the same social-ecological system. Such conclusions suggest that these biocultural practices should be a focal point of biocultural restoration efforts in the 21st century, many of which aim to restore cultural landscapes.

Keywords: alternative regime state; portable biocultural toolkit; social-ecological system theory; Hawaii; *Colocasia esculenta*

1. Introduction

1.1. Social-Ecological Systems and the Application of Ecological Terminology

This paper emphasizes the concept that humans are a part of—not separate from—nature [1], supporting views established by Berkes and Folke [2], and Berkes et al. [3], which hold that the delineation between social systems and natural systems is arbitrary and artificial. Several frameworks for understanding such social-ecological systems have been put forth (e.g., [4–6]), but some have pointed out a disconnect between the frameworks—proposed to understand social-ecological systems—and the biocultural elements that are at the foundation of such systems [7]. This research aims to bridge that gap by (a) presenting theories and methods associated with quantifying biocultural relationships within social-ecological systems; (b) demonstrating how restoring the function of "keystone" components is essential to restoring the structure of social-ecological systems that are observed to be in decline; and (c) demonstrating how restoring the function of "redundant" components is essential to restoring the resilience of such systems. As with our other publications on social-ecological systems, we follow the Walker et al. definition of resilience as the capacity of a system to absorb disturbance and reorganize while undergoing change so as to still retain essentially the same function, structure, identity, and feedbacks [8].

Following Berkes [9], we use the terms "ecological subsystem" and "social subsystem" when discussing particular sides of the social-ecological spectrum, and will do so even when the referenced text uses the term "ecosystem". This allows us to discuss each of the two sides, while maintaining that the two subsystems are not autonomous from one another. We follow others in applying common ecological terms to set up a logical framework for understanding social-ecological systems [10–13]. This paper uses accepted terminology such as 'function', 'functional group', 'diversity', 'keystone', 'redundant', 'regime shift', and 'stable state' of Anderies and Janssen [14] in discussing social-ecological systems. This paper uses the term, 'alternative regime state', to describe alternative stable states that can exist within the same social-ecological system, such as flooded-field agriculture versus rain-fed agriculture practiced by the same culture. An 'alternative regime state' describes different stable states that can exist within the context of the same social-ecological system, whereas a 'regime shift' indicates a stable state that exists within a different social-ecological system altogether, such as a rural agricultural community being transformed into a city and inducing a concurrent shift in the dominant culture. This notion is explored in more detail below.

1.2. Quantifying Biocultural Elements within Social-Ecological Systems

Theories relating to co-evolutionary relationships between human and natural systems are not new [15]. In the 5th century BC, the Greek philosopher Herodotus voiced his observation that events shape both people and nature, and that people and nature interact and evolve together through these events. More recently, Winter and McClatchey [16,17] put forth theories to quantify these co-evolutionary relationships, and established methods for measuring fundamental units of interaction between people and plants—or linked biological-sociocultural relationships (henceforth referred to as biocultural relationships)—in a way that is scalable from simple interactions (one person and one plant) to complex relationships (all of humanity and all plants). Such an approach has been used to address hypotheses about the evolution of interactive relationships [16–18].

The Quantum Co-evolution Unit—or QCU—(Figure 1) is a unit to measure linked, co-evolving relationships such as those observed in social-ecological systems [16–18]. These relationships will henceforth be referred to as "biocultural elements" of such systems. A set of QCUs within a social-ecological system can be quantified (Figure 2, [17]) and assessments of these populations at different times can demonstrate co-evolving biocultural relationships [16,17]. As in many disciplines, units can be considered at different scales for both the ecological component (ecosystem, genus, species, etc.) and the social component (socio-cultural system, community, individual, etc.), to assess the health of diffent aspects of a social-ecological system. The research presented here contends that QCUs can be used as a unit to quantify biocultural elements, such as the following:

- exploring the concept of functional groups within social-ecological systems,
- quantitatively classifying particular elements as either keystone components or redundant components of social-ecological systems,
- quantitatively relating loss of *keystone* components to loss of social-ecological system *structure and function*,
- quantitatively classifying loss of *redundant* components to diminished *resilience* in social-ecological systems,
- identifying alternative regime states within a single social-ecological system,
- quantifying regime shifts between social-ecological systems.

Possessing such an understanding can be key to informing biocultural restoration efforts.

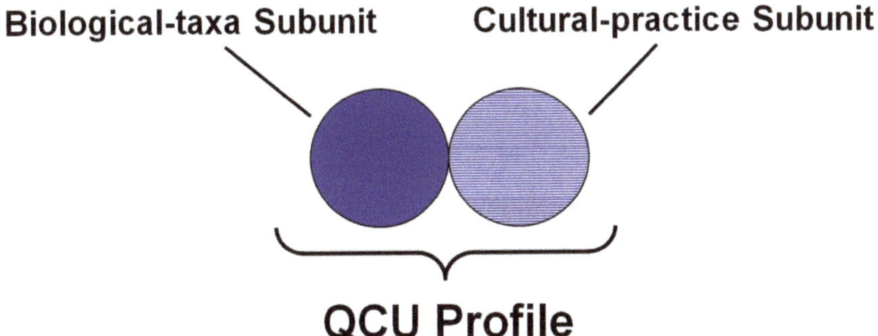

Figure 1. The Quantum Co-evolution Unit (QCU), which relates to the co-evolutionary relationship between biological taxa and human cultures. It is composed of two subunits—the biological-taxa subunit, and the cultural-practice subunit—and is used as a metric for biocultural diversity. The complete unit of the QCU is referred to and described by its QCU profile [17].

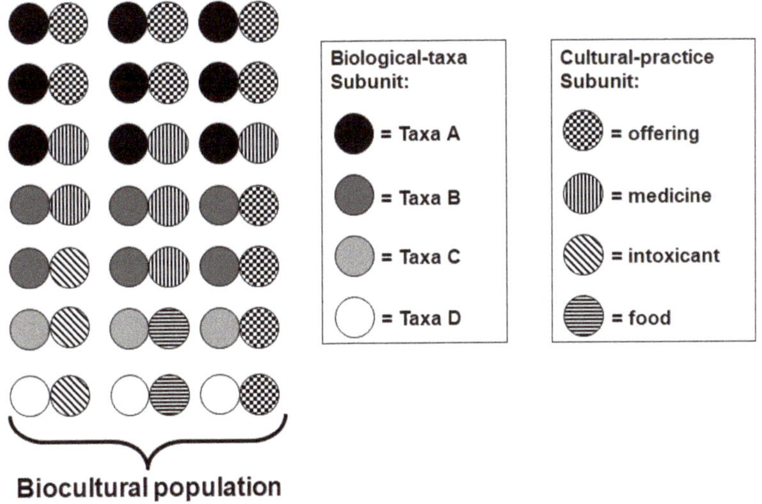

Figure 2. A QCU population. A hypothetical collection of Quantum Co-evolution Units (QCUs) represented, within a social-ecological system, showing proportionality and frequency of various QCUs in relation to one another. The QCU population of a social-ecological system could be sampled over various points in time. Changes could be observed and further quantified. Such changes could include the adoption of new QCUs into the profile, deletion of QCUs from the population, and changes in individual QCU frequency within the population [17].

1.3. The Hawaiian Social-Ecological System

The evolution of social-ecological systems in Hawai'i is uniquely understandable, in part, because the relatively late human colonization allows for tracing the entirety of human history. As such, Hawai'i has been described as a model for the study of social-ecological systems [17,19].

Archaeological evidence indicates settlement no later than 1000 years ago [20], although use of oral history sources indicates initial voyages to the Hawaiian islands may have happened centuries prior [21–24]. In the pre-contact era—prior to contact with Europeans in 1778—the social-ecological system in Hawai'i was intensively managed to maximize resource abundance by attaining a stable state known in Hawaiian as "'āina momona". 'Āina momona is descriptive of a stable state that can exist in alternative forms in Hawaiʻi, and is associated with both flooded-field and rain-fed agriculture [25]. This is a stable state that was brought about via a regime shift (originally from an ecosystem into a social-ecological system that maximized ecosystem services). The social-ecological system in the Hawaiian archipelago associated with the pre-contact era will henceforth be referred to as the "Hawaiian social-ecological system", which is identified by the similarity of linked human-ecological units—being the foundational culture, the available plant species, and the cultural uses of those plants—that existed in the Hawaiian ecoregion in the pre-contact era.

Alternative states of *ʻāina momona* existed in the Hawaiian social-ecological system, and while they had similar functional components with one another, they differed in structure across different regions of the archipelago. As has been described for ecological subsystems [26], it appears that the structures of these alternative regime states were shaped around different keystone components of the same system. Observations about differing structures of alternative regime states associated with either flooded-field or rain-fed agriculture [27] may be related to this phenomenon [28].

The structure of the Hawaiian social-ecological system has been in decline since the 19th century [29]. Gaining an understanding of the biocultural elements, and identifying them as either keystone or redundant components could be beneficial to efforts aimed at restoring the structure, function, and resilience of that system. We utilize theoretical foundations and logical assumptions to explore the the relative importance of biocultural elements in Hawaiian agricultural traditions—a core foundation within the pre-contact social-ecological system in Hawaiʻi.

2. Theoretical Foundations

2.1. The Keystone Concept as Relates to System Structure and Function

Paine [30] first described ecological 'keystone species', as occurring in a situation where patterns of distribution and density of species within an ecological subsystem are disproportionately affected by the activities of a single species. It has since become a major concept within the discipline, but has fueled decades of debate on definitions [31]. Ultimately, this debate stems from disagreements about how to quantify a metaphor. Is the concept of 'keystone' a biological reality, or is it a simple metaphor to understand a complex system? This paper argues the former, and following systems theory [32], asserts that the keystone concept holds true within social-ecological systems. This paper further asserts that it is possible to quantitatively determine a keystone component by assessing functional groups of biocultural elements within a system, and then analyzing the associated diversity within those functional groups.

The disproportionate influence of keystone species suggests there is no functional redundant within the system (Figure 3). If the keystone is removed, then a relatively large number of secondary extinctions would occur [33,34] and the system would reorganize itself with a different structure and function. This process is referred to as a regime shift [35]. Thus, keystone components play a major role in the structure and function of systems.

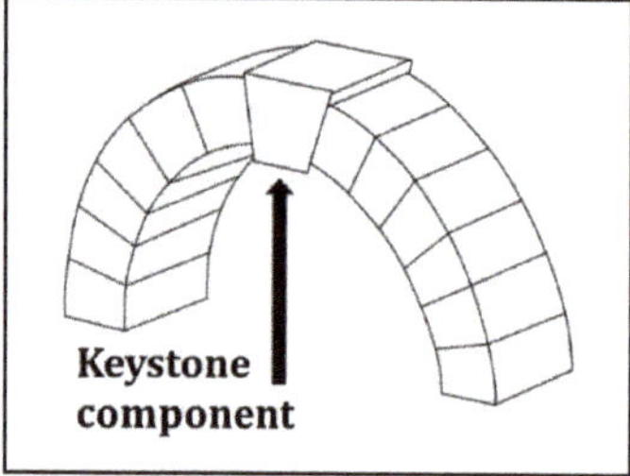

Figure 3. The keystone metaphor. The keystone is the component of a structure that is irreplaceable. Without the keystone, a structure could be reassembled, but could never be the same as if the keystone were present and functioning in its role.

2.2. The Social-Ecological Keystone Concept

The keystone concept of has been applied to various cultural interactions with the biological world [10–12]. Garibaldi and Turner [11] define *cultural* keystone *species* as "the culturally salient *species* that shape in a major way the cultural identity of a people". *Cultural* keystone *practices*, described by Brosi et al. [12], are *traditions* that are so intrinsic to the culture that, if they were to disappear, the culture would be irreversibly altered. Thus, the usage of the keystone concept by researchers working within the social-ecological system paradigm is not new. However, this paper contends that a "social-ecological keystone" is actually a specific biocultural element (i.e., relationship); further, that social-ecological keystones can be quantitatively determined utilizing the theoretical concepts previously established by Winter and McClatchey [16,17].

A relevant concept here is 'functional group'. In ecological subsystems, functional groups are used to lump together species with similar roles—such as top predator, generalist pollinator, nitrogen fixer, and so on. In social-ecological systems, members of functional groups would be biocultural relationships—such as applying herbal medicine, imbibing fermented sugars, weaving baskets, farming complex carbohydrates, and so on. This paper, therefore, refers to "biocultural functional groups", and has identified the QCU as a unit of measure for the biocultural elements within them.

Based in part on Davic's [31] definition of keystone species, this paper submits that a social-ecological keystone is a strongly interacting biocultural element within its functional group, whose top-down effect on biocultural diversity is large relative to all elements within the system. Biocultural elements, and therefore social-ecological keystones, are neither individual taxa nor individual practices, but rather the linked taxa-practice unit. If a social-ecological keystone is severely disabled (or goes extinct), then there would be no substitute without seriously compromising the structure and function of the system—possibly inducing a regime shift. Correspondingly, if a social-ecological keystone were to go extinct it would cause a cascading effect of secondary extinctions of biocultural elements, and subsequently affect the structure of both the ecological and social subsystems. The theories and methods explored in this manuscript could be used to document and understand such processes.

2.3. The Influence of Crop Diversity and Cropping Systems on the Structure of the Hawaiian Social-Ecological System

Functional groups associated with agriculture often determine the structure and function of social-ecological systems managed by agrarian societies because agriculture often dictates the form and hierarchy of the social subsystem, and is the foundation of the economy and politics within the social subsystem [36], which subsequently influences the management of the ecological subsystem. The central role of agriculture in cultural development, political complexity, material economy, and social norms of the Hawaiian social-ecological system has been well explored (see Lincoln and Vitousek [37] for a broad overview and detailed reference list). Based on these concepts, this paper holds the assumption that agriculture is a key biocultural functional group within social-ecological systems managed by agrarian societies.

Hawaiian agriculture manifested in highly diverse forms in the Hawaiian social-ecological system. The most salient division often used in anthropological discussion is the difference between wet (flooded-field, irrigated) and dry (rain-fed) agriculture [38,39]. The state environmental factors that drove the opportunities and constraints of agricultural development are highly organized, but not evenly distributed, in the Hawaiian archipelago. The distribution of geological age and rainfall, which subsequently drive soil fertility and land topography, created a spectrum of agricultural opportunties that spanned almost exclusively rain-fed opportunities on the young island of Hawai'i, to almost exclusively flooded-field opporunities on the oldest of the high islands, Kaua'i [27]. These agricultural forms had different requirements (e.g., levels of organization, infrastructural investment) and offered different effects (e.g., levels of resilience, economic surplus).

In brief, flooded-field agriculture was investment intensive, but low maintainenance while offering low vulnerability to both natural and social perturbances. Consequently, flooded-field agriculture supported socio-political systems on Kaua'i with more diversified social roles and stronger political stability. Conversely, the more labor intensive and vulnerable dryland agricultural systems of Hawai'i Island manifested socio-political systems that were more volatile, saw frequent regime shifts between political leaders, and spawned predatory political ambitions [28]. This implies the existence of "alternative regime states" within the same social-ecological system—a more resilient one built around flooded-field systems of agriculture, and a more vulnerable one built around rain-fed systems of agriculture. These alternative regime states existed even though they are based on the same biocultural elements in the same social-ecological system. The manifestations of the economy and the political systems differed due, at least in part, to the relative and absolute areas of the agricultural systems.

2.4. Social-Ecological System Resilience and the Role of Redundant Components

Resilience [40,41] is a measure of a system's relative ability to absorb disturbance without changing into a different state (i.e., regime shift), such as a different biological community with different ecosystem services [35]. Biological diversity has been shown to be a key factor in the resilience of ecological subsystems [8,42] because it helps to maintain desired states of dynamic regimes in the face of uncertainty and surprise, and also plays a major role in renewing and reorganizing those systems after disturbance [35]. Maintaining such desired states of dynamic regimes (i.e., "stable states") is more specifically dependent on "response diversity", which is the diversity of responses to environmental change among species that contribute to the same function in the system [43]. Therefore, the more nuanced value of diversity is the increased redundancy within functional groups [44]. Such 'functional diversity' refers to the number of species that perform the same function; if a member of a functional group were to be lost (either temporarily or permanently) via a disturbance event, then its function could be replaced by another species. Ecological systems with high response diversity increase the likelihood for reorganization and renewal into a desired state after disturbance [43]; and are, therefore, more resilient. If a set of functionally redundant species does not exhibit any response diversity, then they do not contribute to system resilience [35]. A loss of biodiversity—more importantly a loss of functional diversity and response diversity—is a major contributing factor in regime shifts [35,45]. In the context of ecological subsystems, regime shifts imply a shift in services which that subsystem provides to the socio-cultural subsystem, and are largely irreversible [35].

Throughout this paper we use the term, 'redundant component' to classify a biocultural element of a functional group that could be substituted if removed—that is, a component that is not a keystone. In accordance with systems theory [32], maintaining resilience in social-ecological systems relies on the management of biocultural diversity, including the seemingly redundant components of these systems. Such "social-ecological redundants" contrary to keystones, may not contribute significantly to the structure and function of social-ecological systems individually, however, if they represent response diversity, then such components may contribute significantly to the resilience of social-ecological systems.

We use the QCU (Figure 1) as a unit of measure to quantitatively classify redundant components of social-ecological systems within biocultural functional groups. This paper contends that biocultural functional redundancy exists in instances where a subunit of the QCU (either biological taxa or sociocultural practice) lost via a disturbance or other event can be easily transferred to other corresponding subunits which would ensure the persistence of that biocultural functional group. For example, using the biocultural functional group of "weaving a plant-based fiber", a culture may have five taxa which it uses to weave. If one of the five taxa were to go extinct, the sociocultural practice could continue because of the functional redundancy that exists in that biocultural functional group.

2.5. Theoretical Assumptions

This manuscript builds off of four theoretical assumptions in regards to the keystone concept:

1. Keystone function of a system can be viewed in terms of a functional group.
2. Keystone components of functional groups are dominants within that functional group.

3.   Dominant components of a functional group (i.e., keystone components of a system) are not necessarily dominant components within the overall system.
4.   Shifting dominance within a keystone functional group replaces the keystone component of the system, thus influencing the structure of that system.

3. Testing the Keystone Theory in Social-Ecological Systems

The Hawaiian Social-Ecological System as a Model

There are countless social-ecological systems that could be chosen from around the world to model the theories explored in this paper, but in order to test the conceptual validity of these theories a model social-ecological system is ideal. High islands are excellent examples to discuss system function becuase they are big enough to possess all the biological, ecological, chemical, and physical processes needed for complete system study, yet small enough that the complexity of such systems is perceivable [46,47]. Kirch [19] outlines how both the social and ecological factors of Hawai'i, in particular, lend themselves to serving as a model system for biocultural understanding. State factors influencing ecology are either held constant (e.g., parent material, biota) or are extremely broad yet well organized (e.g., climate, age, topography). Simultaneously, social factors lend themselves to study due to the short timeframe of human colonization, the extreme isolation, and the high level of socio-political complexity achieved.

To exemplify the theories expressed above, we focus on the Hawaiian social-ecological system. At the point of contact with Europeans in 1778, this highly modified system was being managed to maintain a human population on the order of 300,000–800,000 people [48,49]. Furthermore, anthropological discussion has argued that Hawai'i developed high levels of socio-political heirarchy manifested in complex systems of land tenure, resource mangement, and taxation, describing Hawai'i as one of nine civilizations to have independently developed into a state system [50]. The Hawaiian archipelago, therefore, represents an island-bound and intensively managed social-ecological system, with a population size and social structure that makes it comparable to the contemporary period.

4. Methodology

As this is an examination of the structure and function of a system which existed in the past, historical records and archaeological evidence were used to supplement actual observations. The methods described below were used to quantify biocultural diversity, identify key functional groups within systems, as well as quantitatively classify keystone components and redundant components that constitute these functional groups.

4.1. Quantification of Biocultural Diversity

This research considers a unit of biocultural diversity—referred to as the Quantum Co-evolutionary Unit (QCU)—as any one of the human needs, as described by Max-Neef et al. [51,52], which has a satisfier that comes from within the realm of biodiversity. QCUs were generally assessed from the standpoint of viewing them as components within functional groups that are embedded in the entire system, rather than as individual components standing alone within the context of the entire system.

4.2. Assessing Biocultural Functional Groups in Social-Ecological Systems

It is the assumption of this paper that within an agrarian society—such as that associated with the Hawaiian social-ecological system—social-ecological keystones can found within the biocultural functional groups associated with agriculture. Biocultural functional groups were identified by reviewing the seminal literature on ancient agricultural and associated practices [53–59]. From this literature review, commonly occurring categories of agricultural function clearly stood out and were identified to be used to define the biocultural fuctional groups. This literature, while select, forms the broad basis for vast majority of subsequent publications in traditional Hawaiian agriculture.

Once the biocultural functional groups associated with agriculture were determined, their components were then quantified to classify them as a keystone component or a redundant component within each respective functional group. As we attempt to examine a time period in the past, data from historical records and publications were used. Handy et al. [54] is a comprehensively researched tome on ancient Hawaiian agriculture that was produced by the B.P. Bishop Museum in collaboration with anthropologists and a highly-respected, well-published, native-speaking Hawaiian ethnographer; it is widely considered the authoritative volume regarding Hawaiian agriculture. For the purposes of this research, the number of written lines dedicated to each biocultural relationship was used as a proxy for the relative importance of that element (see Table S1). Other sources of knowledge in this area exist [53–59], but those works were not systematically approached from the standpoint of plant-based biocultural relationships as was the work of Handy et al. [54]. The number of lines, therefore, provides the numerical quantification of each element subsequently used to calculate the indexes as described below. While this is an ad hoc approach utilizing only a single, albeit substantial, volume on Hawaiian agriculture, we employ this method to demonstrate the application of the QCU concept in a meaningful way and to provide a starting point from which more intesive analyses can be done in the future.

4.3. Quantitatively Classifying Keystone and Redundant Components

As a means to quantify the relative contribution of elements within functional groups, Davic [31] suggests a variety of community indices could be applied, and uses the dominance index (DI) [60] to determine importance of individual elements within a group:

$$DI_{BP} = (n_{max}/N)$$

where n_{max} represents the number of individuals of the most abundant element, and N is the total number of individuals within the functional group as a whole. In cases where more than one potential keystone is identified within a functional group Davic [31] advocates the community dominance index (CDI) of McNaughton [61]:

$$CDI = (n_1 + n_2/)N$$

such that n_1 and n_2 represent the frequency of the two most abundant species within a functional group. Other ecological measures of dominance may also provide quantitative insights to the relative importance of individual elements to a group. Commonly applied in the field of ecology is the Simpson Domination Index [62]:

$$DI_S = \sum (n_i/N)^2$$

where n_i is the population of each species, and N is the total population. The Simpson method gives greater consideration to diversity within a system; in contrast the Berger-Parker approach [60] does not account for the number of species, but only the total population of the system. However, the Simpson method falls short in that it can only be applied to characterize groups and not individual elements. Our analysis, therefore, applied the Simpson method to the functional groups to provide a more conservative assessment of domination within each group. We then applied the Berger-Parker equation [60] to quantify each species' dominance within the functional groups. Because we applied these indexes to both the groups and their elements, we refer to *domination of* to characterize the inequality within a group, and to *dominance by* to describe the contribution of individual elements to a group. For the purposes of this paper, we have determined that a value of >0.5 for either DI_{BP} or CDI calculations would result in a classification of a biocultural element as a social-ecological keystone, and a value of <0.5 would result in a classification of a biocultural element as a social-ecological redundant.

5. Results

A review of the literature regarding Hawaiian agricultural practices [53–59] yielded three classes of crop systems and eighteen biocultural functional groups that span a range of functions, incuding food production, material resource production, and spiritual/religious practice. Dominance *by* and domination *of* three sets of cropping systems (Tables 1 and 2) and the eighteen biocultural functional groups (Table 3) was calculated.

Table 1. Dominance Index (DI) as calculated for each of the systems of growing crops in the Hawaiian social-ecological system, as documented by Handy et al. [54].

Cropping System	Dominance of Cropping Systems	Domination by Species Assemblage	Associated Crop Species
Rain-fed	0.441	0.263	11
Agroforestry	0.288	0.110	17
Flooded-field	0.272	0.731	7

Table 2. Dominance of crops within three major classes of agricultural systems that existed within the Hawaiian social-ecological system.

Latin Name	Hawaiian Name	Dominance in Rain-Fed Systems	Dominance in Agroforestry	Dominance in Flooded Systems
Aleurites molaccanus	*Kukui*	-	**0.186**	-
Artocarpus altilis	*'Ulu*	-	0.089	-
Broussonetia papyrifera	*Wauke*	0.013	0.066	0.016
Cocos nucifera	*Niu*	-	0.094	-
Colocasia esculenta	*Kalo*	**0.352**	**0.178**	**0.852**
Cordia subcordata	*Kou*	-	0.010	-
Cordyline fruticosa	*Kī*	0.005	0.028	0.014
Curcuma domestica	*'Ōlena*	0.005	0.015	-
Dioscorea alata	*'Uhi*	0.073	0.036	-
Dioscorea bulbifera	*Hoi*	-	0.028	-
Dioscorea pentaphylla	*Pi'a*	-	0.008	-
Ipomoea batatas	*'Uala*	**0.349**	-	-
Lageneria siceraria L. *vulgaris*	*Ipu*	0.086	-	-
Musa ssp.	*Mai'a*	0.023	0.099	0.059
Pandanus tectorius	*Hala*	-	0.084	-
Piper methysticum	*'Awa*	0.040	0.008	0.016
Saccharum offinarum	*Kō*	0.043	0.003	0.024
Schizostachyum glaucifolium	*'Ohe*	-	0.041	-
Tacca leontopetaloides	*Pia*	0.010	0.028	0.019
Count		**11**	**17**	**7**

Table 3. The eighteen biocultural functional groups that embody Hawaiian agriculture traditions as identified in the Handy et al. [54] tome on the topic, dominance index (DI) for each, and associated crop species.

Biocultural Functional Group	Dominance of Hawaiian Agriculture (DI$_S$)	Domination of Group by Crop (DI$_{BP}$)	Number of Associated Crops	Dominant Crop
Complex carbohydrates for food	0.211	0.336	9	*Kalo*
Affiliated with deities	0.129	0.132	12	*Kalo*
Ceremonial plants for religious practice	0.108	0.277	8	*'Awa*
Wood (timber, fuel, vessel, music, misc.)	0.098	0.184	11	*Kukui/Hau*
Famine food for a resilient food system	0.078	0.121	14	*Kalo*
Medicinal applications	0.069	0.105	17	*'Awa*
Leaves for weaving or thatch material	0.059	0.459	4	*Hala*
Fibers for clothing	0.037	1.000	4	*Wauke*
Simple carbohydrate for food	0.037	0.274	6	*Niu*
Mulch for agriculture	0.028	0.378	4	*Kukui*
Relates to the family system	0.029	0.746	2	*Kalo*

Table 3. *Cont.*

Biocultural Functional Group	Dominance of Hawaiian Agriculture (DI$_S$)	Domination of Group by Crop (DI$_{BP}$)	Number of Associated Crops	Dominant Crop
Oil for culinary uses and healing	0.026	0.501	2	*Kukui*
Drink for refreshment and recreation	0.023	0.453	3	*Niu*
Genesis story with the culture	0.020	1.000	1	*Kalo*
Leafy greens for food	0.019	0.600	3	*Kalo*
Fibers for cordage	0.018	0.257	5	*Kalo*
Dye for visual attraction	0.006	0.510	2	*ʻŌlena*
Glue/resin source	0.002	1.000	2	*ʻUlu*

6. Analysis

In examining the dominance of cropping system forms of Hawaiian agriculture (Table 1), 'rain-fed' was the dominant crop system (0.44), followed by agroforestry (0.29), and then flooded-field systems (0.27). However, the split was relatively level, and the Simpson Domination Index (DI$_S$) for Hawaiian Cropping Systems based on these relative abundances yields a moderate value of 0.35. This indicates that a severe loss of any one of the three systems would substantially impact Hawaiian agriculture. Conversely, the domination of each the three cropping systems by their crop components varied significantly, indicating wildly different levels of reliance on critical species. The agroforestry systems, with a very low value of 0.11, could easily absorb the loss of any species, including its dominant species. At the opposite end of the spectrum, flooded-field agriculture with an extremely high value of 0.73 would likely catastrophically fail if its most abundant species were to be removed. Rain-fed agriculture, with a more moderate value of 0.26, would likely struggle but adapt to a removal of the dominant species.

Examining the dominance of crops within these agricultural systems, *kalo* (taro, *Colocasia esculenta*) cultivation was either dominant or co-dominant in all three systems (Table 2). Within flooded-field agriculture *kalo* was highly dominant (DI$_{BP}$ = 0.85), a value that classifies it as a keystone component. Within rain-fed systems *kalo* cultivation was only slightly more dominant (0.352) than *ʻuala* (sweet potato, *Ipomoea batatas*) cultivation (0.349), indicating clear co-dominance. The CDI value of 0.7 classifies the cultivation of these species as a keystone component, which indicates that rain-fed cultivation would likely collapse without either of the two co-dominant species. Within agroforestry systems *kalo* cultivation was also co-dominant (0.18) with *kukui* cultivation (0.19), and was the dominant among non-canopy species. The moderate CDI value of 0.37 suggests that cultivation of either does not play a keystone role, suggesting that agroforestry systems would adapt to the loss of either or both of the co-dominant species.

In examining the eighteen biocultural functional groups (Table 3), cultivating a complex carbohydrate as a food source displayed the highest level of dominance. Other important functional groups (>5% DI$_{BP}$) include religious and ceremonial associations, wood, famine food, thatching, and medicinal uses. *Kalo* again demonstrates significant importance: it is the dominant species of the dominant functional group, in 43% of the important functional groups, and in 33% of all functional groups. Furthermore, it contributes to more functional groups (61%) than any other species.

An additional finding indicates a relationship between dominance *of* functional groups within Hawaiian social-ecological system, and how dominated *by* their species assemblage those functional groups are (Figure 4). This was a highly significant relationship (r^2 0.39, *p* 0.006) described by a log-log function (log(y) = 4.09 − 0.39 × log(x); var(x) = 28.4; var(y) = 903.6; cov[x, y] = −0.5056). Although perhaps intuitive, this indicates that functional groups that make less significant contributions to the social-ecological system are more likely to rely on a smaller assemblage of species. This relationship is important because it indicates that essential functions of a biocultural system will tend to not develop an overly dominant species, likely resulting in increased resilience within the social-ecological system due to functional and responce diversity. In the case of Hawaiʻi, the most dominant functional group—cultivation of complex carbohydrates—has a relatively high DI$_S$ value, indicating a higher reliance on *kalo* than might be expected for such an important function.

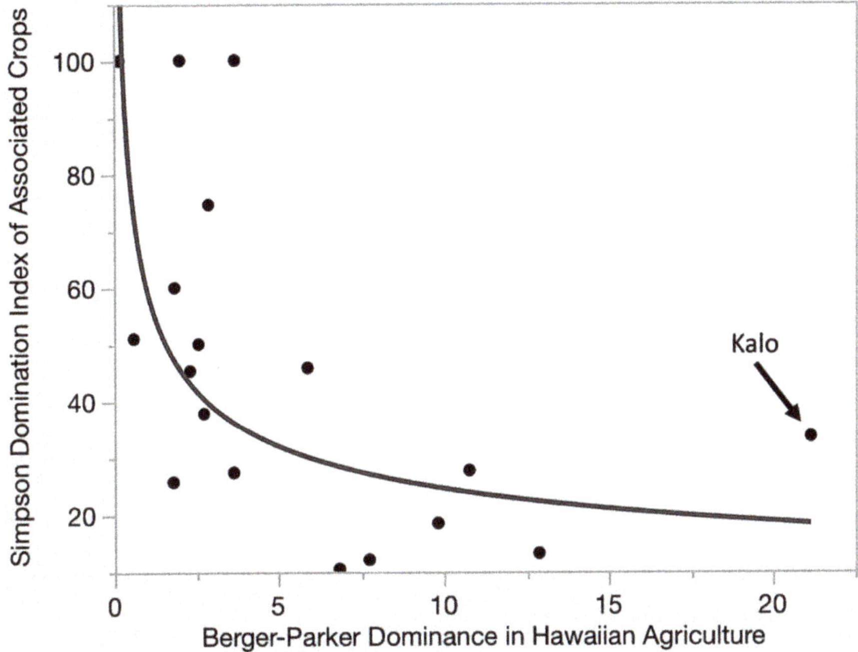

Figure 4. As importance of function groups pertaining to agriculture within the Hawaiian social-ecological systems decreases, the Domination Index within that group tends to increase along a log-log relationship. This indicates the less significant a functional group is, the more heavily it can rely on a smaller species assemblage.

7. Discussion

7.1. Is Kalo Cultivation a Keystone Component of the Hawaiian Social-Ecological System?

Kalo cultivation displayed high dominance within cropping systems, being dominant or co-dominant in all three systems identified, and within biocultural functional groups, being dominant in multiple important groups. Furthermore, *kalo* is dominant within the most important functional group—the cultivation of complex carbohydrates. But do the metrics we've identified adequately capture the importance and impacts of *kalo* within the Hawaiian social-ecological system, especially as relates to structure and function of the system as a whole? These questions are explored in a discussion below.

7.2. Biocultural Relationships between Kalo and Hawaiian Culture

Through the original voyages of Polynesians to what is now known as the Hawaiian Islands a total of at least twenty-seven plant species and six animal species were established in Hawai'i [63], collectively referred to as a "portable biocultural toolkit"—a term we apply to the suite of plants and animals that cultural groups take with them in their diaspora across the globe. The taxa selected were likely chosen because they—through their pre-established links to sociocultural practices—could facilitate the perpetuation of key biocultural functional groups (e.g., eating food, drinking liquid, healing with medicine, clothing production, religious ceremony, storing food, etc.) upon arrival at a new destination. They were also the vehicles for transporting stories, which helped retain lessons about life, family, and culturally-appropriate behaviors; and, therefore, represented important teaching tools for future generations.

As the Hawaiian culture developed, *kalo* came to hold a pre-eminent role on both a practical and philosophical level. This is reflected in a high level of pre-contact diversity with approximately 400 distinct cultivars of *kalo* that the Hawaiian culture co-evolved with [18]. Not only was *kalo* the preferred staple food, but it was also considered to be the original ancestor of the Hawaiian race [54] and the ancient religious system [55,58]. Furthermore, there are many philosophical and symbolic links observed linguistically between the parts and growth form of *kalo*, and the Hawaiian family system and its structure [54].

Evidence for the central importance at which *kalo* was deeply ingrained within the identity, dietary system and religious beliefs of the Hawaiian people is seen in the ancient proverb *"Ola ke kalo, ola ke kanaka"*, which roughly translates to "As long as *kalo* lives, so shall the Hawaiian race"; the converse implication is that if *kalo* were to disappear, so too would the Hawaiian race [18,64]. It is apparent that due to its central importance in the identity, diet, and religion of the ancient Hawaiian society its hypothetical disappearance would drastically alter the cultural subsystem that was built upon it—not only because of its direct importance, but also because of secondary extinctions which would occur due to the breaking of linked biocultural relationships. Thus, it certainly satisfies the Garibaldi and Turner [11] definition of *cultural* keystone *species*, at least within the context of pre-contact Hawaiian culture. One could also argue that the practices associated with *kalo* satisfy the Brosi et al. [12] definition of *"cultural* keystone *practice"* in the same pre-contact period. However, beyond establishing *kalo* as a *"cultural* keystone *species"* and the traditions associated with it as *"cultural* keystone *practices"* via an *a priori* assessment, this paper takes a quantitative approach to determining which specific biocultural relationships can be considered keystone components of the Hawaiian social-ecological system.

7.3. Kalo Is a Dominant Component in Hawaiian Cropping Systems

Kalo was a dominant component in all three cropping systems—rain-fed, agroforestry, and flooded-field—employed in the Hawaiian social-ecological system. 'Uala (sweet potato) was co-dominant in rain-fed systems, which is a strong indicator of functional redundancy in that group. This means that the structure of rain-fed systems could, theoretically, still be maintained in such systems, if *kalo* was not utilized for some reason. The same could not be said for flooded-field systems; the results indicate that *kalo*'s disappearance would result in structural shifts, and potential collapse, in that cropping system. Overall, no cropping system was overbearingly dominant in Hawaiian agriculture when viewed at the scale of the archipelago. However, the spatial distribution of agricultural opportunities was not evenly distributed across the archipelago [27], and the loss of *kalo* would have had different effects based on the local reliance on different cropping systems. Particularly on the oldest island of Kaua'i, where flooded-field agriculture was highly dominant, the theoretical loss of *kalo* would likely drastically disrupt the entire society. This would not necessarily be the case in leeward Hawai'i and Maui Islands, where rain-fed agriculture dominated. It could be argued that the functional redundancy was being enacted in rain-fed systems on the younger islands of Maui and Hawai'i, where shifts in the dominant deity (from Kāne to Lono) and the development of religious sects occurred in areas that, due to environmental limitations of *kalo* cultivation, were dominantly dependent on 'uala.

7.4. Kalo Is a Dominant in a Key Biocultural Functional Group

Of all the biocultural functional groups associated with Hawaiian agriculture, the most important was "cultivating a complex carbohydrate as a food source" (Table 3). Within that biocultural functional group, *kalo* was a strong dominant with no co-dominant (Table 3). Theoretically, this means that if *kalo* was not used in agriculture systems, for whatever reason, the structure of those systems would shift into an altered state, especially in light of the unique agricultural practices associated with *kalo* cultivation as described by Handy et al. [54]. In rain-fed systems, *kalo* could theoretically be replaced by its co-dominant ('uala, sweet potato), but the structure of those systems would shift due to the different agricultural practices associated with the cultivation of each species. For instance, 'uala matures more quickly, and also has a more confined harvest and storage time, indicating that compared to *kalo*, an *'uala*-based system would require a more mobile labor force, and more consistent time in the field to manage the shorter harvest periods. More planning and balance would be needed to manage the short window of harvest for constant food supply, and the systems would likely be more vulnerable to both environmental and social perturbances such as drought or war.

In agroforestry systems, *kalo*'s dominance was due to its ability to grow in low light conditions and, more importantly, be able to store in ground for extended periods and self replicate—all traits that are condusive to its cultivation in these systems. It could be replaced by a redundant component of the sub-canopy layer in that system, but the structure of that system would also shift. In flooded-field systems, there was no redundant component in the pre-contact period that could be grown in such flooded conditions. Therefore, if *kalo* was not cultivated in that system, the lack of a functional redundant would result in an inability of that system to exist, which would result in a structural shift in cropping systems if agriculture was to continue.

7.5. Cropping Systems Associated with Kalo Influenced the Structure of the Hawaiian Social-Ecological System

Although not a native plant, *kalo* played a major role in the highly intensified management of the ecological subsystem, and its cultivation shaped the cultural landscape that is iconically associated with the Hawaiian social-ecological system. *Kalo* is most productively cultivated in terraced and flooded pond-fields ([54], Figure 5), which resemble rice paddies of South-East Asia. Evidence [27,54,65] suggests that nearly all land capable of being irrigated (from the backs of valleys to the alluvial plains bordering the seashore) was converted from its natural (i.e., pre-human) state of lowland wet-to-mesic forest types; to flooded fields for the cultivation of *kalo*. This flooded-field system of agriculture shaped the social-ecological system around it as described below.

Figure 5. Picture of a contemporary flooded- field system in Hawai'i used to cultivate *kalo* (taro, *Colocasia esculenta*). In this style of agricultre, rivers/streams are the central component with its waters being diverted over large areas of adjacent flatlands for the cultivation of *kalo*.

The conversion of large areas of lowland forest into flooded-field systems had three major repercussions which directly influenced the structure and function of the Hawaiian social-ecological system, as well as the outward appearance of its associated cultural landscapes [66]:

1. This conversion induced localized regime shifts in large areas of land (valley floors and alluvial plains) from forest biome to riparian ecotone. This, in essence, expanded and stabilized riparian habitat—a highly productive ecotone—from a relatively limited to a very broad area. Archaeological evidence suggests that such localized regime shifts have occurred [67], and likely extended the range of native water (i.e., riparian) fowl allowing for increases in their populations [68].

2. This conversion theoretically increased the capacity of aquifers (i.e., the islands' ability to retain water). The expansive flooded-field system slowed the flow rate of water on its journey towards the sea, and increased surface area of land covered by water. This cumulatively increased the potential for aquifer recharge. Increasing aquifer recharge potentially increases the level of the aquifer, which could result in more artesian springs at higher elevations than previously existed. The appearance of these springs would further increase the potential for lands—at higher elevation—to be converted to flooded terraces.

3. This conversion likely induced localized regime shifts in estuaries and nearshore reefs from predator-dominated to herbivore-dominated. In theory, this may have been achieved through the development of aquaculture technologies. The emergence of such technologies was likely enabled by the flooded-field system, which mobilized nutrients and then transported them to coastal areas. The water passing through this flooded-field system was presumably enriched due to both a direct and indirect increase in organic matter, and anaerobic soils that mobilized otherwise fixed phosphorous into water systems. This aquaculture system had several classes of fish ponds, including those that walled in large areas of near-shore reef. These walls trapped the enriched water, thus containing algal blooms which allowed for the farming of herbivorous fish within them, while maintaining the health of the reef outside of the walls. This effectively expanded and stabilized estuary habitat—another highly productive ecotone—from a relatively limited to a very broad area. The success of this technology hinged on the management strategies which included methodological removal of top predators.

Each of the above occurrences turned out to be key components to the structure and function of the social-ecological system that existed due to the intensified resource management system of ancient Hawaiian civilization known as the *Moku* System [25]. The ecosystem services provided likely enabled a population boom in the Hawaiian social-ecological system that was sustained until contact with Europeans in 1778 A.D., which subsequently brought unexpected and catastrophic change such as a 90% population collapse resulting from introduced diseases [48].

7.6. Substitution of a Social-Ecological Keystone Alters the Structure of Social-Ecological Systems

An example of resilience is seen in the functional group "crop grown in rain-fed field systems" (Table 2), which has ʻuala as a co-dominant with *kalo*. Rain-fed cropping systems existed in regions in the Hawaiian archipelago that lacked suffient water for large-scale *kalo* cultivation (either through rain or surface water). Areas such as this—as in the *moku* (districts) of Kohala Hema, Kona Akau and Kona Hema on the island of Hawaiʻi—had field systems which were shaped around ʻuala, the co-dominant in rain-fed systems. Lincoln et al. [39] point out that *kalo* was succeeded by ʻuala in regions where *kalo* could not be cultivated on a large scale due to insufficient water availability. In other words, the dominant of a key biocultural functional group (i.e., the keystone of that biocultural functional group) was substituted by one of the redundant components in that functional group.

Theoretically, this would result in a structural shift within the system, and would look different from the one that was shaped around the cultivation of *kalo* in flooded-field systems. Indeed, we argue that an "alternative regime state" existed in different regions within the Hawaiian social-ecological system where ʻuala became dominant. An "alternative regime state" exists when one keystone is succeeded by another in the context of the same social-ecological system. Contemporary analysis of archaeological and historical evidence [27,39,69] supports the notion that this has occurred, and indicates that ʻuala cultivation was the keystone component of social-ecological systems in the Hawaiian islands in regions lacking sufficient water for intensified cultivation of *kalo*. In this case, manifestations were seen in the emergence of religious sects where a shift in primary deities that elevated Lono over Kane and the emergence of new rituals and traditions, such as the *makahiki* festival that originated on the leeward side of Hawaiʻi Island. Further differences have been evidenced in the stability of the political hierarchy and the propensity for predatory warfare.

While these large-scale differences of the structure of the social-ecological system in places where it was shaped around *'uala*—as opposed to *kalo*—are observable, a detailed investigation would likely reveal many more subtle changes at the local level. *Kalo* expresses extreme dominance in key functional groups such as relation to the family system and the genesis of mankind. How did the local biocultural relationships evolve in the absence of *kalo*? The redundancy within the agricultural functional groups allowed for the existence of the Hawaiian social-ecological system in arid regions, albeit one of altered form and structure. In ecological analogy, an "alternative regime state" would be akin to replacing the dominant canopy tree in a forest, with cascading effects on the assemblage of bird, animal and insect species present, while a "regime shift" might be shifting from a forested ecosystem to a shrubland, and all the associated shifts in supported species. Similar regime shifts have also been explored by Scheffer et al. [70].

8. Conclusions

8.1. On Keystone and Redundant Components within Social-Ecological Systems

Social-ecological systems are composed of linked biological-sociocultural relationships, and can be referred to as the "biocultural elements" of these systems. "Keystone" components and "redundant" components exist within the set of biocultural elements that compose social-ecological systems. Theory and methods exist with which to identify and quantify these components, and to correlate them to system health (i.e., system function and resilience), as well as how system health changes over time. Using functional groups to classify and distinguish between keystone components and redundant components is a viable methodology. Such an approach is useful to consider in the context of biocultural restoration of social-ecological systems, as data produced by this approach could be used to influence resource management policies. In accordance with systems theory, this approach could also be applied to other systems—both historical and modern.

However, as in ecology, there are no clearly defined thresholds used to classify components as either a keystone or a redundant, but rather a holistic view of the functional roles must be considered. While quantifying these metics provides insights into the importance of biocultural elements within a social-ecological system, there are no hard cut-off values, or even well-established guidelines, for interpreting data. Therefore, quantification alone cannot define either keystone or redundant components, but qualified assessments can help to illuminate such designations.

Through both quantitative and qualitative methods, we explored functional roles in the Hawaiian social-ecological system, and conclude that *kalo* qualifies as a keystone species for the Hawaiian culture, and further that *kalo* cultivation can be considered a keystone component of the Hawaiian social-ecological system. This suggests *kalo* and its cultivation is vital for the structure and function of the Hawaiian social-ecological system, and that the removal of *kalo* from parts of this system would result in either alternative regime states, or a regime shift resulting in an entirely new social-ecological system. Historical trends over the last two centuries support this notion.

8.2. On Biocultural Diversity and Resilience in Social-Ecological Systems

A loss of biocultural functional groups—due to a lack of both functional redundancy and functional diversity—could induce cascading extinctions on both sides of the social-ecological system. Therefore, increasing biocultural diversity is the most pragmatic way to manage resilience social-ecological systems. Redundancy within biocultural functional groups allows for resilience because of the presence of response diversity in such functional groups. In regions where a particular social-ecological keystone cannot exist, for whatever reason, redundancy in its biocultural functional group can facilitate the existence of an "alternative regime state" within the same social-ecological system, which is built around a successional keystone component. An "alternative regime state" is an altered stable state that is built upon a successional component of a biocultural functional group, one that would likely have a different level of resilience. Such a pehnomenon would also likely result in a cultural landscapes with a different outward appearance occurring within the same social-ecological system. The redundant compontents of Hawaiian agriculture made for a resilient social-ecological system. However, a whole-scale removal of these components would theoretically induce a regime shift and result in an entirely new social-ecological system, and this has been historically observed in Hawai'i.

8.3. On the Model of the Hawaiian Social-Ecological System

Islands are ideal models for system study because they possess all of the biological and physical processes needed to understand complete systems, yet they exist at a scale where complexity is comprehendible. A relatively good understanding of the Hawaiian social-ecological system exists due to a large body of research conducted by native Hawaiians over the last two centuries, and the existence of well-established research institutions. This makes reasonable speculations about structure and function of this system plausible. While rigorous examination of the original Hawaiian social-ecological system can be challenging—as it is largely associated with an era in the past—due to the large body of information associated with it, creative methodologies can be employed to give insights into its structure and function.

8.4. Biocultural Restoration of the Hawaiian Social-Ecological System

In the post-contact era (1778 A.D. onward) Hawaiian culture has been under severe sociocultural pressures such as changes in government, land tenure, religious institutions, economies, language, and others. As a result of these processes native Hawaiians are no longer functioning as the top-level managers of the large-scale social-ecological system that currently exist in Hawai'i. These processes and events are analogous to disturbance, uncertainty, and surprise events [35,43]. Despite these potentially catastrophic events, nearly all aspects of the Hawaiian social-ecological system persist into contemporary times. This may be due, at least in part, to the resilience of key biocultural functional groups associated with Hawaiian culture, particularly those which were shaped around flooded-field system agriculture for the production of *kalo*, waterfowl, and fish. This supports the notion that flooded-field *kalo* cultivation is the foundation of a Hawaiian cultural landscape; and is, therefore, the key to the biocultural restoration of the Hawaiian social-ecological system. Provided that cultural landscapes are the outward appearance of social-ecological systems, focusing on keystone and redundant elements found within a culture's "portable biocultural toolkit" may provide a pathway for maintaining and/or restoring cultural landscapes. We content that while the pre-colonial state cannot be re-created exactly, by looking to the past we can understand and re-create productive and resilient cultural landscapes. Such cultural restoration goes hand-in-hand with ecological restoration.

8.5. Future Research

The theories explored in this paper are not new, but this manuscript puts forth some novel applications of them in the context of social-ecological systems. While the methods presented herein demonstrate some level of credibility to this approach, these notions would need to be assessed in other ways to further test their validity. Some possibilities for future research could include:

- Assessing the percentage of total land area associated with each biocultural functional group to classify between keystone, dominant, and redundant components within social-ecological systems.
- Exploring the functional groups relating to animal husbandry, and assessing dominance in the context of functional groups.
- Expanding these methods to the entire biocultural resource spectrum of a social-ecological unit, which in the Hawaiian archipelago extends from the mountains to sea.
- Assessing the viability of utilizing social-ecological keystones to induce a regime shift back towards the state of abundance known in the Hawaiian language as, "'āina momona" or biocultural resource abundance.

Until more rigorous testing can be done, the concepts explored in this manuscript should still be considered theoretical at best.

Author Contributions: Conceptualization, K.B.W.; Methodology, K.B.W. and N.K.L.; Validation, N.K.L, and F.B.; Formal Analysis, K.B.W., and N.K.L; Investigation, K.B.W., and N.K.L.; Resources, K.B.W.; Data Curation, K.B.W., and N.K.L; Writing—Original Draft Preparation, K.B.W.; Writing—Review & Editing, K.B.W., N.K.L, and F.B.; Visualization, K.B.W., and N.K.L.; Supervision, F.B.; Project Administration, K.B.W.; Funding Acquisition, K.B.W.

Funding: This research received no external funding. The APC was funded by Hawai'i Community Foundation.

Acknowledgments: We express our gratitude to Will McClatchey, Peter Vitousek, Mary Power, and Kim Bridges who have contributed to our research by providing direction for this paper; and to Joshua Diem, who provided the image used in Figure 3. We also thank National Tropical Botanical Garden, and Department of Tropical Plant and Soil Sciences (University of Hawai'i at Mānoa) for supporting this research.

Conflicts of Interest: The authors declare no competing interests as defined by *Sustainability*, or other interests that might be perceived to influence the results and/or discussion reported in this paper. The founding sponsors had no role in the design of the study; in the collection, analyses, or interpretation of data; in the writing of the manuscript and in the decision to publish the results.

References

1. Balée, W. The research program of historical ecology. *Annu. Rev. Anthropol.* **2006**, *35*, 75–98. [CrossRef]
2. Berkes, F.; Colding, J.; Folke, C. (Eds.) *Navigating Social-Ecological Systems: Building Resilience for Complexity and Change*; Cambridge University Press: Cambridge, UK, 2003.
3. Berkes, F.; Folke, C. (Eds.) *Linking Social and Ecological Systems: Management Practices and Social Mechanisms for Building Resilience*; Cambridge University Press: Cambridge, UK, 1998.
4. Ostrom, E. A General Framework for Analyzing Sustainability of Social-Ecological Systems. *Science* **2009**, *325*, 419–422. [CrossRef] [PubMed]
5. McGinnis, M.D.; Ostrom, E. Social-ecological system framework: Initial changes and continuing challenges. *Ecol. Soc.* **2014**, *19*, 30. [CrossRef]
6. Pieter, B.; Schlüter, M.; Sendzimir, J. A Framework for Analyzing, Comparing, and Diagnosing Social-Ecological Systems. *Ecol. Soc.* **2015**, *20*, 18.
7. Sterling, E.J.; Filardi, C.; Newell, J.; Albert, S.; Alvira, D.; Bergamini, N.; Betley, E.; Blair, M.; Boseto, D.; Burrows, K.; et al. Biocultural approaches to sustainability indicators: Bridging local and global scales to foster human adaptive capacity and ecological resilience. *Nat. Ecol. Evol.* **2017**. [CrossRef]
8. Walker, B.; Holling, C.S.; Carpenter, S.R.; Kinzig, A. Resilience, adaptability and transformability in social-ecological systems. *Ecol. Soc.* **2004**, *9*, 5. [CrossRef]
9. Berkes, F. Restoring Unity: The Concept of Marine Social-Ecological Systems. In *World Fisheries: A Social-Ecological Analysis*; Ommer, R.E., Perry, R.I., Cochrane, K., Cury, P., Eds.; Wiley-Blackwell: Oxford, UK, 2011.
10. Cristancho, S.; Vining, J. Culturally defined keystone species. *Hum. Ecol. Rev.* **2004**, *11*, 153–162.
11. Garibaldi, A.; Turner, N. Cultural keystone species: Implications for ecological conservation and restoration. *Ecol. Soc.* **2004**, *9*, 1. [CrossRef]
12. Brosi, B.J.; Balick, M.; Wolkow, R.; Lee, R.; Kostka, M.; Raynor, W.; Gallen, R.; Raynor, A.; Lee, L.D. Quantifying cultural erosion and its relationship to biodiversity conservation: Canoe-making knowledge in Pohnpei, Micronesia. *Conserv. Biol.* **2007**, *21*, 875–879. [CrossRef] [PubMed]
13. Nguyen, M.L.T.; Wieting, J.; Doherty, K.T. Vegetation analysis of urban ethnic markets shows supermarket generalists and Chinatown ethnic-specialist vendors. *Ethnobot. Res. Appl.* **2008**, *6*, 63–85. [CrossRef]
14 Anderies, J.M.; Janssen, M.A. *Sustaining the Commons*; Center for the Study of Institutional Diversity, Arizona State University: Tempe, AZ, USA, 2013.
15. Norgaard, R.B. Sociosystem and ecosystem coevolution in the Amazon. *J. Environ. Econ. Manag.* **1981**, *8*, 238–254. [CrossRef]
16. Winter, K.; McClatchey, W. Quantifying Evolution of Cultural Interactions with Plants: Implications for Managing Diversity for Resilience in Social-Ecological System. In *Ethnobotany: A Focus on Brazil. Functional Ecosystems and Communities*; Albuquerque, U.P., Ed.; Global Science Books Ltd.: Isleworth, UK; Ikenobe, Japan, 2008; Volume 2, pp. 1–10.
17. Winter, K.B.; McClatchey, W. The Quantum Co-evolution Unit: An Example of Awa (*Piper methysticum* G. Foster) in Hawaiian Culture. *Econ. Bot.* **2009**, *63*, 353–362. [CrossRef]
18. Winter, K.B. *Kalo* [Hawaiian *Kalo: Colocasia esculenta* (L.) Schott] Varieties: An Assessment of Ancient and Modern Synonymy and Diversity. *Ethnobot. Res. Appl.* **2012**, *10*, 423–447.

19. Kirch, P.V. Hawaii as a model for human ecodynamics. *Am. Anthropol.* **2007**, *109*, 8–26. [CrossRef]
20. Wilmshurst, J.M.; Hunt, T.L.; Lipo, C.P.; Anderson, A.J. High-precision radiocarbon dating shows recent and rapid initial human colonization of East Polynesia. *Proc. Natl. Acad. Sci. USA* **2011**, *108*, 1815–1820. [CrossRef] [PubMed]
21. Kamakau, S.M. *Ruling Chiefs of Hawaii*; Kamehameha Schools Press: Honolulu, HI, USA, 1961.
22. Kame'eleihiwa, L. *Native Land and Foreign Desires*; Bishop Museum Press: Honolulu, HI, USA, 1992.
23. Cachola-Abad, C.K. Evaluating the orthodox dual settlement model for the Hawaiian Islands: An analysis of artefact distribution and Hawaiian oral traditions. In *The Evolution and Organization of Prehistoric Society in Polynesia*; Graves, M.W., Green, R.C., Eds.; New Zealand Archaeological Association Monograph: Auckland, New Zealand, 1993; Volume 19, pp. 13–32.
24. Cachola-Abad, C.K. An Analysis of Hawaiian Oral Traditions: Descriptions and Explanations of the Evolution of Hawaiian Socio-Political Complexity. Doctoral Dissertation, University of Hawaii at Mānoa, Honolulu, HI, USA, 2000.
25. Winter, K.B.; Beamer, K.; Vaughan, M.B.; Friedlander, A.M.; Kido, M.; Whitehead, A.N.; Akutagawa, M.K.H.; Kurashima, N.; Lucas, M.P.; Nyberg, B. The Moku System: Managing biocultural resources for abundance within social-ecological regions in Hawai'i. *Sustainability* **2018**, in press.
26. Power, M.E.; Tilman, D.; Estes, J.A.; Menge, B.A.; Bond, W.J.; Mills, L.S.; Daily, G.; Castilla, J.C.; Lubchenco, J.; Paine, R.T. Challenges in the quest for keystones. *BioScience* **1996**, *46*, 609–620. [CrossRef]
27. Ladefoged, T.N.; Kirch, P.V.; Gon, S.M.; Chadwick, O.A., III; Hartshorn, A.S.; Vitousek, P.M. Opportunities and constraints for intensive agriculture in the Hawaiian archipelago prior to European contact. *J. Archaeol. Sci.* **2009**, *36*, 2374–2383. [CrossRef]
28. Kirch, P.V.; Zimmerer, K.S. *Roots of Conflict*; School for Advanced Research Press: Santa Fe, NM, USA, 2011.
29. Winter, K.B.; Lucas, M. Spatial modeling of social-ecological management zones of the alii era on the island of Kauai with implications for large-scale biocultural conservation and forest restoration efforts in Hawaii. *Pac. Sci.* **2017**, *71*, 457–477. [CrossRef]
30. Paine, R.T. The Pisaster-Tegula Interaction: Prey, Patches, Predator Food Preference, and Intertidal Community Structure. *Ecology* **1969**, *50*, 950–961. [CrossRef]
31. Davic, R.D. Linking Keystone Species and Functional Groups: A New Operational Definition of the Keystone Species Concept. *Conserv. Ecol.* **2003**, *7*, r11. [CrossRef]
32. Von Bertalanffy, L. *General System Theory*; George Braziller, Inc.: New York, NY, USA, 1968; Volume 41973, p. 40.
33. Pimm, S.L.; Gilpin, M.E. Theoretical issues in conservation biology. In *Perspectives in Ecological Theory*; Roughgarden, J., May, R., Levin, S., Eds.; Princeton University Press: Princeton, NJ, USA, 1989; pp. 287–305.
34. Christianou, M.; Ebenman, B. Keystone Species and Vulnerable Species in Ecological Communities: Strong or Weak Interactors? *J. Theoret. Biol.* **2005**, *235*, 95–103. [CrossRef] [PubMed]
35. Folke, C.; Carpenter, S.; Walker, B.; Scheffer, M.; Elmqvist, T.; Gunderson, L.; Holling, C.S. Regime shifts, resilience, and biodiversity in ecosystem management. *Annu. Rev. Ecol. Evol. Syst.* **2004**, *35*, 557–581. [CrossRef]
36. Rhindos, D. Darwinism and its role in the explanation of domestication. In *Foraging and Farming: The Evolution of Plant Exploitation*; Harris, D.R., Hillman, G.C., Eds.; Unwin Hyman: London, UK, 1989; pp. 27–41.
37. Lincoln, N.K.; Vitousek, P.M. Indigenous Polynesian agriculture in Hawaii. *Oxford Resear. Encycl. Environ. Sci.* **2017**. [CrossRef]
38. Kirch, P.V. *The Wet and the Dry: Irrigation and Agricultural Intensification in Polynesia*; University of Chicago Press: Chicago, IL, USA, 1994.
39. Lincoln, N.K.; Rossen, J.; Vitousek, P.; Kahoonei, J.; Shapiro, D.; Kalawe, K.; Pai, M.; Marshall, K.; Meheula, K. Restoration of 'Āina Malo'o on Hawai'i Island: Expanding Biocultural Relationships. *Sustainability* **2018**, under review.
40. Holling, C.S. Resilience and stability of ecological systems. *Annu. Rev. Ecol. Syst.* **1973**, *4*, 1–23. [CrossRef]
41. Resilience Alliance. 2002. Available online: https://www.resilience.org/about-resilience/ (accessed on 30 June 2018).
42. Holling, C.S. Engineering resilience versus ecological resilience. In *Engineering within Ecological Constraints*; Schulze, P.C., Ed.; National Academy Press: Washington, DC, USA, 1996.
43. Elmqvist, T.; Folke, C.; Nystrom, M.; Peterson, G.; Bengtsson, J.; Walker, B.; Norberg, J. Response Diversity and Ecosystem Resilience. *Front. Ecol. Environ.* **2003**, *1*, 488–494. [CrossRef]
44. Naeem, S. Species Redundancy and Ecosystem Reliability. *Conserv. Biol.* **1998**, *12*, 39–45. [CrossRef]

45. Scheffer, M.; Carpenter, S.; Foley, J.A.; Folke, C.; Walker, B. Catastrophic Shifts in Ecosystems. *Nature* **2001**, *413*, 591–596. [CrossRef] [PubMed]

46. Fosberg, F.R. (Ed.) *Man's Place in Island Ecosystems: A Symposium*; Bishop Museum Press: Honolulu, HI, USA, 1962; 264p.

47. Vitousek, P. *Nutrient Cycling and Limitation: Hawaii as a Model*; Princeton University Press: Princeton, NJ, USA, 2004.

48. Stannard, D.E. *Before the Horror: The Population of Hawaii on the Eve of Western Contact*; University of Hawaii Press: Honolulu, HI, USA, 1994.

49. Swanson, D.A. *The Number of Native Hawaiians and Part Hawaiians in Hawaii, 1778 to 1900: Demographic Estimates by Age, with Discussion*; University of California: Riverside, CA, USA, 2015.

50. Hommon, R.J. *The Ancient Hawaiian State: Origins of a Political Society*; Oxford University Press: Oxford, UK, 2013.

51. Max-Neef, M.; Elizalde, A.; Hopenhayn, M.; Herrera, F.; Zemelman, H.; Jataba, J.; Weinstein, L. Human Scale Development: A Hope for the Future. *Dev. Dial.* **1989**, *1*, 5–80.

52. Max-Neef, M.; Elizalde, A.; Hopenhayn, M. *Human Scale Development: Conception, Application and Further Reflections*; The Apex Press: New York, NY, USA, 1991; 114p.

53. Ii, J.P. *Fragments of Hawaiian History*; Bishop Museum Press: Honolulu, HI, USA, 1959.

54. Handy, E.S.C.; Handy, E.G.; Pukui, M.K. *Native Planters in Old Hawaii: Their Life, Lore, and Environment*; Bishop Museum Press: Honolulu, HI, USA, 1972.

55. Kamakau, S.M. *Ka Hana a Ka Poe Kahiko*; Translated from the Newspaper *Ke Au Okoa*; Bishop Museum Press: Honolulu, HI, USA, 1976.

56. Kamakau, S.M. *Ka Poe Kahiko*; Translated from the Newspaper *Ke Au Okoa*; Bishop Museum Press: Honolulu, HI, USA, 1991.

57. Desha, S.L. *Kamehameha and His Warrior Kekuhaupio*; Frazier, Frances N., Translator; Kamehameha Schools Press: Honolulu, HI, USA, 2000.

58. Malo, D. *Ka Moolelo Hawaii: Hawaiian Traditions*; Chun, Malcolm, Translator; First People's Publications: Honolulu, HI, USA, 2006; 274p.

59. Kepelino. *Kepelino's Traditions of Hawaii*; Beckwith, M.W., Ed.; Bishop Museum Press: Honolulu, HI, USA, 2007; 206p.

60. Berger, W.H.; Parker, F.L. Diversity of planktonic Foraminifera in deep sea sedimanets. *Science* **1970**, *168*, 1345–1347. [CrossRef] [PubMed]

61. McNaughton, S.J. Structure and Function in California Grasslands. *Ecology* **1968**, *49*, 962–972. [CrossRef]

62. Simpson, E.H. Measurement of diversity. *Nature* **1949**, *163*, 688. [CrossRef]

63. Abbott, I.A. *Lāau Hawaii Traditional Hawaiian Uses of Plants*; Bishop Museum Press: Honolulu, HI, USA, 1992.

64. Kelly, M. *Ahupua'a Fishponds and Loi: A Film for Our Time*; Nā Maka o ka Āina: Honolulu, HI, USA, 1992.

65. Handy, E.S.C. *The Hawaiian Planter: His Plants, Methods and Areas of Cultivation*; Bishop Museum Press: Honolulu, HI, USA, 1940; 227p.

66. Molnar, Z.; Berkes, F. Role of traditional ecological knowledge in linking cultural and natural capital in cultural landscapes. In *Reconnecting Natural and Cultural Capital: Contributions from Science and Policy*; Paracchini, M.L., Zingari, P.C., Blasi, C., Eds.; European Union: Luxembourg, 2018; pp. 183–193.

67. Burney, D.A.; Kikuchi, W. A millennium of human activity at Makauwahi Cave, Mahaulepu, Kauai. *Hum. Ecol.* **2006**, *34*, 219–247. [CrossRef]

68. Burney, D.A.; James, H.F.; Burney, L.P.; Olson, S.L.; Kikuchi, W.; Wagner, W.L.; Burney, M.; McCloskey, D.; Kikuchi, D.; Grady, F.V.; et al. Fossil evidence from a diverse biota from Kauai and its transformation since human arrival. *Ecol. Monogr.* **2001**, *71*, 615–641.

69. Kurashima, N.; Jeremiah, J.; Ticktin, T. I ka wa ma mua: The value of a historical ecology approach to ecological restoration in Hawaii. *Pac. Sci.* **2017**, *71*, 437–456. [CrossRef]

70. Scheffer, M.; Carpenter, S.R.; Lenton, T.M.; Bascompte, J.; Brock, W.; Dakos, V.; Van de Koppel, J.; Van de Leemput, I.A.; Levin, S.A.; Van Nes, E.H.; et al. Anticipating critical transitions. *Science* **2012**, *388*, 344–348. [CrossRef] [PubMed]

 *sustainability*

Article

Kū Hou Kuapā: Cultural Restoration Improves Water Budget and Water Quality Dynamics in He'eia Fishpond

Paula Möhlenkamp [1], Charles Kaiaka Beebe [1], Margaret A. McManus [1], Angela Hi'ilei Kawelo [2], Keli'iahonui Kotubetey [2], Mirielle Lopez-Guzman [1,3], Craig E. Nelson [1,3,4] and Rosanna 'Anolani Alegado [1,3,4,*]

[1] Department of Oceanography, University of Hawai'i Mānoa, Honolulu, HI 96822, USA; pmoehlen@hawaii.edu (P.M.); cbeebe@hawaii.edu (C.K.B.); mamc@hawaii.edu (M.A.M.); malg@hawaii.edu (M.L.-G.); craig.nelson@hawaii.edu (C.E.N.)

[2] Paepae o He'eia, Kāne'ohe, HI 96744, USA; hiilei@paepaeoheeia.org (A.H.K.); kelii@paepaeoheeia.org (K.K.)

[3] Daniel K. Inouye Center for Microbial Oceanography: Research and Education, University of Hawai'i Mānoa, Honolulu, HI 96822, USA

[4] Sea Grant College Program, University of Hawai'i, Mānoa, Honolulu, HI 96822, USA

* Correspondence: rosie.alegado@hawaii.edu; Tel.: +01-808-956-0565

Received: 29 September 2018; Accepted: 24 December 2018; Published: 29 December 2018

Abstract: In Hawai'i, the transition from customary subsistence flooded taro agroecosystems, which regulate stream discharge rate trapping sediment and nutrients, to a plantation-style economy (c. the 1840s) led to nearshore sediment deposition—smothering coral reefs and destroying adjacent coastal fisheries and customary fishpond mariculture. To mitigate sediment transport, *Rhizophora mangle* was introduced in estuaries across Hawai'i (c. 1902) further altering fishpond ecosystems. Here, we examine the impact of cultural restoration between 2012–2018 at He'eia Fishpond, a 600–800-year-old walled fishpond. Fishpond water quality was assessed by calculating water exchange rates, residence times, salinity distribution, and abundance of microbial indicators prior to and after restoration. We hypothesized that *R. mangle* removal and concomitant reconstruction of sluice gates would increase mixing and decrease bacterial indicator abundance in the fishpond. We find that He'eia Fishpond's physical environment is primarily tidally driven; wind forcing and river water volume flux are secondary drivers. Post-restoration, two sluice gates in the northeastern region account for >80% of relative water volume flux in the fishpond. Increase in water volume flux exchange rates during spring and neap tide and shorter minimum water residence time corresponded with the reconstruction of a partially obstructed 56 m gap together with the installation of an additional sluice gate in the fishpond wall. Lower mean salinities post-restoration suggests that increased freshwater water volume influx due to *R. mangle* removal. Spatial distribution of microbial bio-indicator species was inversely correlated with salinity. Average abundance of *Enterococcus* and *Bacteroidales* did not significantly change after restoration efforts, however, average abundance of a biomarker specific to birds nesting in the mangroves decreased significantly after restoration. This study demonstrates the positive impact of biocultural restoration regimes on water volume flux into and out of the fishpond, as well as water quality parameters, encouraging the prospect of revitalizing this and other culturally and economically significant sites for sustainable aquaculture in the future.

Keywords: mariculture; aquaculture; community restoration; conservation ecology; Native Hawaiian fishpond; microbes; microbial source tracking

---

1. Introduction

1.1. Native Hawaiian Fishpond Mariculture and Food Security

As the catch rate of our global fisheries levels off due to degradation of the environment and collapse of specific fish populations, the demand for aquaculture production of fish is projected to increase markedly [1]. Concerns over sustainable food production have brought indigenous models of resource management to the fore. Hawai'i currently imports about half of our seafood [2] and local aquaculture is estimated to supply only ~20,000 lbs (9072 kgs) annually [3], but this was not always the case. For centuries, Native Hawaiians developed marine aquaculture that utilized natural enrichments via freshwater from surface and submarine groundwater discharge in managed estuaries, called *loko i'a* (fishponds) [4]. *Loko i'a kuapā* (walled fishponds) were intentionally built in natural embayments at the interface of freshwater streams and the ocean where nutrients from streams promoted the growth of primary producers in constrained brackish ecosystems. The *kuapā* (walls) regulates freshwater inflow to *mākāhā* (size-slotted sluice gates), creates a low wave energy environment within the loko i'a, impedes water volume flux into and out of the loko i'a and ensures that a minimum volume of water is retained in the loko i'a at all times, especially at extremely low tides. Where water volume flux ($m^3\ s^{-1}$) is the volume of water passing through each mākāhā over time. Water volume flux can be in to or out of the loko i'a, depending on mākāhā, tidal stage and other environmental conditions. In this system, unicellular photosynthetic microbes form the base of a complex food web that yield energetically efficient protein production of crustaceans and herbivorous fish species. *Kia'i loko i'a* (fishpond stewards) practiced stock enhancement, leveraging knowledge of juvenile fish migration to trap target species behind mākāhā until reaching maturity and preventing entry of large predators. In addition, kia'i loko regulate water volume flux or harvest fish by blocking mākāhā. It is estimated that loko i'a in Hawai'i could have yielded approximately 2 million pounds of fish per year total historically [5,6].

1.2. The Legacy of Land Use Change and Invasive Species on loko i'a

Physical changes (development, disuse, sedimentation, storm damage) and biological invasions have dramatically altered many loko i'a. Beginning in the 1800s, a shift from subsistence to plantation economy led to erosion and siltation of the nearshore environment. In an attempt to mitigate and stabilize these impacts, mangroves were introduced to Hawai'i in 1902 [7]. Mangroves are highly appreciated in their native habitats for the ecosystem services they provide: shoreline protection and sediment stabilization [8], litterfall subsidy [9] and provision of nursery grounds [8]. Thus, by modifying their environment, mangroves have cascading effects for resident biota, acting as important ecosystem engineers.

However, in Hawai'i, mangroves have caused a variety of negative ecological and economic impacts that motivate their removal [10]. Mangrove's preference for halotypic ecotones favor their growth in estuaries with their root systems obstructing mākāhā, decreasing water volume flux, flushing, and circulation of loko i'a and the streams that feed them [11–13]. Instead of sandy habitats, mangrove vegetated areas have high sedimentation rates and anoxic sediments due to bacterial decomposition of mangrove leaf detritus [11,14]. Moreover, mangrove drawdown of nitrogen and phosphate and decrease dissolved oxygen from overlying waters, potentially inhibiting primary production rates in loko i'a [13]. Importantly, the absence of mangrove feeding specialists in Hawai'i has resulted in the poor assimilation of mangrove-derived nutrients from introduced stands [15] because detritivores native to Hawai'i are not adapted to utilizing mangrove detritus, which tends to be tannin-rich and nitrogen-poor [16].

Post-World War II, a combination of urbanization, the introduction of invasive species, stochastic events (e.g., storms, floods, tsunamis and lava flows) led to deterioration of loko i'a across the state [6]. By 1977, only 28 loko i'a were still in production, and by 1985, merely 7 loko i'a were in commercial or subsistence use [6]. The loss of actively maintained loko i'a exacerbated the spread of invasive mangrove in coastal estuaries [17].

1.3. Revitilization of loko i'a: He'eia Fishpond as a Model

Driven by a desire to re-establish customary practices, provide economic opportunities to local communities and improve production of crustaceans and herbivorous fish, a grassroots movement of loko i'a restoration has gained momentum since the early 2000s [18–20]. Hui Malama Loko I'a is a statewide network of indigenous kia'i loko dedicated to restoring loko i'a for food production [21]. Loko i'a restoration generally entails mangrove removal and dry stacking of basalt with coral/rubble internally. Typical mangrove clearing practices in Hawai'i include the removal of the above-sediment mangrove biomass, leaving intact the prop roots and the root-fiber mat within the sediment. Despite increased loko i'a restoration across the state, we know of no published data on the effects of mangrove removal and loko i'a infrastructure repair on water circulation dynamics and water quality.

Located on the windward side of O'ahu Island, Hawai'i (Figure 1A), He'eia Fishpond (also known as Pihi Loko I'a) is a loko i'a kuapā estimated to have been built 600-800 years ago atop the Malauka'a fringing reef [22] and has been at the forefront of loko i'a restoration in Hawai'i. *Rhizophora mangle* was introduced to the He'eia estuary in 1922 to control runoff from upstream agriculture and stabilize sediments [11,15]. The circulation and water volume flux patterns within He'eia Fishpond were compromised during the Keapuka Flood, which occurred in 1965. The highest discharge rate on record from Ha'ikū and 'Ioleka'a streams occurred during the Keapuka Flood [23] on May 2, 1965. Flood waters first broke the kuapā in the northwestern sector adjacent to He'eia Stream, creating a 183 m opening in the loko i'a. Historical tidal data [24] indicate that the flood likely occurred during a perigean spring tide (a. k. a. King Tide), thus the 56 m break in the kuapā on eastern seaward side as well (Figure 1B, "Ocean Break") likely resulted from build-up of internal pressure within the loko i'a coupled with an extremely low tide outside the loko i'a.

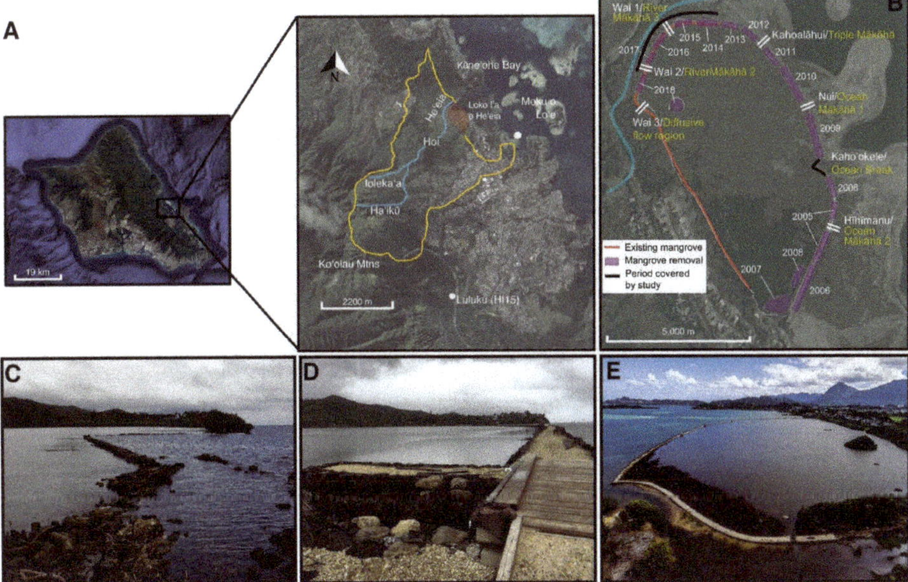

Figure 1. Study site: He'eia ahupua'a and He'eia Fishpond. (**A**) The He'eia *ahupua'a* (social-political governance unit, usually organized along watershed boundaries) is located on the northeast/windward side of O'ahu Island, HI. He'eia ahupua'a is outlined in yellow, He'eia Stream (blue line) originates as Ha'ikū Stream near the ridgeline of the Ko'olau Mountains and converges with Ioleka'a Stream before entering Hoi wetlands and flowing into and past He'eia Fishpond (shaded red) into Kāne'ohe Bay. Weather stations on Moku o Lo'e and Luluku (HI15) rain gauge are indicated by white dots (map

downloaded from USGS National Map Viewer). (**B**) Bio-cultural restoration over the course of this study. Freshwater and marine inputs into He'eia Fishpond via mākāhā (sluice gate) locations and names, yellow: community stewards Paepae o He'eia, white: He'eia Coastal Ocean Observing System; time period of this study (black line) in the context of the chronosequence of mangrove removal and wall rebuilding. From 1965– 2015 a 100 m break in the kuapā (**C**) altered flow patterns in the loko i'a. From 2014–2015, Paepae o He'eia (POH) and community volunteers repaired the kuapā and built a mākāhā (Kaho'okele) (**D**). (**E**) From 2014-2017, POH removed invasive *R. mangle* and repaired kuapā and mākāhā infrastructure on the north quadrant of the loko i'a bordering He'eia Stream (photo courtesy of Samual Kapoi).

As a result of the shift from a constrained to a radically unconstrained system, the fundamental functioning of the loko i'a has changed: the volume became strongly tidally dominated and fish production using customary mariculture techniques could no longer be practiced. A dense mangrove forest around the mouth of He'eia stream expanded into the loko i'a, growing along and eventually obscuring the kuapā and effectively decreasing the amount of water exchange. Sediment loading from He'eia Stream, agriculture and urbanization overwhelmed the original mechanisms by which material was flushed out of the loko i'a [25]. The average loko i'a depth is ~1 m, due to progressive accumulation of terrigenous particulates on the coral benthos, accelerated by a dense mangrove root mass [26]. Increased salinity, organic matter, and turbidity may have facilitated a shift in the biological diversity and composition of the loko i'a away from desirable aquaculture species and toward invasive macroalgae.

Though limited kuapā repair over the last 25 years has enabled conventional net pen aquaculture in the loko i'a, the ecosystem became steadily more eutrophic. In 1988, Mark Brooks leased the property, installing a 0.9 m retaining wall of cement cinder blocks in Ocean Break that reduced the tidal influence and prevented water exchange except at spring tides (Figure 1C). In addition, a previous flood in 1927 deposited a portion of the kuapā into the interior of the loko i'a creating a mangrove stand where introduced cattle egrets *(Bubulcus ibis)* established a rookery (Figure 1B). In 2017 an estimated 2000–3000 cattle egrets dwelled in this mangrove stand. The potential for human and animal health impacts from microbial contamination is a central concern in maintaining an ecologically balanced and productive loko i'a [27,28]. Limited circulation within He'eia exacerbates this issue, particularly given the rich source of guano and nutrients produced by the egret colony.

Since 2001, the Native Hawaiian non-profit organization Paepae o He'eia has sought to foster cultural sustainability and restore and maintain a thriving loko i'a for the local community by linking traditional knowledge and contemporary management practices. As the compromised mākāhā system made regulation of fish migration and recruitment impossible, Paepae o He'eia initially centered their aquaculture activities around high-density cultivation in quarter-acre net pens. From 2006–2009 Paepae o He'eia produced approximately 1.2 metric tons of Pacific threadfin. However two events massive fish mortality events in 2009, prompted a re-evaluation of the use of conventional rearing techniques in He'eia Fishpond. Repairing the kuapā would eliminate the need for net pen aquaculture, enabling fish stock to move throughout the entire loko i'a toward cooler and/or more oxygenated areas in response to future environmental stress. Paepae o He'eia hypothesized that consistent freshwater input and nutrients, via functional mākāhā would increase primary productivity and subsequently increase the biomass of native herbivores in the loko i'a.

1.4. Biocultural Restoration of He'eia Fishpond: 2012–2018

Biocultural restoration from 2012 to 2018 targeted two areas: the gap in the seaward kuapā and the section bordering He'eia Stream (Figure 1B). The restoration phase involving repair of the 56 m kuapā gap (Ocean Break) spanned 2014–2015 and was known as *Pani ka puka* (Shut the door). Kia'i loko used traditional external materials (*pohaku pele*, basalt rock) and a mix of traditional and contemporary internal materials (*ko'a*, coral rubble, and remnant cinder blocks) to coordinate rebuilding of the north and south segments of the broken kuapā to meet in the middle. Rather than rebuild a continuous kuapā spanning the entire seaward side, Paepae o He'eia elected to install a new mākāhā (Kaho'okele) to increase loko i'a circulation, increase oxygenation of the water column, and promote recruitment of marine species (Figure 1C,D). Mākāhā site selection was based on empirical kia'i loko observations of areas with the highest abundance and diversity of marine life (e.g., fish, oysters, macroalgae, sponges).

With the help of over 50,000 community volunteers, Paepae o He'eia has resurrected over 2 km of kuapā along its historical footprint and progressively removed invasive *R. mangle* (Figure 1B–E). Historically, the volume and location of surface water input into He'eia Fishpond from the Hoi wetland and He'eia Stream was confined to water volume flux through mākāhā. After the 1965 Keapuka flood, however, damage to the kuapā and subsequent *R. mangle* growth resulted in an attenuated and diffuse flow of fresh water into the loko i'a. Over this period of this study, Paepae o He'eia commenced kuapā restoration along He'eia Stream and concomitant mangrove removal (Figure 1E) in order to alter the path of surface water into the loko i'a. Kia'i loko posited that restoring the wall and mākāhā would increase the rate of water exchange and flow rate, which might improve fish passage into the estuary. *R. mangle* was initially removed from the remnant kuapā and nearby loko i'a interior by clear-cutting and incineration on site. With the exception of 2014–2015, the mean rate of restoration was 154.84 ± 17.33 m year^{-1}, totaling 619.35 m kuapā (Table S1).

In the present study, we partnered with Paepae o He'eia, kia'i loko of He'eia Fishpond, to assess the impacts of restoration from 2012–2018. We have addressed the following questions: (1) How does kuapā infrastructure repair, including mangrove clearance around the loko i'a periphery, affect circulation dynamics in He'eia Fishpond? (2) How does the potential for increased freshwater and ocean water volume flux alter the overall salinity distribution in the loko i'a? and (3) How do these changes in the physical characteristics of water in the loko i'a alter microbial bioindicators for fecal contamination?

2. Methods and Materials

2.1. Study Site

He'eia Fishpond ($21°26'10.74''$ N, $157°48'28.05''$W) is a 0.356 km^2 embayment located on the windward side of O'ahu Island, Hawai'i (Figure 1A). The loko i'a is completely enclosed by 2.5 km of kuapā and is bordered by Kāne'ohe Bay to the south and east, He'eia Stream to the north, and a remnant irrigation ditch (*auwai*) running longitudinally along its entire west bank. The Ha'ikū Stream near the ridgeline of the Ko'olau Mountains converges with the 'Ioleka'a Stream and becomes He'eia Stream before entering the Hoi wetland. Within the Hoi wetlands, a portion of He'eia Stream is diverted through a network of *auwai*, irrigating taro patches. At the terminus of the watershed, He'eia Stream historically splits, either flowing south in the auwai that parallels He'eia Fishpond or east toward Kāne'ohe Bay. A forest of *R. mangle* occupies the northwest and western periphery of He'eia Fishpond.

Mākāhā are interspersed along the kuapā, connecting the loko i'a to exterior water sources and regulating surface and seawater exchange with the loko i'a (Figure 1B, Table 1). Hereafter, names of mākāhā follow the convention used by Paepae o He'eia in 2018. Designations from previous studies [29,30] are also given. For the past 50 years, mākāhā channels in He'eia Fishpond have had concrete floors with vertical walls composed of basalt and coral rubble with either a semi-permeable barrier fence or grid constructed from wood or plastic (Figure 2). With the exception of Kaho'okele, the floor of the mākāhā are slightly higher than the natural bottom of the loko i'a. All fieldwork was conducted with the permission of Paepae o He'eia and the private landowner, Kamehameha Schools (Joey Char, Land Asset Manager, Kamehameha Schools Community Engagement and Resources Division).

Table 1. Mākāhā names (post-restoration/pre-restoration), latitude and longitude, compass heading, width (m).

Mākāhā	Latitude	Longitude	Heading	Width (m)
Hīhīmanu/Ocean Mākāhā 2	21.4357389	−157.80531	111°/291°	2.00
Kahoʻokele/Ocean Break	21.4372333	−157.80583	80°/260°	3.05
Nui/Ocean Mākaā 1	21.4384222	−157.80675	63°/243°	6.48
Kahoalāhui Kealohi/Triple Mākāhā 1	21.4396667	−157.80993	48°/228°	1.88
Kahoalāhui Koʻa Mano/Triple Mākāhā 2	21.4396667	−157.80993	48°/228°	1.78
Kahoalāhui Kekepa/Triple Mākāhā 3	21.4396667	−157.80993	48°/228°	1.55
Wai 1/River Mākāhā 3	21.4386034	−157.81072	310°/130°	2.18
Wai 2/River Mākāhā 2	21.4379231	−157.80782	290°/110°	1.85
Diffuse flow region/River Mākāhā 1	21.4386583	−157.81077	n/a	n/a

2.2. Water Volume Flux and Volume Change Calculations

To evaluate the current direction (°), water level (m) and water velocity (m s^{-1}) into and out of the loko iʻa, Sontek Argonaut Shallow Water (SW) Profilers (SonTek, San Diego, CA, USA) and battery housings were deployed in each mākāhā for 7 days (Figure 2, Table S1). Each instrument packet was oriented facing into the channel and mounted to 0.7 × 0.7 m metal mooring with ~25 kg weights and placed at the bottom of each mākāhā channel. Measurements were recorded every 20 s with an averaging interval of 10 s. The blanking distance was set to the minimal amount of 0.07 m, as the mean water column was <0.50 m. Over this period, one full neap and spring tide were measured. Water volume flux data and water velocity measurements (m s^{-1}) acquired from the Sontek Argonaut SW Profiler were used to generate rating curves for each mākāhā at (spring flood tide, SF; spring ebb tide, SE; neap flood tide, NF; and neap ebb tide, NE) using the following equation:

$$\phi = wdv \tag{1}$$

where ϕ is the water volume flux, w is the respective mākāhā width (m), d is the water level vector (m) changing over time with the tide, and v is the water velocity (m s^{-1}) through the mākāhā channel [29,30]. Rating curves were fitted using a poly-fit function with a best-fit line and 95% confidence intervals in Matlab (The MathWorks Inc., Natick, MA, USA). To account for bidirectional water flow in the mākāhā due to tidal forcing, water volume flux was determined for an entire tidal cycle at the following tidal stages: SF, SE, NF, and NE. The cycle with the largest tidal amplitude was selected for spring tide, while the cycle with the lowest tidal amplitude was selected for neap tide. The data set was split into flood (from pressure minimum to pressure maximum) and ebb tide (from pressure maximum to pressure minimum) based on tidal stage.

Based on the water volume flux, mean and maximum flow through each mākāhā were calculated for four tidal cycles (SF, SE, NF, NE). Peak water volume flux occurs mid-way between slack tides, thus the water level to water volume flux relationship, the rating curve, typically resembles a "C" curve or vertical sine function. To account for varying tidal cycle length caused by mixed semidiurnal tides in Kāneʻohe Bay, individual mākāhā flow rates were normalized by calculating the total volume of water (m^3) moving through a mākāhā channel at a given tidal cycle and the hourly water volume flux rate. Here, water volume flux values for Kahoalāhui/Triple Mākāhā were calculated by tripling the flow measurements at the northernmost mākāhā channel (Kealohi).

Precipitation, tidal state, wind direction, and wind speed were used as criteria for selecting pre- and post-restoration dates for comparison (Table S3). Daily (cm 24 h^{-1}) and cumulative precipitation over 4 days (cm 96 h^{-1}) were obtained from the NOAA Luluku (HI15) rain gauge station [31]. Mean stream streamflow (mean m^3 s^{-1} 24 h^{-1}) was calculated using data from US Geological Survey discharge station (Haʻikū Station #16275000) obtained from [32]. Wind direction and magnitude was determined from automatic weather station *Moku o Loʻe* (21.4339° N, 157.7881° W), 1.5 km from Heʻeia

Fishpond [33]. A sea level gauge with a water temperature probe, located ~ 10 m offshore of the weather station at a depth of ~ 1 m, was used for tidal data [33].

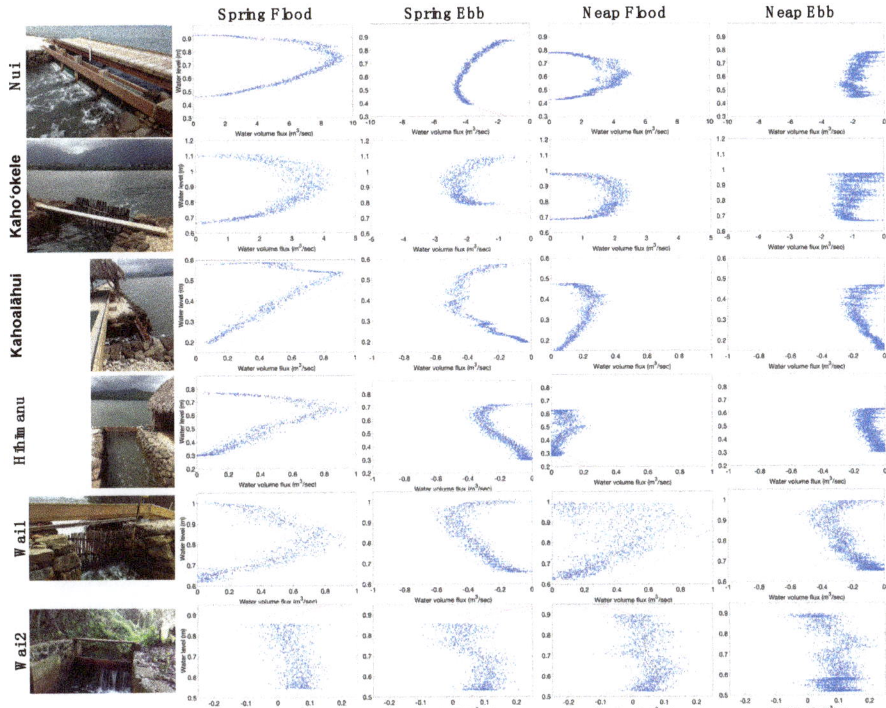

Figure 2. Post-restoration rating curves at each mākāhā over various tidal stages. Water volume passing through mākāhā (sluice gates) as the tide varies (height above the sensor), e.g., during spring flood, water height value is lowest and increaes as tide rises, while during spring ebb water height values are highest and decrease as tidal heigh drops; similarly, for neap flood and ebb. Each point represents the height taken in 20-second intervals over period between successive high and low tides (~6 h). Water volume flux ($m^3\ s^{-1}$) relative to the water level (m) is shown for all 6 mākāhā, 'best fit line' in red, 95% confidence intervals, dashed pink line. Positive values indicate water volume flux into the loko i'a and negative values indicate water volume flux out of the loko i'a.

Loko i'a volume was calculated using 728 bathymetric depth measurements taken in 2007 normalized to mean low low water from a reference HOBO® water level logger (Onset, Bourne, MA, USA) deployed at an interior site (21.43466° N, W 157.80699° W) that recorded tidal fluctuations during bathymetry mapping [26,34]. In 2018, we redeployed a HOBO® water level logger at the same location to recollect reference water level data over a 10-day period. The reference pressure data was corrected for atmospheric pressure fluctuations using a second HOBO logger situated on land to record atmospheric pressure fluctuations reference to adjust for differences in tidal amplitude between pre- and post-restoration.

To calculate post-restoration loko i'a volume, the difference in reference tidal state from pre-restoration (2007) and post-restoration (2018) was applied to the bathymetry dataset at SF, SE, NF, NE tidal states with Station *Moku o Lo'e* as a reference. A rectangular grid with ~1 m spacing and a natural neighbor interpolation was adopted to estimate depths in between measured bathymetry points in Matlab. For each tidal state, a trapezoidal rule was used with no smoothing applied. The small mangrove island located in the northwest quadrant of the loko iʻa was excluded from our calculations. We assume that there is no change in bathymetry over the course of the study.

To derive minimum residence time in Heʻeia Fishpond, the amount of water exchanged during ebb flood transition was calculated for neap and spring tide using the following equations [30]:

$$\tau HFS = \frac{\text{Heeia Fishpond Volume Exchanged (spring high tide − spring low tide)}}{\text{Heeia Fishpond Volume (spring high tide)}} \tag{2}$$

$$\tau HFN = \frac{\text{Heeia Fishpond Volume Exchanged (neap high tide − neap low tide)}}{\text{Heeia Fishpond Volume (neap high tide)}} \tag{3}$$

where τ_{HFS} is minimum residence time during spring tide and τ_{HFN} is minimum residence time during neap tide. To determine residence time, the following assumptions were made: loko iʻa water column is mixed uniformly, all flood and ebb tides are 6 h long, and mākāhā are the only source of water exchange with the following equation:

$$\varphi^x = 0.01 \tag{4}$$

where φ^x is the percentage of water remaining after 1 flushing cycle (12 h) and x is the residence time in flushing cycles to mix the initial water to a 1% dilution.

2.3. Water Quality Sampling Regime

This study utilized on-going efforts by Nā Kilo Honua o Heʻeia (http://nakilohonuaoheeia.org), a Heʻeia coastal ocean observing research collective at the University of Hawaiʻi at Mānoa that has carried out monthly sampling at Heʻeia Fishpond since 2007 [29]. To minimize the variability of physical and chemical characteristics of the loko iʻa due to tidal exchange, all samples were collected during neap tide over a period of 3–4 h The pre-restoration sampling grid was composed of 10 stations within the loko iʻa, P1−P10, whereas the post-restoration sampling grid was composed of 11 stations within the loko iʻa, L01−L11, and one at each of the mākāhā, M01−M06 (Table S3) Pre-restoration sampling dates in 2014 and post-restoration dates from 2017 were selected to minimize variation in precipitation and stream discharge (Table S4). Reference endmembers for oceanic input were taken outside the kuapā at Kahoʻokele/Ocean Break, E01, whereas endmembers for surface freshwater were collected in Heʻeia Stream between the Hoi wetland and Heʻeia Fishpond outside the kuapā, E02. To minimize the disturbance of the water column and benthos prior to measurements, stations were approached against prevailing currents and winds. Salinity was measured using a YSI Professional Plus (ProPlus) multiparameter sonde (YSI Xylem Brand, Yellow Springs, OH). At each station, a measurement was taken ~ 5–10 cm below the water surface ("surface") and 5–10 cm above the benthos ("bottom") by allowing the instrument reading to stabilize for 2–3 minutes before recording values.

Eleven stations were selected for discrete sampling for microbes: Kahoʻokele/Ocean Break, Wai 1, and 9 stations in the loko iʻa interior. Pre-restoration (P01–P10, Ocean Break) and post-restoration (L01–L03, L06–L11, Kahoʻokele, Wai 2) locations differed slightly (Table S4, Figure 5A). At each station, 1L polycarbonate bottles were acid washed and rinsed with ambient surface water three times, before immersion at the surface to fill the bottle completely. Samples were stored at 4 °C and processed within 2 h of collection. Seawater was filtered through a 47 mm diameter, 0.45 μm filter (MCE, Millipore, Sigma, Burlington, MA) and stored at −80 °C prior to DNA extraction.

2.4. Microbial Source Tracking

Total genomic DNA (gDNA) was extracted from filters using the PowerWater DNA Extraction kit (QIAGEN, Germantown, MD, USA) following the manufacturer's instructions. Quantitative PCR (qPCR) was used to determine the abundance of bacterial 16S rRNA genes from mammalian fecal indicator bacteria *Enterococcus* using assay Entero1a [35–37] and Bacteroidales using assay GenBac3 [38–40]. Quantification was performed with the KAPA PROBE FORCE qPCR system (Wilmington, MA, USA) using KAPA PROBE FORCE qPCR Master Mix (20 μL reactions), 400 nM specific Taqman primers (Table 2) and template gDNA diluted 1:5. Standards were run in triplicate using an 8-point, 5-fold serial dilution. Cycling parameters for all assays were: 95 °C for 2 min, 45 cycles of 95 °C for 15 s and annealing/extension at 60 °C for 30 s Ct values were converted to concentrations per 100 mL using the manufacturer's software. The standards used for the Entero1a and GenBac3 assays were genomic DNA extracted from *Enterococcus faecalis* strain V583 (ATCC® 700802D-5™) and *Bacteroides thetaiotaomicron* strain VPI 5482 (ATCC® 29148™), respectively.

Primers previously shown to detect avian fecal contamination in water [41] were tested on *B. ibis* fecal DNA (Table 2). Briefly, fecal material was collected from birds present on the small mangrove island on the loko i'a interior. Total genomic DNA was extracted from avian feces using the DNeasy PowerSoil Kit (QIAGEN, Germantown, MD) following the manufacturer's instructions. qPCR using GFC primers targeting the 16S rRNA gene from *Catellicoccus marimammalium* used the KAPA SYBR FAST qPCR system (20 µL reactions), 400 nM primers, and gDNA diluted 1:5. Cycling parameters were as follows: 95 °C for 3 min for enzyme activation, followed by 40 cycles of 95 °C for 3 s and annealing/extension at 60 °C for 20 s Ct values were calculated as previously described with uncultured *Catellicoccus* sp. 16S rRNA gene, partial sequence (Genbank accession number JN084062) used as a standard.

Table 2. The 16S rDNA oligos used in this study.

Target	Primer	Sequence	References
Enteroccocus	Entero1af	AGAAATTCCAAACGAACTTG	[35–37]
	Entero1ar	CAGTGCTCTACCTCCATCATT	[35–37]
	Entero1ap	6-FAM™/TGGTTCTCT/ZEN™/CCGAAATAGCTTTAGGGCTA/IB®FQ/	[35–37]
Bacteroidales	GenBac3f	GGGGTTCTGAGAGGAAGGT	[38–40]
	GenBac3r	CCGTCATCCTTCACGCTACT	[38–40]
	GenBac3p	6-FAM™/CAATATTCC/ZEN™/TCACTGCTGCCTCCCGTA/IB®FQ/	[38–40]
Catellicoccus marimammalium	GFCf	CCC TTG TCG TTA GTT GCC ATC ATT C	[41]
	GFCr	GCC CTC GCG AGT TCG CTG C	[41]

2.5. Statistics

Statistical significance for pre- and post-restoration events was determined with a pairwise Welch's t-test to account for differences in variance. Mean baseline events pre-restoration and mean baseline events post-restoration for salinity and log-transformed numbers of microbial biomarker abundance were compared with the t-test for statistical significance in R (R Foundation for Statistical Computing) with the p-value for statistical significance set to $p < 0.05$. In addition, correlation of GFC/GenBac3/Entero1a distribution with salinity, date, and location was tested using a generalized additive mixed model (GAMM) in R (R Foundation for Statistical Computing). Mean baseline salinity and log-transformed numbers of microbial biomarker abundance pre- and post-restoration was plotted with a contour plot function in Matlab (The MathWorks Inc., Natick, MA, USA).

3. Results

3.1. Restoration from 2014–2018 Shifted Relative Water Volume Flux Contributions of Each mākāhā

3.1.1. Characterizing mākāhā Water Volume Flux Post-Restoration (2018)

Four mākāhā along the eastern kuapā (Hīhīmanu, Kaho'okele, Nui, Kahoalāhui, Figure 2) were assumed to have bi-directional flow mediated by the semi-diurnal tidal cycle in Kāne'ohe Bay. Three mākāhā in the north and northwest sectors of He'eia Fishpond were documented since the early 1900s to provide conduits for surface water inputs into the loko i'a (Figure 1B). Wai 1 and Wai 2 were restored over the course of this study. Wai 1 is located closest to the mouth of He'eia Stream and allows the bidirectional exchange of fresh and oceanic water [30], whereas Wai 2, located 100 m upstream, has a unidirectional flow of surface water into the loko iʻa. The most upstream mākāhā was destroyed during flood events in 1927 and 1965 and has not yet been restored and, measurements with current meters in this area were not possible.

Precipitation and stream discharge were used as criteria to select water volume flux measurements sampling dates with similar meteorological conditions pre- and post-restoration (Table S3). While daily rainfall ranged from 0.05 cm to 1.32 cm in 2012 (pre-restoration) (mean 0.76 $\pm$ 0.6 s.d. cm), it ranged slightly higher from 0 cm–2.29 cm (mean 1.23 $\pm$ 0.87 s.d. cm) in 2018 (post-restoration). Similarly, Haʻikū Stream discharge ranged from 0.04 m^3 s^{-1}–0.07 m^3 s^{-1} (mean 0.06 $\pm$ 0.013 s.d. m^3 s^{-1}) in 2012 (pre-restoration), and from 0.06 m^3 s^{-1}–0.11 m^3 s^{-1} (mean 0.085 $\pm$ 0.03 s.d. m^3 s^{-1}) in 2018 (post-restoration). Wind direction ranged from E to NE (average wind direction ~50°) with magnitude ranging from 10 to 13 knots pre-restoration and from E to NE (average wind direction ~60°) with magnitudes of 3–13 knots post-restoration.

As each mākāhā was constructed at varying heights from the loko iʻa substratum, flood tide onset and end were defined as low slack water (LSW, water volume flux = 0 m^3 s^{-1}) tide stage and high slack water (HSW, water volume flux = 0 m^3 s^{-1}), respectively. Conversely, ebb tide onset and end were defined as HSW and LSW, respectively. LSW levels range from 0.2 m at Kahoalāhui to 0.65 m at Kahoʻokele and Wai 1. HSW levels range from ~0.5 m at Kahoalāhui to 1.1 m at Kahoʻokele. The consistently high water level at Wai 1 likely due to continuous baseline stream flow into the loko iʻa. We note that Wai 2 exhibits an atypical rating curve as a wooden board in the mākāhā restricts discharge into the loko iʻa only when water levels are higher than the board (Figure 2, Wai 2).

Mean and peak water volume flux were highest during flood tides at all mākāhā. The fastest mean water volume flux (4.18 m^3 s^{-1} at SF and 2.26 m^3 s^{-1} at NF) and peak water volume flux (9.70 m^3 s^{-1} at SF and 5.41 m^3 s^{-1} at NF) were recorded at mākāhā Nui (Table 3). In addition, flood tidal cycle duration was shorter than ebb at all mākāhā at both Spring and Neap, mean tidal duration was 5.23 $\pm$ 1.20 s.d. h and 8.00 $\pm$ 0.84 s.d. h for SF and NF, respectively, whereas mean tidal duration was 6.09 $\pm$ 0.73 s.d. h and 15.67 $\pm$ 1.38 s.d. h for SE and NE, respectively. Taken together, the shorter lag time at high water vs. low water, longer-duration dropping tides and stronger flood than ebb currents suggest that Heʻeia Fishpond is a flood-dominant system.

Table 3. Water volume flux (WVF) dynamics in Heʻeia Fishpond post-restoration (2018).

	Mean WVF (m^3 s^{-1})	Peak WVF (m^3 s^{-1})	Tidal Cycle Length (h)	Cum. Flux per Tidal Cycle (m^3)	WVF Rate (m^3 h^{-1})	Volume Exchanged per Tidal Cycle (m^3)	Relative WVF
Spring Flood				191660	31778	191660	100.00%
Wai 2	0.05	0.16	4.43	840	190	840	0.44%
Wai1	0.40	0.93	4.55	7140	1569	7140	3.37%
Kahoalāhui	1.47	2.76	4.36	24420	5601	24420	12.74%
Nui	4.18	9.70	6.29	97800	15548	97800	51.03%
Kahoʻokele	2.02	4.69	7.29	54380	7460	54380	28.37%
Hīhīmanu	0.39	0.95	5.02	7080	1410	7080	3.69%
Spring Ebb				−174880	−30851	−174880	100.00%
Wai 2	0.07	−0.09	5.50	1560	284	1560	−0.89%
Wai1	−0.32	−0.63	6.32	−7600	−1203	−7600	4.35%
Kahoalāhui	−0.87	−1.86	6.31	−20220	−3204	−20220	11.56%
Nui	−3.60	−4.86	5.53	−76320	−13801	−76320	43.64%
Kahoʻokele	−1.10	−3.12	5.50	−67520	−12276	−67520	38.61%
Hīhīmanu	−0.17	−0.43	7.35	−4780	−650	−4780	2.73%
Neap Flood				141384	16717	141384	100.00%
Wai 2	0.05	0.20	7.41	1300	175	1300	0.92%
Wai1	0.32	0.98	8.29	9720	1172	9720	6.87%
Kahoalāhui	0.51	1.08	7.31	13620	1863	13620	9.63%
Nui	2.26	5.41	9.46	78744	8324	78744	55.70%
Kahoʻokele	1.35	2.52	7.30	36440	4992	36440	25.77%
Hīhīmanu	0.05	0.24	8.20	1560	190	1560	1.10%
Neap Ebb				−159938	−10584	−159938	100.00%
Wai 2	0.88	−0.09	17.46	5640	323	5640	−3.53%
Wai1	−0.17	−0.57	15.50	−9880	−637	−9880	6.18%

Table 3. *Cont.*

	Mean WVF ($m^3\ s^{-1}$)	Peak WVF ($m^3\ s^{-1}$)	Tidal Cycle Length (h)	Cum. Flux per Tidal Cycle (m^3)	WVF Rate ($m^3\ h^{-1}$)	Volume Exchanged per Tidal Cycle (m^3)	Relative WVF
Kahoalāhui	−0.30	−0.9	15.50	−17100	−1103	−17100	10.69%
Nui	−1.60	−3.19	14.09	−81298	−5770	−81298	50.83%
Kahoʻokele	−0.86	−1.80	17.10	−53280	−3116	−53280	33.31%
Hīhīmanu	−0.08	−0.25	14.34	−4020	−280	−4020	2.51%

3.1.2. Changes in Relative Water Volume Flux Post-Restoration

We evaluated the relative contribution of each mākāhā to loko iʻa water exchange during SF, SE, NF, and NE in order to gain insight into how restoration altered circulation in Heʻeia Fishpond. Prior to restoration, Ocean Break, the 0.9 m elbow wall bridging the 56 m gap in the eastern kuapā was lower than the adjoining sections of wall, restricting water exchange to high tidal stages, when the water level exceeded the height of Ocean Break. Restoration resulted in a significant shift in water exchange in the seaward kuapā. The spatial pattern of flushing in Heʻeia Fishpond remains dominated by the mākāhā in the northeast quadrant of the loko iʻa for all tidal stages. Nui, Kahoʻokele, and Kahoalāhui together account for 92% of the water exchanged at spring flood, 94% at spring ebb, 91% at neap flood and 95% at neap ebb tide whereas the southern and eastern edges of the loko iʻa experience relatively low flushing.

When comparing site–specific water volume flux rates pre-restoration (2012) to post-restoration (2018), it becomes evident that the relative magnitude of water volume flux specific to each mākāhā changed due to restoration practices: The total amount of water volume exchanged in a complete tidal cycle decreased from 241,413 m^3 pre-restoration to 194,700 m^3 post-restoration for flood tide and decreased from −241,685 m^3 pre-restoration to −173,080 m^3 post-restoration for ebb tide (Table 4). Pre-restoration, Ocean Break facilitated the largest amount of volume exchange contributing approximately ~80% to total water exchange at both flood and ebb tidal cycles (81.94% for flood, 79.76% for ebb) with mean water velocities of 11.53 $m^3\ s^{-1}$ and −13.55 $m^3\ s^{-1}$ [42]. Pre-restoration, Nui contributed the second largest amount of volume exchange with 12.88% for flood and 11.12% for ebb tide and mean velocities of 1.75 $m^3\ s^{-1}$ and −0.5 $m^3\ s^{-1}$ [42]. While contributing only 10% to water exchange pre-restoration, post-restoration Nui is presently the site with largest water volume exchange. Post-restoration, Nui facilitated about half of the volume exchanged (50.24% at flood tide, 44.1% at ebb tide, Figure 3) with much higher mean water volume flux of 4.18 $m^3\ s^{-1}$ and −3.6 $m^3\ s^{-1}$ (Table 4) than pre-restoration. In contrast to pre-restoration, Kahoʻokele now accounts for the second largest volume exchanged (27.93% and 39.01% for flood and ebb tide respectively, Figure 3) with lower mean water volume flux of 2.02 $m^3\ s^{-1}$ and −1.1 $m^3\ s^{-1}$ compared to pre-restoration. Kahoalāhui is composed of three individual mākāhā post-restoration and together they account for the third largest water volume—roughly 10% of contribution to total water volume flux. The relative contribution in the magnitude of Kahoalāhui increased about six-fold for flood tide and five-fold for ebb tide from pre-restoration to post-restoration (from 1.71% to 12.54% for flood tide and 2.41% to 11.68% for the ebb tide, Table 4). Hīhīmanu did not experience significant changes due to restoration: While accounting for 1.69% at flood and 2.03% for ebb pre restoration, it now accounts for 3.61% and 2.76% at flood and ebb, respectively (Table 4). Mean water volume flux ranged from −0.12 $m^3\ s^{-1}$ to 0.28 $m^3\ s^{-1}$ pre-restoration and is now −0.17 $m^3\ s^{-1}$ to 0.39 $m^3\ s^{-1}$.

Table 4. Change in water volume flux (WVF) rates through mākāhā pre-restoration (2012) and post-restoration (2018).

	Flood Tide				Ebb Tide			
	Pre-Restoration		**Post-Restoration**		**Pre-Restoration**		**Post-Restoration**	
Mākāhā	**Volume Exchange per Tidal Cycle (m³)**	**Relative WVF**	**Volume Exchange per Tidal Cycle (m³)**	**Relative WVF**	**Volume Exchange per Tidal Cycle (m³)**	**Relative WVF**	**Volume Exchanged per Tidal Cycle (m³)**	**Relative WVF**
Wai 2	2057	0.85%	1300	0.67%	−5515	2.28%	5640	−3.25%
Wai 1	2249	0.93%	9720	5.10%	−5791	2.40%	−9880	5.70%
Kahoalāhui	4106	1.71%	24420	12.54%	−5802	2.41%	−20220	11.68%
Nui	31101	12.88%	97800	50.24%	−26886	11.12%	−76320	44.10%
Kahoʻokele/OB	197820	81.94%	54380	27.93%	−192780	79.76%	−67520	39.01%
Hīhīmanu	4081	1.69%	7080	3.61%	−4912	2.03%	−4780	2.76%
Mākāhā Total	241,413	100.00%	194,700	100.00%	−241,685	100.00%	−173,080	100.00%

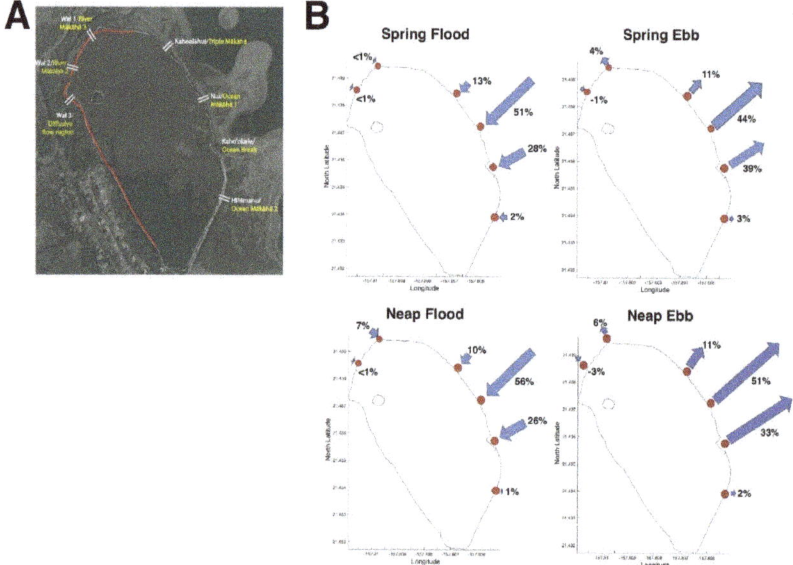

Figure 3. Relative water volume flux post-restoration dominated by Mākāhā Nui, and Mākāhā Kahoʻokele. (**A**) Mākāhā reference map. Pre-restoration names are yellow, post-restoration names are white. (**B**) Relative water flows through each mākāhā during spring flood tide; spring ebb tide; neap flood tide; neap ebb tide. Arrow lengths are visual representations of the relative magnitude of water volume flux at each mākāhā, normalized to the total water volume flux for each respective cycle. Mākāhā location, filled red circles.

In terms of overall volume exchange, the river mākāhā continue to play minor roles in water exchange. In 2018, water volume flux rates measured at Wai 1 were similar to pre-restoration with a relative water volume flux magnitude of 3–7% and low mean flow rates (Figure 3, Table 4). Water passing through Wai 1 increased from 0.93% pre-restoration to 5.1% post-restoration for flood tide, and 2.4% pre-restoration to 5.7% post-restoration for ebb tide. Water volume flux increased from 0.09 m³ s⁻¹ and 0.1 m³ s⁻¹ pre-restoration to 0.4 m³ s⁻¹ and 0.32 m³ s⁻¹ post-restoration. Pre-restoration Wai 2 accounted for 0.85% of water exchange during flood tide and accounts for a slightly decreased water exchange of 0.67% post-restoration for flood tide. For ebb tide, the water exchange reversed from 2.28% pre-restoration to −3.25% post-restoration. Wai 2 displayed unidirectional flow into the loko i'a, regardless of tidal state with solely positive flow velocities and accounting for the lowest water volume flux measured.

3.2. Decrease in loko i'a Volume and Residence Time Post-Restoration

The majority of the loko i'a has relatively uniform and shallow bathymetry of ~0.9 m with the deeper portions around the mangrove island and Ocean Break [26]. Prior to restoration, water exchange along the eastern kuapā only occurred when the water depth exceeded the height of the elbow wall at Ocean Break. Pre-restoration, ~90% of loko i'a water exchange occurred in the northeast corner of the loko i'a via Ocean Break (~80%) and Nui (~10%), suggesting that the eastern half of the loko i'a was better mixed and less stratified than the western side [30,34]. Water volume exchange before restoration was also found to be largely tidally driven, with the greatest volume exchange at mid-tide: ~77% during spring tide and ~42% during neap tide.

Given changes in water volume flux in He'eia Fishpond due to restoration, we determined post-restoration loko i'a volume and residence time for SF, SE, NF, NE. He'eia Fishpond is deepest during SF tide (Figure 4A), averaging 0.89 ± 0.12 m with a minimum water depth of 0.63 m in the center of the loko i'a and a maximum water depth of 1.46 m around the mangrove island in the northwestern corner of the loko i'a. During SF, the maximal volume of the loko i'a is 264,730 m^3 (Figure 4B). The minimum water volume occurs during SE tide when the loko i'a is 48,060 m^3 or 20% of the SF volume (Figure 4B). The mean loko i'a depth at spring ebb tide is 0.17 m ± 0.12 m and ranges from 0 m in the center to 0.74 m around the mangrove island in the northwestern corner of the loko i'a. The NF tidal volume is 149,550 m^3, 56% of the SF tidal volume, with a mean depth of 0.50 m ± 0.12, ranging from 0.25–1.08 m. NE depth ranges from 0–0.79 m, averaging 0.22 ± 0.12 m. NE tidal volume is 63,160 m^3. Restoration regimes resulted in a considerable change of loko i'a volume from pre- (2007) to post-restoration (2018): SE tide loko i'a volume decreased 16,010 m^3, SF volume decreased 17,990 m^3, NE volume decreased 14,890 m^3 and NF volume increased 15,660 m^3 (Figure 4B). Thus, as a result of removing the elbow wall and installing a sixth mākāhā (Kaho'okele), He'eia Fishpond is shallower and has a lower volume at all tidal states except NF.

We calculated that post-restoration, approximately 82% of the loko i'a water is exchanged during the ebb–flood transition at spring tide. During the neap tide ebb-flood transition, 58% of the loko i'a water is exchanged. To be consistent with previous work by Young [30], we defined one flushing cycle as the time that it takes to flush out 82% of loko i'a water during spring ebb tide and to replenish that water again with new Kāne'ohe Bay water during spring flood tide or 12 h Based on the assumption that the incoming water would mix uniformly with the water remaining in the loko i'a during the first flushing cycle (18%), about 3 flushing cycles are required to mix the initial 18% of water to a <1% dilution. Therefore, the post-restoration minimum residence time of He'eia Fishpond is ~ 32 h or under 3 flushing cycles, and occurs during spring tide when water exchange is maximal. In contrast, when water exchange is minimal (e.g., neap tides), the maximum residence time is 64 h More than 5 flushing cycles or 64 h are required to mix the 42% of water retained down to <1% dilution. Water exchange during ebb flood transition experienced a 4.51% increase (from 77.34% pre-restoration to 81.85% post-restoration, Table 4) at spring tide. During neap tide water exchange increased 16.06% (from 41.71% pre-restoration to 57.77% post-restoration, Table 4). As a result, minimum water residence time decreased from 38 h at spring tide pre-restoration to 32 h (~1.5 days) at spring tide post-restoration and maximal residence time during neap tides decreased from 102 h (~8.5 days) at spring tide pre-restoration to 64 h (~5.5 days) at spring tide post-restoration.

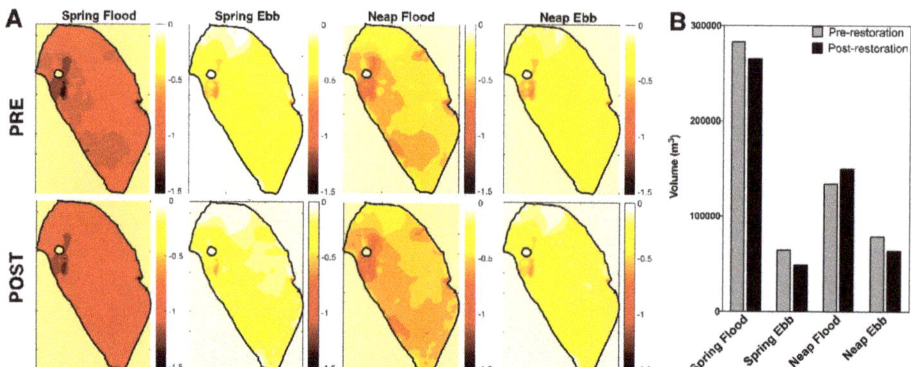

Figure 4. Comparison of He'eia Fishpond depth and volume pre- vs. post-restoration over various tidal stages. (**A**) Loko i'a depth (m) for spring flood, spring ebb, neap flood, neap ebb pre-restoration (top row) vs. post-restoration (bottom row). (**B**) Loko i'a volume (m^3) for each tidal stage pre-restoration (grey) vs. post-restoration (black).

3.3. Spatial Salinity Distribution Significantly Altered due to Restoration

The water column geochemistry of He'eia Fishpond is influenced by the mixing of distinct water masses: surface water from He'eia Stream, whose discharge depends on precipitation; submarine groundwater discharge, composed of a mixture of fresh water from an underground aquifer and recirculated seawater [43]; and seawater from Kāne'ohe Bay that fluctuates with tidal pumping. Built at the interface of He'eia Stream and Kāne'ohe Bay, He'eia Fishpond exhibits a typical vertical salinity gradient—a less dense, freshwater lens atop a more dense, saltier water mass—although mixing of these water masses does occur with increased river flow, winds, and tides. A major motivation for the biocultural restoration of He'eia Fishpond was to increase the freshwater influence in the loko i'a. Kia'i loko hypothesized that brackish conditions would drive primary production of diatoms—a major food source for juvenile mullet, which is a target species. Surface and bottom salinities were measured using a handheld YSI at several locations in He'eia Fishpond (Figure 5A). We selected two pre-restoration sampling events from 2014 and three post-restoration sampling events from 2017 with similar meteorological conditions (Tables S4 and S5). Salinity measurements from pre- and post-restoration work was analyzed as an indicator of loko i'a circulation, mixing, and stratification.

Surface salinity distribution pre- and post-restoration display a strong spatial gradient (Figure 5B, left panels). The highest salinities in both cases were measured along the ocean-ward kuapā near Nui and the Ocean Break/Kaho'okele (station P10), while the lowest salinity was measured along He'eia Stream near Wai 2 (station P3, L07). However, mean pre-restoration salinity was significantly higher than post-restoration salinity, 27.4 ± 4.86 ppt and 20.5 ± 10.41 ppt, respectively (*p*-value < 0.01). With similar meteorological conditions, these data indicate a weaker freshwater influence and stronger salinity gradient pre-restoration. Before restoration, the freshwater wedge did not extend past the western edge of the mangrove island, where salinities ranged from 20–25 ppt (stations P2, P4, P5) and further west, salinities rose to 25–30 ppt (stations P1, P6, P7, P8, P9). Post-restoration however, salinity ranged from 0.10–32.59 ppt with the freshwater wedge from the river extended beyond the mangrove island, which ranged from 15–20 ppt (stations L06, L08, L09), with salinities further west rising to above 20 ppt (station L01 and L05) and 25–30 ppt (stations L02, L03, L04, L11, M03). The presence of strong spatial gradient throughout the restoration process suggests that freshwater from He'eia Stream is more prevalent along the northwestern side of the loko i'a, whereas tidal pumping from Kāne'ohe Bay dominates the southeastern side of the loko i'a.

As expected, bottom waters of the loko i'a had a higher salinity than the surface, however, post-restoration salinity exhibited limited gradient structure post-restoration, whereas the loko i'a bottom pre-restoration was entirely homogeneously mixed with no detectable freshwater influence (Figure 5B, right panels). Mean bottom salinities were significantly higher pre-restoration (31.99 ± 1.82 ppt) as compared to post-restoration (25.17 ± 8.12 ppt), *p*-value < 0.1. Post-restoration, the influence of freshwater from He'eia Stream became more evident, with the majority of the loko i'a salinity ranging from 20–25 ppt (Figure 5B, lower right panel). Similar to the surface salinity spatial distribution, highest measurements were taken near the Kaho'okele and Nui and the lowest measured bottom salinities were taken at Wai 2.

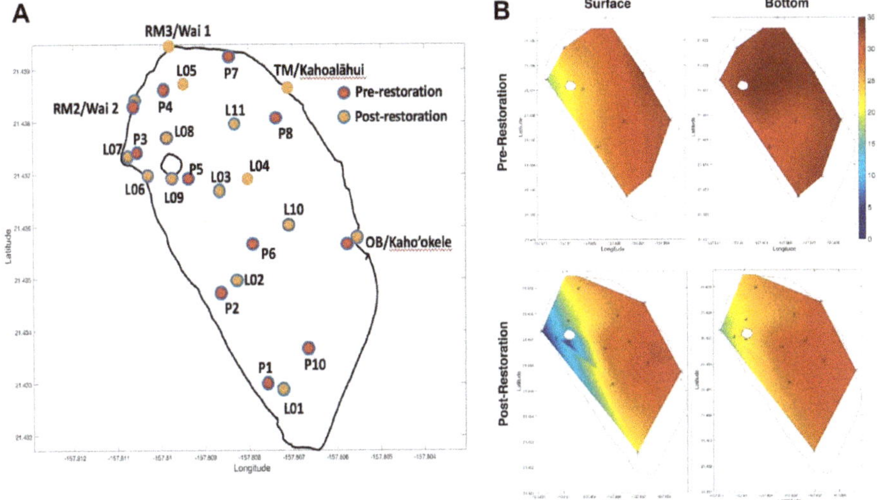

Figure 5. Average salinity of He'eia Fishpond surface and bottom waters decreased due to restoration. (**A**) Discrete sampling sites for microbial indicator species in the water column (blue circles) and/or salinity (pre-restoration, red fill, and post-restoration, orange fill). (**B**) Heat map of salinity as a proxy for the relative proportion of freshwater and ocean water in the loko i'a. Gradient of higher salinity in the eastern sectors of the loko i'a bordering Kāne'ohe Bay and lowest salinity near the diffusive flow region closest to He'eia Stream and the unrestored portion of kuapā is typical of an estuarine saltwater wedge.

3.4. Restoration-Driven Changes to Circulation Altered Microbial Biomarker Spatial Distribution

To understand the consequences of Paepae o He'eia's restoration regime on biological–physical interactions in the loko i'a, we quantified the abundance of microbial biomarkers that have been used previously to track fecal contamination within bodies of water. We focused on 3 specific bacterial groups: *Enterococcus* and Bacteroidales, indicators of contamination from mammals and *C. marimammalium*, an indicator for contamination from avian sources, to investigate how increasing freshwater inputs into the loko i'a potentially affect the biogeography of pathogens.

Discrete samples were collected from a network of stations across the loko i'a along a transect from Wai 2 to Kaho'okele to capture the salinity gradient observed previously (Figure 5A, L03, L06, L07, L09, L10). In addition, we sampled at a higher resolution around the mangrove island on the interior of the loko i'a in order to consider the influence of the large *B. ibis* rookery housed in the *R. mangle* stand. Contrary to expectations, amplification of the 16S rDNA genes from the family Bacteroidales (GenBac3) and the genus *Enterococcus* (Entero 1a) from samples pre- and post-restoration showed no significant difference when averaged across all stations (Figure 6A). We hypothesized that grouping together data may have masked changes in biomarker spatial distribution that occurred due to restoration. We mapped the mean concentration (16S copies 100 mL^{-1}) onto the stations and used a rectangular grid with ~1 m spacing to determine whether the biogeography of *Enterococcus* and Bacteroidales changed from 2014 to 2017 (Figure 6B,C respectively). We found that prior to restoration, the mean concentration of *Bacteroidales* was higher than 10^4 copies per 100 mL across the entire western side of the loko iʻa. In contrast, post-restoration, Bacteroidales concentrations higher than 10^4 copies per 100 mL were restricted to a geographically smaller area of the loko iʻa, adjacent to Wai 2 and the diffuse flow region and lower in the center of the loko iʻa (Fig 6B and 6C, top row). Indeed, when grouped by salinity, freshwater stations showed a statistically significant decrease in Bacteroidales concentration post-restoration (Figure 6D, top row, white). General additive mixed model (GAMM) analysis confirmed that concentration of Bacteroidales negatively correlates with salinity (Figure 6E, top row, Table 5), with the highest concentrations found at stations with the lowest salinity.

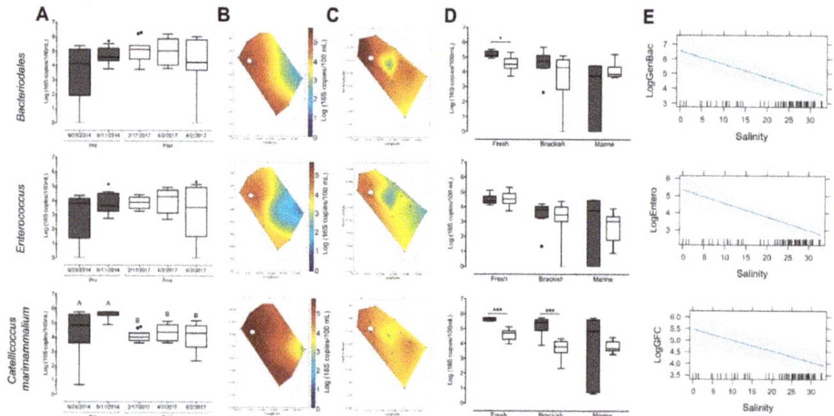

Figure 6. Spatial distribution and mean concentration of *Bacteroidales*, *Enterococcus* and *C. marimammalium* pre- and vs. post-restoration. (**A**) Tukey box–plot diagrams showing concentration in log (16S copies/100 mL) of *Bacteroidales* (GenBac3), *Enterococcus* (Entero1a), and *C. marimammalium* (GFC) before (grey) and after (white) kuapā restoration from all sampling sites. Outliers and 95% confidence intervals are indicated. Heat maps of the averaged abundance of pre-restoration (**B**) and post-restoration (**C**) *Bacteroidales* (GenBac3), *Enterococcus* (Entero1a), and *C. marimammalium* (GFC). Tukey box plot diagrams of Bacteroidales (GenBac3), *Enterococcus* (Entero1a), and *C. marimammalium* (GFC) abundance binned by salinity (freshwater, brackish and marine) of sites pre- and post-restoration, 95% confidence intervals and outliers are indicated. (**D**) Tukey box-plot diagrams showing concentration, log (16S copies/100 mL) of *Bacteroidales* (GenBac3), *Enterococcus* (Entero1a), and *C. marimammalium* (GFC) for before (grey) and after (white) kuapā restoration binned by salinity. Outliers and 95% confidence intervals are indicated, * $p < 0.05$, *** $p < 0.001$. (**E**) Correlation between salinity and biomarker concentration using a generalized additive mixed model.

Table 5. Summary of general additive mixed model (GAMM) analysis. Summary output from the general additive mixed model analyis in R. Input Formula: LogConcentration ~ (1 | Date) + Pre- vs. –postrepair + Salinity; [1] Significance codes: 0 '***' 0.001 '**' 0.01 '*' 0.05.

Bacterial Indicator		Estimate	Std. Error	df	t Value	Pr (>|t|)	Signif. Codes [1]
C. marimammalium	Intercept	4.99266	0.37574	9.16611	13.287	2.69×10^{-7}	***
	Pre-vs. post repair	0.96856	0.4402	4.42676	2.2	0.086085	
	Salinity	−0.04652	0.01226	57.66017	−3.794	0.000357	***
Bacteroidales	Intercept	6.36231	0.48919	10.03302	13.006	1.32×10^{-7}	***
	Pre-vs. post repair	0.45987	0.56039	4.46269	0.821	0.453	
	Salinity	−0.09205	0.01671	57.73214	−5.509	8.75×10^{-7}	***
Enterococcus	Intercept	5.14077	0.36391	12.62958	14.127	4.12×10^{-9}	***
	Pre-vs. post repair	0.45003	0.39462	4.63367	1.14	0.31	
	Salinity	−0.0794	0.01361	57.97529	−5.823	2.66×10^{-7}	***

We also found that when values were grouped across all stations *Enterococcus* concentrations did not change significantly over the course of restoration (Figure 6A, middle row). However, unlike Bacteroidales, the spatial distribution of *Enterococcus* pre- and post-restoration was structured with highest concentrations along the western edge of the loko i'a (10^4 copies per 100 mL) and decreasing concentrations proceeding eastward down to 10^2–10^3 copies per 100 mL (Figure 6B,C, middle row). The lack of difference in pre- vs. post-restoration data was also supported by binning the stations along a salinity gradient (Figure 6D, middle row). As expected, the general additive mixed model (Figure 6E, middle row) confirmed that *Enterococcus* has increased abundance in low salinity environments (Table 5).

To assess *B. ibis* fecal contamination, we first developed microbial source tracking tools by adapting primers specific to the 16S rDNA gene of *C. marimammalium* (Table 2, GFCf and GFCr). These primer pairs had previously been used to detect fecal contamination from gulls, geese, ducks, and chickens [41]. GFC primers specifically amplified fecal DNA from *B. ibis* living at He'eia Fishpond (Figure S1) and were used to determine the extent of contamination from *B. ibis* fecal sources in the loko i'a. Pre-restoration, *B. ibis* fecal contamination was significantly higher across all stations (mean concentrations of 2–4 $\times$ 10^5 copies 100 mL^{-1}) as compared to 10^4 copies 100 mL^{-1} post-restoration (Figure 6A, bottom row), $p < 0.01$. Pre-restoration concentrations of *B. ibis* fecal indicator bacteria were higher across all stations than both *Bacteroidales* and *Enterococcus* (Figure 6B), with greater than 10^3 copies per 100 mL detected at the oceanic stations. In contrast, post-restoration concentrations of *C. marimammalium* decreased by 2 orders of magnitude, and these differences were statistically significant at the fresh and brackish stations (Figure 6D, bottom row). General additive mixed model (GAMM) analysis indicates that while the negative correlation between *B. ibis* fecal indicator bacteria and salinity is not as strong as with Bacteroidales and *Enterococcus*, it does exist (Table 5).

We note two interesting differences in microbial indicator concentrations and biogeography post-restoration. First, we note the appearance of a region where microbial indicator concentrations are low (Figure 6C). We note that differences in the station locations pre- vs. post-restoration may have altered the interpolation of biomarker concentrations. Alternatively, this may suggest that post-restoration, circulation patterns in the center of the loko i'a have resulted in a well-flushed zone. Secondly, spatial variation in 16S copy concentration for all three molecular markers was greater pre-restoration compared to post-restoration (Figure 6D). The variation coefficient (standard deviation divided by mean) pre-restoration was 0.31 (*C. marimammalium*), 0.43 (*Enterococcus*), 0.4 (*Bacteroidales*), while the variation coefficient post restoration was 0.16 (*C. marimammalium*), 0.32 (*Enterococcus*), 0.31 (*Bacteroidales*). We interpret this difference as an indication that pre-restoration, the loko i'a was less homogeneously mixed than post-restoration.

4. Discussion

Embedded between land and sea, He'eia Fishpond is a powerful natural laboratory. We have been provided the unique opportunity to examine how historical land use change has altered the functions of coastal habitats and how biocultural restoration maintains and improves the integrity of these coastal ocean ecosystems in the face of rapid global change. In the current study, we utilized a comprehensive time series dataset of in situ deployments, discrete sampling, and empirical observations to draw a link between restoration efforts and changing loko i'a circulation, as well as water quality dynamics. Specifically, we examined the impact of invasive mangrove removal around the northern loko i'a periphery from 2014–2017 and Pani ka Puka, repair of the Ocean Break in 2015, presenting a comparison of pre- vs. post-restoration ecosystem dynamics along multiple parameters.

4.1. Ho'oniho ka niho (Interlock the Stones [44]): Water Volume Flux Changes due to Kuapā Repair

Generally, understanding the physical environment of He'eia Fishpond advances our knowledge of the dynamic biochemical and physical interactions in Hawaiian estuarine ecosystems. In repairing the physical infrastructure of He'eia Fishpond, Paepae o He'eia has set the stage for the ecology of the loko iʻa to return to the original conditions engineered by *kūpuna* (elders, ancestors) of Heʻeia: a brackish body of water with a consistent volume, maintained by regulated mixing of fresh and marine inputs to facilitate phytoplankton growth. Our study confirms that during baseline conditions, coastal loko iʻa circulation patterns are driven by a combination of either tidal pumping or stream velocity, depending on the location of the mākāhā [30,45]. Water volume flux rates during SF and SE tides from mākāhā bordering Kāneʻohe Bay (Hīhīmanu, Kahoʻokele, Nui, Kahoalāhui), suggest that the loko iʻa is more influenced by oceanic inputs (>95% total mean water volume flux) than freshwater inputs (<5% total mean water volume flux) from Heʻeia Stream during baseline conditions at both pre- and post-restoration (Tables 3 and 4, Figure 3).

Prior to Pani ka Puka, Heʻeia Fishpond acted largely as an unconfined system during spring tides, when the spring flood tide exceeded the height of Ocean Break. In essence, because the Ocean Break was lower in height than the surrounded kuapā, the entire 56 m wide section of Ocean Break functioned like a mākāhā when tidal pumping in Kāneʻohe Bay was higher than the provision elbow wall. Pre-restoration, we observed enormous water volume flux at spring tides, ~80% exchange almost exclusively from Ocean Break (Table 4). However, during neap tides, the loko iʻa was more confined with less exchange and circulation in the southeastern portion of the loko iʻa. In 2015, this expansive section of the wall was repaired and Kahoʻokele was built, shifting relative mākāhā exchange rates at Kahoʻokele to ~30% post-restoration. This dynamic is also reflected in mean water volume flux rates: Pre-restoration, Ocean Break had the highest mean water volume flux rates of ~12–14 m^3 s^{-1} (Table 4), while the Kahoʻokele water volume flux rates post-restoration are dramatically lower (now ~ 1 m^3 s^{-1}, Tables 3 and 4). Mean water volume flux rates at other mākāhā generally increased from pre-restoration to post-restoration, an indication that nearby mākāhā somewhat compensate for the difference in water volume flux between Ocean Break and Kahoʻokele. However, the general "C" shape of rating curves remained similar (Figure 3). In its current state, the addition of Kahoʻokele renders Heʻeia Fishpond a confined system at all tidal states with adequate water exchange in the southeastern region. These findings are supported by Ertekin et al. [46] who modeled circulation patterns at two different Aliʻi loko iʻa on Molokaʻi, which concluded that the number of mākāhā plays a significant role in improving tidal circulation. They concluded that mākāhā distance and location in relation to the physical forces at work (tidal activity, wind, loko iʻa bathymetry, stream location) affected circulation inside the loko iʻa.

Our results suggest that oceanic mākāhā water volume flux is also dependent upon wind forcing, in particular for mākāhā aligned with the trade winds (~70°). Nui and Kahoʻokele account for ~50% and ~30% of total water volume flux respectively, Figure 3. These mākāhā also have the largest cross-sectional areas (Nui: 6.48 m; Kahoʻokele: 3.05 m, Table 1), and are positioned most in–line with the predominant trade wind direction, Nui has a bearing of 63° and Kahoʻokele has a bearing of 80° (Table 1). Wind blowing from the northeast across Kāneʻohe Bay, can accelerate (if the wind aids) or dampen (if the wind opposes) water flow through Nui and somewhat Kahoʻokele, which is aligned with the predominant wind direction of 70°. We also noted that the channel floor of Kahoʻokele is deeper than the adjacent benthos of both the loko iʻa interior and Kāneʻohe Bay. Thus, the mākāhā floor depth may allow slightly higher water volume flux through Kahoʻokele due to lower resistance to water volume flux. In contrast, Kahoalāhui and Hīhīmanu have considerably smaller relative water volume flux (together accounting for ~15%, Figure 3) as the individual channels of Kahoalāhui have small cross–sectional areas and Hīhīmanu has the smallest cross–sectional area (2 m, Table 1), in addition to being positioned at 48° and 111°, respectively. The notion that wind can influence the rate of water flow through mākāhā is supported by a study by Yang [47] who suggested that the rate of water flow through the mākāhā may be altered by wind accelerating or dampening flow when the body of water was large enough. Kāneʻohe Bay and Heʻeia Fishpond are both large enough, and shallow enough to be affected by wind stress in such a way as to act as a secondary driver of water volume flux in this system.

We found that the river mākāhā have significantly lower relative water volume flux rates during base flow conditions (i.e., non-storm) pre- and post-restoration (Wai 1 and Wai 2 together ~5%). Water volume flux through Wai 1, the most seaward mākāhā along the He'eia Stream, is dependent on tidal activity due to its proximity to Kāne'ohe Bay, making it the only freshwater mākāhā that allows bi-directional water flow. Under baseline conditions, the relative water volume flux of water passing through Wai 1 during flood tide is balanced by the amount of water that flows out during ebb tide (Table 3). At flood tides, flow out of Wai 1 is dampened by He'eia Stream, due to flow in the opposite direction into the loko i'a, while He'eia Stream flow is additive during ebb tides. Due to a dam-like structure in the mākāhā (Figure 2), Wai 2 has little to no detectable tidal signal and exhibits exclusive unidirectional flow from He'eia Stream into the loko i'a that is largely dependent stream discharge and precipitation in the He'eia watershed [30,45,47]. During episodic storm events, strong freshwater water volume influx can have pronounced effects on the loko i'a system [30], yet our water volume flux measurements were all conducted at baseline/low flow conditions. We anticipate that the relative contribution of river mākāhā vs. ocean mākāhā, as well as the balance between ebb vs. flood exchange, is likely to change if He'eia Stream discharge increases during storm events. Research comparing baseline to storm conditions to quantify how higher stream velocities affect loko i'a flushing is currently underway and will be the subject of a subsequent contribution.

Assuming the He'eia Fishpond water balance is in steady state, the water volume influx rates should be equivalent to water volume outflux rates. However, we found the difference between spring and neap tidal cycle flow, the sum of flow (m^3) for all mākāhā, to be –16,760 m^3 or ~8% of total flow between SF and SE tide and 18,554 m^3 or ~13% of total flow between NF and NE (Table 4). Post-restoration, this imbalance is most evident in 2 mākāhā: Kaho'okele, which accounts for 28% of water volume influx, and 39% of water volume outflux during spring tide, and Nui, which accounts for 40% of water volume influx and 44% of water volume outflux during spring tide. This pattern is evident at both spring and neap tidal cycles. We posit that trade winds accelerate flow into the loko i'a at Nui during flood tide, which as previously discussed is aligned with the prevailing wind direction during sampling (63°, Table 1). However, during ebb tide, the wind force opposes outflow at Nui, and a small proportion of water volume flux is redistributed to other mākāhā channels thereby compensating for the reduced outflow at Nui (Table 3, Figure 4). However, these site–specific differences do not account for all the discrepancy observed pre- and post-restoration. We attribute discrepancies in water volume flux balances to a number of factors. First, the influence of submarine groundwater discharge (SGD) into He'eia Fishpond is not accounted for in this study. Previous work quantifying SGD at He'eia Fishpond using radon isotope measurements found that the amount of water volume flux from SGD was equal to that of He'eia Stream discharge [43,48]. Second, the water volume flux in the diffuse flow region (Figure 1B), as well as gains or losses of water through small holes in the kuapā, was not quantified and has not been accounted for in our water budget. In addition, though every effort was made to choose tidal cycles similar in length and amplitude for rating curves, rating curves were calculated using in situ data from sequential rather than simultaneous deployments due to the limitation of instruments (Table S1). Some degree of variability in tidal length and amplitude among sites likely exists. Finally, the mixed semidiurnal tides cause large variations in tidal length (Table 3), giving rise to some uncertainty in the final water volume flux rates calculated.

4.2. Paepae ke alo (Raise the Face of the Wall [44]): Volume, Residence Time, and Salinity

Pani ka Puka affected loko i'a volumes and residence times considerably (Figure 4). The addition of a sixth mākāhā (Kaho'okele) led to increased and faster water volume outflux during both NE and SE tides, corresponding to lower volumes post-restoration. Conversely, whereas no water exchange occurred at Ocean Break during neap tides prior to restoration, Kaho'okele allows more water volume influx during NF tide compared to before, resulting in a larger loko i'a volume post-restoration during this tidal stage (Figure 4). These increased water masses cannot be compensated entirely with the flow (1–2 $m^3\ s^{-1}$) through Kaho'okele, which has a smaller cross–diameter, 3.05 m as compared to

Ocean Break, 56 m (Figure 1C,D). We predicted that restoration would result in shorter residence times, particularly during neap tides. Indeed, total exchange rates during spring tides were 5% higher post-restoration, with a 12% shorter minimum residence time of 32 h as compared to 38 h pre-restoration. Changes in post-restoration circulation are more marked during neap tides – water exchange has increased 16% and maximum residence time has decreased 37% from 102 h to 64 h These residence time calculations are tempered by the following assumptions: (1) uniform mixing of the loko i'a water column, (2) all flood and ebb tides are 6 h, (3) mākāhā are the only source of water exchange. However, salinity measurements at the surface and benthos indicate that the water column is sometimes mildly stratified. Furthermore, our data show a large range in tidal cycle duration variability, ranging from 4.43–17.46 h (Table 3). Lastly, submarine groundater discharge and input from the diffuse flow region (Figure 1B) likely are other indirect sources of water exchange. The difference in minimum and maximum residence times emphasizes the importance of differentiating between tidal states when looking at the effects of restoration on the physical environment of the loko i'a.

Concomitant mangrove removal around the stream mouth corresponded with an increase in water volume flowing through Wai 1 from ~1–2% pre-restoration to ~5% post-restoration (Figure 1E and Table 4) and a freshening of the loko i'a post-restoration. At the end of the period of this study, Wai 2 was not fully clear of *R. mangle* and also showed little change in discharge between pre- and post-restoration. We conclude that mangrove removal positively correlates with increased water flow and subsequently improved loko i'a circulation. Increased freshwater volume flux is also reflected in the salinity distribution, which shows a much stronger freshwater signal around the river mākāhā in post-restoration compared to pre-restoration (Figure 5B). We expect that continued removal of mangrove along the loko i'a periphery would increase stream velocity and the mass of freshwater entering He'eia Fishpond. It is also evident that mixing from the ocean is more limited post-restoration, and thus the freshwater coming in may have a greater overall effect on the salinity. Moreover, the temperature of the surface water is often much lower than marine inputs and given concerns about fish stress linked to sea surface warming trends [29], mixing of cooler water may be beneficial to fish survival. In addition, increased freshwater and nutrient input may be beneficial for native macroalgae and phytoplankton to thrive, which is the primary food source for the herbivorous target fish species. While we can only speculate as to the historical biogeochemistry of He'eia Fishpond, the abundance of evidence suggests that increasing freshwater input is necessary for proper management of native marine species. As this is the first study we are aware of that reveals a correlation between mangrove removal and improved loko i'a circulation, we recommend long–term monitoring of fish and phytoplankton diversity and biomass, particularly near the stream so that the connection between mangrove removal, stream flow, and nearshore fishery health can be fully understood.

4.3. Pani hakahaka (Close Gaps/Vacancies [44]): Microbial Indicators as Markers of Watershed Connectivity

To assess water quality and associated human health risk, we used two broad–spectrum microbial bioindicators used by the US Environmental Protection Agency [49,50]. We used primers that targeted the Bacteroidales family (GenBac3) and the *Enterococcus* genus (Entero1a), bacteria that are common in the feces of mammals (Table 2). These non–pathogenic microbes are easy to quantify and have decay rates similar to those of the pathogens of interest [51], hence, they can be strongly associated with the presence of pathogenic microorganisms derived from upstream in the watershed. By performing co-registered sampling of salinity and microbes, we were able to directly correlate fecal indicator concentrations with salinity, an abiotic factor that strongly influences abundance [49,52,53]. We hypothesized that shorter residence time and increased water volume flux would lower the concentration of Bacteroidales and *Enterococcus* in He'eia Fishpond. Instead, we found no significant overall difference in surface mammalian fecal indicator bacteria before and after restoration (Figure 6A). We found coherence between spatial distribution of mammalian fecal indicators with surface salinity (Figure 6B): post-restoration, lower salinity (e.g., more fresh water) in the northwestern sector of the loko i'a corresponded with even higher concentrations of bacterial indicators as compared to pre-restoration whereas higher salinity in the oceanic–dominated areas of the fishpond had even less fecal contamination than pre-restoration. From the spatiatl distrution of each marker (Figure 6D), we attribute the increase in mammalian fecal bacteria in the northwest area of the loko iʻa to increased terrigenous freshwater input from Heʻeia Stream. Because the expansion of freshwater niches is generally more favorable for these microbes to survive [53], these results emphasize the need for enhanced pollution reduction management upstream.

We also evaluated an internal source of fecal pollution deriving from a large colony of *B. ibis* residing on the mangrove island on the loko iʻa interior. In order to quantify *B. ibis* fecal contamination, we optimized primers to *C. marimammalium* (GFC, Table 2), an uncharacterized Gram-positive facultative anaerobe in the order of *Lactobacillales (Fusobacterium)* [54] originally developed to detect fecal contamination from gulls in coastal environments [41,55–57] for cattle egret fecal material (Figure S1). Unlike Bacteroidales and *Enterococcus*, we found a significant decrease in egret fecal bacteria post-restoration, suggesting that increased flushing and decreased residence times had a positive impact on water quality. The pattern of decreasing *C. marimammalium* and consistent abundance of Bacteroidales and *Enterococcus* between the pre- and post–repair periods is intriguing and may be related to differential environmental reservoirs of the two clades targeted by the assays. GenBac3 and Enterola are phylogenetically very broad probes that target a diverse clade of organisms that may contain unknown members with variable salinity tolerances. In contrast, the GFC probes target a specific organism with few environmental isolates having a narrower range of salinity tolerance. As the cattle egret colony on the mangrove island is the primary source of bird fecal contamination to the loko iʻa, eliminating egret habitat by removing the mangrove island is expected to further reduce the amount of contamination from bird feces.

As Hawaiian watersheds are highly interconnected, loko iʻa provide snapshots of ecosystem health for the entire ahupuaʻa. Fecal contamination in our study site confirms the presence of leaking cesspools and/or septic tanks in the Haʻikū and ʻIolekaʻa watersheds. This kind of pollution endangers plans for seafood production as well the public, who participate in numerous educational and cultural activities.

4.4. Pōhaku ka papale (Place the Capstone on the Top [44]): Future Implications of Revitalizing Customary Fishpond Infrastructure

The design of the new kuapā with additional mākāhā represents an innovation of the contemporary kiaʻi loko to mitigate future flooding risk. While deviating from historical photographs from the 1920s, it is likely that over the course of the 800–year existence of Heʻeia Fishpond kuapā infrastructure has been altered in response to hydrological and oceanic conditions. Kelly noted archeological evidence that the kuapā adjacent to Heʻeia Stream has been moved multiple times, potentially due to catastrophic floods [22], suggesting that placement and number of mākāhā were dynamically managed. Paepae o Heʻeia revealed more contemporary evidence of this during the restoration of Nui, when concrete slotted mākāhā, likely built in the 1900s, was found buried in the kuapā interior. Because kiaʻi loko were concerned about future floods and the integrity of a 3 m wall, they reasoned that having a mākāhā would facilitate the release of water pressure during high flow events. The exact location of the mākāhā was based on practitioner knowledge of the circulation and biological diversity of the area. Thus, re–establishment of customary practices encompassed adaptation for increased resilience, as well as future fish recruitment. In support of their hypothesis, kiaʻi loko noted an increase in fish aggregation around Kahoʻokele over the course of Pani ka Puka that has persisted.

A key dimension to restoring Heʻeia Fishpond has been the removal of invasive *R. mangle*, whose roots grow into the kuapā, separating the rock and coral. Furthermore, mangrove roots hold sediment transported from upstream and its leaf litter directly contribute to the organic matter in the pond, changing the chemistry of the benthos and water column. Mangrove canopies acted as a wind block, impeding circulation and oxygenation, creating heterogenous micro–niches within the loko iʻa. Moreover, kiaʻi loko observed that this non–native species also corresponded with the presence of non–native fish, and they speculated that mangrove removal would enable native aquaculture species to compete more effectively in this habitat, potentially by increasing fish passage into the estuary. Examining the rates of sediment transport from the loko i'a out to Kāne'ohe Bay is needed, as well as a more comprehensive understanding of how this introduced species functions in non–native vs native landscapes.

Overall, this study clearly demonstrates the positive impact restoration regimes had on various physical and microbiological components of the loko i'a ecosystem. Our results are encouraging and indicate that there is a significant potential for community–based restoration to revitalize this, and other, culturally and economically significant sites for sustainable aquaculture in the future. More recently, in part because of the ongoing concerted efforts of community organizations like Paepae o He'eia, the coastal area of He'eia was designated as National Estuarine Research Reserve (NERR) in January 2017 to advance research and protection of the He'eia ahupua'a by integrating the traditional Hawaiian ecosystem management approach with contemporary estuarine management practices.

Supplementary Materials: The following are available online at http://www.mdpi.com/2071-1050/11/1/161/s1, Table S1: Kuapā restoration by Paepae o He'eia over the course of this study, Table S2: He'eia Fishpond in situ sampling regime, Table S3: Meteorological conditions pre- and post-restoration water volume flux calculations, Table S4: Discrete sampling station pre –and post-restoration in He'eia Fishpond, Table S5: YSI and discrete sampling meteorological conditions pre –and post-restoration. Figure S1: Positive amplication of 16S rDNA gene from cattle egret feces DNA samples (BF1 and BF2).

Author Contributions: Conceptualization, R.A.A., M.A.M., C.E.N., and P.M.; methodology, M.A.M., M.L.-G., and C.E.N.; software, P.M., C.K.B., and M.L.-G.; validation, M.L.-G.; formal analysis, P.N., M.L.-G., C.E.N., M.A.M., and R.A.A.; investigation, P.N. and C.K.B.; resources, A.H.K., K.K., R.A.A., M.A.M.; data curation, R.A.A.; writing—original draft preparation, P.M.; writing—review and editing, R.A.A., M.A.M., C.E.N., P.B., M.L.-G., C.K.B., A.H.K., and K.K.; visualization, P.M. and R.A.A.; supervision, M.A.M., R.A.A., A.H.K., K.K., and C.E.N.; project administration, R.A.A. and M.A.M.; funding acquisition, M.A.M., R.A.A., A.H.K., K.K.

Funding: This study was funded in part by the University of Hawaii Sea Grant college program, SOEST, under institutional grant NA14OAR4170071 (M.A.M., R.A.A.) from NOAA Sea Grant, Department of Commerce; Water Resources Research Institute Programs Grant No. 2016HI461B and the Hawai'i Department of Health (A.H.K, K.K., R.A.A.). The APC for this publication was cover in part by both Sea Grant Hawai'i and the He'eia National Estuarine Research Reserve. C.K.B. was supported through the Hau'oli Mau Loa Foundation Graduate Fellowship. The views expressed herein are those of the author(s) and do not necessarily reflect the views of NOAA or any of its subagencies.

Acknowledgments: This project would not have been possible without the support and collaboration of Paepae o He'eia, their incredible stewardship of He'eia Fishpond and their support of our work in this wonderfully unique and beautiful coastal system. We thank Conor Jerolmon, Kristina Remple, Christina Comfort, Gordon Walker, Camilla Tognacchini, and Nalani Olguin for assisting with fieldwork and lab analysis and Dr. Kathleen Ruttenberg for the provision of Sontek Argonaut current meters that were invaluable for in situ physical measurements. We are also grateful to the University of Hawaii Sea Grant College Program for continued support of Nā Kilo Honua o He'eia. UNIHI-SEAGRANT-JC-16-26. SOEST publication no. 10622.

Conflicts of Interest: The authors declare no conflict of interest. The funders had no role in the design of the study; in the collection, analyses, or interpretation of data; in the writing of the manuscript, or in the decision to publish the results.

References

1. Mathieson, A.M. The state of world fisheries and aquaculture. In *World Review of Fisheries and Aquaculture*; FAO: Rome, Italy, 2012.
2. Loke, M.K.; Geslani, C.; Takenaka, B.; Leung, P. Seafood consumption and supply sources in Hawaii, 2000–2009. *Mar. Fish. Rev.* **2012**, *74*, 44–51.
3. Keala, G.; Hollyer, J.R. *LOKO I'A*; College of Tropical Agriculture and Human Resources, University of Hawai'i Mānoa: Honolulu, HI, USA, 2007; pp. 1–76.
4. Kikuchi, W.K. Prehistoric Hawaiian fishponds. *Science* **1976**, *193*, 295–299. [CrossRef] [PubMed]
5. Cobb, J.N. *The Commercial Fisheries of the Hawaiian Islands in 1903*; U.S. Government Printing Office: Washington, DC, USA, 1905.
6. Keala, G.; Hollyer, J.R.; Castro, L. *Loko Ia: A Manual on Hawaiian Fishpond Restoration and Management*; University of Hawaii at Mānoa: Honolulu, HI, USA, 2007.

7. Munro, G.C. Island of Moloka ʻi. In *First Report of the Board of Commissioners of Agriculture and Forestry of the Territory of Hawaii for the Period from July*; Board of Commissioners of Agriculture and Forestry: Honolulu, HI, USA, 1904; Volume 1, pp. 94–96.

8. Gedan, K.B.; Kirwan, M.L.; Wolanski, E.; Barbier, E.B.; Silliman, B.R. The present and future role of coastal wetland vegetation in protecting shorelines: Answering recent challenges to the paradigm. *Clim. Chang.* **2011**, *106*, 7–29. [CrossRef]

9. Twilley, R.W.; Lugo, A.E.; Patterson-Zucca, C. Litter production and turnover in basin mangrove forests in southwest Florida. *Ecology* **1986**, *67*, 670–683. [CrossRef]

10. Chimner, R.A.; Fry, B.; Kaneshiro, M.Y.; Cormier, N. Current Extent and Historical Expansion of Introduced Mangroves on Oʻahu, Hawaiʻi. *Pac. Sci.* **2006**, *60*, 377–383. [CrossRef]

11. Allen, J.A. Mangroves as Alien Species: The Case of Hawaii. *Glob. Ecol. Biogeogr. Lett.* **1998**, *7*, 61–71. [CrossRef]

12. Drigot, D.C. *Mangrove Removal and Related Studies at Marine Corps Base Hawaii*; Tech Note M-3N in Technical Notes: Case Studies from the Department of Defense Conservation Program; US Department of Defense Legacy Resource Management Program Publication: Kaneohe Bay, HI, USA, 1999; pp. 170–174.

13. Walsh, G.E. An ecological study of a Hawaiian mangrove swamp. *Estuaries* **1967**, *83*, 420–431.

14. Crooks, J.A. Characterizing ecosystem-level consequences of biological invasions: The role of ecosystem engineers. *Oikos* **2002**, *97*, 153–166. [CrossRef]

15. Demopoulos, A.W.J.; Fry, B.; Smith, C.R. Food web structure in exotic and native mangroves: A Hawaii–Puerto Rico comparison. *Oecologia* **2007**, *153*, 675–686. [CrossRef]

16. Sweetman, A.K.; Middelburg, J.J.; Berle, A.M.; Bernardino, A.F.; Schander, C.; Demopoulos, A.W.J.; Smith, C.R. Impacts of exotic mangrove forests and mangrove deforestation on carbon remineralization and ecosystem functioning in marine sediments. *Biogeosciences* **2010**, *7*, 2129–2145. [CrossRef]

17. Wester, L. Introduction and Spread of Mangroves in the Hawaiian Islands. *Yearb. Assoc. Pac. Coast Geogr.* **1981**, *43*, 125–137. [CrossRef]

18. Force, H.G.M.S.T.; Matsuoka, J.K. *Hawaii. Governor's Molokaʻi Subsistence Task Force Final Report*; Task Force: Washington DC, USA, 1994.

19. Matsuoka, J.K.; McGregor, D.P.; Minerbi, L. *Molokai: A Study of Hawaiian Subsistence and Community Sustainability*; Sustainable Community Development: Studies in Economic, Environmental, and Cultural Revitalizations; CRC Press: Boca Raton, FL, USA, 1998; pp. 25–44.

20. Farber, J.M. *Ancient Hawaiian Fishponds: Can Restoration Succeed on Molokaʻi?* Neptune House Publications: Encinitas, CA, USA, 1997.

21. Declaration of Hui Malama Loko Iʻa. In. 2012. Available online: http://dlnr.hawaii.gov/occl/files/2015/07/Declaration-of-Hui-Malama.pdf (accessed on 19 November 2018).

22. Kelly, M. *Loko Iʻa O Heʻeia: Heeia Fishpond*; Department of Anthropology, Bernice P. Bishop Museum: Honolulu, HI, USA, 1975.

23. Banner, A.H. *A Fresh-Water "Kill" on the Coral Reefs of Hawaii*; Hawaii Institute of Marine Biology, University of Hawaii: Honolulu, HI, USA, 1968.

24. Water Levels—NOAA Tides & Currents. Available online: https://tidesandcurrents.noaa.gov/waterlevels.html?id=1612480&units=standard&bdate=19650101&edate=19651201&timezone=GMT&datum=MLLW&interval=m&action=data (accessed on 15 September 2018).

25. Brooks, M. Heʻeia Fishpond. In *Proceedings of The Governor's Molokaʻi Fishpond Restoration Workshop*; Wyban, C.A., Ed.; Office of Hawaiian Affairs: Hilo, HI, USA, 1991; pp. 20–24.

26. Vasconcellos, S.M.K. *Distribution and Characteristics of a Photosynthetic Benthic Microbial Community in a Marine Coastal Pond*; University of Hawaii Mānoa: Honolulu, HI, USA, 2007.

27. Scott, T.M.; Rose, J.B.; Jenkins, T.M.; Farrah, S.R.; Lukasik, J. Microbial source tracking: Current methodology and future directions. *Appl. Environ. Microbiol.* **2002**, *68*, 5796–5803. [CrossRef] [PubMed]

28. Kirs, M.; Kisand, V.; Wong, M.; Caffaro-Filho, R.A.; Moravcik, P.; Harwood, V.J.; Yoneyama, B.; Fujioka, R.S. Multiple lines of evidence to identify sewage as the cause of water quality impairment in an urbanized tropical watershed. *Water Res.* **2017**, *116*, 23–33. [CrossRef]

29. McCoy, D.; McManus, M.A.; Kotubetey, K.; Kawelo, A.H.; Young, C.; D'Andrea, B.; Ruttenberg, K.C.; Alegado, R.A. Large-scale climatic effects on traditional Hawaiian fishpond aquaculture. *PLoS ONE* **2017**, *12*, e0187951. [CrossRef] [PubMed]

30. Young, C.W. Perturbation of Nutrient Level Inventories and Phytoplankton Community Composition During Storm Events in a Tropical Coastal System: Heeia Fishpond, Oahu, Hawaii. Master's Thesis, University of Hawaii Mānoa, Honolulu, HI, USA, 2011.

31. National Weather Service Honolulu, HI. Available online: http://www.prh.noaa.gov/hnl/hydro/hydronet/hydronet-data.php (accessed on 19 November 2018).

32. USGS Water Data for the Nation. Available online: https://waterdata.usgs.gov/nwis (accessed on 19 November 2018).

33. Weather Observations: Moku o Lo'e, O'ahu | PacIOOS. Available online: http://www.pacioos.hawaii.edu/weather/obs-mokuoloe/ (accessed on 19 November 2018).

34. Timmerman, H.V.; Young, C.; Briggs, R.; D'Andrea, B.; McManus, M.A.; Vasconcellos, S.; Ruttenberg, K.S. Dynamics of Land-Ocean Linkages in a Semi-Enclosed Tropical Coastal System. Unpublished work. 2018.

35. Ludwig, W.; Schleifer, K.H. How quantitative is quantitative PCR with respect to cell counts? *Syst. Appl. Microbiol.* **2000**, *23*, 556–562. [CrossRef]

36. Haugland, R.A.; Siefring, S.C.; Wymer, L.J.; Brenner, K.P.; Dufour, A.P. Comparison of Enterococcus measurements in freshwater at two recreational beaches by quantitative polymerase chain reaction and membrane filter culture analysis. *Water Res.* **2005**, *39*, 559–568. [CrossRef] [PubMed]

37. *Method 1611: Enterococci in Water by TaqMan® Quantitative Polymerase Chain Reaction (qPCR) Assay*; United States Environmental Protection Agency Office of Water (4303T): Washingdon, DC, USA, 2012.

38. Dick, L.K.; Field, K.G. Rapid estimation of numbers of fecal Bacteroidetes by use of a quantitative PCR assay for 16S rRNA genes. *Appl. Environ. Microbiol.* **2004**, *70*, 5695–5697. [CrossRef]

39. Siefring, S.; Varma, M.; Atikovic, E.; Wymer, L.; Haugland, R.A. Improved real-time PCR assays for the detection of fecal indicator bacteria in surface waters with different instrument and reagent systems. *J. Water Health* **2008**, *6*, 225–237. [CrossRef]

40. *Method B: Bacteroidales in Water by TaqMan® Quantitative Polymerase Chain Reaction (qPCR) Assay*; United States Environmental Protection Agency Office of Water (4303T): Washingdon, DC, USA, 2010.

41. Green, H.C.; Dick, L.K.; Gilpin, B.; Samadpour, M.; Field, K.G. Genetic markers for rapid PCR-based identification of gull, Canada goose, duck, and chicken fecal contamination in water. *Appl. Environ. Microbiol.* **2012**, *78*, 503–510. [CrossRef]

42. Moehlenkamp, P. Kū Hou Kuapā: Increase of Water Exchange Rates and Changes in Microbial Source Tracking Markers Resulting from Restoration Regimes at He'eia Fishpond. Master's Thesis, University of Hawaii Mānoa, Honolulu, HI, USA, 2018.

43. Kleven, A. Coastal Groundwater Discharge as a Source of Nutrients to Heeia Fishpond, Oahu, HI. Bachelor's Thesis, University of Hawaii at Mānoa, Honolulu, HI, USA, 2014.

44. Paepae o He'eia. Ho'oniho ka Niho. Unpublished work. 2015.

45. Ertekin, R.C.; Yang, L.; Sundararaghavan, H. *Hawaiian Fishpond Studies: Web Page Development and the Effect of Runoff from the Streams on Tidal Circulation*; University of Hawaii at Mānoa: Honolulu, HI, USA, 1999; pp. 1–53.

46. Ertekin, R.C. *Molokai Fishpond Tidal Circulation Study*; Final Report Submitted to the University of Hawaii Sea Grant College Program; University of Hawaii Sea Grant College Program: Honolulu, HI, USA, 1996.

47. Yang, L. *A Circulation Study of Hawaiian Fishponds*; University of Hawaii, Department of Ocean and Resources Engineering: Honolulu, HI, USA, 2000.

48. Dulai, H.; Kleven, A.; Ruttenberg, K.; Briggs, R.; Thomas, F. Evaluation of Submarine Groundwater Discharge as a Coastal Nutrient Source and Its Role in Coastal Groundwater Quality and Quantity. In *Emerging Issues in Groundwater Resources*; Advances in Water Security; Springer: Cham, Switzerland, 2016; pp. 187–221. ISBN 9783319320069.

49. Noble, R.T.; Lee, I.M.; Schiff, K.C. Inactivation of indicator micro-organisms from various sources of faecal contamination in seawater and freshwater. *J. Appl. Microbiol.* **2004**, *96*, 464–472. [CrossRef]

50. Shanks, O.C.; Kelty, C.A.; Sivaganesan, M.; Varma, M.; Haugland, R.A. Quantitative PCR for genetic markers of human fecal pollution. *Appl. Environ. Microbiol.* **2009**, *75*, 5507–5513. [CrossRef] [PubMed]

51. Murphy, H. Persistence of Pathogens in Sewage and Other Water Types. In *Global Water Pathogens Project Part*; Rose, J.B., Jiménez-Cisneros, B., Eds.; Michigan State University: Lansing, MI, USA, 2017; Volume 4.

52. Ortega, C.; Solo-Gabriele, H.M.; Abdelzaher, A.; Wright, M.; Deng, Y.; Stark, L.M. Correlations between microbial indicators, pathogens, and environmental factors in a subtropical Estuary. *Mar. Pollut. Bull.* **2009**, *58*, 1374–1381. [CrossRef] [PubMed]

53. Shehane, S.D.; Harwood, V.J.; Whitlock, J.E.; Rose, J.B. The influence of rainfall on the incidence of microbial faecal indicators and the dominant sources of faecal pollution in a Florida river. *J. Appl. Microbiol.* **2005**, *98*, 1127–1136. [CrossRef] [PubMed]

54. Sinigalliano, C.D.; Ervin, J.S.; Van De Werfhorst, L.C.; Badgley, B.D.; Ballesté, E.; Bartkowiak, J.; Boehm, A.B.; Byappanahalli, M.; Goodwin, K.D.; Gourmelon, M.; et al. Multi-laboratory evaluations of the performance of Catellicoccus marimammalium PCR assays developed to target gull fecal sources. *Water Res.* **2013**, *47*, 6883–6896. [CrossRef]

55. Ryu, H.; Griffith, J.F.; Khan, I.U.H.; Hill, S.; Edge, T.A.; Toledo-Hernandez, C.; Gonzalez-Nieves, J.; Santo Domingo, J. Comparison of gull feces-specific assays targeting the 16S rRNA genes of *Catellicoccus marimammalium* and *Streptococcus* spp. *Appl. Environ. Microbiol.* **2012**, *78*, 1909–1916. [CrossRef] [PubMed]

56. Cloutier, D.D.; McLellan, S.L. Distribution and Differential Survival of Traditional and Alternative Indicators of Fecal Pollution at Freshwater Beaches. *Appl. Environ. Microbiol.* **2017**, *83*, e02881-16. [CrossRef] [PubMed]

57. Lee, C.; Marion, J.W.; Lee, J. Development and application of a quantitative PCR assay targeting *Catellicoccus marimammalium* for assessing gull-associated fecal contamination at Lake Erie beaches. *Sci. Total Environ.* **2013**, *454–455*, 1–8. [CrossRef]

 *sustainability*

Communication

Linking Land and Sea through Collaborative Research to Inform Contemporary applications of Traditional Resource Management in Hawai'i

Jade M.S. Delevaux [1,2,*], Kawika B. Winter [3,4,5], Stacy D. Jupiter [6], Mehana Blaich-Vaughan [4,7], Kostantinos A. Stamoulis [8], Leah L. Bremer [9,10], Kimberly Burnett [10], Peter Garrod [4], Jacquelyn L. Troller [11] and Tamara Ticktin [1]

[1] Department of Botany, University of Hawai'i at Mānoa, Honolulu, HI 96822, USA; ticktin@hawaii.edu
[2] School of Ocean and Earth Science and Technology, University of Hawai'i at Mānoa, Honolulu, HI 96822, USA
[3] Hawai'i Institute of Marine Biology, University of Hawai'i at Mānoa, Kaneohe, HI 96744, USA; kwinter@ntbg.org
[4] Department of Natural Resources and Environmental Management, University of Hawai'i at Mānoa, Honolulu, HI 96822, USA; mehana@hawaii.edu (M.B.-V.); garrod@hawaii.edu (P.G.)
[5] Limahuli Garden and Preserve, National Tropical Botanical Garden, Hā'ena, HI 96714, USA
[6] Wildlife Conservation Society, Melanesia Program, Suva, Fiji; sjupiter@wcs.org
[7] Sea Grant College Program & Hui 'Āina Momona, University of Hawai'i at Mānoa, Honolulu, HI 96822, USA
[8] School of Molecular and Life Sciences, Curtin University, Perth, WA 6102, Australia; kostanti@hawaii.edu
[9] University of Hawai'i Water Resources Research Center, University of Hawai'i, Honolulu, HI 96822, USA; lbremer@hawaii.edu
[10] University of Hawai'i Economic Research Organization, University of Hawai'i, Honolulu, HI 96822, USA; kburnett@hawaii.edu
[11] Oceantroller LLC, Honolulu, HI 96819, USA; jackie.troller@gmail.com
* Correspondence: jademd@hawaii.edu

Received: 22 July 2018; Accepted: 28 August 2018; Published: 3 September 2018

Abstract: Across the Pacific Islands, declining natural resources have contributed to a cultural renaissance of customary ridge-to-reef management approaches. These indigenous and community conserved areas (ICCA) are initiated by local communities to protect natural resources through customary laws. To support these efforts, managers require scientific tools that track land-sea linkages and evaluate how local management scenarios affect coral reefs. We established an interdisciplinary process and modeling framework to inform ridge-to-reef management in Hawai'i, given increasing coastal development, fishing and climate change related impacts. We applied our framework at opposite ends of the Hawaiian Archipelago, in Hā'ena and Ka'ūpūlehu, where local communities have implemented customary resource management approaches through government-recognized processes to perpetuate traditional food systems and cultural practices. We identified coral reefs vulnerable to groundwater-based nutrients and linked them to areas on land, where appropriate management of human-derived nutrients could prevent increases in benthic algae and promote coral recovery from bleaching. Our results demonstrate the value of interdisciplinary collaborations among researchers, managers and community members. We discuss the lessons learned from our culturally-grounded, inclusive research process and highlight critical aspects of collaboration necessary to develop tools that can inform placed-based solutions to local environmental threats and foster coral reef resilience.

Keywords: ridge-to-reef; groundwater; land-use; nutrients; bleaching; scenario; resilience; collaboration; scientific tools; management

---

1. Introduction

Pacific Islands are ideal systems to understand land-sea links in the context of social-ecological system resilience [1–3], defined as the capacity of the system to cope with disturbances without shifting to an alternative state while maintaining its functions and supporting human uses [4,5]. Around the Pacific Islands [6,7], local knowledge and associated management practices (e.g., agroforestry, fisheries management) have been recognized to play a key role in building resilience to disturbances [8–10]. These local ecological knowledge systems (henceforth LEK) are customary knowledge-practice-belief systems passed down orally over generations, through adaptive management [8]. This knowledge is formed through historical resource-use practices and long-term, qualitative observations over a restricted geographical area. LEK continues to be modified under rapidly changing social, economic and ecological contexts. Where indigenous peoples depend on local environments for resources, they have also adopted conservation practices, which in some cases can enhance abundance or/and biodiversity [11]. For instance, traditionally managed community fisheries in Hawai'i have exhibited equal or higher biomass than even no-take marine protected areas [9,12]. Because of the long-term and place-based understanding embodied in LEK systems, there is increasing recognition of the importance of integrating LEK into management strategies to build resilience [13–15], especially in Pacific Islands, where environments are unpredictable and highly vulnerable to climate change [16].

Awareness of natural resource decline has contributed to a cultural renaissance across the Pacific Islands, where local communities seek to revive local customary and place-based management approaches [17], such as customary *moku* (ridge-to-reef) management approaches [18,19], *kapu* (traditional closures) and *pono* (sustainable) practices to protect biocultural resources and foster social-ecological resilience [17,20]. These social-ecological systems can be defined as Indigenous and community conserved areas (ICCA), where "natural and/or modified ecosystems containing significant biodiversity values, ecological services and cultural values, are voluntarily conserved by indigenous, mobile and local communities, through customary laws and other effective means" [21]. In ICCAs, local people, who are intimately connected to the environment, culturally and/or through their livelihoods make decisions over how resources are used and have the capability to enforce regulations, which can lead to effective conservation outcomes (even if conservation is not the primary objective) [22,23]. Ridge-to-reef management systems that integrate LEK can enhance social-ecological resilience through reducing impact from climate disturbances and strengthening governance systems with capacity to quickly organize and act [2]. These types of ICCAs offer lessons in integrating traditional knowledge and management practices into sustainability and conservation planning but require national level legal and policy changes to accommodate and empower the ICCAs operating at the watershed-reef level [8,24]. The restoration of local management is challenging, because users have often grown in number and shifted in character from small, homogenous resident populations using resources for subsistence, to transient, global tourist populations using the same resources for recreation [25,26].

After nearly two centuries of decline of the Hawaiian biocultural resource management system, there has been a resurgence of interest—from within academia and the policy realm, as well as at the community level—in reviving that system to restore biocultural resource abundance [27]. This renaissance has inspired an attempt to align traditional Hawaiian biocultural resource management with contemporary frameworks of ecosystem-based management that re-establish the cohesive links between terrestrial and marine systems, encompassing integrated ecological and social processes from ridge-to-reef [28–30]. There has been a growing focus on a land-division scale known as *moku* to revive traditional resource management in a localized context as a means for communities to engage in biocultural restoration. *Ahupua'a* are social-ecological communities nested within *moku*, which are delineated as land-divisions that often extend from the mountains to the sea and exist within the context of the Hawaiian system of governance and biocultural resource management [27]. Motivations by Hawaiian communities to employ contemporary ICCAs include access to and restoration of biocultural resources, security of land and resource tenure, security from outside threats, financial benefit from resources or social-ecological system functions, participation in management, empowerment, capacity building and cultural identity and cohesiveness [8,11,31]. Perpetuating ancestral practices related to food systems provides roles for community members of all ages, while maintaining relationships and balance with the natural world in specific areas [32,33].

Despite management challenges, declining resource health and conflicts over access, two *ahupua'a* (social-ecological communities) embody this cultural renaissance [34]. Hā'ena on the windward side of Kaua'i Island and Ka'ūpūlehu on the leeward side of Hawai'i Island have successfully maintained control of a critical component of their food system by enhancing the management of their coastal resources through the creation of innovative ICCA's (Figure 1). Both places have become the first, officially sanctioned ICCAs in the U.S. State of Hawai'i. Coral reef fish caught near shore with nets, pole and line, or spears, are a very important component of community food systems [6,35]. Therefore, both communities in this study initiated marine closures of different sizes to protect fishery species, many of which are also known to feed on algae (herbivorous fishes). Without these herbivorous species, algae blooms can cover the reef when excess nutrients flow into the sea from the land. By eating the algae, these protected fishes create space for new corals to settle and ensure the persistence or resilience of the reefs.

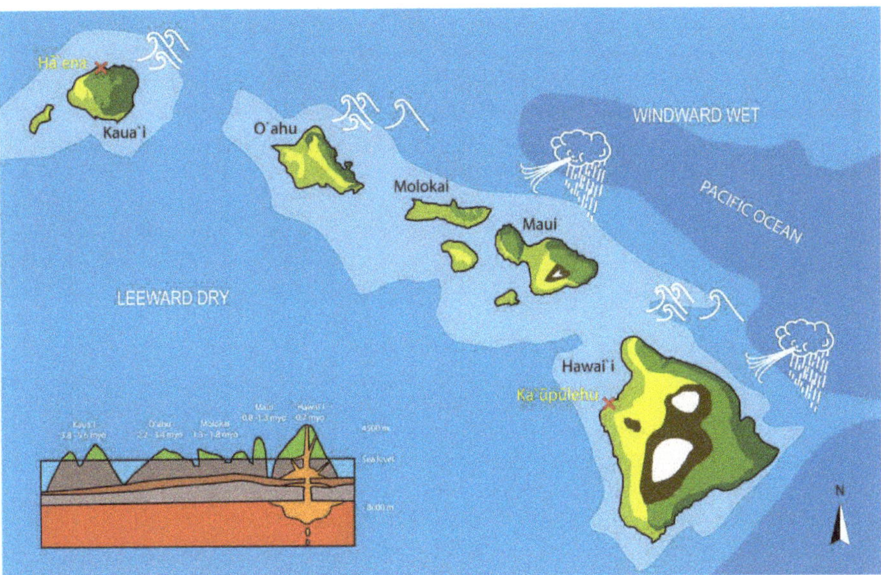

Figure 1. Locations of Hā'ena and Ka'ūpūlehu *ahupua'a* on Kaua'i and Hawai'i along the main Hawaiian Island chain, with island age and the direction of the prevailing north-east trade winds and ocean swell indicated.

These local communities are also interested in reviving the *ahupua'a* approach by better understanding how land-based sources of pollutants from golf courses, lawns and cesspools affect their marine ecosystems to inform alternative land-use options [36]. Even with healthy herbivorous fish populations, these pollutants take a toll on coral reefs, especially with increases in ocean temperature and acidity as a result of climate change. Therefore, it is important to these communities and the health of all marine ecosystems, to ensure that future coastal planning takes land-based impacts into account. Effective ridge-to-reef management requires improved understanding of land-sea linkages and tools to evaluate the effects of land (e.g., nutrients carried through groundwater) and marine (e.g., wave power and reef topography) drivers on coral reefs to inform resilience management in the face of climate change. In response to these gaps, we adopted the traditional *ahupua'a* framework to study the effect of coastal development on coral reefs under projected climate impacts and identified place-based management actions that can boost system resilience. Here, we provide an overview of: (1) the renaissance of the traditional resource management system of Hawai'i, with a focus on two communities with ICCAs; (2) how applied collaborative science can support management; and (3) the development of decision support tools grounded in place-based management.

2. The Renaissance of Traditional Resource Management of Hawai'i

2.1. The Story of How Hā'ena Became a Marine ICCA

Recognizing the importance of customary Hawaiian management and subsistence fishing, Hawai'i enacted legislation in 1994 that allows the Department of Land and Natural Resources (DLNR) to designate community based subsistence fishing areas (CBSFAs) for "reaffirming and protecting fishing practices customarily and traditionally exercised for purposes of Native Hawaiian subsistence, culture and religion" [37]. This created a pathway to designate marine ICCA's in Hawai'i. Achieving a CBSFA designation allows community members to assist DLNR to develop and enforce place-specific management strategies/laws that regulate resources they depend on from the shoreline to one mile out to sea, or the edge of the coral reef, based on Native Hawaiian values and ancestral practices [37]. This designation allows residents to work with the state DLNR Division of Aquatic Resources (DAR) to develop and enforce laws (S.B. 2501, 23rd Leg., Reg. Sess. Hawai'i 2006) [6,25]. Through traditional Hawaiian values, the CBSFA designation emphasizes the connection between the environment and communities, whereby if you care for the environment, the environment will care for you. CBSFAs represent an agency-recognized avenue for local community groups to assert their indigenous rights by proposing management measures informed by customary fishing and management practices to sustain the health and abundance of marine resources for generations in the Hawaiian Islands [38].

Like many other places in Hawai'i, land privatization, along with coastal development of vacation and luxury homes, fragmented the land in the 1960s [39], which led many long-time families to move from the area [6]. Today, the rural *ahupua'a* is mostly owned by the State of Hawai'i and the non-profit organization, National Tropical Botanical Garden (NTBG), with ~140 private residences along the coast (Figure 2A). The NTBG, *Kānaka Maoli* (indigenous Hawaiian) and *kama'āina* (place-based community) of Hā'ena (henceforth Hā'ena community) persisted in the creation of rules guided by ancestral norms: *hō'ihi* (respectful reciprocity), *konohiki* (inviting ability) and *kuleana* (rights based on responsibilities) [33,40]. In 2006, the State of Hawai'i designated Hā'ena as its first CBSFA [25]. After this designation, the community was empowered to work with the state resource management agency to co-develop fishing regulations and secure their approval through the same onerous public process as any administrative rules promulgated by state government agencies [33].

In August 2015, after nearly ten years of planning and negotiation, over seventy meetings, fifteen rule drafts, three public hearings and multiple studies undertaken to document visitor impacts, user groups, fishery health and the importance of locally caught fish within and beyond the Hā'ena community, these rules became law [6,33] (Figure 2B). The community of Hā'ena managed to restore local-level management of their near-shore fishery by co-creating CBSFA rules to govern fishing and all coastal uses, including recreational activities based on customary practices and customary norms for the area [37]. The significance of this event cannot be understated. This was the first time in the state of Hawai'i that local-level fisheries management rules, based on indigenous Hawaiian practices, were recognized. Passage of these rules made Hā'ena the first coastal area in Hawai'i to be permanently governed by community developed, local-level rules based on ancestral knowledge and practices [33]. As the first site to work with DAR to co-create rules formally adopted as state law, Hā'ena set a precedent for at least 19 other Hawai'i communities pursuing co-management of local fisheries [37]. Many communities across Hawai'i view this effort as a larger community movement to increase self-sufficiency and restore formal local-level control over ocean resources as a food source [33].

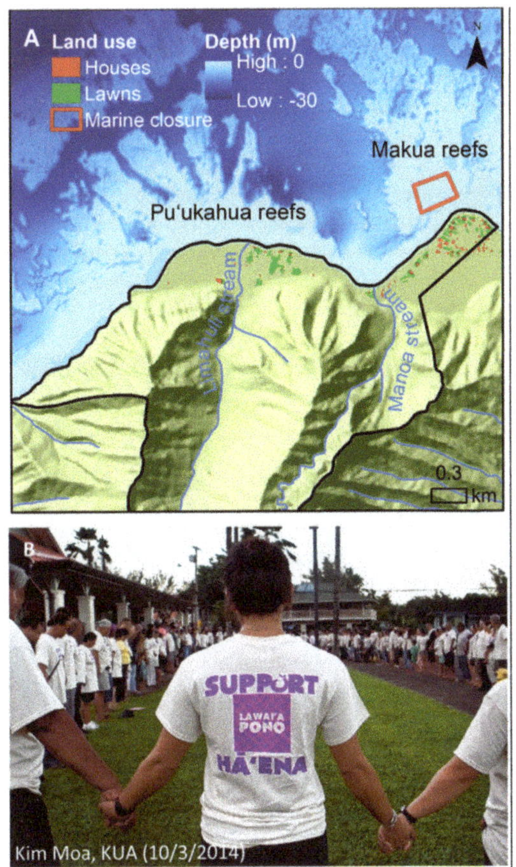

Figure 2. Haʻēna study site. (**A**) Haʻēna land use and community based subsistence fishing area (CBSFA) marine refuge boundaries and (**B**) The opening *pule* (prayer) prior to the public hearing in Hanalei for the Haʻēna CBSFA package rules.

Traditional coastal management in Hāʻena relied on protecting key spawning and feeding areas for fishes [25,40]. One of the largest fringing reef systems in the main Hawaiian Islands is found in Hāʻena, where a large lagoon formed by the back-reef provides wave sheltered nursery habitat for culturally and economically important soft and hard bottom target fish species [3,33,41]. The reefs provide daily fish protein for many Hawaiian and other local families, as well as for *ʻahaʻaina* (feasts commemorating events including weddings, birthdays, funerals and graduations) and other celebrations on Kauaʻi [6]. Therefore, among these rules, a marine refuge (*Makua Puʻuhonua*) was designated in the sheltered lagoon of Makua to protect a key fish nursery area (see Figure 2A). By closing this area to fishing and all recreational use, the community successfully created a refuge grounded in indigenous practices and knowledge [11]. This closure protects culturally important fish species from being captured by fishers or disturbed by snorkelers, kite boarders, stand-up paddlers and others during vulnerable life stage (spawning) and behavior (feeding) [25].

2.2. The Story of How Kaʻūpūlehu Became a Marine ICCA

Kaʻūpūlehu is both commercially and residentially more developed than Hāʻena, with two large luxury resorts, a golf course and several private residences along the southern end of the coast (see

Figure 3A). Few lineal descendants and longtime residents of Ka'ūpūlehu live within the *ahupua'a* but many live nearby, maintaining strong connections to their ancestral lands [14,36]. The entire *ahupua'a* is owned by the largest private landowner in the state of Hawai'i, Kamehameha Schools (KS–an indigenous Hawaiian educational trust and the State of Hawaii's largest private landowner), which was established for the benefit of *Kānaka Maoli* [34,36]. KS seeks to balance multiple economic, educational, cultural and environmental goals [34,36,42,43]. Cultural and place-based values are of high priority for KS and *kama'āina* of Ka'ūpūlehu, (henceforth Ka'ūpūlehu community) are involved in resource management advisory councils, educational programs and cultural restoration projects in the *ahupua'a* [36]. Environmental outcomes, including groundwater recharge and restoring abundant nearshore fisheries, are also highly valued for cultural and economic purposes, as groundwater is the main water source statewide and fisheries are used for subsistence [44].

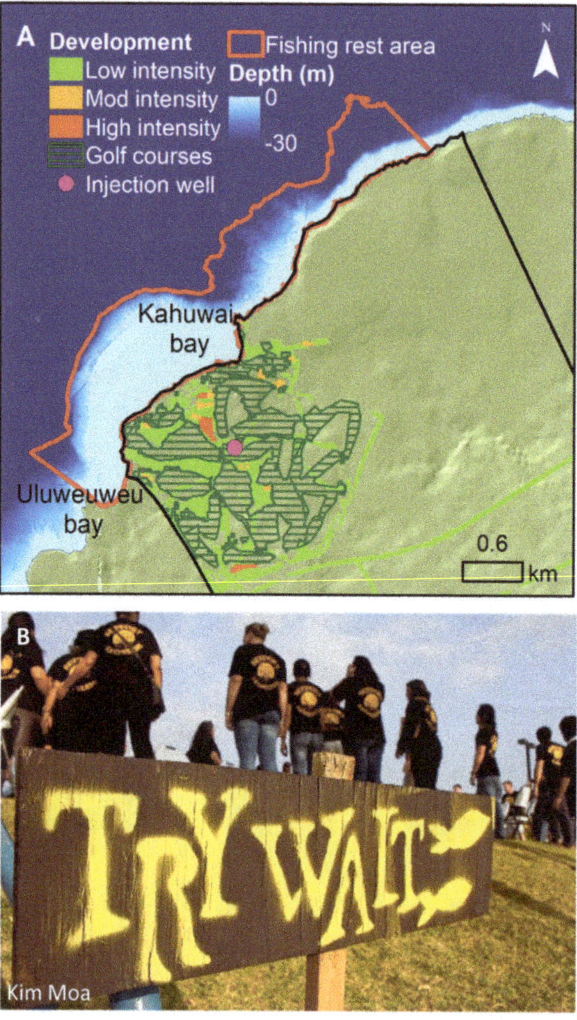

Figure 3. Ka'ūpūlehu study site. (**A**) Ka'ūpūlehu land use and marine reserve map; (**B**) Gathering for the opening *pule* prior to the public hearing for the 'Try Wait' fishing rest area in Ka'ūpūlehu.

Given that the scarcity of water resources limits agriculture, marine resources gathered historically from the ocean played a vital role in the diet of Ka'ūpūlehu families and their subsistence practices [45]. The families that lived (and live) in Ka'ūpūlehu were experts in their resources and knew how to survive within these rugged lands [45]. However the community has observed drastic declines in their coastal resources within the past 40 years, with the opening of the resort and the Ka'ahumanu Highway in 1975 thereby requiring and providing easier access to the isolated waters of Ka'ūpūlehu [45]. In response, the community sought to maintain and restore coastal and marine life health along with their interconnected traditions, before the system could no longer recover and all knowledge was forgotten. In 2015, after nearly ten years of planning and negotiation and over 350 community meetings and multiple studies undertaken to document fishing impacts and coral reef health, the community of Ka'ūpūlehu initiated a law implementing a 10-year fishing rest period known as 'Try Wait' (see Figure 3B), which adopted a term in the local pidgin language meaning, "Let's wait a moment." The protected area extends out to 120 feet deep (or 36.6 m) along a large portion of the coastline. This resulted in the protection of the entire fringing reef (see Figure 3A). Providing full protection of the nearshore reef for a 10 year period while the community develops their long-term management plan is a management strategy grounded in indigenous practices [11].

3. Developing Scientific Tools through Collaboration Grounded in a Hawaiian Approach

Collaborative research among scientists and local communities has the potential to overcome limitations of often-practiced 'expert' driven, narrowly focused scientific research. Collaborative research incorporates the dynamic interactions between people and nature, rather than viewing people only as "managers" or "stressors" [26], and positive outcomes for social-ecological management have been documented (e.g., [46]). Processes to define research questions and objectives based on collaborative approaches can also empower indigenous people and communities [26] and generate possibilities for complementary use of scientific and traditional knowledge [13,15,26]. This type of research requires understanding linkages and feedback loops between nature and people to inform local management in those particular places [26].

Our research process included five main steps and involved managers, scientists and the stewards of the land at different stages: (1) Problem formulation; (2) scenario design; (3) conceptual and model development; (4) scenario modeling and analysis; and (5) informing land-sea planning (Figure 4 and Table 1). Both communities were interested in restoring a ridge-to-reef approach to address contemporary environmental issues, including coastal development and fishing pressure impacts on coral reefs combined with bleaching from climate change. In collaboration with local landowners and communities, we developed a decision support framework grounded in Native Hawaiian culture by adopting the traditional *ahupua'a* lens to assess the impact of coral reefs under projected land use and climate change scenarios, combined with the marine closures. Key collaborators included local community members (e.g., landowners, care takers and active nonprofits), managers with jurisdiction across the ridge-to-reef ecological unit and local experts and scientists. Through local leaders (e.g., K.B.W. and M.B-.V. at Hā'ena, who are co-authors on this paper) and previous work with community members, we identified environmental concerns, ground-truthed models and identified solutions to mitigate local threats. Managers at the state level included the Hawai'i Department of Health (HDOH), which manages water quality from ridge-to-reef and ensures compliance with the Clean Water Act. Scientists and local experts from multiple disciplines, including terrestrial and marine ecologists, social scientists, economists, modelers, hydrogeologists and geographers were involved at different stages of the process to identify and link all the key processes and components that are important in the decision-making process spanning the top of the mountains to the sea and the community in between.

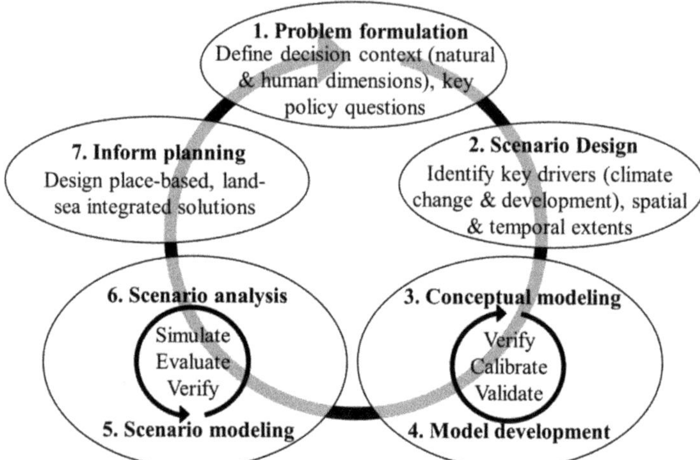

Figure 4. Collaborative science process. The collaborative science process involves stewards/care takers, managers and scientists, or a combination at multiple stages (see Table 1 for roles fulfilled by each group of actors).

Table 1. Roles of multiple actors in a collaborative science process. Stewards/care takers, resource managers and scientists, or a combination play multiple roles at multiple stages.

Stages	Actors	Roles
1. Problem formulation	Stewards/care takers	Community members, land owners and non-profits
	Scientists	Geographer, ecologist, hydro-geologist & planners
	Resource managers	Hawai'i Department of Health
2. Scenario design	Stewards/care takers	Preferences, vision & concerns
	Scientists	Compile data & map scenarios
	Resource managers	Share data
3. Conceptual modeling	Stewards/care takers	Determine key system components, indicators and processes
	Scientists	
	Resource managers	
4. Model development	Scientists	Measure indicators, design & build models, develop user friendly outputs
5. Scenario modeling	Scientists	Model indicator changes per scenario
	Resource managers	Assess & ground-truth outputs
	Stewards/care takers	
6. Scenario analysis	Scientists	Perform indicator analysis & assess potential risk for each scenario
7. Inform planning	Stewards/care takers	Guide place-based management
	Scientists	Synthesize & communicate scenario results
	Resource managers	Guide policy-making

First, we formulated the problem and key policy questions by consulting community members (e.g., landowners, caretakers and active nonprofits), managers with jurisdiction across the ridge-to-reef ecological unit and local experts and scientists to define the decision contexts. Second, we designed scenarios in partnership with local communities to capture their concerns, which included increases in coastal development and climate change impacts on coral reef habitat (e.g., corals and turf) and associated culturally important fisheries (e.g., surgeonfishes, parrotfishes and jacks) and the potential recovery from the recently enacted marine closures. We reviewed zoning documents produced by the County of Kaua'i of Hawai'i and the Office of Planning related to coastal zone planning to determine where coastal development was allowable and feasible to project future land-use change. At the same time, we compiled all the existing data at both sites to calibrate the land-sea models. The database from the HDOH was used to inform the calibration of land-use nutrient loadings rates (e.g., the wastewater injection well loading rate). A local non-profit (The Nature Conservancy) and research group at the University of Hawai'i (Fisheries Ecology & Research Lab) provided empirical data to calibrate the coral reef models at Ka'ūpūlehu and Hā'ena, respectively. Wedetermined the impact of co-occurring human drivers on coral reefs by coupling the human driver scenario analysis (climate change, coastal development and marine closures) with the development of a novel linked land-sea modeling framework for Hā'ena and Ka'ūpūlehu *ahupua'a*. Local ecological and expert knowledge about the coral reef benthic habitat and key fish distributions was used to ground-truth our coral reef indicator maps under present conditions, resulting from the model development phase. For example, the first version of the models provided some outputs that were not consistent with local observations, which led to revisions of the modeling framework until consistency was reached. Subsequently, the downstream fate of nutrients from upstream sources was modeled and projected impacts on coral reefs was assessed under the different scenarios to identify areas on land where managing human-derived nutrients can promote coral reef resilience [3,47]. The modeled scenario outputs were evaluated against the local communities' observations about the location of re-occurring algae blooms and bleaching impacts. Based on the community and managers' feedback, our findings are currently being used to shape place-based management solutions grounded in a ridge-to-reef approach and the indicators can be monitored to track the policy effectiveness. The HDOH also funded the dissemination of these research findings through a statewide conference in 2018 (July–August).

4. A Novel Linked Land-Sea Decision Support Tool for Local Management

The framework links land to sea through groundwater and tracks changes in abundance and distribution of multiple benthic and fish indicators under each scenario (Figure 5) (see [3] for more details). For each site, natural driver data, including topography and bathymetry, and rainfall and wave patterns, were included in the ridge-to-reef modeling framework to represent the natural disturbance regimes specific to each place (Figure 5A). The terrestrial drivers modeled included groundwater flow and nutrient fluxes, incorporating natural and human-derived nutrient flux. The marine drivers characterized the marine habitat conditions and were derived from the SWAN wave model and LiDAR bathymetry data with GIS-based models (Figure 5F). The coral reef predictive models were calibrated on local coral reef survey data [41,48]. To measure proxies of ecological resilience, which also represented important cultural resources to the local communities, the coral reef models focused on four benthic groups, known to change under land-based runoff and bleaching impacts, and four fish indicator groups subject to fishing pressure. The benthic groups were crustose coralline algae (CCA), hard corals, turf and macroalgae (Figure 5G). CCA and corals are active reef builders which provide habitat for reef fishes. CCA also stabilize the reef in high-wave environments. Abundant benthic algae can be a sign of high nutrients and/or low numbers of herbivorous fish, and can harm coral health through competition for space. Herbivorous and piscivorous fish identified as important by the communities (e.g., surgeonfishes, parrotfishes and jacks) were modeled based on their feeding modes and ecological role: (1) browsers; (2) grazers; and (3) scrapers; along with (4) piscivores, which are key fishery species and indicators of fishing pressure [49] (Figure 5H).

The human driver scenarios included coral bleaching, coastal development and marine closures [50]. Two future coastal development scenarios were based on current land zoning from the Hawai'i State Office of Planning and utilized the three commonly used types of wastewater treatment systems in Hawai'i (cesspools, septic tanks and injection wells) (Figure 5B). Nitrogen and phosphorus fluxes were modeled under each coastal development scenario and diffused in the ocean using a GIS-based coastal discharge model (Figure 5E). Two coral bleaching scenarios were derived from projected coral bleaching impacts for the region (Figure 5C). The marine closure scenario assumed removal of fishing pressure within the marine closure boundaries (Figure 5D) [38,44]. The climate change scenarios were applied in combination with the coastal development and marine closure scenarios [51,52]. Under each scenario, our land-sea models predicted the change in nutrient flux and associated abundance of the coral reef indicators (Figure 5G,H). Based on predicted changes, this approach informs place-based solutions rooted in the *ahupua'a* approach, by identifying priority areas on land where management can promote coral reef resilience to climate change (Figure 5I). The development of this new technology necessitated a collaborative process, which leveraged both scientific and local knowledge by involving scientists, community members and resource managers.

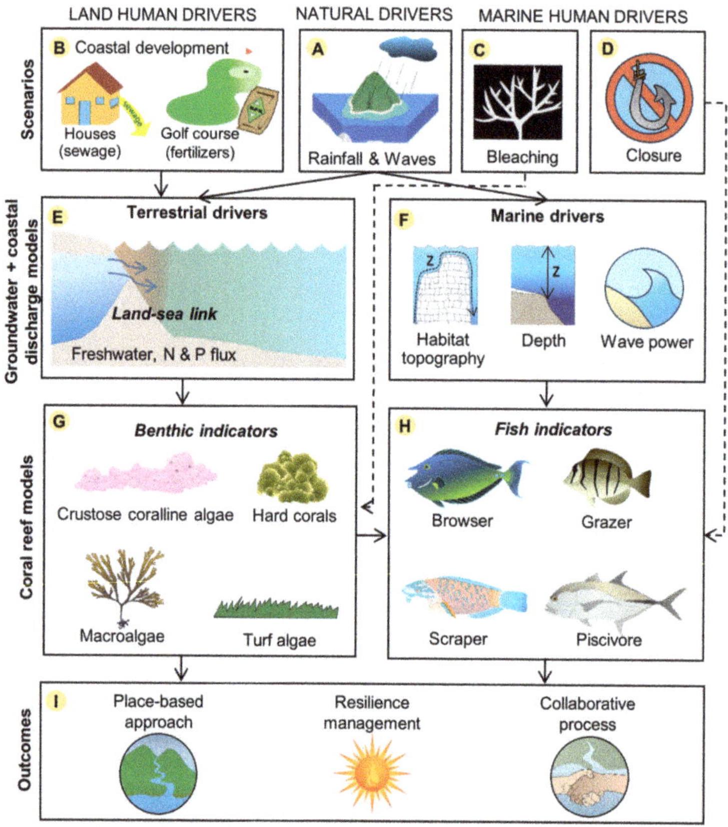

Figure 5. Linked land-sea modeling framework. The framework accounted for (**A**) natural and human drivers of coral reefs. Human drivers consisted of (**B**) land-based (coastal development) and (**C,D**) marine-based (bleaching and closure) scenarios. (**E**) The terrestrial drivers included submarine groundwater and nutrient discharge. (**F**) The marine drivers characterized the marine habitats. Under each scenario, coral reef models track changes in (**G**) benthic and (**H**) fish indicator abundance. This approach identified (**I**) priority areas on land where management can promote coral reef resilience to climate change through a collaborative process. Adapted from [3,47,53].

4.1. Place-Based Models

Due to direct exposure to the prevailing trade winds, Hā'ena *ahupua'a* receives very high rainfall (4040 mm·year^{-1}), resulting in large fluvial and groundwater inputs [54] (see Figure 6A). Dominated by steep cliffs, the Hā'ena *ahupua'a* is 7.3 km^2 and spans 1006 m elevation from the summit of Ali'inui Mountain to the sea, with two flowing perennial streams in the Limahuli and Mānoa valleys. On the other hand, Ka'ūpūlehu *ahupua'a* receives much less precipitation (ranging from 1350 to 260 mm·year^{-1} from ridge-to-reef) due to its location in the rain shadows of Mauna Loa and Mauna Kea mountains [55]. Geologically young, the surface is less eroded with poorly developed ephemeral stream channels and groundwater seeping along the coast [56] (see Figure 6B). The *ahupua'a* covers 104 km^2 and spans 2518 m elevation from the summit of Hualalai Mountain to the sea. High rainfall in Hā'ena results in nearly three times more groundwater discharge (10,279 m^3/year/m) compared to Ka'ūpūlehu (3085 m^3/year/m), which also means that nutrients are more diluted (less concentrated) than Ka'ūpūlehu, which is much drier. Our groundwater models showed that groundwater in Ka'ūpūlehu has higher levels of nitrogen from natural sources (38,900 kg/year or 7.08 kg/m/year) compared to Hā'ena (29,200 kg/year or 6.02 kg/m/year). Hā'ena is rural with limited development and agriculture, so most of the nutrients come from natural processes, with the exception of land areas to the east of the *ahupua'a* where nutrients are largely human-derived (human-derived nutrients: N: 7.8% and P: 5.5%), compared to more developed Ka'ūpūlehu (human-derived nutrients: N: 24% and P: 35%). The key sources of human-derived nutrients were wastewater from houses on cesspools at Hā'ena and the golf course and wastewater from the injection well at Ka'ūpūlehu.

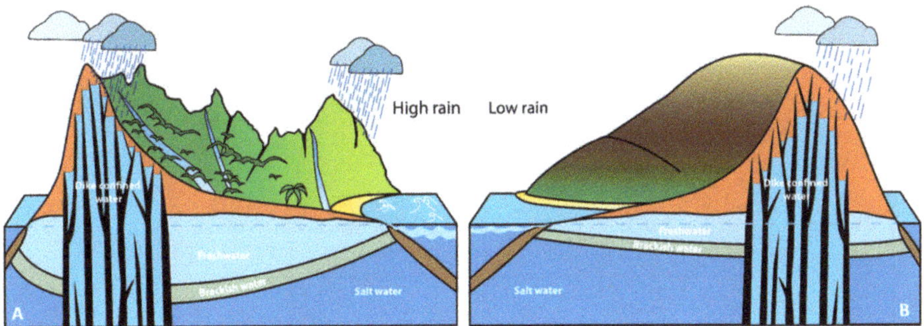

Figure 6. Illustration of the groundwater system at Hā'ena and Ka'ūpūlehu. (**A**) Hā'ena is located on old, wet, wave exposed coast of Kaua'I; (**B**) Ka'ūpūlehu is young, dry and wave sheltered.

Due to its older geological age and exposure to marine erosion from oceanic swells at Hā'ena (nearly one order of magnitude higher than Ka'ūpūlehu) has over time carved wider and shallower reef flats and produced shallow lagoons protected from the swell by well-developed reef crests [57]. The back-reef areas form lagoons that are protected from wave power by well-developed reef crests and support a benthic community dominated by corals and macroalgae [41]. The benthic community on the wave-exposed fore-reef is dominated by crustose coralline algae (CCA) and turf algae [58,59]. Our coral reef models showed that high wave power at Hā'ena has shaped the living community of the reefs, which are dominated by CCA and turf algae with many grazers and less scrapers (see Figure 7A). The Makua lagoon area is an exception where corals are able to grow, sheltered from powerful waves by a well-developed reef crest. In comparison, the coral reefs of Ka'ūpūlehu are younger and form a relatively narrow fringe on the steep slope of that island [57]. Because the reef is sheltered from large winter waves, its slopes are dominated by corals and have high habitat complexity, which supports higher fish biomass, particularly scrapers, while the shallow reef flats are dominated by turf algae with some CCA and support lower fish biomass (see Figure 7B) [48]. Browser abundance was low at both sites. Our coral reef models also showed that land-based nutrients from groundwater can increase benthic algae, suppress coral and CCA and decrease numbers of locally important fish at both sites.

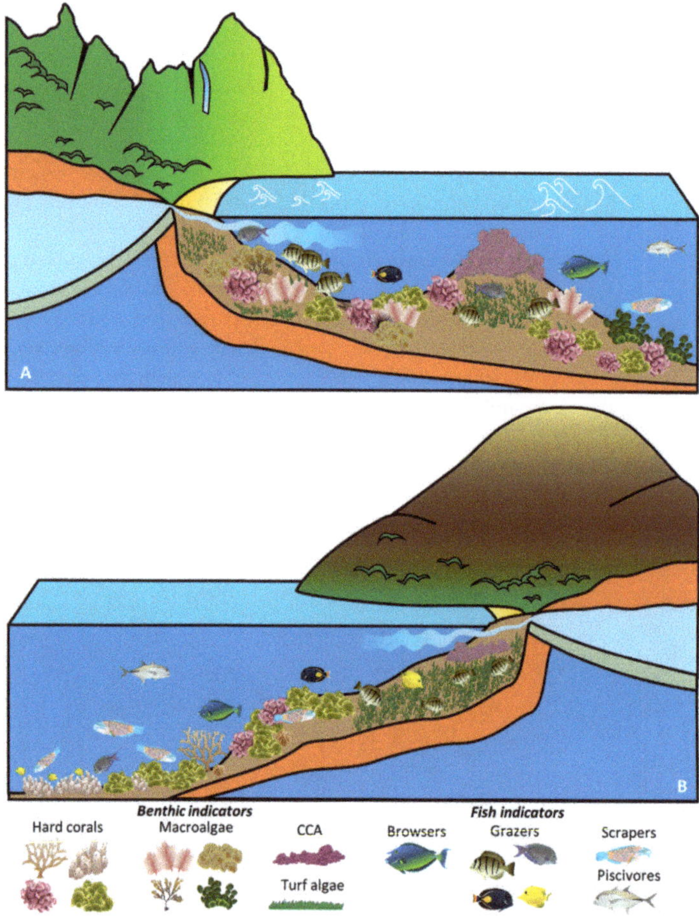

Figure 7. Illustrations of the coral reefs. (**A**) Coral reefs in Hā'ena are characterized by a reef crest dominated by crustose coralline algae (CCA) and turf algae and back reef with abundant corals and macroalgae with many grazers and less scrapers and (**B**) Coral reefs in Ka'ūpūlehu are dominated by corals on the slopes and turf algae on the reef flats with many scrapers.

4.2. Place-Based Solutions

Using this framework, we located coral reefs vulnerable to local and global human stressors and linked them to areas on land where limiting sources of human-derived nutrients could prevent increases in benthic algae and promote chances of coral recovery from bleaching. Under the high coastal development scenario, most of the total nutrient increase (>2000 kg) occurs to the east and center of the *ahupua'a*, where flushing and mixing from waves is limited by the reef crests of Makua and Pu'ukahua reefs. Some of these areas that contribute high levels of human-nutrients lie upstream from the protected reef fish nursery at Makua. Coral reefs in Hā'ena may appear less susceptible to nutrient inputs from coastal development because they benefit from dilution and mixing from high freshwater and wave power. However, we showed that the back-reef of Makua is vulnerable to algae blooms (habitat area loss: 8.2%; shift in fish biomass composition, marked in pink in Figure 8A) and coral bleaching (habitat area loss: 13%, coral percent cover loss: 0–9%, fish biomass loss: −3.3%; marked in yellow in Figure 8A) due to the nearness of human-derived nutrient sources, limited mixing due to shallow depths and low wave power, and abundant corals and algae. Under the high coastal development scenario at Ka'ūpūlehu, most of the nutrient increase (>8000 kg) occurs to the north of the *ahupua'a*, downstream from the proposed development. On the other hand, coral reefs appear more vulnerable to nutrient inputs from more coastal development, combined with higher levels of background nitrogen in the groundwater and limited dilution and mixing from low rainfall and wave power (habitat area loss: 14%, fish biomass loss: 0.6%; marked in pink in Figure 8B). Additionally, Ka'ūpūlehu's plentiful coral cover is prone to coral bleaching (habitat area loss: 13%, coral cover loss 0–13%, fish biomass loss: −1.5%; in yellow in Figure 8B).

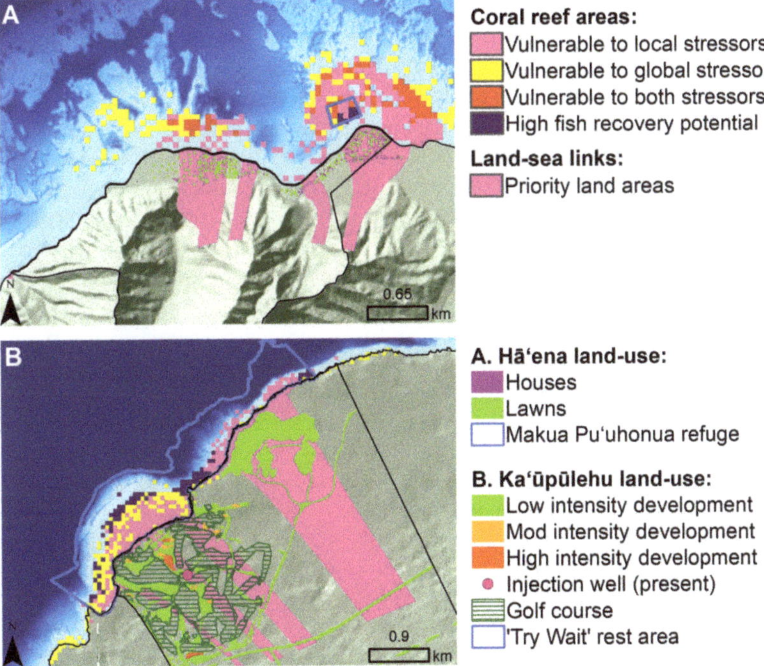

Figure 8. Coral reef areas vulnerable to local and global human stressors (i.e., nutrients and bleaching), coral reef areas with high fish recovery potential and priority land areas for management. Coral reef areas vulnerable to local and global human stressors (i.e., nutrients and bleaching), coral reef areas with high fish recovery potential and priority land areas, where local management actions can target wastewater and fertilizer practices at (**A**) Hā'ena and (**B**) Ka'ūpūlehu. Projected high coastal development land use/cover and marine closure/fishing rest areas are also shown.

Although the extent to which nutrient levels interact with elevated SST to affect the outcome of bleaching events remains poorly understood, it is increasingly recognized that water quality plays a complex role in the fate of nearshore coral reefs under climate change [60–62]. This seems to be the case since excess nutrients have been shown to impact coral reefs by promoting benthic algae growth and reducing coral's ability to recover from bleaching impacts [63,64]. When combining the effects of future coastal development and climate change on coral reefs, the impact worsens at both sites. Coral reefs vulnerable to both (coral reef areas marked in red in Figure 8A,B) do not overlap at Ka'ūpūlehu (habitat area loss: 20.8%, fish biomass loss: −1.6%), while the shallow back-reef of Makua at Hāʻena is vulnerable to both stressors due to limited wave mixing (habitat area loss: 21.1%, fish biomass loss: −3.3%).

Given that climate change and coastal development occur simultaneously, these results suggest that adopting local management can benefit both places. Land-based management can improve the benthic habitat conditions by preventing increases in benthic algae, which promotes coral recovery from bleaching within & outside the marine closures (habitat gain: 8% at both sites). Therefore, to promote coral reef resilience to climate change, the Hāʻena community may benefit from upgrading cesspools in the priority areas we identified, located upstream from Makua (located in pink zone in Figure 8A). Based on our findings, the Kaʻūpūlehu community could focus on minimizing phosphorus inputs from the wastewater injection well by increasing the nutrient removal through treatment (located in the pink zone below the injection well in Figure 8B) to reduce the vulnerability of coral reefs located downstream. In addition, the community could help foster resilience of their coral reefs by ensuring that environmentally sound practices are continued when fertilizing the golf course, particularly in the land areas located upstream from Uluweuweu bay and Kahuwai bay (located in pink zone in Figure 8B). This may also help to protect the water quality of a culturally important groundwater spring (*Wai a Kāne*) that was identified by the Kaʻūpūlehu community in Kahuwai bay (Figure 3B). While marine-based management increases the herbivore population within the reserves, which can supplement adjacent reef through spillover (fish biomass gain within the marine closure boundaries: +13% at Hāʻena and +2.6% at Kaʻūpūlehu). Overall, this research supports the communities' concerns and provides evidence that more coastal development can potentially negatively impact culturally important fisheries at both sites.

4.3. Application and Transferability

Ridge-to-reef management that integrates LEK has been widely advocated because it can improve social-ecological resilience. Watershed units have commonly defined the ecological systems in traditional management systems, which have been found in the Pacific north west, Asia, Africa and Oceania [65]. Among the richest set of ridge-to-reef, social-ecological system approaches to natural resources management is found in Oceania [8,65]. Examples include the *tambak* in Indonesia [66], the *puava* in the Solomon Islands, the *tabinau* in Yap, the *vanua* in Fiji [67] and the *moku* in Hawaiʻi [27]. Through this research, we show that place-based solutions that integrate land and sea processes are critical for addressing local environmental threats. We demonstrate that culturally grounded and inclusive research can guide management actions with multiple benefits such as improved groundwater and coastal water quality and foster the resilience of coral reefs, which are important food production systems for local communities. The lessons learned from this process highlight the critical aspects of collaboration necessary to develop scientific tools that can inform these practical and appropriate management actions. Managing ICCAs requires taking into account interests at all levels, evaluating trade-offs and finding win-win solutions [23].

There is a strong need for planning tools that can prioritize local management actions at relevant spatial scales for decision makers, which are simple to interpret and implement [68,69]. These decision support tools can easily be updated as more data becomes available, or model components of the framework can be substituted or added based on management objectives. For example, in another application, we substituted the groundwater models with the open source Integrated Valuation of Ecosystem Services and Tradeoffs (InVEST) spatially-explicit Sediment Delivery Ratio Model (SDR version 3.2) [70] to model the impact of sediment runoff on coral reefs in Fiji [71]. To support Fijian communities currently working with government, NGOs and the private sector to design and implement Integrated Coastal Management plans [72], we applied the modified land-sea modeling framework with scenario planning in Kubulau District (Fiji), where logging and commercial agriculture expansion competes with forest conservation and potentially fisheries livelihoods, to identify where forest conservation or restoration actions could benefit coral reefs [71]. In addition to fostering collaboration, this approach offers a flexible, transferable, data-driven, place-based model that is spatially-explicit and relies on increasingly available free remote sensing imagery and bathymetry data (i.e., Worldview III, GEBCO).

5. Conclusions

These research findings suggest that different environmental conditions make place-based solutions essential [23], because one-size-fits-all kinds of management ignore issues of place and scale [26]. Rodgers et al. [73] provided the first quantitative statewide evidence that watershed and adjacent coral reef health are significantly interconnected in Hawai'i, with the exception of ridge-to-reef systems on the windward side due to exposure to high rainfall and wave power, implying that reefs in these locales are less vulnerable to land-based activities. Consistent with Rodgers et al. [73], our impact assessment showed that Ka'ūpūlehu is vulnerable to local land-use change, as well as climate change impacts. Our findings also revealed that coral reefs on the windward side are vulnerable to local land-use change at the local-scale, especially back-reef systems, like Makua back-reef. In addition, our results show that managing local human drivers can foster coral reef resilience to global human drivers. Although the marine closures can promote reef recovery, they are not always able to offset the impacts from coastal development and other land-based activities on coral reefs, especially beyond their boundaries. Due to the risk that coastal development can undermine local marine conservation efforts, it is essential to manage upstream land-use change.

Therefore, local-scale and place-based solutions are particularly important in Hawai'i, where locally sourced food is socially and culturally important and food systems are vulnerable to coastal development and climate change impacts [18,33]. Our research provides place-based case studies of the interaction of researchers, community members, resource managers and policy makers to inform future planning. ICCAs, such as Hā'ena and Ka'ūpūlehu, can help redefine co-management and the role of local communities and institutions throughout Hawai'i. Although this management approach may be more suited for communities with strong ancestral ties to the place, as larger and more heterogeneous communities, such as Maunalua Bay on East O'ahu, will require early onset and more efforts to build consensus [74] and necessitate creative strategies to engage the various members [75]. However, this type of research can help coordinate and facilitate reaching agreements across different community groups by testing policies prior to implementation and bridging gaps between managers and communities by visualizing synergies and trade-offs on maps.

Author Contributions: Conceptualization, J.M.S.D., K.B.W., S.D.J., M.B.-V., L.L.B., K.B., T.T.; Methodology, J.M.S.D., K.B.W., S.D.J.; Software, J.L.T.; Validation, J.M.S.D. and K.B.W.; Formal Analysis, J.M.S.D.; Investigation, K.B.W., S.D.J., M.B.-V., K.B., T.T.; Resources, J.M.S.D., J.L.T.; Data Curation, J.M.S.D.; Writing—Original Draft Preparation, J.M.S.D. and K.B.W.; Writing—Review & Editing, K.B.W., S.D.J., M.B.-V., L.L.B., K.A.S., K.B., P.G., J.L.T., T.T.; Visualization, J.M.S.D. & J.L.T.; Supervision, K.W., S.D.J., T.T.; Project Administration, K.B.W.; Funding Acquisition, K.B.W., S.D.J., M.B.-V., K.B., T.T.; J.M.S.D., J.L.T., K.A.S. revised the illustrations to reflect the specificity of each place.

Funding: This research was funded by the NSF Coastal SEES project number 1325874 and the APC was funded by Hawai'i Community Foundation.

Acknowledgments: Many thanks to a collaborating artist, Sophie Eugène, for the illustrations and the Integration and Application Network, University of Maryland Center for Environmental Science (ian.umces.edu/symbols/) for the marine symbols. The authors are very grateful for Robert Whittier from the Hawai'i Department of Health for embarking on this collaborative research that allowed us to produce a story that goes beyond the sum of its parts. The authors would like to acknowledge the Principal Investigators (Alan M. Friedlander, Heather McMillen), the research collaborators (Greg Guannel, Natalie Kurashima, Lisa Mandle, Robert Toonen, Anders Knudby, Jonatha Giddens, Cheryl Geslani) and data providers (Tracy Wiegner, Chad Wiggins, & Eric Conklin) on this research project. Finally, we are grateful to our community members, key landowner partners, Kamahameha schools in Ka'ūpūlehu and, Limahuli Gardens and Preserve in Hā'ena, who inspired this research and made this project possible.

Conflicts of Interest: The authors declare no competing interests as defined by *Sustainability*, or other interests that might be perceived to influence the results and/or discussion reported in this paper. The founding sponsors had no role in the design of the study; in the collection, analyses, or interpretation of data; in the writing of the manuscript and in the decision to publish the results.

References

1. Maris, K.A. Under What Circumstances do People Put Unsustainable Demands on Island Environments?: Evidence from the North Atlantic. Ph.D. Thesis, University of Edinburgh, Edinburgh, UK, 2007.
2. Jupiter, S.D.; Wenger, A.; Klein, C.J.; Albert, S.; Mangubhai, S.; Nelson, J.; Teneva, L.; Tulloch, V.J.; White, A.T.; Watson, J.E. Opportunities and constraints for implementing integrated land–sea management on islands. *Environ. Conserv.* **2017**. [CrossRef]
3. Delevaux, J.M.S.; Whittier, R.; Stamoulis, K.A.; Bremer, L.L.; Jupiter, S.; Friedlander, A.M.; Poti, M.; Guannel, G.; Kurashima, N.; Winter, K.; et al. A linked land-sea modeling framework to inform ridge-to-reef management in high oceanic islands. *PLoS ONE* **2018**, *13*, e0193230. [CrossRef] [PubMed]
4. Folke, C.; Carpenter, S.R.; Walker, B.; Scheffer, M.; Chapin, T.; Rockström, J. Resilience Thinking. *Ecol. Soc.* **2010**, *15*, art20. [CrossRef]
5. Nyström, M.; Folke, C.; Moberg, F. Coral reef disturbance and resilience in a human-dominated environment. *Trends Ecol. Evol.* **2000**, *15*, 413–417. [CrossRef]
6. Vaughan, M.B.; Vitousek, P.M. Mahele: Sustaining communities through small-scale inshore fishery catch and sharing networks. *Pac. Sci.* **2013**, *67*, 329–344. [CrossRef]
7. Barnett, J.; Campbell, J. *Climate Change and Small Island States: Power, Knowledge, and the South Pacific*; Earthscan: London, UK, 2010.
8. Berkes, F. Indigenous ways of knowing and the study of environmental change. *J. R. Soc. N. Z.* **2009**, *39*, 151–156. [CrossRef]
9. Friedlander, A.M.; Shackeroff, J.M.; Kittinger, J.N. Customary marine resource knowledge and use in contemporary Hawai'i. *Pac. Sci.* **2013**, *67*, 441–460. [CrossRef]
10. Ticktin, T.; Whitehead, A.N.; Fraiola, H. Traditional gathering of native hula plants in alien-invaded Hawaiian forests: Adaptive practices, impacts on alien invasive species and conservation implications. *Environ. Conserv.* **2006**, *33*, 185–194. [CrossRef]
11. Gadgil, M.; Berkes, F.; Folke, C. Indigenous Knowledge for Biodiversity Conservation. *Ambio* **1993**, *22*, 151–156.
12. Poepoe, K.K.; Bartram, P.K.; Friedlander, A.M. The use of traditional knowledge in the contemporary management of a Hawaiian community's marine resources. In *Fishers' Knowledge in Fisheries Science and Management*; Haggan, N., Neis, B., Baird, I.G., Eds.; UNESCO Publishing: Paris, France, 2005.
13. Berkes, F.; Colding, J.; Folke, C. Rediscovery of traditional ecological knowledge as adaptive management. *Ecol. Appl.* **2000**, *10*, 1251–1262. [CrossRef]
14. McMillen, H.; Ticktin, T.; Friedlander, A.; Jupiter, S.; Thaman, R.; Campbell, J.; Veitayaki, J.; Giambelluca, T.; Nihmei, S.; Rupeni, E.; et al. Small islands, valuable insights: Systems of customary resource use and resilience to climate change in the Pacific. *Ecol. Soc.* **2014**, *19*, 44. [CrossRef]
15. Berkes, F.; Folke, C.; Colding, J. *Linking Social and Ecological Systems: Management Practices and Social Mechanisms for Building Resilience*; Cambridge University Press: Cambridge, UK, 2000; ISBN 978-0-521-78562-4.
16. Keener, V. *Climate Change and Pacific Islands: Indicators and Impacts: Report for the 2012 Pacific Islands Regional Climate Assessment*; Island Press: Washington, DC, USA, 2013.
17. Johannes, R.E. The renaissance of community-based marine resource management in Oceania. *Annu. Rev. Ecol. Syst.* **2002**, *33*, 317–340. [CrossRef]
18. McGregor, D.P.; Morelli, P.; Matsuoka, J.; Minerbi, L.; Becker, H.A.; Vanclay, F. An ecological model of well-being. In *The International Handbook of Social Impact Assessment Conceptual and Methodological Advances*; Edward Elgar Publishing: Cheltenham, UK, 2003; pp. 109–126.
19. Minerbi, L. Indigenous management models and protection of the ahupua'a. *Soc. Process Hawai'i* **1999**, *39*, 208–225.
20. Foale, S.; Cohen, P.; Januchowski-Hartley, S.; Wenger, A.; Macintyre, M. Tenure and taboos: Origins and implications for fisheries in the Pacific. *Fish Fish.* **2011**, *12*, 357–369. [CrossRef]
21. IUCN. *Indigenous and Community Conserved Areas*; IUCN: Gland, Switzerland, 2008.
22. Pathak, N.; Bhatt, S.; Balasinorwala, T.; Kothari, A.; Borrini-Feyerabend, G. Community Conserved Areas: A Bold Frontier for Conservation. TILCEP/AIUCN, CENESTA, CMWG and WAMIP: Tehran. Available online: http://cmsdata.iucn.org/downloads/cca_briefing_note.pdf (accessed on 3 September 2018).

23. Berkes, F. Community conserved areas: Policy issues in historic and contemporary context. *Conserv. Lett.* **2009**, *2*, 20–25. [CrossRef]

24. Oviedo, G. Community-conserved areas in South America. *Parks* **2006**, *16*, 49–55.

25. Vaughan, M.B.; Ardoin, N.M. The implications of differing tourist/resident perceptions for community-based resource management: A Hawaiian coastal resource area study. *J. Sustain. Tour.* **2014**, *22*, 50–68. [CrossRef]

26. Berkes, F. Rethinking Community-Based Conservation. *Conserv. Biol.* **2004**, *18*, 621–630. [CrossRef]

27. Winter, K.B.; Beamer, K.; Vaughan, M.B.; Friedlander, A.M.; Kido, M.; Akutagawa, M.K.H.; Kurashima, N.; Nyberg, B. The Moku System: Managing biocultural resources for abundance within social-ecological regions. *Sustainability* **2018**. under review.

28. Bremer, L.L.; Delevaux, J.M.; Leary, J.J.; Cox, L.J.; Oleson, K.L. Opportunities and Strategies to Incorporate Ecosystem Services Knowledge and Decision Support Tools into Planning and Decision Making in Hawai'i. *Environ. Manag.* **2015**, *55*, 884–899. [CrossRef] [PubMed]

29. Jokiel, P.L.; Rodgers, K.S.; Walsh, W.J.; Polhemus, D.A.; Wilhelm, T.A. Marine resource management in the Hawaiian Archipelago: The traditional Hawaiian system in relation to the Western approach. *J. Mar. Biol.* **2010**, *2011*, 151682. [CrossRef]

30. Derrickson, S.A.K.; Robotham, M.P.; Olive, S.G.; Evensen, C.I. Watershed management and policy in Hawaii: Coming full circle. *J. Am. Water Resour. Assoc.* **2002**, *38*, 563–576. [CrossRef]

31. Jupiter, S.D.; Cohen, P.J.; Weeks, R.; Tawake, A.; Govan, H. Locally-managed marine areas: Multiple objectives and diverse strategies. *Pac. Conserv. Biol.* **2014**, *20*, 165–179. [CrossRef]

32. McGregor, D. *Na Kua'aina: Living Hawaiian Culture*; University of Hawaii Press: Honolulu, HI, USA, 2007; ISBN 978-0-8248-2946-9.

33. Vaughan, M.B.; Ayers, A.L. Customary Access: Sustaining Local Control of Fishing and Food on Kaua'i's North Shore. *Food Cult. Soc.* **2016**, *19*, 517–538. [CrossRef]

34. Pascua, P.; McMillen, H.; Ticktin, T.; Vaughan, M.; Winter, K.B. Beyond services: A process and framework to incorporate cultural, genealogical, place-based, and indigenous relationships in ecosystem service assessments. *Ecosyst. Serv.* **2017**, *26*, 465–475. [CrossRef]

35. Kittinger, J.N.; Teneva, L.T.; Koike, H.; Stamoulis, K.A.; Kittinger, D.S.; Oleson, K.L.L.; Conklin, E.; Gomes, M.; Wilcox, B.; Friedlander, A.M. From Reef to Table: Social and Ecological Factors Affecting Coral Reef Fisheries, Artisanal Seafood Supply Chains, and Seafood Security. *PLoS ONE* **2015**, *10*, e0123856. [CrossRef] [PubMed]

36. Bremer, L.L.; Mandle, L.; Trauernicht, C.; Pascua, P.; McMillen, H.L.; Burnett, K.; Wada, C.A.; Kurashima, N.; Quazi, S.A.; Giambelluca, T.; et al. Bringing multiple values to the table: Assessing future land-use and climate change in North Kona, Hawai'i. *Ecol. Soc.* **2018**, *23*. [CrossRef]

37. Vaughan, M.B.; Caldwell, M.R. Hana Pa'a: Challenges and lessons for early phases of co-management. *Mar. Policy* **2015**, *62*, 51–62. [CrossRef]

38. DAR. *Management Plan for the Hā'ena Community-Based Subsistence Fisheries Area, Kauai*; Division of Aquatic Resources, Hawaii Department of Land and Natural Resources: Honolulu, HI, USA, 2016.

39. Andrade, C. *Ha'ena: Through the Eyes of the Ancestors*; University of Hawaii Press: Honolulu, HI, USA, 2008; ISBN 978-0-8248-3119-6.

40. Vaughan, M.B.; Thompson, B.; Ayers, A.L. Pāwehe Ke Kai a'o Hā'ena: Creating State Law based on Customary Indigenous Norms of Coastal Management. *Soc. Nat. Resour.* **2017**, *30*, 31–46. [CrossRef]

41. Goodell, W. Coupling Remote Sensing with In-Situ Surveys to Determine Reef Fish Habitat Associations for the Design of Marine Protected Areas. Master's Thesis, The University of Hawai'i, Honolulu, HI, USA, 2015.

42. Goldstein, J.H.; Caldarone, G.; Duarte, T.K.; Ennaanay, D.; Hannahs, N.; Mendoza, G.; Polasky, S.; Wolny, S.; Daily, G.C. Integrating ecosystem-service tradeoffs into land-use decisions. *Proc. Natl. Acad. Sci. USA* **2012**. [CrossRef] [PubMed]

43. Kamehameha Schools. *Kūhanauna: A Generation on the Rise*; Kamehameha Schools Strategic Plan 2015–2020; Kamehameha Schools: Honolulu, HI, USA, 2016.

44. TNC. *Ka'ūpūlehu Conservation Action Plan*; The Nature Conservancy: Arlington County, VA, USA, 2015.

45. McMillen, H.; Ticktin, T.; Springer, H.K. The future is behind us: Traditional ecological knowledge and resilience over time on Hawai'i Island. *Reg. Environ. Chang.* **2017**, *17*, 579–592. [CrossRef]

46. Olsson, P.; Folke, C. Local Ecological Knowledge and Institutional Dynamics for Ecosystem Management: A Study of Lake Racken Watershed, Sweden. *Ecosystems* **2001**, *4*, 85–104. [CrossRef]

47. Delevaux, J.M.S.; Stamoulis, K.; Whittier, R.B.; Jupiter, S.D.; Bremer, L.L.; Friedlander, A.M.; Kurashima, N.; Vaughan, M.B.; Winter, K.B.; Giddens, J.L.; et al. Local place-based management can promote coral reef resilience to climate change. *Ecol. Appl.* **2018**. under review.

48. Minton, D.; Conklin, E.; Friedlander, A.; Most, R.; Pollock, K.; Stamoulis, K.; Wiggins, C. *Establishing the Baseline Condition of the Marine Resources: Results of the 2012 and 2013 Ka'ūpūlehu, Hawai'i Marine Surveys*; The Nature Conservancy: Arlington County, VA, USA, 2015; pp. 1–46.

49. Sandin, S.A.; Smith, J.E.; DeMartini, E.E.; Dinsdale, E.A.; Donner, S.D.; Friedlander, A.M.; Konotchick, T.; Malay, M.; Maragos, J.E.; Obura, D.; et al. Baselines and Degradation of Coral Reefs in the Northern Line Islands. *PLoS ONE* **2008**, *3*, e1548. [CrossRef] [PubMed]

50. The Integration and Application Network. Symbols. Available online: ian.umces.edu/symbols/ (accessed on 3 September 2018).

51. Hoeke, R.K.; Jokiel, P.L.; Buddemeier, R.W.; Brainard, R.E. Projected changes to growth and mortality of Hawaiian corals over the next 100 years. *PLoS ONE* **2011**, *6*, e18038. [CrossRef] [PubMed]

52. Lovell, E.; Sykes, H.; Deiye, M.; Wantiez, L.; Garrigue, C.; Virly, S.; Samuelu, J.; Solofa, A.; Poulasi, T.; Pakoa, K.; et al. Status of Coral Reefs in the South West Pacific: Fiji, Nauru, New Caledonia, Samoa, Solomon Islands, Tuvalu and Vanuatu. *Status Coral Reefs World* **2004**, *2*, 337–362.

53. Liu, Y.; Gupta, H.; Springer, E.; Wagener, T. Linking science with environmental decision making: Experiences from an integrated modeling approach to supporting sustainable water resources management. *Environ. Model. Softw.* **2008**, *23*, 846–858. [CrossRef]

54. Calhoun, R.S.; Fletcher, C.H. Measured and predicted sediment yield from a subtropical, heavy rainfall, steep-sided river basin: Hanalei, Kauai, Hawaiian Islands. *Geomorphology* **1999**, *30*, 213–226. [CrossRef]

55. Izuka, S.K.; Engott, J.A.; Bassiouni, M.; Johnson, A.G.; Miller, L.D.; Rotzoll, K.; Mair, A. *Volcanic Aquifers of Hawai'i—Hydrogeology, Water Budgets, and Conceptual Models*; Scientific Investigations Report; U.S. Geological Survey: Reston, VA, USA, 2016; p. 172.

56. Knee, K.L.; Street, J.H.; Grossman, E.E.; Boehm, A.B.; Paytan, A. Nutrient inputs to the coastal ocean from submarine groundwater discharge in a groundwater-dominated system: Relation to land use (Kona coast, Hawaii, U.S.A.). *Limnol. Oceanogr.* **2010**, *55*, 1105–1122. [CrossRef]

57. Fletcher, C.H.; Bochicchio, C.; Conger, C.L.; Engels, M.S.; Feirstein, E.J.; Frazer, N.; Glenn, C.R.; Grigg, R.W.; Grossman, E.E.; Harney, J.N.; et al. Geology of Hawaii Reefs. In *Coral Reefs of the USA*; Riegl, B.M., Dodge, R.E., Eds.; Springer: Cham, The Netherlands, 2008; pp. 435–487. ISBN 978-1-4020-6846-1.

58. Jokiel, P.L.; Brown, E.K.; Friedlander, A.; Rodgers, S.K.; Smith, W.R. Hawai'i Coral Reef Assessment and Monitoring Program: Spatial Patterns and Temporal Dynamics in Reef Coral Communities. *Pac. Sci.* **2004**, *58*, 159–174. [CrossRef]

59. Friedlander, A.M.; Brown, E.K.; Jokiel, P.L.; Smith, W.R.; Rodgers, K.S. Effects of habitat, wave exposure, and marine protected area status on coral reef fish assemblages in the Hawaiian archipelago. *Coral Reefs* **2003**, *22*, 291–305. [CrossRef]

60. Anthony, K.R.N. Enhanced energy status of corals on coastal, high-turbidity reefs. *Mar. Ecol. Prog. Ser.* **2006**, *319*, 111–116. [CrossRef]

61. Wenger, A.S.; Williamson, D.H.; da Silva, E.T.; Ceccarelli, D.M.; Browne, N.K.; Petus, C.; Devlin, M.J. Effects of reduced water quality on coral reefs in and out of no-take marine reserves. *Conserv. Biol.* **2015**, *30*, 142–153. [CrossRef] [PubMed]

62. Morgan, K.M.; Perry, C.T.; Johnson, J.A.; Smithers, S.G. Nearshore Turbid-Zone Corals Exhibit High Bleaching Tolerance on the Great Barrier Reef Following the 2016 Ocean Warming Event. *Front. Mar. Sci.* **2017**, *4*, 224. [CrossRef]

63. Fabricius, K.E. Effects of terrestrial runoff on the ecology of corals and coral reefs: Review and synthesis. *Mar. Pollut. Bull.* **2005**, *50*, 125–146. [CrossRef] [PubMed]

64. Smith, J.E.; Brainard, R.; Carter, A.; Grillo, S.; Edwards, C.; Harris, J.; Lewis, L.; Obura, D.; Rohwer, F.; Sala, E.; et al. Re-evaluating the health of coral reef communities: Baselines and evidence for human impacts across the central Pacific. *Proc. R. Soc. B* **2016**, *283*, 20151985. [CrossRef] [PubMed]

65. Berkes, F.; Kislalioglu, M.; Folke, C.; Gadgil, M. Minireviews: Exploring the basic ecological unit: Ecosystem-like concepts in traditional societies. *Ecosystems* **1998**, *1*, 409–415. [CrossRef]

66. Johannes, R.; Lasserre, P.; Nixon, S.; Pliya, J.; Ruddle, K. Traditional Knowledge and Management of Marine Coastal Systems. *Biol. Int. Spec.* **1983**. Special Issue No. 4. Available online: iubs.org/pdf/publi/BISI/SPECIAL%20ISSUE%204a.pdf (accessed on 30 August 2018).

67. Ruddle, K.; Akimichi, T. *Maritime Institutions in the Western Pacific*; Serie Ethnological Studies; National Museum of Ethnology: Osaka, Japan, 1984; Volume 17.

68. Melbourne-Thomas, J.; Johnson, C.R.; Fung, T.; Seymour, R.M.; Chérubin, L.M.; Arias-González, J.E.; Fulton, E.A. Regional-scale scenario modeling for coral reefs: A decision support tool to inform management of a complex system. *Ecol. Appl.* **2011**, *21*, 1380–1398. [CrossRef] [PubMed]

69. Guerry, A.D.; Ruckelshaus, M.H.; Arkema, K.K.; Bernhardt, J.R.; Guannel, G.; Kim, C.-K.; Marsik, M.; Papenfus, M.; Toft, J.E.; Verutes, G. Modeling benefits from nature: Using ecosystem services to inform coastal and marine spatial planning. *Int. J. Biodivers. Sci. Ecosyst. Serv. Manag.* **2012**, *8*, 107–121. [CrossRef]

70. Hamel, P.; Chaplin-Kramer, R.; Sim, S.; Mueller, C. A new approach to modeling the sediment retention service (InVEST 3.0): Case study of the Cape Fear catchment, North Carolina, USA. *Sci. Total Environ.* **2015**, *524*, 166–177. [CrossRef] [PubMed]

71. Raudsepp-Hearne, C.; Peterson, G.D.; Bennett, E.M. Ecosystem service bundles for analyzing tradeoffs in diverse landscapes. *Proc. Natl. Acad. Sci. USA* **2010**, *107*, 5242–5247. [CrossRef] [PubMed]

72. Delevaux, J.M.S.; Jupiter, S.D.; Stamoulis, K.A.; Bremer, L.L.; Wenger, A.S.; Dacks, R.; Garrod, P.; Falinski, K.A.; Ticktin, T. Scenario planning with linked land-sea models inform where forest conservation actions will promote coral reef resilience. *Sci. Rep.* **2018**, *8*, 12465. [CrossRef] [PubMed]

73. Ku'ulei, S.R.; Kido, M.H.; Jokiel, P.L.; Edmonds, T.; Brown, E.K. Use of integrated landscape indicators to evaluate the health of linked watersheds and coral reef environments in the Hawaiian Islands. *Environ. Manag.* **2012**, *50*, 21–30.

74. Ayers, A.L.; Kittinger, J.N. Emergence of co-management governance for Hawai'i coral reef fisheries. *Glob. Environ. Chang.* **2014**, *28*, 251–262. [CrossRef]

75. Lowry, K.; Adler, P.; Milner, N. Participating the Public: Group Process, Politics, and Planning. *J. Plan. Educ. Res.* **1997**, *16*, 177–187. [CrossRef]

 *sustainability* 

Article

Ma Kahana ka 'Ike: Lessons for Community-Based Fisheries Management

Monica Montgomery [1,*] **and Mehana Vaughan** [1,2,3]

1   Department of Natural Resources and Environmental Management, College of Tropical Agriculture and Human Resources, University of Hawai'i at Mānoa, Honolulu, HI 96822, USA; mehana@hawaii.edu
2   Hui 'Āina Momona, University of Hawai'i at Mānoa, Honolulu, HI 96822, USA
3   University of Hawai'i Sea Grant College Program, University of Hawai'i at Mānoa, Honolulu, HI 96822, USA
*   Correspondence: mongomer@hawaii.edu

Received: 3 August 2018; Accepted: 17 October 2018; Published: 20 October 2018

Abstract: Indigenous and place-based communities worldwide have self-organized to develop effective local-level institutions to conserve biocultural diversity. How communities maintain and adapt these institutions over time offers lessons for fostering more balanced human–environment relationships—an increasingly critical need as centralized governance systems struggle to manage declining fisheries. In this study, we focus on one long-enduring case of local level fisheries management, in Kahana, on the most populated Hawaiian island of O'ahu. We used a mixed-methods approach including in-depth interviews, archival research, and participation in community gatherings to understand how relationships with place and local governance have endured despite changes in land and sea tenure, and what lessons this case offers for other communities engaged in restoring local-level governance. We detail the changing role of konohiki (head fishermen) in modern times (1850–1965) when they were managing local fisheries, not just for local subsistence but for larger commercial harvests. We also highlight ways in which families are reclaiming their role as caretakers following decades of state mismanagement. Considerations for fisheries co-management emerging from this research include the importance of (1) understanding historical contexts for enhancing institutional fit, (2) enduring community leadership, (3) balancing rights and responsibilities, and (4) fostering community ability to manage coastal resources through both formal and informal processes.

Keywords: traditional resource management; konohiki; co-management; institutional fit; social-ecological systems; biocultural restoration; fisheries; Hawai'i

1. Introduction

Societies settled in a particular place for an extended period of time tend to co-evolve with their environment, adjusting resource use to ecological variability and social changes [1–3]. Accumulated knowledge of place informs collective decision-making, which over time shapes institutions, or rules-in-use, that are compatible with local social–ecological systems [4,5]. Local level institutions and the generations of knowledge that inform them are critical to achieving sustainable fisheries [6–8]. Due to colonial and economic influences, these time-tested systems have become fragmented or displaced in many parts of the world [7,9,10]. However, many rural and Indigenous communities, with growing support from governing bodies and resource managers, are reviving and adapting Indigenous and place-based resource management systems to fit contemporary contexts, often through collaborative partnerships [11–14].

Community-based collaborative management, or co-management, is the sharing of management authority and responsibility, often between local communities and government agencies [15]. Through co-management, community groups can exercise greater autonomy and decision-making power, support short-staffed and underfunded government agencies, apply traditional and place-based knowledge, and tailor management to local social–ecological contexts [16–19]. Co-management is an ongoing process that evolves through practice and social learning [19]. This joint learning-by-doing process offers a promising institutional framework through which communities can exercise greater influence while enhancing the adaptive capacity of coastal social–ecological systems [19,20].

Hawai'i's *konohiki* system is an example of a community-based co-management institution that adapted to social–ecological change. Understanding how locally managed *konohiki* fisheries continued to operate under the territory and later state of Hawai'i can inform community efforts to restore local level fisheries governance within contemporary centralized state management systems. In the *ahupua'a* of Kahana, *konohiki* fishing rights, which were terminated throughout most of Hawai'i in 1900, were officially recognized through the mid-1960s. As one of the longest lasting *konohiki* fisheries in Hawai'i, Kahana offers insights into how features of this Hawaiian biocultural resource management system endured in modern times (1850–1965). Current community efforts to adapt features of this system to contemporary contexts also provide important considerations for collaborative management and biocultural restoration. Using a mixed-methods approach, we explore how relationships with place and local governance endure despite changes in land and sea tenure, and what lessons this case offers for other communities engaged in restoring local level fisheries governance.

2. Background

2.1. History of Hawaiian Fisheries Management

Hawaiian fisheries management is based on familial and reciprocal relationships with marine resources [21,22]. The land and sea were not owned but communally accessed and cared for at the local level, within *moku* (districts) or *ahupua'a*, defined as "culturally appropriate, ecologically aligned, and place specific [land divisions] with access to diverse resources" [23] (p. 71). Within *ahupua'a*, residents and *konohiki* shared both stewardship responsibilities and exclusive harvest rights [24].

Konohiki, or local headmen, were traditionally appointed by ruling chiefs to oversee the well-being of *ahupua'a* resources and residents [24,25]. *Konohiki* held extensive knowledge of the local ecology and natural cycles in order to effectively monitor fishery health [24–26]. Through consultation with local elders and expert fishermen, *konohiki* determined when it was appropriate to place restrictions on certain species or areas to protect their replenishment [26]. Adherence to these restrictions was motivated by strict enforcement as well as shared cultural, social, and spiritual values [26,27]. *Konohiki*, which translates to "to invite ability," also had to earn and maintain the respect of *ahupua'a* residents in order to mobilize participation in communal caretaking and harvesting efforts [28]. If *konohiki* did not treat the people fairly, residents, who tended the land and sea, were free to move to a different *ahupua'a* [24,25]. Thus '*āina momona*, or abundant lands, was an indication of balance and harmony between *konohiki* and *ahupua'a* residents [24].

Konohiki fishing rights were first written into law in 1839 with Hawai'i's Declaration of Rights, and later codified with the passage of the Civil Code of 1859. These laws designated fishing grounds for the exclusive access of *konohiki* and *ahupua'a* residents, "but not for others" [29] (p. 2). The *konohiki* could legally regulate the fishery by either placing a restriction each year on one species for personal use, or by prohibiting all fishing during certain months, then exacting one-third of the catch upon reopening the fishery [29].

Western encroachment transformed Hawaiian land and sea tenure in the mid- to late-19th century. Transition from a communal land tenure system to a private property regime began with Hawai'i's land division process in 1846–1855, which apportioned the land among the King, ruling chiefs or *konohiki*, and *ahupua'a* residents. Although intended to secure Native Hawaiian rights to ancestral lands and resources, creation of fee simple land titles instead paved the way for dispossession [30]. Following the overthrow of the Hawaiian monarchy, the Organic Act of 1900 established Hawai'i as a USA territory and required registration of all *konohiki* fisheries. Only 101 of the estimated 300 to 400 known *konohiki* fisheries were successfully registered [29]. From 1900 through the 1970s, the Hawai'i territorial and later state governments sought to systematically condemn even these remaining local fisheries, opening them for public access [29,31].

Today, the State of Hawai'i's Department of Land and Natural Resources exercises authority over Hawai'i's 750 m of coastline and 1.3 million acres of state lands and coastal waters, which extend 3 m offshore [32]. Fisheries regulations based on size and bag limits, seasonal closures, and gear restrictions are administered statewide through the department's Division of Aquatic Resources. This centralized, top-down management contrasts the community-based *konohiki* system that maintained fishery abundance in past generations [26,33]. Hawai'i's fisheries have considerably declined under state management [34] due to insufficient funds, staffing, and place-based knowledge tailored to address local social-ecological complexities [26,33,35]. In response, communities across Hawai'i, including fishing families of Kahana on the island of O'ahu, are working to restore local level governance through reviving traditional knowledge and place-based caretaking.

2.2. Study Site: Kahana, Ko'olauloa, O'ahu

The *ahupua'a* of Kahana is located within the *moku* (district) of Ko'olauloa on the windward side of O'ahu, Hawai'i's most developed island (Figure 1). This rural *ahupua'a* encompasses a single watershed (5228.7 acres) that empties into Kahana Bay, an important spawning and nursery ground for many native aquatic species. Kahana's nearshore fishery sustained a thriving Native Hawaiian population prior to Western contact [36], and continues to be important for the cultural identity and subsistence of area families. Kahana has a long history of Native Hawaiian presence, with archaeological research indicating habitation for over 800 years [37]. Historical continuity is further evidenced by Kahana's many *mo'olelo* (oral stories) of Hawaiian deities dating from time immemorial, and by its cultural landscapes and seascapes [38–41], including Huilua Fishpond, a stone wall enclosure traditionally used for aquaculture.

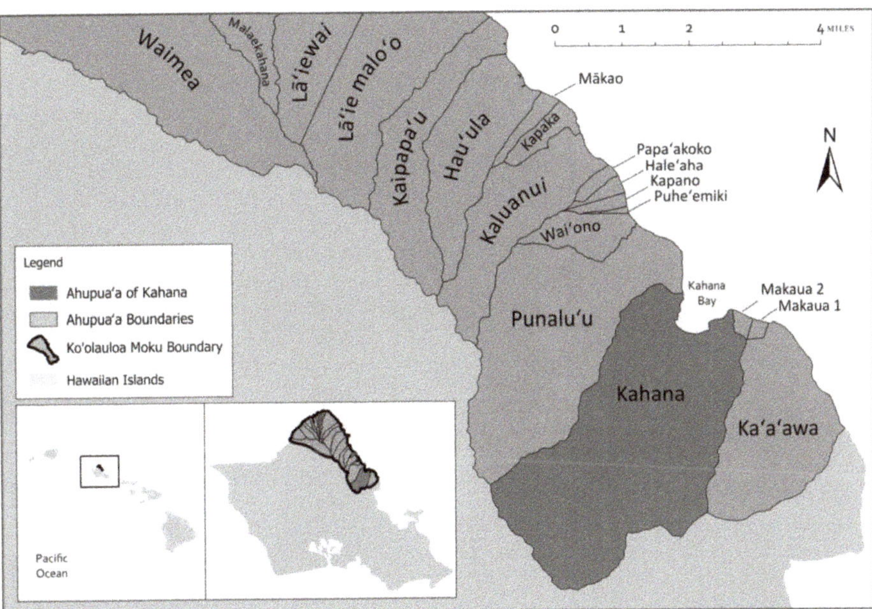

Figure 1. Location of the *ahupua'a* of Kahana within the *moku* (district) of Ko'olauloa on the island of O'ahu. Geographic Information System (GIS) boundaries and names are delineated by the "Ahupua'a" layer provided by the Office of Hawaiian Affairs and distributed by the Hawai'i Statewide GIS Program.

Konohiki fishing rights were officially recognized in Kahana until 1965–1969, when the State of Hawai'i acquired the *ahupua'a* and opened its fishery for public access. Today, the *ahupua'a* is established as a recreational state park offering ten beach campsites, two designated hiking trails, and interpretive programs for visitors to learn about Hawaiian culture. Kahana families, many with multigenerational and genealogical ties to the area, continue to reside in the park on long-term leases with the state. Despite nearly half a century of planning efforts, there is no master plan or shared state and community vision for managing the *ahupua'a* and its resources, including the nearshore fishery.

3. Methods

For this study, the lead author conducted 19 in-depth interviews, 9 with elders and fishers of Kahana and 10 with state resource management personnel, along with informal discussions and time spent between June 2015 and April 2017 participating in community events and workdays. Using a snowball sampling method [42], community members identified elders and fishers' (subsistence and commercial) extensive knowledge of the history related to Kahana's fishery. Semi-structured interviews focused on stories of place, stewardship practices, perceived changes and threats to coastal resources, and recommendations for improving management, capturing personal recollections from as early as the 1930s. State resource management personnel representing different agencies in the Department of Land and Natural Resources were identified by community members, as well as other personnel [42], based on having worked with the Kahana community or their familiarity with community concerns and caretaking efforts. Semi-structured interviews with state resource management personnel addressed management approaches in Kahana, as well as agency efforts and experiences working with community.

Secondary data include Kahana-related research publications such as scientific studies, state government documents including master plans and environmental impact statements, and transcripts from previous collections of interviews [21,43–46] with 23 community members from across the district of Ko'olauloa (Table 1). The lead author also searched archival records such as English language newspapers, correspondence by historic land owners, photos and maps from the Hawai'i State Archives, Bishop Museum, and Brigham Young University–Hawai'i. Newspaper articles were also accessed from online repositories, dating from as early as 1852. A constructivist grounded theory approach guided collection and analysis of qualitative data in this study [47]. Codes derived from interview transcripts, field notes, and secondary data were used to develop conceptual categories related to *konohiki* and community caretaking.

Table 1. Number of individuals interviewed (primary interviews, N = 19; secondary interviews, N = 23).

Primary Interviews	Total: 19
State Resource Management Personnel	
Department of Land and Natural Resources	2
Division of State Parks	3
Division of Aquatic Resources	3
Division of Boating and Ocean Recreation	1
Division of Conservation and Resource Enforcement	1
Community Members	
Elders (Born 1930s–1940s)	7
Younger fishers (Born 1960s)	2
Secondary Interviews	**Total: 23**
Community Members	
Elders (Born 1900–1951)	23

4. Results

Although Kahana formally maintained local level *konohiki* fishing rights through the mid-1960s, over half a century longer than most communities in Hawai'i, the modern *konohiki* system detailed in this case study operated within a very different context than during pre-contact times (Table 2). The *konohiki*, who are remembered by community elders, operated between the 1920s and mid-1960s within a changed system of Western land privatization, commercial use of the nearshore fishery, and the ability to lease *konohiki* fishing rights. Transition to a private property regime following the land division process of 1846–1858 brought new interpretations of the role and responsibilities of *konohiki* [26]. Once a position held by well-respected individuals appointed to oversee the well-being of *ahupua'a* residents and resources, *konohiki* of the late 19th and early 20th centuries enjoyed the benefits of owning particular lands while no longer accountable for ensuring their provision of abundance. According to certified deeds, purchase of Kahana's *konohiki* land came with the rights to the *konohiki* fishery (and fishponds) [48]. Thus, Kahana's *konohiki* fishing rights exchanged hands with new landowners, from Caesar Kapa'akea and Chiefess Keohokālole in 1850, to three different Chinese businessmen, and finally to Ka Hui Kū'ai i Ka 'Āina o Kahana (Hui of Kahana, or Hui) in 1874 [36,48]. The Hui was a group of 95 mostly Native Hawaiians from Kahana and surrounding *ahupua'a* who collectively purchased back the land and divided it among 115 shares [36]. However, with Hawai'i's increasingly foreign-controlled government and new laws enabling wealthy foreigners to acquire Hawaiian lands, Hui shares and other land parcels were eventually sold outside of the Kahana community. By 1903, the majority of Hui shares were owned by Mary Foster, a wealthy part Hawaiian businesswoman with no known connection to Kahana [36], and by 1930, she owned 99% of the *ahupua'a* [49].

Between 1905 to around 1965, Kahana's konohiki fishery appears to have operated under Mary Foster or her estate (having majority rule on Hui decisions) for commercial production, while ahupua'a residents continued to exercise their legal right to the fishery for subsistence. Court documents under the territory of Hawai'i reveal successful registration of Kahana's *konohiki* fishery in 1905, to the Hui of Kahana as vested owners [50]. Within the same year, archival records indicate that Mary Foster and her estate, acting on behalf of the Hui, began leasing the fishing rights for commercial operation [51]. Although newspaper advertisements announced lease of the fishing rights for the highest bidder, community interviews reveal that lessees were still expected to fulfill the traditional *konohiki* responsibility of leading communal fishing efforts. To invite community ability and bring people together for collective harvests, lessees still had to earn and maintain the respect of community members.

4.1. Persisting Role of Konohiki to Invite Ability

Although *konohiki* of the 19th and 20th centuries acquired their positions through new means, the role of *konohiki* remained an important aspect of life in Kahana. Community interviews recall three *konohiki* operating in Kahana, some at the same time (Table 2). The first *konohiki* people remember is Samuel Pua Ha'aheo, who held the fishing rights possibly as early as 1924, when he became the caretaker of Huilua Fishpond, through to 1946 [40]. According to interviews, Pua was not originally from Kahana but married a woman from the area. Archival records show that he rented property from Mary Foster [52]. Pua's responsibilities were to monitor the fishery, determine when it was time to collectively surround schools of fish, guide fishers from the shoreline, and oversee distribution of the catch. Interviewees expressed that Pua was beloved and well-respected because he was perceived as taking care of everyone, especially the elders.

Table 2. Changes in land and sea tenure in Kahana, Oʻahu (1200–1969).

Year	Event	Citation
≈1200	Kahana's most recent continuous settlement began around A.D. 1200.	[37]
1778	The first European, British explorer James Cook, arrives in the Hawaiian Islands; Kahana is a thriving farming and fishing community of ≈600–1000 Native Hawaiians.	[36,49]
1839–1840	First written recognition of *konohiki* fishing rights in the Kingdom of Hawaiʻi's Declaration of Rights and later Constitution.	[29]
1846–1855	Land division process across Hawaiʻi results in the award of ≈5050 acres of *konohiki* land in Kahana to Chiefess Keohokālole, and ≈200 acres of other lands to 34 Kahana residents.	[36]
1857–1872	Keohokālole sells Kahana's *konohiki* land, with the rights to the *konohiki* fishery and fishpond, to AhSing, a Chinese businessman, who later sells to J.A. Chuck, then H. Ahmee.	[48]
1874	Ka Hui Kūʻai i ka ʻĀina O Kahana (Hui of Kahana) initiates purchase of Kahana's *konohiki* land from H. Ahmee, with 95 mostly Native Hawaiian members from the area holding 115 shares.	[36]
1887	The first share of the Hui of Kahana is sold outside of the Hui.	[36]
1900	The Organic Act establishes Hawaiʻi as a USA territory; repeals *konohiki* laws except for registered fisheries.	[29]
1905	Kahana's *konohiki* fishery is successfully registered to the Hui of Kahana, March 30.	[50]
1924	Pua Haʻaheo steps into the role of fishpond caretaker (and likely also *konohiki*).	[40]
1925	Nick Peterson becomes the foreman for Mary Foster's property holdings in Kahana.	[53]
1930	Mary Foster passes away with 99% ownership of Kahana (less six parcels), turning it over to her estate.	[49]
1946	April 1 tidal waves take the lives of Pua Haʻaheo's grandchildren and Pua leaves his roles as *konohiki* and fishpond caretaker; the Kamakeʻeāina family assumes the role of *konohiki*.	2015–2016 Interviews
1959	Hawaiʻi is admitted as the 50th state of the United States.	[30]
1960	Nick Peterson passes away.	[53]
Mid-1960s	The last collective community harvest for *akule* (big eye scad, *Selar crumenopthalmus*) is led by the Kamakeʻeāaina family; Kahana's nearshore fishery begins to decline.	2015–2016 Interviews
1965–1969	Public is notified of condemnation proceedings for Kahana's *konohiki* fishery and *ahupuaʻa*; State purchases the *ahupuaʻa* and Kahana's *konohiki* fishery is opened to the public.	[36,49]

Following Pua, most community members remember the Kamakeʻeāina family, and in particular, Uncle or Papa ʻĀina Pahumoa Kamakeʻeāina from Lāʻiemaloʻo, a nearby *ahupuaʻa* within the same district (Figure 1). The Kamakeʻeāinas are considered to be the last *konohiki* of Kahana, maintaining this role through the mid-1960s. One of the eldest interviewees described how the first time the Kamakeʻeāinas surrounded *akule* (big eye scad, *Selar crumenopthalmus*) in Kahana, not a single resident went to the beach to help. However, younger interviewees only remember doing collective harvests with the Kamakeʻeāinas and recall Uncle ʻĀina as a person who always gave fish. Together, these varied accounts suggest the Kamakeʻeāinas were able to build their relationships with the Kahana community over time.

During both Pua's and the Kamake'eāinas' tenures as *konohiki*, Nicholas Peterson, who was part Hawaiian and originally from the south side of O'ahu, served as "the caretaker of Kahana". However, as one elder clarified, "Uncle Nick wasn't really the People's *konohiki*, so they say, but [he worked] for Mary Foster … He had to make sure that he made the bucks out of the fishing and make sure that she had revenues coming in." According to newspaper articles, Peterson worked as a foreman for Mary Foster from around 1925 through to 1960 [53]. Community interviews suggest that Peterson collected rent from the residents, made sure people were taking care of upland resources, enforced Kahana's restriction on harvesting *akule*, and sold the fish surrounded from communal harvests. Whether responsible for leading the fishing effort or ensuring reliable harvests, the 20th century *konohiki* of Kahana continued to mobilize community efforts around cultivating and harvesting abundance.

4.2. Managing Land-Sea Connectivity

One way in which Kahana's *konohiki* maintained the health of the nearshore fishery was by managing the land and streams, which affect coastal resources. According to community interviews, it was Peterson, who primarily oversaw the whole *ahupua'a*. As one long-time resident describes, Peterson would make sure people were "keeping their yards clean, make sure they came out and did what they're supposed to, make sure the rivers were clean, the stream beds, and just [responsibilities] that belong to our people anyway." Elders remember Kahana being well cared for when there was a *konohiki*, commenting in particular on how Kahana Stream was cleared of vegetation. Many interviewees recalled how the *hau* (*Hibiscus tiliaceus*), an invasive tree that has overgrown the stream in recent decades, never used to reach the water's surface. The importance of managing land–sea connectivity is emphasized in the Hui's requirement of the Kamake'eāinas, as non-residents, to cultivate one *lo'i* (flooded-field agriculture) in Kahana in order to hold the *konohiki* fishing rights. Residents also cared for their own *lo'i* and home gardens, and recall how "from time to time, everyone went up the valley to clean the [irrigation ditches]. We never let the [irrigation ditches] get dirty or blocked with rubbish. We all cleaned it together" [54] (p. 13). In these various ways, Kahana's *konohiki* and the community managed the land and sea as an integrated unit, recognizing that fisheries management begins on land.

4.3. Protecting Spawning Behavior

Kahana's *konohiki* also protected natural processes to replenish the fishery, such as spawning behavior. Kahana Bay provides important spawning and nursery habitat for many of Hawai'i's native aquatic species, including hammerhead sharks, manta rays, *moi* (*Polydactylus sexfilis*; Pacific threadfin), *āholehole* (*Kuhlia xenura*; Hawaiian flagtail), and *'ama'ama* (striped mullet, *Mugil cephalus*). However, Kahana is most famed for the large schools of *akule* (big eye scad, *Selar crumenopthalmus*) that aggregated in the center of its bay to spawn [39]. From as early as 1852, lasting into the mid-1960s, Kahana's *konohiki* actively claimed exclusive rights to harvest *akule* [50,55]. This restriction on harvesting *akule* additionally meant that no one except the *konohiki* could even enter the center of the bay. As one elder explained, "If you want to go fishing, we got to stick to the side of the bay and then go out. They don't like us in the middle, because we're going to chase the *akule* away. Chase the fish away." *Ahupua'a* residents could still exercise their right to the *konohiki* fishery by accessing the edges of the bay and its fringing reefs to fish and gather for home consumption. Elders attribute a productive nearshore fishery—with schools of fish visiting the bay more frequently, in greater numbers, and with larger individuals—to *konohiki* management. Kahana Bay's establishment as a *konohiki* fishery protected it from overuse by prohibiting entry of non-residents, and the restriction on harvesting *akule* further protected replenishment of fish stocks by ensuring that spawning behaviors were not disrupted.

4.4. Sharing Responsibility

Another enduring feature of *konohiki* was the sharing of responsibility, exemplified through the continued practice of *hukilau*. *Hukilau* is a communal surround net fishing method in which *lau* (ti leaf) is intertwined with rope and used to guide schools of fish towards the shore. In modern practice, nets were also used and attached to the rope. Nearly every community member interviewed in this study, and in previous studies, discussed *hukilau* for *akule*. This Hawaiian fishing method required the help of the whole community to contribute various skills and labor including spotting the school of fish, rowing the boat, diving to check the net, pulling the net in, and loading baskets full of fish to sell at the market. Once a school of fish was surrounded, the work continued with everyone helping to clean the boats and nets, patch up holes in the nets, cut and gather wood to build racks for the nets to dry on, and store everything back in Kahana's net and boat houses, ready for the next surround. All of this shared responsibility and work was coordinated by the *konohiki*.

4.5. Supporting Communal Benefit

Only by sharing in the work, the entire community also could share in the benefits. The *konohiki* who led the fishing effort was also responsible for overseeing the distribution of catch from each surround, making sure that everyone—workers, community members, and even visitors—had fish to bring home to their families. One resident, born and raised in Kahana, shared of Pua Ha'aheo, "No matter how big the school or how small the school … each resident had their share of fish to go home with. That's how it was. So that's how he maintained the fishing rights over here" [56]. In addition to their shared distribution of the catch, everyone who helped also received a share of the money earned from selling the fish. Kahana elders describe how Mary Foster and the *konohiki* would split the money fifty-fifty, then the *konohiki* "would split the money up so everybody got money … [Your share] would depend on how old you were and how much work you did." Pua and the Kamake'eāinas kept their roles as *konohiki* through facilitating collective ability to harvest then distributing the share to see that the community, especially elders, were well provided for.

4.6. Limited State Capacity to Manage Ahupua'a Resources

Kahana's elders shared how their nearshore fishery, even as it was being used for both subsistence and commercial purposes, was well cared for and abundant under *konohiki* management. However, with the loss of *konohiki* fishing rights, interviewees began to see the fishery decline. Between 1965 and 1969, the State of Hawai'i acquired Kahana through eminent domain, establishing the *ahupua'a* as a recreational state park and opening its *konohiki* fishery for public access. Facing eviction, Kahana families protested and successfully lobbied the state legislature to remain on the land [49]. Today, just 28 households, many with multi-generational and genealogical ties to Kahana, have secured 65-year leases with the state [57]. Per lease terms, residents are required to provide 25 hours each month of cultural activities and interpretive services for park visitors, as part of the state's "living park" concept. Since Kahana became a state park, residents have endured decades of meetings, interviews, and surveys involved in state-funded studies, legislative reports, environmental impact assessments, at least eight master planning efforts and numerous park program proposals [49,58]. Yet there is no master plan for Kahana and the majority of State Parks and community interviewees agree the "living park" concept is no longer viable.

Community interviewees also expressed that the state has poorly addressed threats to coastal resources, including intensive commercial fishing, disruptive recreational uses such as jet skis, invasive species introductions, and habitat change. State Parks interviewees expressed they lack the capacity and expertise to manage ecological function. Though State Parks is tasked to manage Kahana's natural and cultural resources, their efforts largely focus on creating public recreational opportunities. Additionally, State Parks does not have jurisdiction over the nearshore fishery. Interviews with personnel from Aquatic Resources, Boating and Ocean Recreation, and Conservation and Resources

Enforcement highlight their responsibility to the whole archipelago, and therefore their inability to focus limited resources on any one community's concerns. Following decades of failed planning efforts, mismanagement and fishery decline, the Kahana community, including residents and fishers from across the district of Ko'olauloa, is strengthening local governance to improve coastal resource health.

4.7. Reviving and Strengthening Local Fisheries Institutions

Kahana families are finding creative ways to manage their coastal resources through reviving *ahupua'a* health and food systems, restoring customary harvesting practices, teaching across generations, and building community relationships and management capacity. With the help of younger residents and Ko'olauloa families, elders have restored and continue to maintain *lo'i* (flooded-field agriculture) that were cared for within their families for generations. These farmers also work together to maintain the traditional irrigation ditch that feeds their *lo'i* with stream water. These traditional food systems have become features of the park, where visitors from schools and organizations across the island visit to learn and contribute through workdays.

Huilua Fishpond, which once functioned as a community food source, also fell into disuse under state ownership. In 2015, young leaders worked with State Parks to obtain a master permit, in compliance with state and federal regulations, to begin restoring the fishpond's 1000-foot wall. Although the permit determines the extent of activities that can be performed, fishpond caretakers lead the restoration effort and design best-management practices based on traditional environmental and cultural protocol [59].

Other young community members created the community-based nonprofit organization, Kahana Kilo Kai, in 2014 to steward the bay. They formally enrolled in the "Adopt-a-Harbor" program their first year to restore area facilities long neglected by Boating and Ocean Recreation. They also worked with the agency to post signage reminding jet skiers of Hawai'i's "slow-no-wake" laws, to minimize disturbance to fish spawning behavior. Along with these efforts, three youth completed training in identifying and reporting violations of fishing regulations through the Conservation and Resources Enforcement agency's Makai Watch program. Over the course of this study, Kahana families gathered at the pier regularly to monitor recreational activities, record fish catches and spawning times, build relationships with fishers, and educate the public about both state regulations and local values for using area resources responsibly. In recognition of these collective efforts by Kahana's young leaders, one state resource manager stated, "They don't just suggest things, they don't just plan things. They're doing it. They're working in the taro patches, they're carrying [rocks] at the fishpond . . . they can walk their talk." Through restoring traditional agriculture and aquaculture, and promoting responsible use of the nearshore fishery, the Kahana community is reintegrating land and sea management, while reasserting their role as caretakers.

Kahana fishing families are also reviving other customary nearshore management and fishing practices, such as *hukilau* or communal surround net fishing. Though *hukilau* ended in Kahana in the mid-1960s, community members revived the practice for *'ama'ama*, a native mullet, with guidance from one particular elder over the past ten years. While *hukilau* requires substantial effort, this harvesting practice, in which everyone who participates can contribute, strengthens community through relearning, reconnecting, and working together.

Community interviews also highlight potential rules for future local-level fisheries management that draw upon customary practices. The majority of community interviews expressed interest in introducing a new state law to place an island-wide seasonal restriction on harvesting *akule*, as well as placing a ban on disruptive recreational uses in Kahana, such as jet skis, to protect fish spawning behaviors. Many, including commercial fishers within the community, also share a desire to limit commercial fishing. Guided by elder knowledge and recommendations, the Kahana community is relearning harvesting and caretaking practices of *konohiki* to improve management of their nearshore fishery.

Another way in which Kahana families are strengthening local influence is through teaching across generations. Over the past several years, young community members have organized fishing camps, held for multiple days each summer, to educate area families and especially youth about ecosystem health and function, caretaking values, and responsible harvesting practices. The camps integrate Indigenous and Western knowledge systems through activities such as sewing and patching fishing nets, monitoring changing ecological conditions according to the Hawaiian moon calendar, and analyzing fish gonads to document spawning times. Many community elders and longtime fisher men and women informally teach at the camps, but also, learn. One Kahana elder shared, "A lot of the things that I've been taught today [through fishing camps and community gatherings] was never ever taught to our young ones or even to our old ones." Logistically, hosting the camps in Kahana requires Special Use permits to be filed with State Parks at least 45 days in advance. Aside from general rules for using the area, these permits secure space for families, many of whom no longer live in Kahana, to gather, teach, and learn together on the land. These camps also strengthen relationships with nonprofit and state resource management personnel, who are also invited to participate in and help conduct activities.

Kahana families are also building their capacity to manage coastal resources through participating in local and global community partnerships and knowledge sharing networks. As a member of the nonprofit organization, Kua'āina Ulu 'Auamo (KUA), Kahana families are connected to more than 33 communities across Hawai'i working to protect and steward their lands and fisheries, including the restoration of 38 fishponds. In 2016, Kahana families expanded this network internationally, hosting a fishpond workday with over 100 Indigenous and community leaders, practitioners, and supporters from 35 nations. The global gathering was organized by KUA to build solidarity among individuals engaged in community grassroots efforts prior to attending the IUCN (International Union for Conservation of Nature) World Conservation Congress in Honolulu.

Although the Kahana community lacks formal government recognition of local management rights or co-management agreements, these informal efforts demonstrate community action and commitment. State resource management personnel across agencies have begun to take notice and lend their support. Within the timeframe of this study, state personnel attended fishing camp activities to give talks to the youth, helped to pull *hukilau* nets in to shore, and passed rocks at the fishpond alongside community members. These means of informal engagement are beginning to repair state–community relationships, while also building community confidence, connections, and ability. Here, informal caretaking efforts hold promise for collaborative management driven by community objectives and needs.

5. Discussion

Understanding how Indigenous and place-based institutions historically operated and adapted to social–ecological change, and how they can fit within contemporary contexts is essential for biocultural restoration [6,60,61]. We use a case study approach focused on one Hawai'i fishing community reclaiming their role as caretakers despite changes in land and sea tenure, governance, access, and use. Emerging from this research are key considerations for community-based collaborative management. These include: (1) understanding historical context for enhancing institutional fit, (2) enduring community leadership, (3) balancing rights and responsibilities, and (4) fostering community ability to manage coastal resources through both formal and informal processes.

5.1. Understanding Historical Context for Enhancing Institutional Fit

Understanding social-ecological systems from a historical perspective is critical in designing effective collaborative fisheries management institutions. Institutions are systems of rights, formal and informal rules, and decision-making procedures that guide human–environment interactions [62]. Institutional fit refers to how well institutions match a particular social–ecological system, often in terms of their spatial, temporal, and functional contexts [63–65]. Findings from this research emphasize that design of fisheries institutions must also factor in the historical contexts of a social–ecological system, specifically the traditional and place-based institutions historically in place. *Konohiki* management was tailored to local socio–cultural contexts and ecological systems, including prime spawning habitat and the impacts of land-based activities on the nearshore fishery. The experience of coastal resource decline is not unique to Kahana, but is shared across Hawai'i [21,26], the Pacific [7,11], and other parts of the world [5,9] where Western models and concepts of resource management have replaced Indigenous and place-based institutions. Still, a wealth and diversity of traditional and place-based knowledge systems endure, and can be adapted within contemporary contexts to improve coastal resource management [3,5,11–13,66]. Rather than simply replicating specific practices of these time-tested institutions, it is important to understand their foundational principles, the key functions they fulfilled, and how they can be adapted to foster future biocultural restoration.

5.2. Enduring Community Leadership

Hawai'i's *konohiki* system is an example of a community-based collaborative management institution that endured by maintaining key features within a changing social–ecological context. One enduring feature was *konohiki* facilitation of community caretaking and harvesting efforts, which depended upon in-depth knowledge of the local ecology, exceptional fishing skills, and ability to earn and maintain the respect of the community [28,67]. Such leadership roles are increasingly recognized as important attributes contributing to sustainable fisheries [20,68–70], specifically when leadership is perceived to be legitimate and highly engaged [71]. A recent analysis of 130 co-managed fisheries worldwide identified leadership—the presence of at least one respected individual with entrepreneurial skills, who is driven by collective interests and committed to the co-management process—as the most important feature for successful co-management [68]. The Kahana case study provides an example in which leadership emerges through a community as a collective in the context of perceived government inaction and resource decline. In this case, leadership is dispersed among elders as knowledge holders and young adults who mobilize community around various caretaking activities. These collective efforts demonstrate how multiple sources of leadership can be complementary, interact through mutual support, and coexist within community, adding to a pluralistic conceptualization of environmental leadership [70].

5.3. Balancing Rights and Responsibilities

Maintaining collective benefit through balancing rights and responsibilities was another enduring feature of *konohiki* management. *Konohiki* carried unique responsibility to oversee harvests, facilitating a system in which everyone who contributed to collective work would in turn receive the benefits of reliable harvests [24,25,67]. Rights and responsibilities for all resource users need to be balanced and based upon contributions to collective efforts and caretaking [60]. One key challenge to fisheries co-management in Hawai'i is that traditional management rested upon reserving distinct rights for area residents [13,29]. Under the state, open access allows the public to use and harvest coastal resources with no expectation to care for them, cultivate their abundance, or give back in any way. This decoupling of rights and responsibility has resulted in coastal resource decline worldwide [7,14,72]. Displacement of Indigenous people throughout the world for the purpose of establishing national parks, uninhabited wilderness [9], and marine protected areas [7] violates their customary rights and prevents them from exercising their distinct responsibilities to care for their homelands. Managing natural areas for public benefit limits community ability to continually interact with, eat from, perpetuate knowledge of, and govern coastal resources for which they are responsible. Emerging literature on co-management emphasizes the need to maintain a balanced distribution of rights and responsibilities, obligations, and benefits amongst all resource users [60,61,67].

5.4. Fostering Community Ability to Manage Coastal Resources through Both Formal and Informal Processes

This case study also highlights the power of informal fisheries co-management as a means to foster community ability to lead caretaking efforts. Co-management comprises a variety of institutional arrangements shaped by different goals, partners, knowledge systems, and degrees of power sharing [15–20]. Such arrangements are negotiated and acknowledged through a formal (e.g., officially recognized by law) or informal (e.g., verbally accepted) agreement among partners [18]. In Kahana, the fishery has been minimally managed with archipelago-wide species-specific regulations administered by the state. Formal state park leases focus only on community service to educate visitors. However, within the last decade, Kahana families have been reclaiming their role as caretakers through informal means of managing coastal resources. Many of these caretaking activities require state approval (e.g., through special permits) and increasingly engage state resource management personnel, for example, to give invited talks to youth at fishing camps.

Through informal co-management, communities have the flexibility to self-organize, determine their own management objectives, and act upon them within their own time frames. By working within informal co-management systems, collective governance of fisheries, along with feelings of empowerment and shared responsibility, can be achieved and lead to effective management practices [73]. Informal caretaking efforts in Kahana also highlight the importance of creating protected spaces for families and community to spend time together and build upon generational knowledge of place and practice. Traditional activities provide not only a gathering space for renewing relationships with family, community, and place, but also a foundation for cultural resurgence and resilience [74]. Adaptive community-led management can guide progress towards more just and effective conservation solutions, restoring coastal resources along with Indigenous and place-based communities' rights and responsibilities [60].

Still, this study recognizes the need for higher levels of governance to formally complement and support community caretaking efforts [19,20,69]. Informal co-management can be limited in confronting unsustainable fishing and recreational activity, and may require legal backing [19]. Long-term sustainability requires place-specific rules, the ability of communities to recognize and respond to change, and support from higher levels of organization [12,69]. In this case, as in others, informal efforts can strengthen relationships that pave the way for more formal co-management agreements [16,67].

6. Conclusions

Understanding how Indigenous and place-based institutions historically operated and adapted to social–ecological change, and how they can be reinvigorated within contemporary contexts, is essential for biocultural restoration [6,60,61]. This study provides a historical perspective of fisheries governance within one rural Hawaiian fishing community, covering the transition from local to state level fisheries management, to an emerging collaborative arrangement led by community. This study also demonstrates how informal community initiative *ma ka hana ka 'ike*, to learn by doing the work [75], can be more powerful than formal co-management arrangements for building community ability [67,74]. Findings also emphasize the value of understanding historical institutions that adapted to local socio-cultural and ecological contexts, building collective leadership that fulfills traditional functions, and balancing rights and responsibilities amongst all resource users. Effective fisheries governance requires true partnerships, formal and informal, that value Indigenous and place-based knowledge systems and while creating the space for communities to build enduring relationships among people, place and practice.

Author Contributions: Both authors were involved in the design of this research. M.M. wrote the initial draft manuscript and carried out the field research. Both authors contributed to analysis and writing the paper. Both authors read and approved the final manuscript.

Funding: The APC was funded by Hawai'i Community Foundation.

Acknowledgments: Mahalo first and foremost to the Kahana elders and fishers who shared their stories, knowledge of place, and recommendations for Kahana's people, fishery, and future. We also thank the state resource management personnel who took time to share their thoughts and efforts moving forward. A special thanks to Shaelene Kamaka'ala, who has been a tremendous support and guiding light in this work. We also thank Hō'ala 'Āina Kūpono, the Kū'oko'a Strategic Initiative, the Hawai'i Community Foundation, Malia Akutagawa, Kirsten Oleson, Heather McMillen, and the Ko'olauloa families who continue to care for this place. Credit goes to Kalani Quiocho for the title, meaning "Through Kahana, one learns," a word play on the Hawaiian proverb, "Ma ka hana ka 'ike."

Conflicts of Interest: The authors declare no conflict of interest.

References

1. Turner, N.J.; Ignace, M.B.; Ignace, R. Traditional Ecological Knowledge and Wisdom of Aboriginal Peoples in British Columbia. *Ecol. Soc. Am.* **2000**, *10*, 1275–1287. [CrossRef]
2. Janssen, M.A.; Anderies, J.M.; Ostrom, E. Robustness of social-ecological systems to spatial and temporal variability. *Soc. Nat. Resour.* **2007**, *20*, 307–322. [CrossRef]
3. Berkes, F. *Sacred Ecology*, 3rd ed.; Routledge: New York, NY, USA, 2012.
4. Berkes, F.; Turner, N.J. Knowledge, learning and the evolution of conservation practice for social-ecological system resilience. *Hum. Ecol.* **2006**, *34*, 479–494. [CrossRef]
5. Ostrom, E. *Governing the Commons*; Cambridge University Press: Cambridge, UK, 1990.
6. Colding, J.; Folke, C.; Elmqvist, T. Social institutions in ecosystem management and biodiversity conservation. *Trop. Ecol.* **2003**, *44*, 25–41.
7. Ruddle, K.; Hickey, F.R. Accounting for the mismanagement of tropical nearshore fisheries. *Environ. Dev. Sustain.* **2008**, *10*, 565–589. [CrossRef]
8. Ruiz-Mallén, I.; Corbera, E. Community-based conservation and traditional ecological knowledge: Implications for social-ecological resilience. *Ecol. Soc.* **2013**, *18*. [CrossRef]
9. Stan, S. *Indigenous Peoples, National Parks, and Protected Areas: A New Paradigm Linking Conservation, Culture, and Rights*; University of Arizona Press: Tucson, AZ, USA, 2014.
10. Turner, N.J.; Gregory, R.; Brooks, C.; Failing, L.; Satterfield, T. From invisibility to transparency: Identifying the implications. *Ecol. Soc.* **2008**, *13*, 1–7. [CrossRef]
11. Johannes, R.E. The Renaissance of Community-Based Marine Resource Management in Oceania. *Annu. Rev. Ecol. Syst.* **2002**, *33*, 317–340. [CrossRef]
12. Stephenson, J.; Berkes, F.; Turner, N.J.; Dick, J. Biocultural conservation of marine ecosystems: Examples from New Zealand and Canada. *Indian J. Tradi. Knowl.* **2014**, *13*, 257–265.
13. Vaughan, M.B.; Thompson, B.; Ayers, A.L. Pāwehe Ke Kai a'o Hā'ena: Creating State Law based on Customary Indigenous Norms of Coastal Management. *Soc. Nat. Resour.* **2016**, *30*, 1–16. [CrossRef]
14. Friedlander, A.M. Marine conservation in Oceania: Past, present, and future. *Mar. Pollut. Bull.* **2018**, *135*, 139–149. [CrossRef] [PubMed]
15. Pomeroy, R.S.; Berkes, F. Two to tango: The role of government in fisheries co-management. *Mar. Policy* **1997**, *21*, 465–480. [CrossRef]
16. Borrini-Feyerabend, G.; Pimbert, M.; Farvar, M.T.; Kothari, A.; Renard, Y. *Sharing Power, Learning by Doing in Co-Management of Natural Resources Throughout the World*; IIED and IUCN/CEESP/CMWG: Cenesta, Tehran; Earthscan: London, UK, 2004.
17. Nielsen, J.R.; Degnbol, P.; Viswanathan, K.K.; Ahmed, M.; Hara, M.; Abdullah, N.M.R. Fisheries co-management—An institutional innovation? Lessons from South East Asia and Southern Africa. *Mar. Policy* **2004**, *28*, 151–160. [CrossRef]
18. Pomeroy, R.S.; Rivera-Guieb, R. *Fishery Co-Management: A Practical Handbook*; CABI Publishing: Cambridge, MA, USA, 2005.
19. Berkes, F. *Coasts for People: Interdisciplinary Approaches to Coastal and Marine Resource Management*; Routledge: New York, NY, USA, 2015.
20. Olsson, P.; Folke, C.; Berkes, F. , Adaptive comanagement for building resilience in social–ecological systems. *Environ. Manag.* **2004**, *34*, 75–90. [CrossRef] [PubMed]

21. Maly, K.; Maly, O. *Volume I: Ka Hana Lawai 'Aa Me Na Ko 'Ao Na Kai 'Ewalu, A History of Fishing Practices and Marine Fisheries of the Hawaiian Islands*; Prepared for The Nature Conservancy; Kumu Pono Associates LLC: Hilo, HI, USA, 2003.

22. Poepoe, K.K.; Bartram, P.K.; Friedlander, A.M. The use of traditional Hawaiian knowledge in the contemporary management of marine resources. In *Conference Proceedings: Putting Fishers' Knowledge to Work*; UBC Fisheries Centre, University of British Columbia: Vancouver, BC, Canada, 2003.

23. Gonschor, L.; Beamer, K. Toward in Inventory of Ahupua'a in the Hawaiian Kingdom: A Survey of Nineteenth-and early Twentieth-Century Cartographic and Archival Records of the Island of Hawai'i. *Hawaii. J. Hist.* **2014**, *48*, 53–87.

24. Akutagawa, M.; Williams, H.; Kamaka'ala, S. *Traditional and Customary Practices Report for Mana'e, Moloka'i: Traditional Subsistence Uses, Mālama Practices and Recommendations, and Native Hawaiian Rights Protections of Kama'āina Families of Mana'e Moku, East Moloka'i, Hawai'i*; Prepared for Office of Hawaiian Affairs; Office of Hawaiian Affairs: Honolulu, HI, USA, 2016.

25. Steele, C. He Ali'i Ka 'Āina; He Kauwā Ke Kanaka (The Land is Chief; Man is its Servant): Traditional Hawaiian Resource Stewardship and the Transformation of the Konohiki. Master's Thesis, University of Hawai'i at Mānoa, Honolulu, HI, USA, 2015.

26. Jokiel, P.L.; Rogers, K.S.; Walsh, W.J.; Polhemus, D.A.; Wilhelm, T.A. Marine Resource Management in the Hawaiian Archipelago: The Traditional Hawaiian System in Relation to the Western Approach. *J. Mar. Biol.* **2011**, *2011*. [CrossRef]

27. Titcomb, M. *Native Use of Fish in Hawaii*; University of Hawai'i Press: Honolulu, HI, USA, 1972.

28. Andrade, C. *Hā'ena: Through the Eyes of the Ancestors*; University of Hawai'i Press: Honolulu, HI, USA, 2008.

29. Kosaki, R.H. *Konohiki Fishing Rights*; Hawai'i Legislature: Honolulu, HI, USA, 1954.

30. McGregor, D.P.; Mackenzie, M.K. *Mo'olelo Ea O Na Hawai'i History of Native Hawaiian Governance in Hawai'i*; Prepared for the Office of Hawaiian Affairs; Office of Hawaiian Affairs: Honolulu, HI, USA, 2014.

31. Meller, N. *Indigenous Ocean Rights in Hawai'i*; Sea Grant Marine Policy and Law Report, UNIHI-SEAGRANT-MP-86-01; University of Hawai'i Sea Grant College Program: Honolulu, HI, USA, 1985.

32. Department of Land and Natural Resources About DLNR. Available online: http://dlnr.hawaii.gov/about-dlnr/ (accessed on 31 July 2018).

33. Friedlander, A.M.; Shackeroff, J.M.; Kittinger, J.N. Customary Marine Resource Knowledge and Use in Contemporary Hawai'i. *Pac. Sci.* **2013**, *67*, 441–460. [CrossRef]

34. Friedlander, A.M.; Rodgers, K.S. Coral reef fishes and fisheries of south Moloka'i., Chapter 7. In *The Coral Reef of South Moloka'i, Hawai'i; Portrait of a Sediment-Threatened Fringing Reef*; Field, M.E., Cochran, S.A., Logan, J.B., Storlazzi, C.D., Eds.; US Geological Survey Scientific Investigations: Denver, CO, USA, 2007; Volume 5101, pp. 59–66.

35. Schemmel, E.M.; Friedlander, A.M. Participatory fishery monitoring is successful for understanding the reproductive biology needed for local fisheries management. *Environ. Biol. Fishes* **2017**, *100*, 171–185. [CrossRef]

36. Stauffer, R.H. *Kahana: How the Land Was Lost*; University of Hawai'i Press: Honolulu, HI, USA, 2004.

37. Beggerly, P.P. Kahana Valley, Hawai'i, a Geomorphic Artifact: A Study of the Interrelationships among Geomorphic Structures, Natural Processes, and Ancient Hawaiian Technology, Land Use, and Settlement Patterns. Doctoral Dissertation, University of Hawai'i at Mānoa, Honolulu, HI, USA, 1990.

38. Sterling, E.P.; Summers, C.C. *Sites of Oahu*; Bishop Museum Press: Honolulu, HI, USA, 1978.

39. Handy, E.S.; Handy, E.G.; Pukui, M.K. *Native Planters in Old Hawai'i*; Bishop Museum Press: Honolulu, HI, USA, 1972.

40. Kelly, M. *Background History of Huilua Fishpond, Kahana Bay, Ko'olau Loa, O'ahu*; Bishop Museum Press: Honolulu, HI, USA, 1979.

41. Wyban, C.A. *Interpretive Materials for Huilua Fishpond Kahana Valley State Park*; Prepared for Department of Land and Natural Resources, State Parks Division; Department of Land and Natural Resources: Honolulu, HI, USA, 1992.

42. Biernacki, P.; Waldorf, D. Snowball sampling: Problems and techniques of chain referral sampling. *Sociol. Methods Res.* **1981**, *10*, 141–163. [CrossRef]

43. Maly, K.; Maly, O. *He Wahi Moʻolelo No Kaluanui Ma Koʻolauloa, Mokupuni ʻO Oʻahu, A Collection of Traditions, Historical Accounts and Kamaʻāina Recollections of Kaluanui and Vicinity, Koʻolauloa, Island of Oʻahu*; Prepared for The Nature Conservancy; Kumu Pono Associates LLC: Hilo, HI, USA, 2004.

44. *Kenneth Baldridge Oral History Collection, 1971–2004*; Joseph, F. Smith Library Archives and Special Collections; Brigham Young University: Lāʻie, HI, USA.

45. *Clinton Kanahele Collection*; Joseph, F. Smith Library Archives and Special Collections; Brigham Young University: Lāʻie, HI, USA, 1970.

46. Kodama-Nishimoto, M.; Nishimoto, W.S.; Oshiro, C.A. (Eds.) *Talking Hawaiʻi's Story: Oral Histories of an Island People*; University of Hawaiʻi Press: Honolulu, HI, USA, 2009.

47. Charmaz, K. *Constructing Grounded Theory: A Practical Guide through Qualitative Analysis*, 2nd ed.; Sage: Thousand Oaks, CA, USA, 2014.

48. *Ahupuaʻa of Kahana Title Deeds: Certified Copies, 1856–1881*; Box 1, Folder 1, Mary E. (Robinson) Foster Papers 1844–1930, Collection M-433; Hawaiʻi State Archives: Honolulu, HI, USA, 1856.

49. Jaworowski, S. *Kahana: What Was, What Is, What Can Be*; Legislative Reference Bureau: Honolulu, HI, USA, 2001.

50. *Kahana Fishery, 1902–1942*; Box 10, Folder 109, Mary E. (Robinson) Foster Papers 1844-1930, Collection M-433; Hawaiʻi State Archives: Honolulu, HI, USA, 1844.

51. *Hui of Kahana: Receipts for Expenses, 1901–1930*; Box 8, Folder 89, Mary E. (Robinson) Foster Papers 1844-1930, Collection M-433; Hawaiʻi State Archives: Honolulu, HI, USA.

52. *Reports of Revenues from Leases and Rentals, 1910–1920*; Box 9, Folder 93, Mary E. (Robinson) Foster Papers 1844–1930, Collection M-433; Hawaiʻi State Archives: Honolulu, HI, USA.

53. Mr. N. Peterson Dies at Kahuku. *The Honolulu Advertiser*. 11 April 1960. Available online: http://www.newspapers.com (accessed on 14 April 2017).

54. ʻOhana Unity Council. The Living Park Plan of Kahana's People. Kahana, HI, USA, 1979. Available online: https://docs.google.com/viewer?a=v&pid=sites&srcid=ZGVmYXVsdGRvbWFpbnxrYWhhbmFwbGFufGd4Ojk5NTA0MjRhYTM1ZTNiMA (accessed on 16 October 2018).

55. NOTICE. *The Polynesian*. 10 January 1852. Available online: http://www.chroniclingamerica.com/ (accessed on 27 March 2017).

56. Soga, B. *Transcript of an Oral history Conducted by Kenneth W. Baldridge, 21 August 1992, Kenneth Baldridge Oral History Collection, 1971–2004*; Joseph F. Smith Library Archives and Special Collections; Brigham Young University: Lāʻie, HI, USA.

57. Townscape, Inc. *Ahupuaʻa ʻO Kahana State Park*; Phase 1A Planning Draft Progress Report; Prepared for Division of State Parks, State of Hawaiʻi; Townscape, Inc.: Honolulu, HI, USA, 2017.

58. Hopkins, C. *Status Report on Kahana Valley State Park*; Prepared for Twelfth Legislature of the State of Hawaiʻi, Regular Session; Office of Hawaiian Affairs: Honolulu, HI, USA, 1984.

59. Watson, T.; Cain, M.; Lemmo, S.; Doktor, L.; Asuncion, B.; Mossman, L.; Lyles, J.; Farinbolt, N.; Kittinger, J. *Hoʻāla Loko Iʻa: Permit Application Guidebook*; Department of Land and Natural Resources, Office of Conservation and Coastal Lands: Honolulu, HI, USA, 2016.

60. Gavin, M.C.; McCarter, J.; Mead, A.; Berkes, F.; Stepp, J.R.; Peterson, D.; Tang, R. Defining biocultural approaches to conservation. *Trends Ecol. Evol.* **2015**, *30*, 140–145. [CrossRef]

61. Aswani, S.; Ruddle, K. Design of realistic hybrid marine resource management programs in Oceania. *Pac. Sci.* **2013**, *67*, 461–476. [CrossRef]

62. Young, O.R. *The Institutional Dimensions of Environmental Change: Fit, Interplay, and Scale*; MIT Press: Cambridge, MA, USA, 2002.

63. Cumming, G.; Cumming, D.H.; Redman, C. Scale mismatches in social-ecological systems: Causes, consequences, and solutions. *Ecol. Soc.* **2006**, *11*. [CrossRef]

64. Folke, C.; Pritchard, L., Jr.; Berkes, F.; Colding, J.; Svedin, U. The problem of fit between ecosystems and institutions: Ten years later. *Ecol. Soc.* **2007**, *12*. [CrossRef]

65. Epstein, G.; Pittman, J.; Alexander, S.M.; Berdej, S.; Dyck, T.; Kreitmair, U.; Rathwell, K.; Sergio, V.; Vogt, J.; Armitage, D. Institutional fit and the sustainability of social–ecological systems. *Curr. Opin. Environ. Sustain.* **2015**, *14*, 34–40. [CrossRef]

66. Vaughan, M. *Kaiāulu: Gathering Tides*; Oregon State University Press: Corvallis, OR, USA, 2018.

67. Winter, K.B.; Beamer, K.; Vaughan, M.; Friedlander, A.M.; Kido, M.H.; Whitehead, A.N.; Akutagawa, M.K.H.; Kurashima, N.; Lucas, M.P.; Nyberg, B. The Moku System: Managing Biocultural Resources for Abundance within Social-Ecological Regions in Hawai'i. *Sustainability* **2018**, *10*, 3554. [CrossRef]

68. Gutiérrez, N.L.; Hilborn, R.; Defeo, O. Leadership, social capital and incentives promote successful fisheries. *Nature* **2011**, *470*, 386–389. [CrossRef] [PubMed]

69. Ostrom, E. A general framework for analyzing sustainability of social-ecological systems. *Science* **2009**, *325*, 419–422. [CrossRef] [PubMed]

70. Case, P.; Evans, L.S.; Fabinyi, M.; Cohen, P.J.; Hicks, C.C.; Prideaux, M.; Mills, D.J. Rethinking environmental leadership: The social construction of leaders and leadership in discourses of ecological crisis, development, and conservation. *Leadership* **2015**, *11*, 396–423. [CrossRef]

71. Crona, B.; Gelcich, S.; Bodin, O. The importance of interplay between leadership and social capital in shaping outcomes of rights-based fisheries governance. *World Dev.* **2017**, *91*, 70–83. [CrossRef]

72. Costanza, R.; Andrade, F.; Antunes, P.; van den Belt, M.; Boersma, D.; Boesch, D.; Catarino, F.; Hanna, S.; Limburg, K.; Low, B.; et al. Principles for sustainable governance of the oceans. *Science* **1998**, *281*, 198–199. [CrossRef] [PubMed]

73. Hauzer, M.; Dearden, P.; Murray, G. The effectiveness of community-based governance of small-scale fisheries, Ngazidja island, Comoros. *Mar. Policy* **2013**, *38*, 346–354. [CrossRef]

74. Corntassel, J. Re-envisioning resurgence: Indigenous pathways to decolonization and sustainable self-determination. *Decol. Indig. Educ. Soc.* **2012**, *1*, 86–101.

75. Pūku'i, M.K. *'Olelo No'eau: Hawaiian Proverbs & Poetical Sayings*; Bishop Museum Press: Honolulu, HI, USA, 1983; Volume 71.

Article

Restoration of 'Āina Malo'o on Hawai'i Island: Expanding Biocultural Relationships

Noa Kekuewa Lincoln [1,*], Jack Rossen [2], Peter Vitousek [3], Jesse Kahoonei [4], Dana Shapiro [5], Keone Kalawe [6], Māhealani Pai [7], Kehaulani Marshall [8] and Kamuela Meheula [4]

1 Department of Tropical Plants and Soil Sciences, College of Tropical Agriculture and Human Resources, University of Hawai'i at Mānoa, 3190 Maile Way, St. John 102, Honolulu, HI 96822, USA
2 Department of Anthropology, Ithaca College, Ithaca, NY 14850, USA; jrossen@ithaca.edu
3 Department of Biology, Stanford University, Stanford, CA 94305, USA; vitousek@stanford.edu
4 Kahalu'u Kūāhewa, Kailua-Kona, HI 96740, USA; jkkahoonei@gmail.com (J.K.); akmeheula@gmail.com (K.M.)
5 Māla Kalu'ulu Cooperative, Captain Cook, HI 96704, USA; malakaluulu@gmail.com
6 Independent Scholar, Keauhou, HI 96739, USA; kkalawe@hushmail.com
7 Kamehameha Schools 'Āina Pauahi o Kona, Kailua-Kona, HI 96740, USA; mpai@ksbe.edu
8 Ulu Mau Puanui, Kamuela, HI 96743, USA; kehaulanimarshall@yahoo.com
* Correspondence: nlincoln@hawaii.edu; Tel.: +1-808-956-6498

Received: 31 July 2018; Accepted: 23 October 2018; Published: 31 October 2018

Abstract: Before European contact, Native Hawaiian agriculture was highly adapted to place and expressed a myriad of forms. Although the iconic lo'i systems (flooded irrigated terraces) are often portrayed as traditional Hawaiian agriculture, other forms of agriculture were, in sum, arguably more important. While pockets of traditional agricultural practices have persevered over the 240 years since European arrival, the revival of indigenous methods and crops has substantially increased since the 1970s. While engagement in lo'i restoration and maintenance has been a core vehicle for communication and education regarding Hawaiian culture, it does not represent the full spectrum of Hawaiian agriculture and, on the younger islands of Hawai'i and Maui in particular, does not accurately represent participants' ancestral engagement with 'āina malo'o (dry land, as opposed to flooded lands). These "dryland" forms of agriculture produced more food than lo'i, especially on the younger islands, were used to produce a broader range of resource crops such as for fiber, timber, and medicine, were more widespread across the islands, and formed the economic base for the powerful Hawai'i Island chiefs who eventually conquered the archipelago. The recent engagement in the restoration of these forms of agriculture on Hawai'i Island, compared to the more longstanding efforts to revive lo'i-based cultivation, is challenging due to highly eroded knowledge systems. However, their restoration highlights the high level of place-based adaptation, demonstrates the scale and political landscape of pre-European Hawai'i, and provides essential elements in supporting the restoration of Hawaiian culture.

Keywords: traditional agriculture; indigenous agriculture; biocultural; restoration; Hawai'i

---

1. Introduction

Biocultural restoration relies upon understanding specific cultures and practices within an ecological context [1,2]. Increasingly, several fields are asserting the importance of using an approach that recognizes the intertwined nature of people and place to develop adaptive management strategies [3,4]. Doing so requires a holistic understanding of both the sociocultural and ecological systems, and in particular, the relationships and feedbacks that are encompassed within socioecological systems. It has been suggested that islands in general are well suited to the study of coupled natural and human systems and Hawai'i in particular has been hailed as a model system for the exploration of these complex human-environment dynamics [5]. This is because unique attributes of both the natural and social environments present ideal combinations of complexity and tractability. For instance, Hawai'i encompasses an extraordinary range of variation in climate and substrate age in a small area, but the resulting diversity in soils and ecosystems is highly organized and predictable in its distribution [6]. Similarly, Hawai'i reached a high state of political complexity and state governance prior to the arrival of Europeans, yet represents a very short timeline of human occupation and a relatively closed social system at the archipelago level [5]. Consequently, unique opportunities for understanding the development and diversity of biocultural relationships exists in Hawai'i.

Of particular interest to both research and restoration have been Hawaiian agricultural systems. Before European contact, Native Hawaiian agriculture was highly adapted to place and expressed a myriad of forms [7]. The development of unique resource management practices evolved to local environments maximized efficient productivity [8–14]. This led to highly specific, place-adapted indigenous knowledge that powered the political evolution of the ancient Hawaiian state. In the past few decades there has been an increasing recognition of the high value of this knowledge in contemporary resource management and land stewardship, and growing efforts to preserve and revitalize such knowledge. Restoration of these place-adapted systems has proven challenging due to significant loss of traditional ecological knowledge.

In reviewing efforts to revitalize traditional agricultural techniques, we see several commonalities between organizations that are undertaking these efforts; in particular, the application of highly interdisciplinary, non-linear approaches that rely on strong relationships between players across different disciplines and epistemologies. We suggest that oversimplification of the diverse, place-based requirements and practices associated with traditional Hawaiian agriculture has impeded in-depth understanding of traditional Hawaiian agriculture, and consequently, has also impeded the restoration of these systems. Furthermore, we suggest that examining these agricultural efforts within a landscape-level socioecological context is essential to understanding their function and roles in both the past and the present. While exploring these theoretical underpinnings, we also discuss practical components of conducting biocultural restoration.

Understanding Environmental and Social Adaptation in Hawaiian Agriculture

In illustrating the adaptive nature of socioecological systems, this paper presents a novel treatment of the evolution and function of one core biocultural coupling—agriculture. As concepts of biocultural management grow, it is important to exemplify how form and function of socioecological couplings are a product of both the environmental and the social landscape. Even common and essential elements, in this case of agriculture, manifested differently within the larger socioecological landscape in the past, and similarly manifest differently within the contemporary efforts to restore these systems.

In this paper, we first present a review of the evolution of agricultural form in Hawai'i based on local environments, and illustrate the knowledge specificity and system functionality that existed in the past. We then consider recent efforts to expand the form of agricultural restoration from that which has dominated the last 30 years of effort by describing several organizations and their efforts. These organizations were selected as, to the authors' expert knowledge, the leading efforts on Hawai'i Island to expand the form of traditional agriculture restoration. We explore common elements of the efforts, how they differ from previous efforts, and how these differences are manifested within contemporary social and political movements.

Importantly, we utilize indigenous methodology that directly engages those intimately involved in the efforts. This includes participants from both western and indigenous science perspectives. All participants are highly experienced experts that, in a traditional ethnographic study, would be treated as human subjects within a study rather than given the opportunity to tell their own story directly. It is important to note that all the authors have been intimately involved with the restoration of these systems in different capacities and speak from immersed experience in the process; 11 of the

15 participants and six of the nine authors are native Hawaiians. All involved are highly experienced in traditional agriculture in a myriad of forms.

2. Evolution and Restoration of Traditional Hawaiian Agricultural Forms

The unique and highly diverse biophysical landscape of Hawai'i compared to the southern Polynesian islands supported the development of new agricultural practices that were not found elsewhere in the Pacific. Hawai'i is the only Polynesian group north of the equator, the only islands with mountain peaks over 4000 m, and the only islands with constant and current volcanic activity, resulting in a denser and more diverse array of soils and ecosystems. The most salient division in land types, recognized even by ancient agriculturalists, were 'āina wai (inundated, wet lands) and 'āina malo'o (non-flooded, dry lands).

'Āina wai and 'āina malo'o supported distinctly different forms of agriculture. In particular, 'āina wai primarily supported lo'i—flooded, irrigated agriculture akin to rice paddies but focused on kalo (taro, *Colocasia esculenta*) (Throughout this paper we use the Hawaiian crop names to emphasize that, although they are common tropical species found over broad ranges, the landraces with which the Hawaiian culture coevolved are unique Hawaiian cultivars. The names and encoded knowledge that accompany this specific group of cultivars sets it apart from the species as a whole; cultural restoration arguably could not occur, and certainly could not occur to the full extent, with just any taro species, but only the specific cultivars that are appropriately referred to as kalo.) Hawaiians deliberately created and altered 'āina wai specifically for lo'i cultivation through the building of terraces, excavating of lands, and construction of dams and canals. In contrast to the relatively tight coupling of 'āina wai to lo'i cultivation, 'āina malo'o supported a broad range of agricultural strategies that included home gardens (kīhāpai), agroforestry (mahi 'ulu lā'au), intensive dryland farming (mahi 'ai), and a range of other strategies (see Lincoln and Vitousek 2017 for an overview). Additionally, there were "hybrid" systems that developed from diverting water from wet areas to irrigate dry lands intermittently, or dry lands that were intermittently wet on their own through seasonal rivers and springs.

In sweeping terms, archaeologists and anthropologists consider lo'i agriculture to be high in *landesque capital* [15], requiring significant infrastructural investment to construct terraces and canals to control the flow of water. Following construction, lo'i agriculture produced a significant surplus, as the flowing water reduced labor demands for weeding, fertilization, and watering of crops while supporting high productivity [16]. This form of production was also resilient against natural perturbations, such as drought, and social disturbances, such as war. In contrast, agriculture practiced on dry lands is presented as having less infrastructural development and higher labor costs, resulting in lower surplus production and, therefore, higher vulnerability to social disturbances [17]. Such systems, being dominantly rainfed, are also inherently more variable in their production, both spatially and temporally; therefore agriculture on 'āina malo'o has been considered more vulnerable to natural disturbances as well [12]. However, these generalizations are built upon sparse investigations into traditional agricultural systems, and none of them on operational systems in Hawai'i. As described, the forms of agriculture on 'āina malo'o are too diverse for easy generalization. Recent research by the authors identifies such systems that have minimal infrastructural investment and minimal labor requirements [10], extremely high infrastructural investment and moderate labor requirements [18], and moderate infrastructural development with high labor requirements [13].

Although the iconic lo'i systems are dominantly portrayed as traditional Hawaiian agriculture, agriculture on 'āina malo'o was, arguably, more important (especially on the younger islands): they produced more food than lo'i, at least on the younger islands; they were used to produce a broader range of crops with resource crops for fiber, timber, and medicine grown almost exclusively in dryland conditions; they were more widespread across the archipelago, occurring everywhere Hawaiians inhabited; and they formed the economic base for the powerful Hawai'i Island chiefs who eventually conquered the archipelago [17,19–21]. Documentation and modeling of rainfed agriculture in the state conservatively indicates that, in terms of land area, agriculture on 'āina malo'o exceeded lo'i agriculture at least five times [19,20].

2.1. Traditional Agriculture on 'Āina Malo'o

While lo'i agriculture is primarily based on kalo and is relatively consistent in its form, agriculture on 'āina malo'o was much more diverse, utilizing a range of cropping systems including small heavily-managed gardens near house sites, large-scale intensive multi-crop systems, mixed agroforestry, swidden or shifting agriculture, and arboriculture (Figure 1) [7,10,16,19–55]. The diverse forms of cropping systems reflected the differing ecosystems and topography that 'āina malo'o occupied. The drivers of agricultural form appear to dominantly be the soil depth and fertility, the slope of the land, and rainfall, although other local variations likely played a role as well [7,19,20,54].

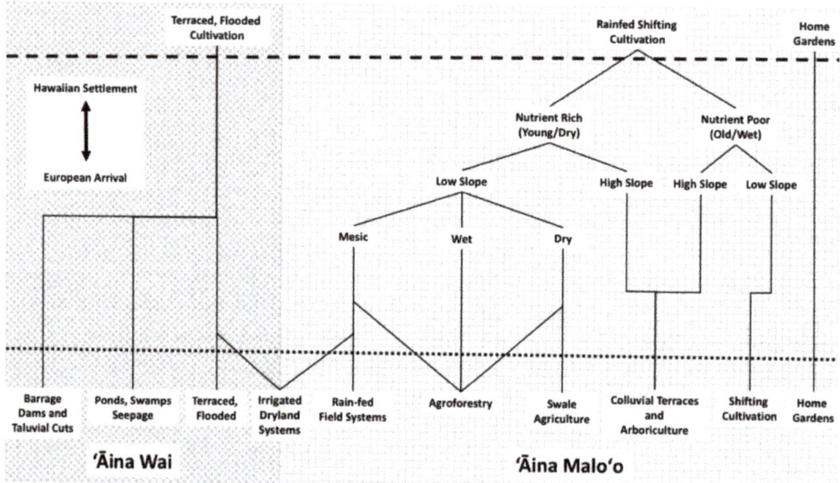

Figure 1. A rough typology for agricultural production systems in pre-contact Hawai'i; the general categories are not exhaustive and represent a spectrum of practices. The dashed horizontal line is intended to demarcate systems that the Polynesian discoverers of Hawai'i had in mind when they arrived; the lower dotted line indicates the range of techniques employed at the time of European arrival. There is no implication that cropping systems are invariant over time along a given line; to the contrary, we know that systems of lo'i expanded over time, rain-fed field systems underwent infilling and intensification, and shifting cultivation systems began to manage the fallow as well as the cropping phase intensively; other systems no doubt developed as well.

A considerable portion of the area devoted to rainfed cultivation occurred in vast, intensively developed "systems," such as the Leeward Kohala Field System (Figure 2). These intensive systems were confined to areas with high natural soil fertility [9,56] and adequate rainfall, mostly on the young islands of Hawai'i and Maui [19]. These field systems are defined by common elements of agricultural infrastructure, including long linear embankments and built stone mounds, although considerable diversity in form and application of the infrastructure is evident. Embankments were planted with taller crops such as kō (sugarcane, *Saccharum officinarum*), mai'a (plantain/banana, *Musa* spp.), and kī (ti, *Cordyline fruticosa*), and bordered cleared fields containing the primary staples of kalo, 'uala (sweet potato, *Ipomoea batatas*), and 'uhi (greater yam, *Dioscorea alata*). These continuous systems occupied vast areas on the younger islands; the largest of which was likely the Ka'ū system that may have covered over 50,000 acres [11,19]. Portions of these systems were likely farmed seasonally based on patterns of rainfall and temperature [12,13].

Figure 2. The infrastructural remnants of the Leeward Kohala Field System—a vast, dense network of rainfed farming plots.

Agroforestry and other forms of tree agriculture represented another significant fraction of agriculture on 'āina malo'o (Figure 3). Tree crops, such as 'ulu (breadfruit, *Artocarpus altilis*), kukui (candlenut, *Aleurites moloccanus*), niu (coconut, *Cocos nucifers*), hala (*Pandanus tectorius*), and 'ōhi'a'ai (mountain apple, *Syzygium malaccense*) were employed extensively by Hawaiians, primarily in places that were too dry, too rocky, too steep, too salty, too infertile, or too small for the "system" form of agriculture discussed above (e.g., Reference [20]), although several extensive agroforestry systems were developed in fertile areas (e.g., Reference [10]). Agroforestry in ancient times included mono-cropped arboricultural stands, multi-tiered diversified agroforestry, and the alteration and tending of native forests (e.g., References [21,28,34]).

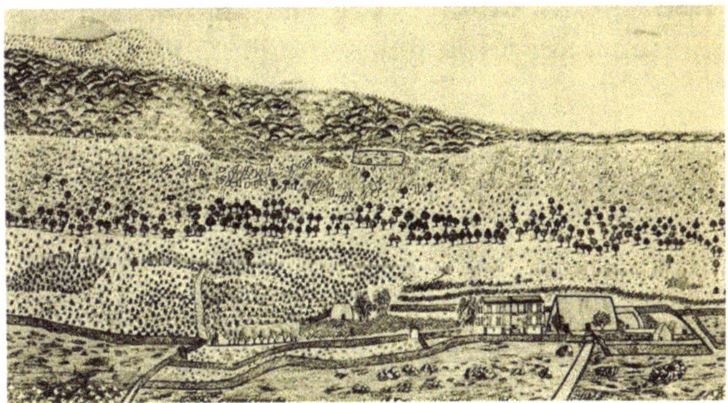

Figure 3. An 1836 drawing by Persis Goodale Thurston depicts the different rainfed farming zones within Kona, Hawai'i Island. The *kula* lands in the foreground represented opportunistic agriculture and home gardens in the dry lowlands, the *kalu'ulu* arboriculture appears as a distinct band of breadfruit trees across the landscape, the *'āpa'a* planting zone follows with its intensive stone infrastructure depicted, and finally the *ama'u* zone as managed native forest.

Non-flooded agriculture also includes a range of miscellaneous techniques that were smaller in scale and scope, but collectively were applied to a large area. These were highly diverse forms of practice and infrastructure at a micro-habitat scale; they included check dams (pa'amua), water holes (nā loko wai), terraces (kīpapa), intermittent water manipulation, stone and earth mounds (pu'u), swales, built soils, and other innovations that took advantages of local topography and environment. These developments ranged in intensity, scale, and productivity, but were highly place-adapted to maximize the scope of agriculture given the local opportunities and constraints. These variable developments often occurred adjacent to, or even embedded within, the intensive systems as the landscape shifted into more marginal environments in terms of water or soil fertility. However, some regions without potential for intensive systems of agriculture applied these alternative techniques extensively.

On Hawai'i Island, the vast majority of agriculture was of non-flooded forms, although a few opportunities existed for lo'i agriculture in older, windward areas. Moving clockwise around the island, we generalize its agricultural opportunities (Figure 4). Starting at the northern point of the Kohala peninsula, small valleys were developed for limited lo'i with rainfed agriculture (probably shifting cultivation) occurring between the valleys, and in a late pre-contact development, tunnels and canals were constructed to irrigate interfluvial areas [57]. On the northeastern coast a series of large valleys offered ideal locations for lo'i with agroforestry conducted on the colluvial valley slopes. Moving south into Hamakua vast areas of agroforestry were employed, and unique swidden and arboriculture systems established, along with sparse lo'i opportunities in the many small streams and rivers. Larger rivers flow into Hilo Bay, and relatively large systems of lo'i were established there. In the very young but wet regions of Puna, vast areas of multi-tiered agroforestry existed along with multiple forms of agricultural gardens such as planting pits and built soils. Surrounding the southern point of Ka'u was perhaps the largest intensive rainfed field system. The southwest coast, being very dry and young, offered limited opportunities for agriculture that took advantage of microsite development. Along the Kona coast, another large, intensive field system existed. Moving north along the west coast the landscape again becomes dry and provides only for limited development of agriculture at opportunistic sites. At the inlet just south of the Kohala peninsula, two intensive hybrid systems that intermittently irrigated dryland areas existed. Finally, along the western coast of Kohala, a final, intensive dryland field system (the Leeward Kohala Field System) existed inland. While this captures the large-scale patterns of agricultural developments around the island, it is important to note that a substantial amount of variation occurred within these generalizations.

2.2. The Decline and Rise of Traditional Agriculture

Following European colonization, with the decline of the native population, the privatization of lands, the introduction of plantation agriculture, and the control of water resources, native Hawaiian agriculture diminished substantially [58]. In particular, the rainfed agricultural systems, which were both more vulnerable and had land more conducive to plantation agriculture, declined very rapidly. While lo'i systems also declined precipitously, their physical infrastructure and continued practice were sustained at a much higher rate.

While pockets of traditional agricultural practices have persevered over the 240 years since European arrival, the revival of indigenous methods and crops has significantly increased since the "Hawaiian Renaissance" of the 1970s (e.g., see Kagawa-Viviani et al., this issue). Since then, hundreds of individual lo'i terraces in dozens of districts have been restored, both into commercial and subsistence production; often through the efforts of nonprofit organizations focused on cultural and environmental restoration and education. Conversely, restoration on 'āina malo'o remained largely non-existent. That lo'i have been prevalent in initial restoration efforts could be expected for several reasons. First, it follows the ancient temporal pattern, in which wetland areas with abundant freshwater resources were developed first by the original settlers of the island [7,59]. Furthermore, lo'i are common throughout Polynesia and therefore represent a knowledge system with more common and recoverable knowledge from other areas. As indicated above, lo'i infrastructure and practices have been better preserved and thus represented a more accessible starting point for restoration efforts; this is coupled with the fact that labor requirements of lo'i are typically lower on a per area basis, allowing restoration to occur with a relatively small cohesive group and therefore more easily obtaining a "critical mass" to power the efforts. Finally, and not to be understated, lo'i terraces are used to grow kalo (taro, *Colocasia esculenta*) to the near exclusion of all other crops [60]. Kalo is a piko, both spiritually and physically, of the Hawaiian people; its importance as "the staff of life" gives it a central role in any efforts of biocultural restoration. Piko literally refers to the navel and umbilical cord, or a summit, and symbolically refers to a connection to the world. In Hawaiian epistemology, a person has three piko that connected one to the spiritual and physical world—the fontanel that connected one to theirhis/her ancestors, the navel that connected one to the present world, and the genitals that connected one to theirhis/her future. This worldview recognizes that humans are a product of genetic and environmental history, that they are intimately connected to everything in the present, and that their being will impact everything to come in the future. Kalo is seen as a manifestation of this connection, as it is connected to mankind through ancestoryancestral cosmology, connected to mankind by reciprocal sustenance (humans farming kalo and kalo feeding humans), and connected to mankind in their relationship into the future. More practically, kalo was the preferred staple of the people and the gods, and therefore central to the diet of the people and religious and ceremonial practices.

3. Reviving 'Āina Malo'o

While engagement in lo'i restoration and maintenance has been a core vehicle for communication and education regarding Hawaiian culture, it does not represent the full spectrum of Hawaiian agriculture and, on the younger islands of Hawai'i and Maui in particular, does not accurately represent participants' ancestral engagement with 'āina malo'o. To facilitate discussion of biocultural restoration stemming from the revived cultivation of 'āina malo'o multiple representatives from each of five identified organizations (Figure 4) contributed: Ulu Mau Puanui, Maluaka, Māla Kalu'ulu, Hui Mālama i ka 'Ala 'Ūlili, and Ho'o'ulu'ulu Kahalu'u. The contributors represent a wealth of experience regarding efforts on 'āina wai and 'āina malo'o, and are among the leading organizations conducting agricultural restoration on 'āina malo'o. We present a brief case study on three of the organizations to exemplify key aspects of the efforts. The three were selected not only because they are the most developed of the organizations, but represent the most substantially different pathways to the restoration that is occurring.

3.1. Ulu Mau Puanui

While researching the leeward Kohala slopes, using one of the most striking rainfall gradients on the planet to study soils and ecosystems, Peter Vitousek and colleagues had the opportunity to collaborate with archaeologists studying the rain-fed Leeward Kohala Field System, a 6500-hectare area that was once farmed intensively by Hawaiians. The region, which is mostly used for cattle today, retains the imprint of Hawaiian agricultural practices, with the infrastructure still etched on the landscape (Figure 2). Together, ecologists and archaeologists developed an understanding of why the systems exist where they are, eventually demonstrating that the location of the field systems related to soil development and thresholds of soil properties that change with age and rainfall. The interdisciplinary team also studied the Hawaiian populations that lived in leeward Kohala and how their societies functioned and evolved [19,39,61–72]. However, for all the research that situated the development of agriculture within environmental and social context, they did not understand *how* rain-fed agricultural systems worked, namely how people grew crops and how they sustained the productivity of that land for centuries under conditions where most people worldwide practiced much less efficient slash-and-burn agriculture. Recognition of this shortcoming led to the founding of Ulu Mau Puanui, a community-based non-profit organization that established three permanent garden plots spanning the rainfall gradient in an effort to rediscover the agricultural practices associated with the field system.

Figure 4. Location of the five organizations focused on restoration of traditional dryland agriculture and a depiction of the general patterns of agricultural reliance of Hawai'i Island estimates by ethnographic sources, archaeological surveys, and biogeochemical models. It is important to note that the forms of agriculture presented are only the broad categories of the dominant forms applied, and many nuanced variations within any area occurred, including areas that do not depict any agriculture. For instance, in Kona, although dominantly rainfed areas of spring-fed, flooded or irrigated cultivation occurred, along with areas of agroforestry.

Puanui is one of thirty-three *ahupua'a* in leeward Kohala that make up the Kohala Field System. (*Ahupua'a* is a traditional land division system, generally considered the smallest land division in Hawai'i that still retained strong political oversight, which, for the most part, coincides with the concentric geography of islands, in which the divisions extend from an upland interior to the ocean, encompassing a range of ecosystems and resource types [73].) While large ranching landowners now own most of the Kohala Field System, Kamehameha Schools owns the narrow *ahupua'a* of Puanui (but has leased it to Parker Ranch for grazing for many years). (The Kamehameha Schools is a private educational trust endowed by the will of Hawaiian Princess Bernice Pauahi Bishop (1831–1884) that provides preferential admittance to Native Hawaiian students. They are the largest private landholder in Hawai'i and a very substantial organization in the state. As the literal and metaphorical descendant of the Hawaiian monarchy, they are in the critical eye of the Native Hawaiian population, which insists that they be leaders and advocates for Hawaiian culture and well-being.) When approached, both Kamehameha Schools and Parker Ranch were highly supportive of an effort to bring Hawaiian crops, and the Hawaiian community, back to the Kohala Field System at Puanui. Ulu Mau Puanui's efforts focused on outreach and education to the broader community. The gardens attracted substantial local interest, including groups from several schools that made multiple repeat visits. It quickly became clear that there is no substitute for experience and experimentation in this landscape; the system itself must have evolved that way, and as we seek to understand it, we find that experience and experiments unlock knowledge in the community as well as providing scientific information.

Ulu Mau Puanui manages the gardens of Puanui, provides access to the land for schools and community members, and encourages groups to come and work on the rain-fed agricultural system and to contribute to the process of discovery. The mission of Ulu Mau Puanui is to "engage in hands-on, land-based learning and culturally-centered science with learners, educators, families, and community in order to revitalize and better understand the Kohala Field System". The vision is that,

> when we are successful, our communities will appreciate the scope, diversity, and global significance of Hawaiian agriculture as it was practiced before European contact. We will understand that Hawaiian agriculture arose from a populous, organized and innovative society, and that the society in turn was shaped by its interactions with the land. We will build on that understanding to create an innovative and dynamic modern society that has a deep understanding and connection to its land. It is our hope that this transformation will spread across the Archipelago, and across the Pacific to produce a transformed modern agricultural system that draws from the wisdom and sustainability of the past, the knowledge and experience of local farmers and ranchers, and the best agricultural practices of the wider world that provides Hawaii with most of our food.

At Ulu Mau Puanui, a sustained interaction between scientists, respected cultural leaders, and community and student members has resulted in trust building and mutual exchange that has caused all parties to ask deeper questions and examine their own biases and assumptions in new ways. Culturally-centered science—the integration of Hawaiian ways of knowing that helps inform and inspire scientific inquiry—has been the cornerstone of activities. This has led to an inquiry-based framework that promotes creativity while practicing cultural values such as *kilo*—observation, *pili*—relationship to the land and others, *hōʻihi*—respect, and *kuleana*—responsibility and privilege. This process has impacted the participants and the researchers alike and brought two, often disparate, perspectives much closer together. Since established, the restoration at Puanui has led to multiple publications that have directly examined elements of biocultural restoration [8,13,14,74].

3.2. The Maluaka Project

The Maluaka Project was born from the joining of forces between a series of service-learning anthropology classes taught by Jack Rossen during academic winter intersession and the mapping and restoration of the ten-acre parcel Maluaka parcel of the North Kona agricultural field system by Keone Kalawe and Māhealani Pai. The collaborative archaeological project involves excavation and intensive water flotation to examine field engineering and to recover plant remains. The work is conducted in collaboration with Kamehameha Schools and involves linkages with lineal descendant of that land, elementary, intermediate, and high school students, and at-risk youths, teaching all of them the complexity and genius of Hawaiian agricultural systems and combatting the negative stereotypes of ancient Hawaiians created and maintained by foreigners (e.g., stupid, lazy, etc.). The long-term goal is to revitalize the ancient agricultural terraces and platform system, utilizing Native Hawaiian knowledge and fine-grained archaeological and archaeobotanical data to understand the site in terms of spirituality, technology, layout, and plant patterns.

Over the years, a relationship was formed through sustained interaction. Community-based clearing, restoration, mapping, and utilization of the site was underway by Māhealani Pai and Keone Kalawe. Courses designed for New York college students to experience the culture and history of Hawaiʻi Island, led by Jack Rossen, performed service at a wide range of venues, including Maluaka. Each year, the group would spend more and more time at Maluaka, contributing labor while learning about the system. Efforts at the site grew in scale and scope, clearing and restoring more of the site while increasing community engagement and education. Each year the college returned, Māhealani, Keone, and their students had cleared and mapped more of the site. At one point, Māhealani wondered what type of research would be needed to understand the specific agricultural patterns of the planting platforms and pits we were uncovering; Jack Rossen mentioned that from the perspective of an archaeobotanist, it was a matter of excavating with an emphasis on water flotation recovery and microscope analysis of plant remains, along with starch, pollen, and phytolith studies. This conversation led to the devotion of courses to the Maluaka site and the creation of on-site field schools. During four summer field seasons (2015–2018), excavations in various sectors and elevations of Maluaka occurred, recovering numerous artifacts, and more importantly, discovering high levels of infrastructural development such as an extensive underground canal system, the system of firepits cut into pahoehoe lava, and three to five meter high mounds used as observation points. Unlike many areas of Kona, Maluaka has seasonal water sources in the form of groundwater and spring-fed wells.

How did the Maluaka Project develop from philosophical and intellectual perspectives? It began with the foundation of long-term relationships of friendship and trust. As visitors to the island, New York researchers and students gave volunteer labor over several years with interest and respect for Native culture and history. The Native Hawaiian counterparts gave welcome, cultural perspectives and indigenous practices. From an intellectual standpoint, both parties knew they wanted to combine Native wisdom and knowledge with Western science to understand Hawaiian agricultural systems from a more powerful perspective than could be accomplished by either approach alone. This combination means understanding how archaeology and archaeobotany can provide carefully collected systematic data, and how the long-term site mapping and contemporary usage and observations contribute to a fine-grained understanding. From the start, all understood that Hawaiian agriculture must be understood in terms of Native Hawaiian concepts of land and social organization. Everyone involved has endeavored to understand agriculture as part of broader, integrated, and aligned sacred landscapes. That means understanding the agricultural configuration and observation points (pānānā) of Maluaka in relation to sacred sites (heiau) at the coast below at Keauhou-Kahaluʻu, and other major nearby sites such as the Kāneaka holua (land-sledding) slide. Most importantly, all agreed that the research must have practical applications: to understand the modern potential of the agricultural system and rebuild with our eyes toward the future food sovereignty of Hawaiʻi.

3.3. Māla Kaluʻulu

Māla Kaluʻulu Cooperative (MKC) was born out of a desire to restore the *kaluʻulu*—a nine square mile band of traditional breadfruit (*Artocarpus altilis*) agroforestry that stretched 20 miles across the Kona landscape. (The origin of the term *kaluʻulu* is uncertain but appears in early historical land claim records describing the breadfruit zone. Some elders have indicated the term should be, or is a contraction of, *ka ulu ʻulu*, literally meaning "the breadfruit grove".) Research by Noa Lincoln into the extent and productivity of the breadfruit belt in this area suggested that it produced between 20 and 50 million pounds of breadfruit annually [10], and that the establishment of the breadfruit belt appears to be suited to the unique biogeochemical factors of the region [9]. Inspired by this research, a group of local farmers, entrepreneurs, and educators, in partnership with Noa Lincoln, formed MKC and applied for, and won, the 2015 Mahiʻai Matchup, a farming business plan competition supported by the Pauahi Foundation and Kamehameha Schools which provided a 4-acre land parcel in the heart of the ancient breadfruit belt.

During its establishment, MKC's founding members discussed at length the model of incorporation, ultimately settling on a worker cooperative venture for multiple reasons. Foremost, it was agreed that it was of the utmost importance that the restoration provides people livelihoods and opportunities. It was argued that the cooperative model reflected the traditional social system, in which *kuleana* dominated. *Kuleana* is often defined simply as "responsibility," but in reality was a reciprocal function of rights that were based on one's responsibilities; in ancient times, a person's ability to access resources related to his or her contribution to maintaining those resources, just as the rights of cooperative members to access benefits is based on their relative contributions, or in cooperative terms, patronage. Furthermore, in restoring traditional agriculture, MKC wanted to establish that the systems were viable and relevant in the modern world and therefore wanted to develop in a way that could ultimately be self-sufficient. Enrollment in the cooperative is open to all, and the leadership donates time to ensure the success of the program.

Since its inception, MKC has worked to transform its own and other Kona parcels away from dominance by invasive species or mono-cropped agriculture back into the traditional breadfruit-based systems that once existed. These multi-layered agroforestry systems produced environmental benefits, were biodiverse and resilient, reduced the need for chemicals and inputs, were highly productive, and were culturally relevant ways of producing food [75]. The mission of MKC is "to enhance our understanding, appreciation, and utilization of traditional Hawaiian land use practices focusing on food production, and through food production and distribution, research, and education, to enhance sustainability and self-sufficiency in the Hawaiian Islands." Guided by Native Hawaiian values and practices, MKC aims to work with others across the State to collectively re-learn and enliven the techniques our predecessors used to subsist on the *'āina* (land), and through proper engagement increase engagement in, and awareness of, Hawaiian cultural perspectives on environmental health and stewardship. Through this work, MKC helps to advance understanding of innovative farming principles developed by early Hawaiians and how these principles can apply to contemporary cropping systems. In addition to restoration and research of the traditional system, MKC develops an "adapted" version of the system to meet today's market demand better; for instance, by planting a wider variety of crop sub-species for year-round production. The concomitant restoration of the traditional system and development of a modern version may help to demonstrate the viability and relevance of traditional farming practices in today's socioeconomic environment.

In addition to ongoing restoration of the traditional agroforestry system, MKC engages in holistic agricultural production, research, and education. The restoration plan is based on extensive ethnohistorical testimony that described in detail, from both western and native perspectives, the form and function of the *kalu'ulu*. Research interests have since emerged in how the traditional agroforestry system interacts with the young soils of south Kona, and in documenting the impacts of ecosystem services, nutrient cycles, and biodiversity. Multiple research partnerships have been established to pursue these interests. On-site research has been participatory, with researchers participating in the farming activities, co-forming research concepts, and leveraging research studies to create more opportunities for engagement and collaboration. From the beginning of this project, an emphasis was placed on community outreach, initially by sharing the ethnohistory and previous research that has been conducted about the *kalu'ulu* and growing to include sharing results from the restoration activities and related emerging research projects hosted at MKC. Since it was formed, MKC has hosted dozens of educational programs, farmer trainings, researchers and interns, and events at the farm site.

4. Form and Function of Biocultural Restoration on 'Āina Malo'o

4.1. He loa ka 'imina—Long is the Search

A key challenge to restoring traditional dryland systems is the lack of working reference systems to serve as models for the restoration and research efforts. This relative lack of knowledge requires highly interdisciplinary approaches that triangulate agricultural form and function. Multiple lines of evidence are explored in each of the restoration efforts, drawing upon archaeology, archaeobotany, biogeochemistry, agronomy, ethnographic and ethnohistorical accounts, and living culture to develop models of each system. This process is far from linear, but rather is an iterative and interactive learning process, similar to descriptions of learning in adaptive co-management settings (Figure 5) [76]. Investigations of archaeology and ethnohistory influence the design of scientific field experiments and restoration; these trials further inform practices. Outreach activities at all the sites share research and experiential findings while engaging visitors in ways that enable inputs of local and traditional knowledge. The knowledge input from the community also feeds into future activities and experiments. Through this iterative process, refinement of our understanding occurs while simultaneously powering awareness, connections, and, ultimately, cultural revitalization.

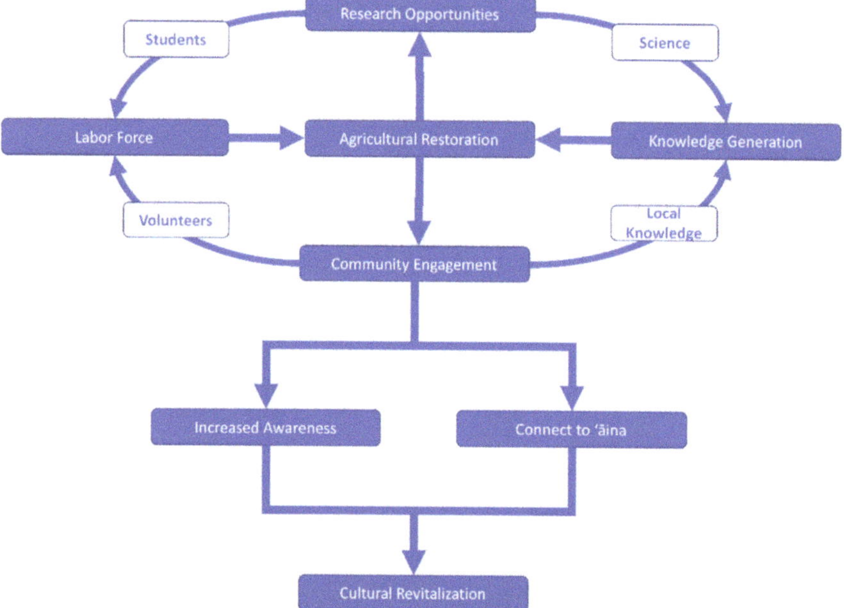

Figure 5. A conceptual diagram that outlines the multiple pathways used to drive agricultural restoration, and how that further leads to cultural revitalization. Each organization applies each of the pathways, although to different extents. The upper loop represents a feedback mechanism in which the agricultural restoration drives opportunities for research and engagement, which further drives opportunities for restoration. This feedback loop powers cultural revitalization through the engagement of the community.

Despite significant variations in their starting points, each organization leverages its disciplinary strengths while pursuing multiple methods. Extensive intact archaeology at Puanui [77–80] has provided for rediscovery based on archaeology and archaeobotany; however, the scantily recorded ethnohistory at Puanui resulted in an agricultural system that was well understood from a theoretical standpoint but poorly understood in practice. In contrast, Māla Kaluʻulu has minimal physical infrastructure associated with its agroforestry and therefore minimal archaeological or archaeobotanical data. However, as one of the primary points of European contact in the 18th century, their region has exceptionally well-recorded ethnographic and historical testimony detailing many practical aspects of the agriculture. Despite considerable losses in knowledge and practice, each effort has managed to create a sharp picture of what life and agriculture were like before decline.

4.2. He aliʻi ka ʻāina—The Land is Chief

Much of the awe which dryland systems inspire is due to their scale. At the parcel level, rainfed agricultural remains are often not conspicuous, and also lack the particular serene beauty of loʻi with flowing water and rich mud that seems inherently more attractive on a small scale. Each organization expressed how this has led to initial challenges of attracting interest and support, and have utilized a shared strategy of connecting to a landscape level scale to overcome this issue. Puanui, the oldest of the organizations, relies directly on the visible scale of the Kohala Field System as the best preserved and most observable example, owing to the lack of plantation agriculture that destroyed traditional infrastructure in many places, and the contemporary cattle grazing that maintains visibility of the agricultural features (Figure 2). Engagement activities at Puanui include a hike up Puʻu Kehena (a cinder cone) that offers visitors a panoramic view at the vast extent of the agricultural remains blanketing the landscape. From this exceptional vantage, many truths become apparent to the observer: the vastness of the system speaks to the ancient population, the political organization needed, the ecological gradients encompassed, and the integrated socio-political features. This powerful view was a pivotal aspect to building support for the effort and continues to be a major tool of engagement and impact. Although the extent of the traditional agricultural systems is not directly observable at the other sites, each develops aspects of scale to paint a picture of ancient agriculture within a broader landscape. Māla Kaluʻulu emphasizes how the interactions of politics, ecology, and breadfruit defined the extent of the *kaluʻulu* system and describes the ancient productivity in terms of modern populations to demonstrate the scale of the development. Maluaka connects its plot to other major landscape features within its land division to elucidate how the more monumental developments near the coast were integrally connected to and supported by the agricultural developments in the uplands.

With their scale, dryland systems carry a story of political development, innovation, and complexity that may have many parallels to continued developments for Hawaiian sovereignty and self-determination today (e.g., References [81,82]). In contrast, loʻi tend to situate agriculture within the spiritual and family realm, owing to the strong connections of kalo to these aspects of Hawaiian culture and loʻi restoration has played a central role in the revival of cultural pride and practices (e.g., Kagawa-Viviani et al., this issue). The themes of scale and community food systems in the strategy of each of these organizations tends to place non-flooded agriculture within the socio-political history of the islands, emphasizing the importance of food, and in particular, the social movements powered by the vast rainfed agricultural areas. Indeed, these areas are what separated the Hawaiian archipelago from the rest of the Pacific, allowing the development of the most complex political systems [5]. The scale of rainfed systems is also what separated the young islands of Hawaiʻi and Maui from the rest of the archipelago, eventually powering the conquest of the archipelago by Kamehameha, a Hawaiʻi Island chief who was born and grew to power in the Leeward Kohala Field System, and moved to the Kona system once he achieved paramountcy in the archipelago [17]. Pulling the parallel into the present, engagement at these sites tends to raise discussion of the importance of large-scale sustainable food systems for security and self-determination of communities today.

In this way, the restoration of Hawaiian agricultural systems could be seen to parallel the larger cultural movement of the islands. Starting with loʻi, early focus of the renaissance movement focused on building internal strength and cohesiveness within the Hawaiian community, reconnecting to values and practices that were lost or hidden away. The use of kalo as a spiritual and family center was a pivotal symbol to revive the basic units of the Hawaiian social fabric. Subsequently, Hawaiian activism engaged more expansive goals, revitalizing the fight for sacred spaces locally, such as the pushback against further telescope construction on Mauna Kea Volcano [83], and sovereignty in the international arena, such as appeals to the United Nations that have formally recognized Hawaiʻi's status as a sovereign country under military occupation by the United States [84]. These more recent activities have coincided with the greater awareness and restoration of the large-scale dryland agricultural systems that were essential to the development of the high levels of political complexity represented by the Hawaiian Kingdom [5]. This coincidence of activities may suggest a growing resurgence of Hawaiian cultural activity beyond the individual family and community units to a broader political framework.

4.3. He māʻona moku—A Satisfaction with the Land

At all sites an emphasis is placed on place-based adaptations of Hawaiian agriculture to environmental variation and microclimates. At Maluaka, the young lava flow creates a highly diverse topographic landscape, with many localized high and low points that direct the flow of water and the accumulation of soils; correspondingly, very high heterogeneity of infrastructure is apparent, demonstrating extreme adaptation to microhabitats and topography at a scale that, until recently, was not well documented or even widely known. At Puanui, although the infrastructure is simple and regular, engagement at the site emphasizes the adaptation of the cropping systems to the local environment and to the huge environmental gradient encompassing the three restoration sites. Similarly, at Māla Kaluʻulu, while the agroforestry system could be described as a whole, the small-scale variation of planting based on light gaps and water distribution is substantial and dynamic. Here, planting trials have demonstrated different niches for the rhizome-based crops as a function of light availability: kalo will grow only in the drip lines at the edge of the canopy of larger trees, while ʻawapuhi (shampoo ginger, *Zingiber zerumbet*) will only grow under canopied areas.

The organizations here push beyond the "keystone" biocultural relationship with kalo [60] by expanding the range of crops and cropping systems and by reviving place specific knowledge and practices. This is important to the overall resilience and diversity of culture and practice, expanding the suite of biocultural couplings to strengthen the larger socioecological system of the modern Hawaiian culture. This is particularly important on Hawaiʻi Island, where the history of Native Hawaiians dominantly consisted of interactions with ʻāina maloʻo and a broad range of staple and supplementary crops. The need to recognize and revitalize that range of interaction is a critical element to the identity and practice of Hawaiʻi Island culture. Following 120 years of colonial occupation and often de facto banning of Hawaiian culture (e.g., banning of ʻawa (*Piper methysticum*) drinking; Kagawa-Viviani et al., this issue), the mental health and well-being of Hawaiians have declined precipitously, as it has done globally for indigenous peoples (e.g., References [85–88]). Engagement with the land and agriculture, a central component to Hawaiian culture in which social and family values are encoded, is critical in restoring identity and wellbeing to Hawaiian people [89,90]. By providing a broader range of relationship between people, plants, and places, well-being follows (e.g., References [91–93]). This is particularly relevant for the many locations and individuals where the practice of loʻi agriculture does not appropriately address their genealogical connection to the land.

While Hawaiian and other indigenous place-based adaptations in resource management is often proclaimed to be exceptional [2,94–96], the more consistent practices associated with loʻi do not capture the diversity of adaptive strategies in the way that agriculture on ʻāina maloʻo does. This is largely a function of necessity. Loʻi systems occur in lowland valleys with flowing water that regulates multiple aspects of the cropping system—they therefore do not have to deal with the same level of spatial and temporal variation in environment that agriculture on ʻāina maloʻo does. The extensive loss of the detailed knowledge and practices associated rainfed systems (e.g., References [21,35,36]) requires a revival of knowledge, which, despite significant information from investigation, can only be regained through active restoration on the land.

What we now see as the archaeological infrastructure is only the remaining physical manifestation of the diverse practices and cropping systems employed within the sites. A simple comparison of the sites themselves offers a case-study of adaptation, with vastly different forms consisting of different suites of crops, temporal patterns, and practices seen (e.g., References [8,10,12,14,18,97]). One thing that has been clear in the restoration of these systems is the diverse and innovative methods for managing water. In Kohala, research has demonstrated how the long walls running perpendicular to the wind were a vital mist-trap, enhancing and concentrating soil moisture to facilitate better growing of sweet potatoes [8,14]. In Kona, highly diverse infrastructure appears adapted to the changing water situations. Infrastructure, such as cut canals to move, store, and disperse intermittent water flows [18], mounds to take advantage of areas where water pools (J. Kahoonei, pers. com.), and a host of strategies to prevent water from evaporation in lower elevations, illustrate the adaptive hydrological strategies (e.g., Reference [97]). Conversely, the restoration also highlights how much is still unknown. For instance, the striking difference in wall orientation between agricultural areas, with Leeward Kohala alignments perpendicular to the slope and wind while Kona alignments are parallel to them, has not been adequately explained.

4.4. E hoʻohuli ka lima i lalo—Turn the Hands Down

Community engagement at these sites is multifaceted but active. A "service learning" component is associated with education all the sites. For the host, it provides a vital source of labor while simultaneously driving a particular and important experience. Everyone works at these sites—be he or she world-class researchers or delinquent students, elder or child—and they work side-by-side. The action of collectively getting one's hands dirty has a substantial effect on bringing people to the same level in a way that is not easily replicated, driving informal sharing through which knowledge transfer that may otherwise be difficult can occur. Here real bi-directional learning occurs between generations and ways of knowing. Subsequently, the restoration of each site is presented in mission statements as a vehicle through which to engage and grow communities of people. These restoration efforts are not solely restoration of physical infrastructure or ecosystems, but the biocultural systems of food and culture. Each mission statement includes phrases such as "space for the community to connect" and "regenerate responsibilities," recognizing that these agricultural systems are socio-ecological landscapes that rely on the physical environment, biota, and human knowledge and practice to function correctly. Each of these organizations expressed that such relationships are difficult to teach but can be learned; educating people about a biocultural relationship through talk provides little adoption of practice but conducting activities that actively place individuals in direct contact with the environment allows the opportunity for those biocultural connections to be formed.

Although people might believe that it would be easy to restore these systems, the participants were clear about the difficulties. It is not just the physical clearing and planting of an ancient agricultural system, but it is the revival of place and history, the healing and building of relationships, and the cultivation of interaction. All of these take considerable time and commitment, influential leaders, and collaboration. It takes dedication to a cause in which importance and impact are not necessarily immediately seen. The sustained partnerships for restoration, research, and education requires working across academic and rural community partners, which is not always easy. The success of both the research and engagement hinges on positive and productive relationships. These relationships are bound by a common interest in understanding the sustainability of the respective systems in both environmental and socio-cultural terms, which must emerge from a diversity of knowledge sources.

4.5. Pupukahi i holomua—Unite in Order to Progress

The success of each effort has relied heavily on collaboration at multiple levels. Collaboration is essential between the leaders and the landowners, with the community, with scientists, and with the broader public. With many different stakeholders in each of the efforts, there exist many different lenses on the value and purpose of the restoration. Finding a way to move everyone together in a productive way is a crucial part of the process. Discussions emphasized the critical importance of trust-building between stakeholder groups. Particularly in Hawaiʻi where there has been a long history of science focused on a still-living culture subjected to recent (and active) traumas, there have been many conflicts between scientists and communities. The groups have different timelines, where scientists are often driven by short time frames dictated by grants and careers, while communities have a long-term perspective on value and multi-generational outcomes. Recognizing and managing the different time-scales is an integral part of the process. There have also been issues of knowledge ownership and give-back. Often scientists "mine" local knowledge that is used to guide research, and report that local knowledge directly by themselves, thereby representing a taking of community knowledge. Simultaneously many researchers do not make the time and effort to conduct strong outreach and reporting of results to the community of interest. Historically, this has represented a one-way flow of knowledge away from locations, leading to burnt bridges and an overall distrust of many scientists by local communities.

There were several common strategies employed by the organizations to overcome relationship barriers. First and foremost is addressing the past issues by forging better relationships with communities, acknowledging their knowledge as central, committing to long-term engagement, and conducting appropriate sharing of scientific knowledge back to the locales. Aiding in this is that each site has a "kanaka scholar," typically a native Hawaiian or local scientist who served as a liaison between outside researchers and communities to ensure proper acknowledgment of contributions, to translate ground-based activities and observations into academic speak, and to prioritize communicating science back to the community. These individuals also help to forge stronger relationships with researchers coming from the outside. Often this is as simple as "translation," helping people from different backgrounds and perspectives speak the same language. The quality of these relationships is directly related to perceived value on both sides; win-win activities are essential to continuing and growing relationships. At Maluaka, an active program has been developed that engages local at-risk youth in the archaeological investigations through field schools, rather than solely utilizing college archaeology majors as is often the case. This has allowed direct and indirect benefits to both the community and the archaeology. Ultimately, this is an indigenous approach that pursues multiple benefits, values, and emphasizes relationships and impact, rather than solely the science itself.

4.6. He aha ka mea nui?—What Is the Most Important Thing?

The approach by these organizations also recognizes that these systems are not just agriculture, but are deeply integrated into a more extensive, sacred, socio-ecological system that was intact until relatively recently; they represent a vital part of a living culture and embody history, religion, livelihoods, environment, and culture as a whole. Each organization attempts to recreate this aspect of a whole. The emphasis on relationships extends beyond those between people to also include relationships among people and the places, the plants, and the history. This recognizes in multiple ways the importance of interactions between people and elements of natural and social environments. At Māla Kalu'ulu, the emphasis on rights to the land based on one's inputs revives the ancient cornerstone of *kuleana*. At Puanui, each visit asks that participants introduce themselves to the place, and to envision their role in the future of the place, and at all sites engagement in protocol that grounds participants in the moment, in the place, at to each other is consistently practiced. The different strengths, yet common goals and outcomes, of the restoration efforts highlighted in this article are a testimony to the interconnected nature of the biocultural system as a whole. This is a distinctly native perspective that does not separate the daily activities from identity, values, and spirituality. The mix of knowledge sources that form the foundation of these restoration efforts each offer their opportunities and challenges, but engaging with all sources is a crucial element to the initial success of these operations.

Following the political shift with the illegal overthrow of the Hawaiian monarchy [98] and subsequent ongoing military occupation by the United States [84], Hawaiian people have been drastically impacted. In virtually every socio-economic statistic measured, from education to income to domestic violence to incarceration, Hawaiian people score the lowest of all groups within their ancestral home of Hawai'i [99]. While the creation of a new socioeconomic system has benefitted many non-Hawaiian immigrants to Hawai'i, Native Hawaiians have constantly struggled for rights and opportunities. By reviving Hawaiian relationships with the land, with their history, and with each other, significant gains have been made in the education, pride, and organization of the Hawaiian people [90]. In our experience, lo'i cultivation has had a critical impact on reconnecting people to the land, revitalizing ancestral responsibilities, and generating awareness of Hawaiian accomplishments and excellence prior to being displaced. This has strengthened a desire to return to traditional knowledge and epistemology to support and maintain Hawaiian communities, practices, and land stewardship. Furthermore, the expansion of restoration from lo'i systems of agriculture to systems on 'āina malo'o appears to parallel larger shifts in the Hawaiian community that seek to increase self-sufficiency, expand land stewardship, and increase activity in the realms of policy and activism. Each organization inadvertently, perhaps unconsciously, contributes to these movements by sharing the extent of historical scale and political power associated with the vast agricultural developments on

Hawai'i Island. To see and hear of the complexity and sophistication of the Hawaiian society in the past simultaneously emphases to some participants what was lost.

5. Conclusions

Even though non-flooded agriculture was, in ancient times, much more widespread and likely more important than lo'i, today lo'i restoration outweighs restoration of agriculture on 'āina malo'o. Restoration of 'āina malo'o agriculture by several organizations on Hawai'i Island has important biocultural consequences, particularly when compared to the more widespread restoration of lo'i. The use of highly interdisciplinary work to triangulate a more complete understanding of the social and physical aspects of the systems is crucial, particularly where knowledge systems have been severely eroded. This triangulation includes scientific investigation, use of historical resources, effectively tapping into local ecological knowledge, and conducting practical, experiential learning through active practice. This approach requires strong relationships and appropriate engagement with the community and culture at all levels; it is essential to building the complex relationships that make these efforts work. Only through strong engagement and mutual respect have these efforts been made possible and successful, and their success is often facilitated by someone with a foot in both Western and indigenous worlds. Organizations must leverage their strengths in this process and situate their connections in an appropriate socio-cultural role. This may require creative framing, such as how these organizations found ways to connect to the large scale of the systems even if the specific restoration plots are small. These essential aspects support previous example and case studies within the field of biocultural restoration.

Biocultural restoration of agriculture on 'āina malo'o highlights the oversimplification in the treatment of ancient Hawaiian agriculture by the dominant narratives told largely in the fields of archaeology and anthropology and perhaps too readily adopted culturally. In particular, there is a very high level of diversity of form of traditional agriculture on 'āina malo'o, and it cannot appropriately be lumped into "rainfed agriculture" as it has previously. High levels of place-specific knowledge are being uncovered through interdisciplinary and multi-epistemological restoration teams. Understanding agriculture on 'āina malo'o broadens the scope of biocultural relationships by engaging a more significant range of crops and therefore assortment of associated practices. Finally, agriculture on 'āina malo'o grounds itself in the scale and scope of the younger islands, and in the political processes that the vast agricultural areas powered prior to European arrival. We suggest that underlying socioecological functions that underpin agricultural types differ substantially, with lo'i and kalo focusing on family and spirituality, while agriculture on 'āina malo'o with its range of crops and systems emphasizing socio-political complexity. These same biocultural themes could parallel larger movements within the revitalization of Hawaiian culture.

Author Contributions: Conceptualization, N.L.; Methodology, N.L. and J.R.; Formal Analysis, N.L.; Investigation, N.L., J.R., J.K., K.M. (Kehaulani Marshall), K.K., M.P., D.S., P.V., and K.M. (Kamuela Meheula); Resources, N.L.; Data Curation, N.L.; Writing—Original Draft Preparation, N.L. and J.R.; Writing—Review and Editing, N.L., P.V., J.R., and D.S.; Visualization, N.L.; Supervision, N.L., J.R., and P.V.; Project Administration, N.L.

Funding: Maluaka and Puanui were funded by Kamehameha Schools. Manuscript preparation was supported in part, by Hawai'i Community College, a seed grant from the University of Hawai'i at Hilo, and the USDA National Institute of Food and Agriculture, Hatch project HAWU8035 administered by the College of Tropical Agriculture and Human Resources.

Acknowledgments: We acknowledge the many individuals who contributed to the organizations engaged, and who provided input to this paper. Contributors who were not co-authors include Aurora Kagawa-Viviani, Natalie Kurashima, Leon Nō'eau Peralto, and Anissa Lucerno.

Conflicts of Interest: The authors declare no conflict of interest. The funders had no role in the design of the study; in the collection, analysis, or interpretation of the date/in the writing of the manuscript, or in the decision to publish the results.

References

1. Pretty, J.; Adams, B.; Berkes, F.; De Athayde, S.F.; Dudley, N.; Hunn, E.; Sterling, E. The Intersection of Biological Diversity and Cultural Diversity: Towards Integration. *Conserv. Soc.* **2009**, *7*, 100–112.
2. Berkes, F.; Colding, J.; Folke, C. Rediscovery of Traditional Ecological Knowledge as Adaptive Management. *Ecol. Appl.* **2000**, *10*, 1251–1262. [CrossRef]
3. Briggs, J. The Use of Indigenous Knowledge in Development: Problems and Challenges. *Prog. Dev. Stud.* **2005**, *5*, 99–114. [CrossRef]
4. Hicks, C.C.; Levine, A.; Agrawal, A.; Basurto, X.; Breslow, S.J.; Carothers, C.; Charnley, S.; Coulthard, S.; Dolsak, N.; Donatuto, J. Engage Key Social Concepts for Sustainability. *Science* **2016**, *352*, 38–40. [CrossRef] [PubMed]
5. Kirch, P.V. Hawaii as a Model System for Human Ecodynamics. *Am. Anthr.* **2007**, *109*, 8–26. [CrossRef]
6. Vitousek, P.M. The Hawaiian Islands as a Model System for Ecosystem Studies. *Pac. Sci.* **1995**, *49*, 2–16.
7. Lincoln, N.K.; Vitousek, P.M. Indigenous Polynesian Agriculture in Hawai'I. *Oxf. Res. Encycl. Environ. Sci.* **2017**. [CrossRef]
8. Lincoln, N.K.; Kagawa-Viviani, A.; Marshall, K.; Vitousek, P.M. Observations of Sugarcane and Knowledge Specificity in Traditional Hawaiian Cropping Systems. In *Sugarcane*; Murphy, R., Ed.; Nova Science Publishers: Hauppauge, NY, USA, 2017.
9. Lincoln, N.; Chadwick, O.; Vitousek, P. Indicators of Soil Fertility and Opportunities for Precontact Agriculture in Kona, Hawai'i. *Ecosphere* **2014**, *5*, 1–20. [CrossRef]
10. Lincoln, N.; Ladefoged, T. Agroecology of Pre-Contact Hawaiian Dryland Farming: The Spatial Extent, Yield and Social Impact of Hawaiian Breadfruit Groves in Kona, Hawai'i. *J. Archaeol. Sci.* **2014**, *49*, 192–202. [CrossRef]
11. Quintus, S.; Lincoln, N.K. Integrating Local and Regional in Pre-Contact Hawaiian Agriculture at Kahuku, Hawai'i Island. *Environ. Archaeol.* **2018**, 1–16. [CrossRef]
12. Kagawa-Viviani, A.K.; Lincoln, N.K.; Quintus, S.; Lucas, M.P.; Giambelluca, T.W. Spatial Patterns of Seasonal Crop Production Suggest Coordination Within And Across Dryland Agricultural Systems of Hawai'i Island. *Ecol. Soc.* **2018**, *23*, 20. [CrossRef]
13. Kagawa, A.K.; Vitousek, P.M. The Ahupua'a of Puanui: A Resource for Understanding Hawaiian Rain-Fed Agriculture. *Pac. Sci.* **2012**, *66*, 161–172. [CrossRef]
14. Marshall, K.; Koseff, C.; Roberts, A.L.; Lindsey, A.; Kagawa-Viviani, A.K.; Lincoln, N.K.; Vitousek, P.M. Restoring People and Productivity to Puanui: Challenges and Opportunities in the Restoration of an Intensive Rain-Fed Hawaiian Field System. *Ecol. Soc.* **2017**, *22*, 23. [CrossRef]
15. Brookfield, H. Intensification Intensified: Prehistoric Intensive Agriculture in the Tropics. *Archaeol. Ocean.* **1986**, *21*, 177–180. [CrossRef]
16. Kirch, P.V. *The Wet and the Dry*; University of Chicago Press: Chicago, IL, USA, 1994.
17. Kirch, P.V.; Zimmerman, K.S. *Roots of Conflict*; School for Advanced Research Press: Santa Fe, NM, USA, 2011.
18. Rossen, J.; Pae, M. Archaeological Investigations at Maluaka. In Proceeding of Society for Hawaiian Archaeology Annual Conference, Honolulu, HI, USA, 29 September 2017.
19. Ladefoged, T.N.; Kirch, P.V.; Gon, S.M.; Chadwick, O.A.; Hartshorn, A.S.; Vitousek, P.M. Opportunities and Constraints for Intensive Agriculture in the Hawaiian Archipelago Prior to European Contact. *J. Archaeol. Sci.* **2009**, *36*, 2374–2383. [CrossRef]
20. Kurashima, N. Hō'ulu'ulu: The Biocultural Restoration of Indigenous Agroecosystems in Hawai'i. Doctoral Dissertation, Department of Botany University of Hawai'i at Mānoa, Honolulu, HI, USA, 2016.
21. Kelly, M. *Na Mala o Kona: Gardens of Kona*; Department of Anthropology Report: Honolulu, HI, USA, 1983.
22. McCoy, M.D.; Graves, M. Small Valley Irrigated Taro Agriculture in the Hawaiian Islands: An Extension of the 'Wet and Dry' Hypothesis. *Senri Ethnol. Stud.* **2012**, *78*, 115–133.
23. Kirch, P.V.; Holson, J.; Legacy, P.; Cleghorn, P.; Chadwick, O. Five Centuries of Dryland Farming and Floodwater Irrigation at Hōkūkano Flat, Auwahi, Maui Island. *Hawaii Archaeol.* **2013**, *13*, 70–102.
24. Vitousek, P.M.; Chadwick, O.A.; Hilley, G.; Kirch, P.V.; Ladefoged, T.N. Erosion, Geological History, and Indigenous Agriculture: A Tale of Two Valleys. *Ecosystems* **2010**, *13*, 782–793. [CrossRef]
25. Tomonari-Tuggle, M.J. *Archeological Data Recovery Investigations for the Northern Portion of Captain Cook Ranch*; International Archaeological Research Institute Inc.: Honolulu, HI, USA, 2006.

26. Schilt, R. *Subistence and Conflict in Kona, Hawai'i: An Archeological Study of the Kuakini Highway Realignment Corridor*; Bishop Museum Press: Honolulu, HI, USA, 1984.

27. Allen, M.S. Bet-Hedging Strategies, Agricultural Change, and Unpredictable Environments: Historical Development of Dryland Agriculture in Kona, Hawaii. *J. Anthr. Archaeol.* **2004**, *23*, 196–224. [CrossRef]

28. Allen, M.S. *Gardens of Lono: Archaeological Investigations at the Amy BH Greenwell Ethnobotanical Garden, Kealakekua, Hawai'i*; Bishop Museum Press: Honolulu, HI, USA, 2001.

29. Haun, A.E.; Henry, D. *Archeological Inventory Survey (TMK: 3-7-9-05:76, 77 and 78) Lands of Honalo, Maihi 1-2 and Kuamo'o 1, North Kona District, Island 39 of Hawai'i*; Haun and Associates Inc.: Kailua-Kona, HI, USA, 2010.

30. Escott, G.; Spear, R.L. *A Data Recovery Report for Hokulia: A Study of a 1540 Acre Parcel in the Ahupua'a of Honua'ino, Hokukano, Kenaueue, Haleki'i, Ke'eke'e, 'Ilikahi, Kanakau, Kalukalu, and 'Ono'uli, in the Districts of North and South Kona Island of Hawai'i, Hawai'i*; SCS: Kailua-Kona, HI, USA, 2003.

31. Kirch, P.V.; Sahlins, M. *Anahulu: The Anthropology of History in the Kingdom of Hawaii, Volume 2: The Archaeology of History*; University of Chicago Press: Chicago, IL, USA, 1992.

32. Kirch, P.V. Valley Agricultural Systems in Prehistoric Hawaii. *Asian Persepct.* **1977**, *20*, 246–280.

33. Newman, T.S. Hawaiian Agricultural Zones Circa A.D. 1832: An Ethnohistory Study. *Ethnohistory* **1971**, *18*, 335–351. [CrossRef]

34. Meilleur, B.A.; Jones, R.R.; Tichenal, C.A.; Huang, A.S. *Hawaiian Breadfruit: Ethnobotany, Nutrition, and Human Ecology*; College of Tropical Agriculture and Human Resources, Univeristy of Hawaii at Mānoa: Honolulu, HI, USA, 2004.

35. Handy, E.S.; Handy, E.G.; Pukui, M.K. *Native Planters in Old Hawai'i: Thier Life, Lore, and Environment*; Bishop Musem Press: Honolulu, HI, USA, 1972.

36. Handy, E.S.C. *The Hawaiian Planter: His Plants, Methods and Areas of Cultivation, Bulletin 161*; Bishop Museum: Honolulu, HI, USA, 1940.

37. Hammatt, H.H.; Clark, S.D. *Archaeological Testing and Salvage Excavations of a 155 Acre (Ginter) Parcel in Na Ahupua'a Pahoehoe, La'aloa, and Kapala'alaea, Kona, Hawai'i Island*; Cultural Surveys Hawaii: Lawai, HI, USA, 1980.

38. Hammatt, H.H.; Borthwick, D.F.; Colin, B.L.; Masterson, I.; Wong-Smith, H. *Archaeological Inventory Survey and Limited Subsurface Testing of a 1540-Acre Parcel in the Ahupua'a of Honua'ino, Hokukano, Kanaueue, Haleki'i, Ke'eke'e, 'Ilikahi, Kanakau, Kalukalu, and 'Ono'uli, Districts of North and South Kona, Island of Hawai'i*; Cultural Surveys Hawaii: Honolulu, HI, USA, 1997.

39. Vitousek, P.M.; Ladefoged, T.N.; Kirch, P.V.; Hartshorn, A.S.; Graves, M.W.; Hotchkiss, S.C.; Tuljapukar, S.; CHadwick, O.A. Soils, Agriculture, and Society in Precontact Hawai'I. *Science* **2004**, *304*, 1665–1669. [CrossRef] [PubMed]

40. Palmer, M.A.; Michael, G.; Ladefoged, T.N.; Chadwick, O.A.; Ka'eo Duarte, T.; Stephen, P.; Vitousek, P.M. Sources of Nutrients to Windward Agricultural Systems in Pre-Contact Hawai'i. *Ecol. Appl.* **2009**, *19*, 1444–1453. [CrossRef] [PubMed]

41. Ellis, W. *Journal of William Ellis: A Narrative of a Tour through Hawai'i 1823*; Hawaiian Gazette Company: Honolulu, HI, USA, 1917.

42. Menzies, A. *Hawaii Nei 128 Years Ago*; Archibald Menzies: Honolulu, HI, USA, 1920.

43. Cordy, R.; Tainter, J.; Regner, R.; Hitchcock, R. *An Ahupua'a Study: The 1971 Archaeological Work at Kaloko Ahupua'a North Kona, Hawai'i: Archaeology at Kaloka-Honokokau National Historical Park*; National Park Service, Department of the Interior: Tuscon, AZ, USA, 1991.

44. Horrocks, M.; Rechtman, R.B. Sweet Potato (Ipomoea Batatas) and Banana (Musa Sp.) Microfossils in Deposits from the Kona Field System, Island of Hawaii. *J. Archaeol. Sci.* **2009**, *36*, 1115–1126. [CrossRef]

45. Rechtman, R.; Maly, K.; Clark, M.; Dougherty, D.; Maly, O. *Archaeological Inventory Survey of the Ki'ilae Estates Development Area (TMK:3-8-5-05:19, 22, 26, 27), Ki'ilae and Kauleolī Ahupua'a, South Kona District, Island of Hawai'i*; Rechtman Consulting: Keeau, HI, USA, 2001.

46. Mccoy, M.D. The Development of the Kalaupapa Field System, Moloka'i Island, Hawai'i. *J. Polyn. Soc.* **2005**, *114*, 339–358.

47. Henry, D.; Wolforth, T. *Settlement and Agriculture in the Central Kona Field System Kula: Archaeological Inventory Survey of the Ginter Parcel, Lands of Kapala'alaea 1st, La'aloa 1st and 2nd, and Pahoehoe 4th, North Kona District, Island of Hawai'i*; CSH: Hilo, HI, USA, 1998.

48. Burthchard, G.C. *Population and Land-Use of the Keauhou Coast: The Mauka Land Inventory Survey, Keauhou, North Kona, Hawai'i Island*; International Archaeological Research Institute Inc.: Honolulu, HI, USA, 1996.

49. Monahan, C.M.; Thurman, D.W.; Thurman, R.R. *Intensive Mapping & Archaeological Condition Assessment 24-Acre Portion of Kahalu'u Field System Kahalu'u Ahupua'a, North Kona District Hawai'i Island, TMK (3) 7-8-002:010 (Por.) & 7-8-008:078*; Prepared for Kamehameha Schools: Honolulu, HI, USA, 2015.

50. Vitousek, P.; Chadwick, O.; Matson, P.; Allison, S.; Derry, L.; Kettley, L.; Luers, A.; Mecking, E.; Monastra, V.; Porder, S. Erosion and the Rejuvenation of Weathering-Derived Nutrient Supply in an Old Tropical Landscape. *Ecosystems* **2003**, *6*, 762–772. [CrossRef]

51. McElroy, W.K. *Wailau Archaeological Research Project 2005 and 2006 Results Wailau and Hālawa Ahupua'a, Ko'olau District, Moloka'i, Hawai'i*; Department of Anthropology, Univeristy of Hawai'i: Honolulu, HI, USA, 2007.

52. Coil, J.; Kirch, P. An Ipomoean Landscape: Archaeology and the Sweet Potato in Kahikinui, Maui, Hawaiian Islands. *Ocean. Monogr.* **2005**, *56*, 71–84.

53. Kirch, P.V.; Hartshorn, A.S.; Chadwick, O.A.; Vitousek, P.M.; Sherrod, D.R.; Coil, J.; Holm, L.; Sharp, W.D. Environment, Agriculture, and Settlement Patterns in a Marginal Polynesian Landscape. *Proc. Natl. Acad. Sci. USA* **2004**, *101*, 9936–9941. [CrossRef] [PubMed]

54. Kurashima, N.; Kirch, P.V. Geospatial Modeling of Pre-Contact Hawaiian Production Systems on Moloka'i Island, Hawaiian Islands. *J. Archaeol. Sci.* **2011**, *38*, 3662–3674. [CrossRef]

55. Kirch, P.V.; Coil, J.H.; Hartshorn, A.S.; Jeraj, M.; Vitousek, P.M.; Chadwick, O.A. Intensive Dryland Farming on the Leeward Slopes of Haleakala, Maui, Hawiian Islands: Archaeological, Archaeobotanical, and Geochemical Perspectives. *World Archaeol.* **2005**, *37*, 240–258. [CrossRef]

56. Vitousek, P.M.; Chadwick, O.A.; Hotchkiss, S.C.; Ladefoged, T.N.; Stevenson, C.M. Farming the Rock: A Biogeochemical Perspective on Intensive Agriculture in Polynesia. *J. Pac. Archaeol.* **2014**, *5*, 51–61.

57. Mccoy, M.D.; Graves, M.W. *An Archaeological Investigation of Halawa and Waiapuka Ahupua'a, North Kohala District, Hawai'i Island*; Report on File with the Hawaii State Historic Preservation Division: Kapolei, HI, USA, 2008.

58. Cordy, R.H. The Effects of European Contact on Hawaiian Agricultural Systems-1778–1819. *Ethnohistory* **1972**, *19*, 393–418. [CrossRef]

59. Kirch, P.V. The Evolution of Sociopolitical Complexity in Prehistoric Hawaii: An Assessment of the Archaeological Evidence. *J. World Prehistory* **1990**, *4*, 311–345. [CrossRef]

60. Winter, K.B.; Lincoln, N.K.; Berkes, F. The Social-Ecological Keystone Concept: A Quantifiable Metaphor for Understanding the Structure, Function, and Resilience of a Biocultural System. *Sustainability* **2018**, *10*, 3294. [CrossRef]

61. Lee, C.T.; Tuljapurkar, S.; Vitousek, P.M. Risky Business: Temporal and Spatial Variation in Preindustrial Dryland Agriculture. *Hum. Ecol.* **2006**, *34*, 739–763. [CrossRef]

62. Kirch, P.V.; Asner, G.; Chadwick, O.A.; Field, J.; Ladefoged, T.; Lee, C.; Puleston, C.; Tuljapurkar, S.; Vitousek, P.M. Reprint: Building and Testing Models of Long-Term Agricultural Intensification and Population Dynamics: A Case Study from the Leeward Kohala Field System, Hawai'i. *Ecol. Model.* **2012**, *241*, 54–64. [CrossRef]

63. Field, J.S.; Kirch, P.V.; Kawelu, K.; Ladefoged, T.N. Households and Hierarchy: Domestic Modes of Production in Leeward Kohala, Hawai'i Island. *J. Isl. Coast. Archaeol.* **2010**, *5*, 52–85. [CrossRef]

64. Cachola-Abad, C.K. An Analysis of Hawaiian Oral Traditions: Descriptions and Explanations of the Evolution of Hawaiian Socio-Political Complexity. Doctoral Dissertation, Department of Anthropology, University of Hawaii at Mānoa, Honolulu, HI, USA, 2000.

65. Field, J.S.; Ladefoged, T.N.; Kirch, P.V. Household Expansion Linked to Agricultural Intensification during Emergence of Hawaiian Archaic States. *Proc. Natl. Acad. Sci. USA* **2011**, *108*, 7327–7332. [CrossRef] [PubMed]

66. Field, J.S.; Ladefoged, T.N.; Sharp, W.D.; Kirch, P.V. Residential Chronology, Household Subsistence, and the Emergence of Socioeconomic Territories in Leeward Kohala, Hawai'i Island. *Radiocarbon* **2011**, *53*, 605–627. [CrossRef]

67. Ladefoged, T.N.; Lee, C.T.; Graves, M.W. Modeling Life Expectancy and Surplus Production of Dynamic Pre-Contact Territories in Leeward Kohala, Hawai'i. *J. Anthr. Archaeol.* **2008**, *27*, 93–110. [CrossRef]

68. Ladefoged, T.N.; Graves, M.W. Evolutionary Theory and the Historical Development of Dry-Land Agriculture in North Kohala, Hawai'i. *Am. Antiq.* **2000**, *65*, 423–448. [CrossRef]

69. Ladefoged, T.N.; Preston, A.; Vitousek, P.M.; Chadwick, O.A.; Stein, J.; Graves, M.W.; Lincoln, N. Soil Nutrients and Pre-European Contact Agriculture in the Leeward Kohala Field System, Island of Hawai'I. *Archaeol. Ocean.* **2018**, *53*, 28–40. [CrossRef]

70. Meyer, M.; Ladefoged, T.N.; Vitousek, P.M. Soil Phosphorus and Agricultural Development in the Leeward Kohala Field System, Island of Hawai'i. *Pac. Sci.* **2007**, *61*, 347–353. [CrossRef]

71. Mulrooney, M.A.; Ladefoged, T.N. Hawaiian Heiau and Agricultural Production in the Kohala Dryland Field System. *J. Polyn. Soc.* **2005**, *114*, 45–67.

72. Ladefoged, T.N.; Graves, M.W.; McCoy, M.D. Archaeological Evidence for Agricultural Development in Kohala, Island of Hawai'i. *J. Archaeol. Sci.* **2003**, *30*, 923–940. [CrossRef]

73. Gonschor, L.; Beamer, K. Toward an Inventory of Ahupua'a in the Hawaiian Kingdom: A Survey of Nineteenth- and Early Twentieth-Century Catographic and Archival Records of the Island of Hawai'i. *Hawaii. J. Hist.* **2014**, *48*, 53–87.

74. Vitousek, P.M.; Beamer, K. Traditional Ecological Values, Knolwedge, and Practices in Twenty-First Century Hawai'I. In *Linking Ecology and Ethics for a Changing World*; Rozzi, R., Pickett, S.T., Palmer, C., Armesto, J.J., Callicott, J.B., Eds.; Springer: New York, NY, USA, 2013; pp. 63–70.

75. Langston, B.; Lincoln, N.K. The Role of Breadfruit in Biocultural Restoration and Sustainability in Hawai'i. *Sustainability* **2018**, *10*, 3965.

76. Armitage, D.; Marschke, M.; Plummer, R. Adaptive Co-Management and the Paradox of Learning. *Glob. Environ. Chang.* **2008**, *18*, 86–98. [CrossRef]

77. Rosendahl, P.H. Aboriginal Agriculture and Residence Patterns in Upland Lapakahi, Island of Hawaii. Doctoral Dissertation, Department of Anthropology, University of Hawai'i at Mānoa, Honolulu, HI, USA, 1972.

78. Tuggle, H.D.; Griffin, P.B. *Lapakahi, Hawaii: Archaeological Studies (No. 5)*; University of Hawaii Press: Honolulu, HI, USA, 1973.

79. Ladefoged, T.N.; Graves, M.W. Variable Development of Dryland Agriculture in Hawai'i: A Fine-Grained Chronology from the Kohala Field System, Hawai'i Island. *Curr. Anthr.* **2008**, *49*, 771–802. [CrossRef]

80. Ladefoged, T.N.; McCoy, M.D.; Asner, G.P.; Kirch, P.V.; Puleston, C.O.; Chadwick, O.A.; Vitousek, P.M. Agricultural Potential and Actualized Development in Hawai'i: An Airborne LiDAR Survey of the Leeward Kohala Field System (Hawai'i Island). *J. Archaeol. Sci.* **2011**, *38*, 3605–3619. [CrossRef]

81. Trask, H.K. Native Social Capital: The Case of Hawaiian Sovereignty and Ka Lahui Hawaii. In *Social Capital as a Policy Resource*; Springer: Boston, MA, USA, 2001; pp. 149–159.

82. Kauainui, J.K. *Hawaiian Blood: Colonialism and the Politics of Sovereignty and Indigeneity*; Duke University Press: Durham, UK, 2008.

83. Witze, A. Mountian Battle. *Nature* **2015**, *526*, 24–28. [CrossRef] [PubMed]

84. United Nations Acknowledges Occupation of the Hawaiian Kingdom. Available online: http://hawaiiankingdom.org/blog/united-nations-acknowledges-the-occupation-of-the-hawaiian-kingdom/ (accessed on 10 September 2018).

85. Yuen, N.Y.; Nahulu, L.B.; Hishinuma, E.S.; Miyamoto, R.H. Cultural Identification and Attempted Suicide in Native Hawaiian Adolescents. *J. Am. Acad. Child Adolesc. Psychiatry* **2000**, *39*, 360–367. [CrossRef] [PubMed]

86. Stephens, C.; Porter, J.; Mettleton, C.; Willis, R. Disappearing, Displaced, and Undervalued: A Call to Action for Indigenous Health Worldwide. *Lancet* **2006**, *367*, 2019–2028. [CrossRef]

87. Mokuau, N.; Matsuoka, J. Turbulence among a Native People: Social Work Practice with Hawaiians. *Soc. Work* **1995**, *40*, 465–472. [CrossRef]

88. Benham, M.K.A.; Heck, R.H. *Culture and Educational Policy in Hawai'i: The Silencing of Native Voices*; Routledge: New York, NY, USA, 2013.

89. Yamashiro, A.; Goodyear-Ka'opua, N. *The Value of Hawai'i: Ancestral Roots, Oceanic Visions, Volume 2*; University of Hawaii Press: Honolulu, HI, USA, 2014.

90. Goodyear-Ka'ōpua, N. Rebuilding the 'Auwai: Connecting Ecology, Economy and Education in Hawaiian Schools. *Alternative* **2009**, *5*, 46–77. [CrossRef]

91. Chinn, P.W. Decolonizing Methodologies and Indigenous Knowledge: The Role of Culture, Place and Personal Experience in Professional Development. *J. Res. Sci. Teach.* **2007**, *44*, 1247–1268. [CrossRef]

92. McCoy, K.; Tuck, E.; McKenzie, M. (Eds.) *Land Education: Rethinking Pedagogies of Place from Indigenous, Postcolonial, and Decolonizing Perspectives*; Routledge: New York, NY, USA, 2017.

Sustainability **2018**, *10*, 3985

93. McMullin, J. The Call to Life: Revitalizing a Healthy Hawaiian Identity. *Soc. Sci. Med.* **2005**, *61*, 809–820. [CrossRef] [PubMed]
94. Jokiel, P.L.; Rodgers, K.S.; Walsh, W.J.; Polhemus, D.A.; Wilhelm, T.A. Marine Resource Management in the Hawaiian Archipelago: The Traditional Hawaiian System in Relation to the Western Approach. *J. Mar. Biol.* **2011**, *2011*, 151682. [CrossRef]
95. Smith, M.K.; Pai, M. The Ahupua'a Concept: Relearning Coastal Resource Management from Ancient Hawaiians. *Naga Iclarm Q.* **1992**, *15*, 11–13.
96. Feinstein, B.C. Learning and Transformation in the Context of Hawaiian Traditional Ecological Knowledge. *Adult Educ. Q.* **2004**, *54*, 105–120. [CrossRef]
97. Lincoln, N.K.; Vitousek, P. Nitrogen Fixation during Decomposition of Sugarcane (Saccharum Officinarum) Is an Important Contribution to Nutrient Supply in Traditional Dryland Agricultural Systems of Hawai'i. *Int. J. Agric. Sustain.* **2016**, *14*, 214–230. [CrossRef]
98. *Apology Act*; United States of America Public Law 103–150, 107 Stat. 1510; The United States Constitution: Washington, DC, USA, 23 November 1993.
99. Kamehameha Schools. *Ka Huaka'i: 2014 Native Hawaiian Education Assessment*; Kamehameha Schools Press: Honolulu, HI, USA, 2014.

 *sustainability*

Article

Biocultural Restoration of Traditional Agriculture: Cultural, Environmental, and Economic Outcomes of Lo'i Kalo Restoration in He'eia, O'ahu

Leah L. Bremer [1,2], Kim Falinski [2,3], Casey Ching [4], Christopher A. Wada [1], Kimberly M. Burnett [1], Kanekoa Kukea-Shultz [3,5], Nicholas Reppun [5], Gregory Chun [6,7], Kirsten L.L. Oleson [4] and Tamara Ticktin [8]

[1] University of Hawai'i Economic Research Organization, University of Hawai'i at Mānoa, Honolulu, HI 96822, USA; lbremer@hawaii.edu (L.L.B.); cawada@hawaii.edu (C.A.W.); kburnett@hawaii.edu (K.M.B.)

[2] Water Resource Research Center, University of Hawai'i at Mānoa, Honolulu, HI 96822, USA; kim.falinski@tnc.org

[3] The Nature Conservancy, Hawai'i Marine Program, Honolulu, HI 96817, USA; kanekoaks@gmail.com

[4] Department of Natural Resources and Environmental Management, University of Hawai'i at Mānoa, Honolulu, HI 96822, USA; caseymc@hawaii.edu (C.C.); koleson@hawaii.edu (K.L.L.O.)

[5] Kāko'o 'Ōiwi, Kāne'ohe, HI 96744, USA; nick@kakoooiwi.org

[6] Social Science Research Institute, University of Hawai'i at Mānoa, Honolulu, HI 96822, USA; gchun711@hawaii.edu

[7] Hawai'inuiākea, School for Hawaiian Knowledge, University of Hawai'i at Mānoa, Honolulu, HI 96822, USA

[8] Botany Department, University of Hawai'i at Mānoa, Honolulu, HI 96822, USA; ticktin@hawaii.edu

* Correspondence: lbremer@hawaii.edu; Tel.: +1-808-956-2325

Received: 31 July 2018; Accepted: 25 November 2018; Published: 29 November 2018

Abstract: There are growing efforts around the world to restore biocultural systems that produce food while also providing additional cultural and ecological benefits. Yet, there are few examples of integrated assessments of these efforts, impeding understanding of how they can contribute to multi-level sustainability goals. In this study, we collaborated with a community-based non-profit in He'eia, O'ahu to evaluate future scenarios of traditional wetland and flooded field system agriculture (lo'i kalo; taro fields) restoration in terms of locally-relevant cultural, ecological, and economic outcomes as well as broader State of Hawai'i sustainability goals around food, energy, and water. Families participating in the biocultural restoration program described a suite of community and cultural benefits stemming from the process of restoration, including enhanced social connections, cultural (re)connections to place, and physical and mental well-being, which inspired their sustained participation. We also found benefits in terms of local food production that have the potential to provide economic returns and energy savings over time, particularly when carried out through a hybrid non-profit and family management model. These benefits were coupled with potential changes in sediment and nutrient retention with implications for water quality and the health of an important downstream fish pond (loko i'a) and coral reef social-ecological system. Compared with the current land cover (primarily invasive grasses), results suggest that full restoration of lo'i kalo would decrease sediment export by ~38%, but triple nitrogen export due to organic fertilizer additions. However, compared with an urban scenario, there were clear benefits of agricultural restoration in terms of reduced nitrogen and sediment runoff. In combination, our results demonstrate that a biocultural approach can support the social and financial sustainability of agricultural systems that provide multiple benefits valued by the local community and non-profit while also contributing to statewide sustainability goals.

Sustainability **2018**, *10*, 4502

Keywords: biocultural restoration; food energy water; ecosystem services; cultural services; sustainable agriculture; Hawai'i; taro; wetland agriculture; flooded field systems; lo'i kalo; sediment; nutrients

---

1. Introduction

Across the globe, multiple factors have stimulated a growing interest in biocultural approaches to ecological restoration [1–5], including the strengthening of indigenous cultural revitalization movements [6,7], increasing acknowledgement among conservation and restoration professionals of the importance of social-ecological linkages [8–11], and growing recognition that restoration success often depends on community engagement [12–14]. Biocultural approaches to restoration focus on both ecological outcomes, such as biodiversity restoration and erosion control, as well as cultural outcomes, such as restoration of culturally important species and the traditions associated with them and community (re)connection to place [15–17]. Building on broader theories of social-ecological systems that emphasize the links between humans and the environment [18,19], biocultural approaches place further emphasis on a place-based approach and explicitly recognize that cultural and biological outcomes are interlinked and mutually reinforcing [20]. For example, the restoration of indigenous food production systems can strengthen cultural identity and improve nutrition and food self-sufficiency, while also restoring habitat for native plants and animals [2]. Restored habitats can then reinforce other cultural traditions that are linked to them [20]

Whereas industrial monocultures are often framed in terms of tradeoffs among food production and other outcomes such as water quality [21], traditional agricultural systems are typically characterized as potentially sustainable systems with synergies among desired ecological, cultural, and economic outcomes [22–24]. A social-ecological systems or biocultural approach to restoring traditional agriculture explicitly focuses on the links between the ecological and socio-cultural processes that underpin these systems [3]. Garnering broad support for biocultural restoration can be challenging, however, as short-term revenue is typically lower than monoculture agriculture or alternative land uses like urbanization, and conservation efforts often prefer complete restoration of 'natural' systems to maximize ecological benefits. While a growing body of literature points to a wide array of cultural and ecological benefits of biocultural restoration that can contribute to multi-level sustainability goals [7,16,20], there are few studies that evaluate these benefits (as well as tradeoffs) in a holistic and inclusive way. A framework for evaluating synergies amongst and tradeoffs across objectives of biocultural restoration could facilitate inclusion of these approaches in multi-scale restoration planning and facilitate adaptive management [9].

There are several existing approaches that can be adapted to guide an evaluation framework for biocultural restoration projects. Considering the environmental dimensions, the food-energy-water (FEW) nexus has emerged as an important framework for illuminating hidden synergies and tradeoffs in agriculture of great relevance to local, regional, and international sustainability initiatives [25], such as the United Nations Sustainable Development Goals (SDGs). While the FEW nexus has the potential to shed light on biocultural restoration approaches, it has generally focused on large-scale agriculture and largely ignores cultural aspects that can play critical roles in long-term societal sustainability and adaptive management. As an important complement to address cultural dimensions, a growing theory and emerging examples of inclusive valuation of land-use futures within the ecosystem services or "nature's contributions to people" literature sheds light on strategies to bring together diverse methodologies to assess benefits and tradeoffs in terms of locally relevant and linked ecological, cultural, and socio-economic concerns [16,26,27]. This work has been furthered by indigenous and place-based perspectives on cultural ecosystem services, often framed as reciprocal human-environment relationships well-aligned with biocultural restoration approaches [17,20]. We propose that combining and adapting the FEW nexus and ecosystem services (including place-based, indigenous cultural ecosystem services) frameworks has important potential to contribute to inclusive assessments of biocultural restoration.

Pacific islands can be considered as model systems to study biocultural restoration of traditional agriculture due to a combination of the socio-cultural significance of local food systems and their geographic isolation, which makes reduced dependence on imported food and energy sources an important part of building resilience. Pacific islands are also characterized by relatively small watersheds and tightly linked land-sea resources, making terrestrial agricultural practices and marine ecosystem health (also important for local food production) intricately linked [28,29]. In Hawai'i, elevated interest in the biocultural values of these traditional agricultural systems is demonstrated by a growing communities working to restore traditional terrestrial agriculture as well as nearshore aquaculture for linked cultural, economic, and environmental benefits [3,30]. The State of Hawai'i through the Aloha + challenge has also committed to interconnected sustainability goals around food, energy, and water (as well as links to local ecosystems and culture) and is recognized as an example of local implementation of the SDGs [31]. These commitments include: doubling food production in 20 years (Hawai'i currently imports nearly 90% of its food [32]); achieving 100% renewable energy by 2040; protecting watersheds and linked marine ecosystems; and facilitating re-connection to place and community-based management [31]. Achieving multiple sustainability goals simultaneously will require re-establishing community-based diversified agricultural systems with low energy requirements and few environmental tradeoffs. Yet, across the State, large tracts of agricultural land are currently fallow or used for high value export crops, largely due to the challenging economics of small-scale farming for local food production [33]. In this context, it is imperative to understand how biocultural approaches to traditional agricultural restoration can contribute towards achieving both local objectives and formal State of Hawai'i sustainability goals around food, energy, and water.

In order to contribute towards a better understanding of the multi-level outcomes of biocultural restoration projects, we collaborated with Kāko'o 'Ōiwi, a community-based non-profit in He'eia on O'ahu at the forefront of biocultural restoration of traditional agriculture, to evaluate the likely future benefits and tradeoffs of their vision around restoring a degraded and invaded wetland to lo'i kalo and native wetland plant communities. We specifically identified key ecological, cultural, and socio-economic outcomes of interest to Kāko'o 'Ōiwi and the local community who participate in the biocultural restoration efforts, and assessed those in combination with the broader State of Hawaii's sustainability goals around food, energy, and water [31]. For evaluating the various outcomes we used mixed methods, including participatory methods using an indigenous cultural ecosystem service process and framework [17]; sediment and nutrient retention ecosystem service modeling [34,35]; and food, energy, and water tradeoff analysis [25,36–39].

This study integrates diverse theories into a practical framework for evaluating biocultural restoration initiatives, with the aim of facilitating their on-the-ground planning, adaptive management, and assessment. Our research team built strong relationships with community members and non-profit staff to develop and apply the framework, ensuring that evaluation outcomes and approaches are reflective of local concerns while also linking to broader statewide sustainability objectives. Applying the framework collaboratively, we addressed the following research questions: (i) what are the locally-relevant cultural, environmental, and economic outcomes of biocultural restoration of lo'i kalo in He'eia, Hawai'i?; (ii) to what extent can biocultural restoration of lo'i kalo in He'eia, Hawai'i contribute to statewide sustainability goals around food, energy, and water?; and (iii) what are the synergies and tradeoffs among these outcomes and how can this inform design of biocultural restoration projects in He'eia and beyond?

2. Methods

2.1. Site Description

Our study site He'eia, O'ahu is an ahupua'a (traditional Hawaiian political-ecological boundary), where several NGOs are committed to a biocultural approach of restoring traditional social-ecological systems from mauka to makai (land-to-sea), which include wetland and marine fish ponds (loko i'a), upland agroforestry, and lo'i kalo. The area has recently been designated a Natural Estuarine Research Reserve, and the first of such reserves to explicitly focus on restoration for socio-cultural benefits [40]. The full He'eia watershed (11.5 km^2) spans from the top of the Ko'olau mountain range to Paepae o He'eia (traditional fish pond) and then to Kāne'ohe Bay, a 45 km^2 sheltered bay and highly valued conservation and subsistence and commercial fishing area. Average rainfall is 3020 mm at the summit and 1205 mm at the coast [41]. The watershed has two smaller basins, Ha'iku and 'Ioleka'a, that contribute to streamflow in He'eia stream (Figure 1).

We focus on a ~800,000 m^2 wetland managed by the community-based non-profit Kāko'o 'Ōiwi who seek to restore the now primarily invasive vegetation to the lo'i kalo systems present until the 1930s. The mission of Kāko'o 'Ōiwi is to "perpetuate the cultural and spiritual practices of Native Hawaiians," of which restoring lo'i kalo through a biocultural approach is a central part. They have worked to restore lo'i kalo since 2008 and with the help of volunteers and a family program have successfully restored ~7,500 m^2 of lo'i alongside other managed areas, and seek to restore another ~500,000 m^2 to lo'i and other intercropped species (see below) in the next 20 years alongside a series of other restoration activities, including restoration of native wetland plant communities and agroforestry [42] (See SI Methods). They also aim to restore an additional ~100,000 m^2 of the area as retention basins (including areas restored to loko i'a (fish ponds) and native wetland plant communities) with the remaining area including, streams, channels, access roads and small buildings, such as a commercial kitchen, poi (a traditional Hawaiian food made from mashed taro corms) mill, and community gathering place [42].

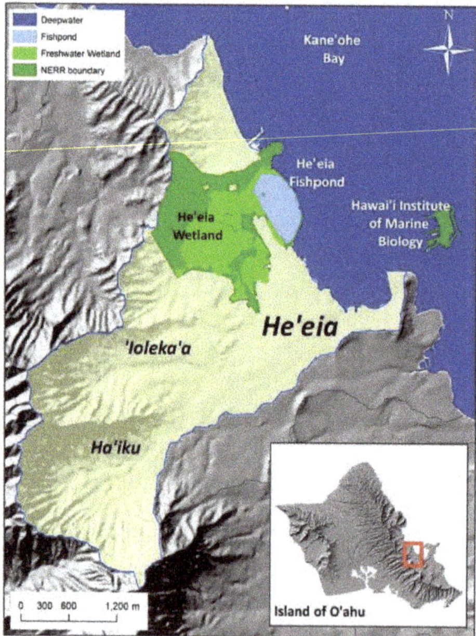

Figure 1. Location of He'eia wetland within the He'eia watershed, O'ahu. NERR = National Estuarine Research Reserve.

2.2. Outcomes Evaluated

Emerging theory on biocultural restoration suggests that evaluations of biocultural projects should include place-based outcomes and indicators defined by and relevant to local actors [20,43]. At the same time, for biocultural approaches to gain traction in broader land-use planning initiatives, demonstrating their value to broader state, regional, and global sustainability goals and objectives is important [9,43] In the context of this study, we defined key outcomes of importance to Kākoʻo ʻŌiwi's and the local community [42] as well to the achievement of statewide sustainability goals around food, energy, and water [31] (Table 1). An important *first* goal of Kāko'o 'Ōiwi is to build capacity of farmers and provide local families with the opportunity to learn to farm and care for loʻi kalo and (re)connect to the land. Among several initiatives to achieve this (see SI Methods), the organization began a family or 'ohana program where local families care for and maintain loʻi. There are currently 11 families participating in the project and scaling this program is an important part of the non-profit's growth strategy. *Second*, in line with the State's goal to double local food production, Kākoʻo ʻŌiwi strives to produce traditional and diversified food crops. While financial return is not a central goal, the organization sees financial sustainability as an important part of their ability to sustain food production. *Third*, in line with the State of Hawai'i's goal to be 100% renewable by 2045, the organization seeks to understand how renewable energy could be incorporated into their long-term plans and how traditional agriculture can reduce energy inputs associated with food production and imports. *Fourth*, in line with the State's goal of protecting watersheds and marine ecosystems, there is a strong interest in understanding how loʻi kalo and broader wetland restoration influences the sediment and nutrient retention functions of the wetland given links to the downstream fish pond and critical coral reef habitat of Kāneʻohe Bay, which both have high ecological, economic, and cultural value (Table 1).

Table 1. Outcomes of interest to Kāko'o 'Ōiwi and to the broader State of Hawai'i Aloha + Challenge sustainabilty goals [31].

Outcome	Why Important?
Community and cultural outcomes	Central to Kāko'o 'Ōiwi's mission and important for local community members
Traditional and diversified crop production	Traditional food production a key management goal of Kāko'o 'Ōiwi; economic returns important for long-term restoration success; contributes to statewide sustainability goal to double local food production by 2030;
Energy savings	Potential cost savings for Kāko'o 'Ōiwi with renewable energy; contributes to statewide sustainability goal to be 100% renewable by 2040
Sediment and nutrient retention (water quality)	Important for broader He'eia social-ecological system and food production downstream; contributes to statewide sustainability goals around watershed and marine conservation.

2.3. Assessment Methods

We provide a brief overview of methods used to assess each outcome listed in Table 1. Detailed methods can be found in the Supplementary Information.

2.3.1. Community and Cultural Outcomes

To understand participant families' perspectives and experiences with the family or 'ohana program to date, we conducted semi-structured, in-depth interviews with 8 of 11 families in the pilot project following an informal gathering between the researchers, participants, and Kākoʻo ʻŌiwi staff. Interviews were conducted while working with the families in the lo'i for 2–3 hours, which helped to build relationships and facilitate conversations (see SI Methods for guiding questions and further methods).

Our interview approach and analysis adapted an existing process and framework developed by Pascua et al. [17] to assess cultural ecosystem services from an indigenous Hawaiian perspective. This framework was developed through participatory methods within other indigenous Hawaiian communities and proved to be a more appropriate classification for cultural services than western frameworks such as the Millenium Ecosystem Assessment cultural ecosystem service categories [44]. Rather than framing cultural ecosystem services as uni-directional benefits from nature to people, Pascua et al. [17] focus on reciprocal relationships between people and ʻāina (land, literally that which feeds), which resonated with the place-based, indigenous and local communities with whom she worked. The cultural service categories encompass the Hawaiian language that contain nuanced meaning not reflected within their English translations. As such, it was important to use a Hawaiʻi-based categorization method to document the results and maintain the authenticity of outcomes mentioned within each interview.

Interview quotes and themes were coded according to Pascua et al. [17]'s Hawaiʻi-based cultural ecosystem service framework that includes four main categories: ʻIke (knowledge); Mana (spiritual landscapes); Pilina Kānaka (social connections); and Ola Mau (physical and mental well-being) as well as subsets of those categories. Where themes did not fit into existing categories, a new category was created (see Table 4 and SI Methods).

2.3.2. Future Scenario Analyses

In order evaluate ecological and economic outcomes important both to Kākoʻo ʻŌiwi and of relevance to State of Hawaiʻi sustainability goals around food, energy, and water, we developed a spatially explicit future agricultural restoration scenario based on Kākoʻo ʻŌiwi's conceptual plan and input from the farm manager and executive director (Figure 2; Table 2; SI Methods). The scenario is over a 20-year period (2018–2037) in line with the timeline of Kākoʻo ʻŌiwi's conceptual plan [42]. As a point of comparison for the sediment and nutrient retention analyses only, we included an urban scenario based on development patterns of neighboring urban areas (Table 2; SI Methods).

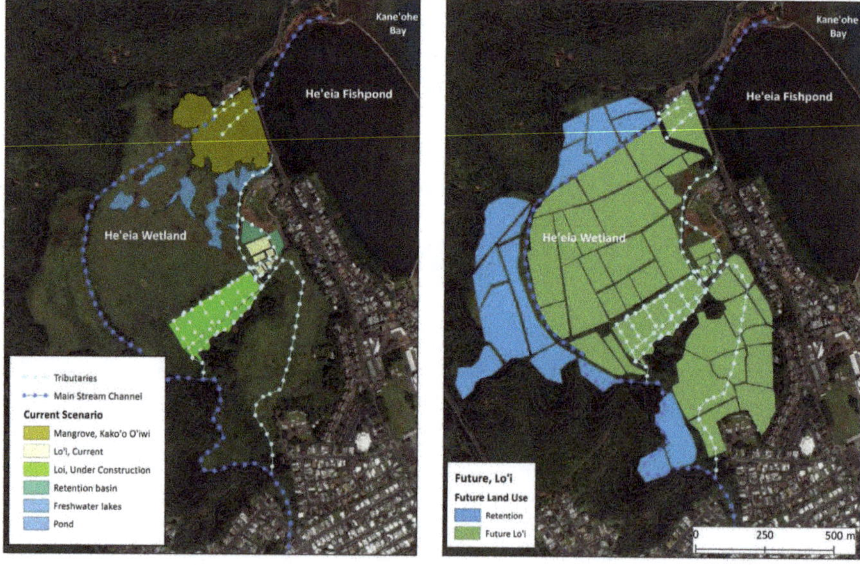

Figure 2. (left) Current land cover in Heʻeia wetland. Note that mangroves (an invasive species in Hawaiʻi) are currently being removed, but that the effect of mangrove removal was not considered in this study; **(right)** restored agriculture scenario in Heʻeia wetland. Retention areas are re-planted with wetland native plants or restored to loko iʻa (fish ponds).

Table 2. Descriptions of current, restored agriculture, and urban scenarios of He'eia wetland.

Scenario	Description
Current land use	Wetland area (~800,000 m^2) mainly dominated by invasive guinea grass (*Megathyrsus maximus*) and Job's tears (*Coix lacryma-jobi*); 7300 m^2 of lo'i kalo already restored
Restored agriculture	~500,000 m^2 restored to mix of kalo (taro; *Colocasia esculenta*), ulu (breadfruit; *Artocarpusaltilis altilis*) and mai'a (banana; *musa sp.*); ~100,000 m^2 to retention basins including loko i'a (fish ponds) and native wetland plants; additional areas for waterways, roads, educational and community buildings, and other infrastructure
Urban	The full 1,600,000 m^2 parcel managed by Kāko'o 'Ōiwi is converted to mid-intensity urban development. On-site disposal systems, similar to the surrounding communities are assumed to be dominant.

2.3.3. Traditional and diversified crop production and economic returns

We estimated potential crop production (local food production) and economic returns (total revenues, total costs, and total profits) over the future agricultural restoration scenario under varying assumptions of banana (*musa sp.*), breadfruit (*Artocarpusaltilis altilis*), and taro (*Colocasia esculenta*) productivity [45–53], with the percent area of each crop defined by Kāko'o 'Ōiwi staff. (Table 3; SI Methods). While kalo (taro) is the dominant crop, we included additional crops as Kāko'o 'Ōiwi's production model includes diversified cropping systems (Table 2).

Table 3. Assumptions underlying estimation of economic returns from food production.

Crop	Area Added Annually (m^2)	Area in Production by Year 20 (m^2)	Area in Year 20 as % of Total Farm Area	Yield (kg/m^2)	Farm Price ($/kg)	Production cost ($/kg)
Banana	2428	50,181	10%	0.58–2.39	2.34	1.50
Breadfruit	4856	50,181	10%	0.20–2.43	2.58	0.32–0.83
Taro	6475	125,453	25%	1.12–1.79	5.51–10.34	6.33–10.14

2.3.4. Energy Savings

In order to understand how utilizing solar power could increase Kāko'o 'Ōiwi's financial returns while contributing to the State's goal to be 100% renewable by 2045, we evaluated the use of renewable energy for food production as well as avoided energy use for food imports in the restored agriculture scenario (SI Methods).

On-farm Solar Energy for Food Processing

Kāko'o 'Ō'iwi is hoping to realize cost savings over the long run by operating the proposed taro processing facility using solar power. Commercial poi production is similar to wet milling of other starches. Assuming that energy requirements for wet milling corn are comparable to those for poi production, we calculated the required daily energy and corresponding photovoltaic (PV) system power capacity to produce poi for taro production scenarios 1 and 2 [54,55]. PV installation costs were then estimated assuming a cost of $3.01 per watt [56]. Utility bill savings and revenue from solar energy sales back to the grid were estimated using Hawaiian Electric Company's (HECO) Schedule 'G' General Service rate of $0.27/kWh and the HECO Customer Grid-Supply rate of $0.15/kWh (SI Methods).

Avoided Energy Inputs

One of the main reasons for higher energy efficiency in the case of organic farming is the lack of input of synthetic nitrogen (N) fertilizers, which require high energy consumption for production and transport. Kāko'o 'Ō'iwi currently applies 151 kg ha^{-1} of organic fishmeal N-fertilizer for taro production and would continue this practice under both future scenarios. Indirect energy savings—i.e., fossil fuel energy inputs that would have been required to synthetically produce the required amount of N-fertilizer—were calculated for the 20-year management period [57].

We also estimated the fuel needed to ship taro to Hawai'i as a conservative estimate of the energy offset from locally produced taro (SI Methods).

Sediment and Nutrient Retention

Sediment and nitrogen retention (accumulation) within the wetland and export from He'eia stream were estimated for the current, restored agriculture, and urban scenarios. These are important metrics of success for Kāko'o 'Ōiwi as changes in water quality through sediment and nutrients directly links wetland management to the fish pond and coral reef which have high socio-economic, cultural, and ecological significance for the broader He'eia system. For both parameters, we estimated the input and export based on existing data or model estimates to determine the amount retained by the restoration area.

The calculation for sediment retention in the current scenario employed a simple box model (retention equals sediment input minus export) based on data from the United States Geological Survey (USGS) for import into the wetland [58] and Hawai'i Department of Health (HDOH) [59] and USGS [60] for export out of the wetland (see SI Methods for further details). Sediment retention in the restored agriculture scenario was estimated through literature values of annual retention rates of rice paddies (a similar system to lo'i) [61]. Total sediment retained in the restored scenario was estimated by multiplying annual retention rate (kg m^{-2}) by the retention area. Sediment export from the restored agriculture scenario was estimated as the net current input [58] from above the wetland less the total retained. To estimate sediment export from He'eia wetland in the urban scenario, we used the InVEST Sediment Delivery Ratio model [34,62] (See SI Methods).

We used a box model approach to estimate the amount of nitrogen input to and export from the wetland. We modeled nutrient input into the wetland from the broader He'eia watershed with the InVEST Nutrient Delivery Model (InVEST NDR) [34] (see SI Methods). For the current scenario, we calculated nitrogen retention in the wetland using USGS discharge data [60] and HDOH average nutrient concentrations at the He'eia stream mouth [59] (SI Methods). For the restored agriculture scenario, we conservatively estimated fertilizer rates within the restored lo'i areas as approximately 15 g m^{-2} based on manager input and plant (taro) uptake at 23% of fertilizer N applied [63]. For the urban scenario we assumed there was no retention capacity within the wetland. Given that there are no centralized wastewater treatment options, we estimated a similar input of N (as from the surrounding neighborhoods) in wastewater from on-site sewage disposal systems [64] (SI Methods).

3. Results

3.1. Community and Cultural Benefits

The families participating in the 'ohana program spoke of motivations for participation that went far beyond the direct benefit of producing kalo, spanning all inter-related categories of Pascua et al. [17]'s framework ('Ike—knowledge; Mana—spiritual landscapes; Pilina Kānaka—social connections; and Ola Mau—mental and physical well-being) and beyond (see Table 4 for examples and quotes). Families consistently mentioned that they most value the intangible benefits gained from the opportunity to maintain their own lo'i such as developing a reciprocal relationship between and among kānaka (Indigenous Hawaiians) and 'āina (land):

> "My family gets to eat, mentally, physically, spiritually. The place that we're working at becomes more abundant and healthier and restored. The thriving factor increases as we work not only for my own family but for the place."

Families talked of an ability to connect with landscape or experience a sense of "feeling the living, breathing, 'āina" while working in the place (Mana; Table 4). Many spoke of the opportunity to mālama

Hāloa, referencing the kin relationship between kānaka and kalo (taro) with roots in creation described in the Hawaiian creation chant, the Kumulipo (Table 4). In many cases, the program was their first opportunity to have access to land and water that allowed them to fulfill this kuleana (responsibility) to their ancestors by following in their footsteps and perpetuating the traditional practice (Table 4).

Social connections (Pilina kānaka; Table 4) were also an important motivation and perceived benefit of the program. Goods produced through the traditional agricultural practice were primarily discussed in the context of having kalo to take home or share. They valued building an ʻohana through common experience with like-minded individuals, sharing of the work, and establishing trusting relationships and aloha for each other over all else (Table 4). There was also strong interest in passing down knowledge, passions, and environmental knowledge to new generations, keiki (children), and other families that may enter the program (Table 4).

While participants emphasized that kalo, per se, was not their primary motivation for participation, they pointed to the importance of beliefs and cultural practices around food cultivation and preparation:

> "I think most of us in our hui have experience with food and we know that if you're working around food and you're working around something you're going to eat later on, you've got to have good thoughts and say good words and put good intentions into something you're going to put into your own body later."

They emphasized that the program helps to build a sense of pride and accomplishment, creating a sense of belonging (safe space), and providing the opportunity to progress as kānaka (Hawaiians). They gain a sense of pride, accomplishment, and mental healing from coming to the loʻi, putting in the work, and watching their kalo progress. In the words of a participant:

> "Just creating this safe space where people can feel like they're okay and they don't have to be judged. They can just come here and feel aloha. Creating a sense of belonging and that they belong to something bigger, and something healthy and something that's aloha."

However, there was also concern about how to deal with potential changes to the current positive atmosphere as the program grows; the trusting relationships, the heart of the ʻohana, may dissipate with a larger number of families and loʻi:

> "The one thing that I think about as we see all these families here is that we really have to continue to create this space of positivity, so that we don't have to deal with the things like stealing or...you know what I mean? As you get hundreds and hundreds, as organizations get big, they tend to make these operational rules and that's when people start to get...when attitudes change. And so on..."

Table 4. Cultural outcomes discussed by community participants in the ʻOhana program categorized by Pascua et al. [17]'s cultural ecosystem service framework developed through participatory methods with several indigenous and local communities in Hawaiʻi.

ʻIke (knowledge)	
Ma ka Hana ka ʻike (learn place-based practices by actually doing them)	
<ul><li>Learning mahiʻai kalo (wetland taro farming).</li><li>Experience builds on previous knowledge/experience and adapting practices to a particular landscape.</li><li>Practice promotes other place-based cultural practices such as kuʻiʻai (traditionally pounding poi with a board and stone).</li></ul>	*"The number one thing I value about this experience is, just the opportunity and just the fact that I was asked to participate and given the opportunity to be a part of it and to bring my family along with me ... this opportunity is unique because it was like 'Okay, go take care of a particular loʻi by yourself, with our families..we'll give you guys some tools, we'll give you guys some direction, but go...go at it. You're going to help us increase the abundance of this area."*

Table 4. *Cont.*

Nānā i ke Kumu (observe familiar natural processes and seasonal occurrences)

- Observation of rain intensity, stream flow, etc. and learning corresponding responses in the lo'i.
- Understanding of physical aspects of the landscape needed to maintain lo'i (soil, water flow, different types of mud, nutrient cycles, etc.).
- Expansion of ecological knowledge and interactions within the ecosystem (presence of different animals with different seasons).
- Learning of human impacts on landscape.
- Ability to observe how natural processes influence each other on an ahupua'a scale.

"It's definitely personal growth out here because you're learning science and you've got to learn the moons. You've got to know when it's going to rain too hard. We live in Kāne'ohe, so when we have heavy rain it's like 'Uh oh'...we're thinking about the lo'i. We've had it when the water was up to here, on my hips, but I think we're kind of getting smart about adjusting the water and water flow. With these (lo'i), the makawai go this way so we have to be careful. We're at the top so it feeds into the next one and the next one."

Hālau 'Ike (diverse formal and informal learning)

- 'Āina-based/culture-based education for keiki.
- Informal learning through working in the place and alongside other people.
- Opportunities for teaching moments for communities and the broader world about cultural practice.
- Enhancement of environmental awareness and sustainability.
- Ability to draw links between technology and scientific study to place-based observations.
- Gaining perspective of environmental processes operating on a wider scale.
- Learning how people fits into the broader picture of the environment (reciprocal relationships between kanaka and 'āina, being part of the solution).

"They appreciate it a lot more . . . They understand the kuleana. And I like that about . . . this opportunity, the fact that we have our own place and we have the ability to come and take the kids and do this with them. Eventually, I'm going to lose that, my kids are going to go to sports on Saturdays; it's just good to have this time with them. You know, and it's not a chore to try to get them to do this. I have two daughters so my youngest daughter is like, 'I don't want to get muddy dad...' but once she gets in, she has fun."

Mana (Spiritual landscapes)

Ho'omana/Mauli Ola (spiritual beliefs and practices allowing people to interact with mana of a landscape)

- Physical connections with the landscape facilitate mental connections (feeling the living, breathing 'āina); and cultural identity.
- Gaining an awareness of sacredness of landscape.
- Fulfillment of a kuleana.
- Connections with ancestral presence by following in their footsteps with cultural practice.

"So there's that saying I always go back to, 'you don't grow the kalo, the kalo grows you.' And what I've gained is just a different perspective in the different aspects of life really. From the relationship side of how man fits into the larger picture of the environment and the 'āina and seeing the 'āina as an actual living breathing thing, because when you come out here on Sundays and it's just you and it's quiet like this, you can actually hear and feel and see how it's moving. You can see all the life in the water, you can almost feel and hear the 'āina speaking to you, too."

Wahi Pana (appropriate access to, and understanding of place-specific practices associated with storied landscapes)

- Promoting an understanding of and respect for the value of the practice within the place.
- Learning of the stories of the landscape.
- Understanding the meanings of place/weather event names.
- Opportunity and access to engage in cultural practice within the place. Access to place to mālama hāloa.

"That's another thing that's changed for me too, is we notice that we're so excited to talk about the different stories . . . . And the kids are starting to talk about it too and for us it's like 'oh, this is great, I'm hearing the children talk about Keahiakahoe' and there's so much. This is like an open classroom because there's always something to learn

---

Table 4. *Cont.*

'Aumakua/Kinolau (presence/recognition of familial gods/ancestors)

• Deepened connection with Hāloa; recognizing kalo as a family member. • Presence and significance of names of rain, winds, etc., to various aspects of the landscape.	*"Something so fundamental, from who we are as kanaka, what are we supposed to do? We're supposed to mālama Hāloa, we're supposed to mālama our older brother. 'You have a chance to mālama Hāloa? Awesome!' Because I don't know how many of us would love to have a plot of land and water where we could potentially do that. That's so many Hawaiian families' dream."*

Hō'ailona (presence of environmental signs/indicators and the ability to recognize them)

• Ability to recognize different species of plants and animals indicating different states of the lo'i and different seasons. • Ability to recognize bioindicators of various cycles within the ahupua'a system. • Listening to omens from weather dictating when to/ not to work (during heavy rainfall, difficult conditions, etc.).	*"It just feels really comfortable there and there's some real specific things . . . .that you get to know all the sounds of the birds that are going to be around and you're working and you're like 'oh, you can hear the ae'o.' You don't even have to look up but you know exactly and you can picture where they must be because you can hear their direction. You know the buzzing of the pinao when they're around, it's just a lot of specifics. You even get a sense of what's normal weather for that time of morning and what's unusual. Things are shifting, it's a different season. You're so much more present with the environmental setting there than I know I am at other places."*

Pilina Kānaka *(social interactions)*

Ho'olako (perpetuation of practices/skills allowing individuals to provide for their families)

• Kalo for the home/family. • Kalo for community and sharing.	*"We're part of the family program, but my intentions for this particular one, is mainly for our school and also for my family because this is more than enough kalo to go around. So we can be a part of the family program and also be able to provide 'school' with whatever kalo they need. Families can come and eat better, learn about how and why what foods work and which ones don't, and try to get Hawaiians back to what they ate before."*

'Ike Aku, 'Ike Mai (share traditional/local knowledge and values)

• Passing down knowledge and passions to new generations and keiki. • Creating an ono (taste) for the traditional practice. • Setting/shifting/sharing an awareness, intention, and purpose while working in the landscape; cultural continuity in life. • Promoting new, young leaders. • Ways to connect mindsets to entire ahupua'a.	*"We have so much to teach, not only our own keiki but other communities. If you look at how our island is a microcosm for the entire planet, hopefully it can be used on a global level. With Hawaiians, and kilo, and the way they would be so in tune that when one thing is blossoming they know which fish are running. And the wet and dry season, there was just so much detail. And when you're that in tune and everything, you're actively engaging all those nodes of connectivity between the different aspects of life. Especially when you name things, that's such a Hawaiian thing to do. Or just labeling something wind, or something rain. If you look at Hawaiian culture and Hawaiian language, there's thousands of names for rain and names are descriptive, same thing with winds, clouds "*

Kōkua Aku, Kōkua Mai (presence of strong social ties/social networks)

• Establishing of trusting relationships and aloha for each other. • Building community. • Building an 'ohana through common experience with like-minded individuals. • Sharing of the work. • Expanding social networks. • Facilitation of goods exchanges not within monetary means.	*"It's a prioritization of our relationships with that 'āina and with each other. If that's the driving force, then, people aren't caring about who got how much and when and what and all that stuff. Among our hui, and I'm sure others as well are similar this way, there is a genuine wanting to take care of each other and take care of that place. If that stays the priority, I don't think there would be any problems."*

Table 4. *Cont.*

Ola Mau (Physical and mental wellbeing)

Lako/Momona (availability and access to subsistence resources rich enough for people to thrive)

• Access/presence of water; water coming up from below creating potential for lo'i. • Adequate stream flow processes and nutrient cycling, weather, mud, etc. • Presence of ae'o and pinao. • Access to restoration of whole ahupua'a.	*"I've always wanted to do this. But I just didn't have the resources. So once this came about, it was perfect, honestly to me. The water is the main thing; the water was the source of life for Hawaiians. "Water is life" without the water, you can't do anything."*

Ho'oikaika Kino (active lifestyle to support the physical demands of specialized practices)

• Building strength from work. • Opportunities for self, family, keiki to be outdoors from time required to do the work. • Encourages appreciation and patience for the work.	*"It's more than just a lo'i, it's like being on a big farm, being outside, and stepping on sleeping grass, getting little cuts and scrapes and bruises, those are all wins for me. Those are all little victories in our family because those are the kinds of memories that I have as a kid growing up so, it just makes me feel good that I'm able to pass those kinds of experiences on to my kids."*

'Oihana (engaging in family roles and occupations)

• Perpetuation of the practice for passing to new generations of kalo farmers. • Preservation of mahi'ai kalo occupation. • Presence of strong family role models. • Encouragement of future mahi'ai kalo. • Roles within the mahi'ai kalo process. • Mahi'ai roles as providers.	*"It's been a bit of an adjustment not fully being in control of the amount of water coming in, or the taro that we plant, the types and variety of kalo that we plant, or the planting style. It's okay to kind of sit back and do it the Kāko'o way. I've been more focused on building relationships and just getting our kids outside and having them develop more of that relationship to 'āina, has been more of the gem for me … Usually I'd be in more control of what I'd want to see, but I'm kind of okay in not being in control and just putting my hands in and just doing the work."*

Personal and mental wellbeing

• Provides a physical and mental sanctuary (place to recharge). • Strengthening sense of or reconnection with cultural identity (becoming pa'a. • Sense of belonging (safe space). • Cultivates a sense of pride and accomplishment; opportunity to aloha or mālama something. • Brings about a sense of awareness of intention, purpose, and mindset within the landscape. • Building a reciprocal relationship with land (kalo grows you). • Encourages an appreciation and patience for the practice. • Sense of joy in the family.	*"It's a sense of pride, just enjoying and getting enjoyment. The luana of this place is maika'i. Just creating this safe space where people can feel like they're okay and they don't have to be judged. They can just come here and feel aloha. Creating a sense of belonging and that they belong to something bigger, and something healthy and something that's aloha. And strengthening their sense of identity as a Hawaiian, living in Hawai'i. Those two things, I think, create confidence … like I said, you make decisions for the greater good."*

3.2. Traditional and Diversified Crop Production and Economic Returns

Over the 20-year management period (2018–2037), the range of estimated profit for breadfruit and banana was positive (Table 5). Profit from raw taro (without sales of additional products) was negative for both scenarios. However, including additional taro products such as poi increased profits, as did allowing for volunteer family effort (15% reduction in costs), although the lower bound of the estimated profit range remained negative (see Figures S1–S4). When considering the three crops collectively and assuming no volunteer effort, combined profit ranged from –$6.33 million to $14.22 million (Figure 3). Under the family program, the estimated range of profits increased to –$4.09 to $16.46 million (Figure 3).

Table 5. Estimated total production and profits over 20 years for banana, breadfruit, and taro.

Crop	Total Production over 20 Years (Million kg)	Total Profit over 20 Years (Million $)
Banana	0.31–1.26	0.26–1.06
Breadfruit	0.14–1.65	0.24–3.68
Taro	1.48–2.36	−6.84–9.48
Taro (with family program)	1.48–2.36	−4.59–11.72
All crops	1.93–5.57	−6.33–14.22
All crops (with family program)	1.93–5.57	−4.09–16.46

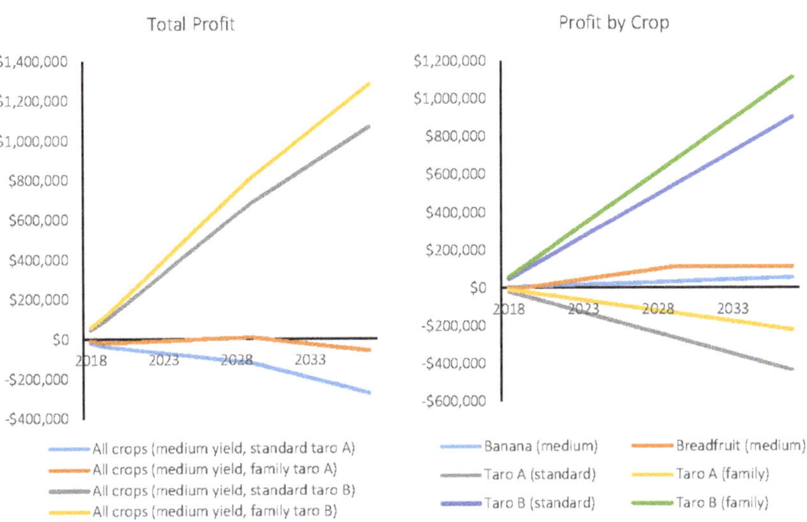

Figure 3. Total profits in different crop scenarios (**left**) and profit by crop type (**right**). Note that *taro A* and *B* are used to differentiate the two yield-cost-price scenarios developed in collaboration with Kāko'o 'Ōiwi (see SI Methods for details) and *family* denotes implementation of the family volunteer farming program.

3.3. Energy Savings

Estimated annual energy savings, owing to the use of organic N-fertilizer in place of synthetic N-fertilizer in taro production, ranged from 0.16 to 0.25 kWh/m^2. Over the 20-year management period, this amounted to a cumulative energy savings in the range of 0.20–0.33 million kWh. In terms of energy offsets, the low taro yield scenario would offset 46–48 tons of heavy fuel oil, while the high yield scenario would offset 115–120 tons over the 20-year project time horizon. This is equivalent to 5.3 million kWh (low) to 13.6 million kWh (high).

Aside from interest in supporting the State's renewable energy goals, Kāko'o 'Ōiwi is interested in becoming more self-sufficient in their on-farm energy use to increase the profitability of their programs. The energy required for poi milling in taro production scenario A (1.12 kg/m^2 yield) totaled 37,361 kWh over 20 years. Powering this process with solar energy would require a 3410-watt PV system at an installation cost of $10,327. Cumulative utility bill savings of $9908 combined with $5094 in revenue from selling excess energy back to the grid resulted in a payback period of 15 years. For taro production scenario B (1.79 kg/m^2 yield), energy input totaled 59,778 kWh over 20 years. The higher input would require a larger (5450-watt) PV system at an installation cost of $16,523. This resulted in cumulative utility bill savings and revenue from grid supply sales equal $15,853 and $8151 respectively. The payback period remained unchanged at 15 years.

3.4. Sediment and Nutrient Retention

Overall, we found that the full agriculture restoration scenario decreased sediment export by 38% compared with the current scenario, but that nutrient export could increase by as much as 240% due to fertilizer inputs. However, nutrient export in the restored agriculture scenario was still less than half of export in the urban development scenario (Figure 4). Specifically, sediment exported to Kāne'ohe bay from the wetland was estimated at 1070, 668 and 2365 tons year^{-1} for the current, agriculture restoration and urban scenarios, respectively (Table 2). Nutrient export was 2500, 8515, and 17,995 kg year^{-1} for the current, agriculture restoration, and urban scenarios.

Sediment export at He'eia stream mouth during baseflow for the current scenario was based on average TSS concentration exported from the wetland (18.5 mg L^{-1} from 2013 to 2017; [59]) and combined mean daily discharge of Ha'iku and 'Ioleka'a streams (1.8 2.87 cfs; [60]), translating to approximately 75 tons year^{-1} of sediment. We assumed this was 7% of the sediment budget given that much of export is not associated with baseflow [65], which translated into a total current export of 1070 tons year^{-1}. Thus current net input and export (2335 tons input and 1070 tons export), translated to an accumulation of 1265 tons of sediment in the wetland per year (see SI Methods).

For the restored agricultural scenario, Slaets et al. [61]'s accumulation rate for lo'i retention in the agricultural restoration scenario translated to a future accumulation of 1670 tons of sediment per year (for the ~600,000 m^2 of retention space available in this scenario). Sediment retention was null in the urban scenario and conversion of the upland areas within Kāko'o 'Ōiwi from non-native vegetation to urban also increased sediment export by 31 tons year^{-1}.

Nitrogen

Predicted N export from upper He'eia watershed into the wetland using the InVEST NDR model under the current scenario was 4990 kg year^{-1}. Using the Department of Health [59] average N concentrations and baseflow estimates, we calculated that 2500 kg year^{-1} of this import is currently retained by the wetland, resulting in an export of ~2500 kg year^{-1}.

For the restored agriculture scenario, we estimated that an additional 6015 kg year^{-1} of N (from fertilizer) would be added to the system that may not be taken up by crops (based on 23% uptake by plants), resulting in an export of 8515 kg year^{-1}. Urban development contributed a wastewater derived nitrogen load of 12,960 kg year^{-1} in addition to the current predicted export, resulting in a total export of 17,955 kg year^{-1} (Figure 4).

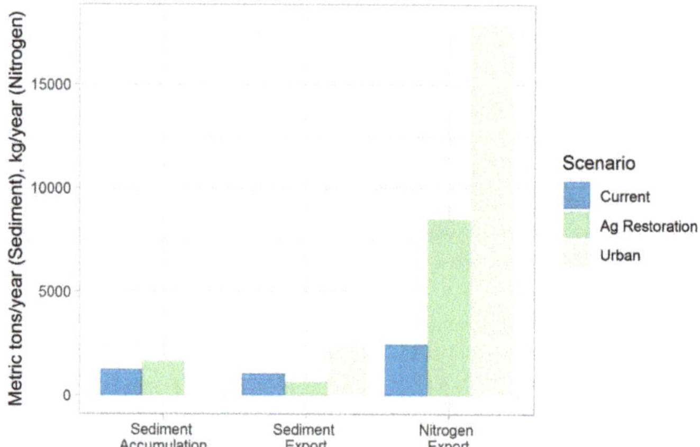

Figure 4. Summary results of sediment accumulation (tons/year), sediment export (tons/year) and nitrogen export (kg/year) under the current, restored agriculture, and urban scenarios.

4. Discussion

Interest in biocultural approaches to restoration are growing across the world [2,9,20,43], and Hawai'i is emerging as a hotspot for biocultural restoration of traditional agriculture [7,15]. This can be attributed to an indigenous Hawaiian cultural renaissance around food and ahupua'a and broader moku (district or region) management [3,66] coupled with the State of Hawai'i's recent commitment to doubling local food production [31]. Statewide efforts to increase local food while also moving towards renewable energy, watershed and marine ecosystem protection, and local (re) connection to place situate these systems-level biocultural restoration efforts as transformative projects for sustainability and resilience at multiple scales.

A defining principle of biocultural approaches to conservation and restoration is the need to start from a cultural, place-based approach while also acknowledging diverse actors and multiple goals across spatial scales [9,43]. In this study we set out to assess future scenarios of restoration in terms of outcomes valued by the local community and a community-based non-profit as well as the State of Hawai'i as formalized in sustainability goals around food, water, and energy [31]. In the context of biocultural restoration of lo'i kalo in He'eia, O'ahu, we identified a diverse set of locally relevant environmental, cultural, and economic goals, many of which align with statewide sustainability goals (see Table 1). We found a number of benefits, including those stemming from the biocultural restoration process, as well as key challenges facing this and similar biocultural restoration projects as they begin to scale.

From the perspective of participating families in the 'ohana program, the most important benefits associated with cultivating kalo were associated with the *process* of restoring lo'i kalo rather than just the cultural or economic benefits of the end product. In particular, families valued the opportunity to (re)connect to important biocultural landscapes and build social connections among like-minded people. That cultural benefits emerge from the process of restoring reciprocal relationships to place rather than just the end products of biocultural restoration has been noted elsewhere [7,20,67]. The attention to restoring ritual and the cultural protocols alongside ecological systems is an important theme emerging from this study and other articles in this issue [7,20]. While work on "relational values" [68] and to some extent "nature's contribution to people" [27] has acknowledged this, the vast majority of literature in ecosystem services has focused on the value of the end state rather than the process of getting there.

In regards to the process of restoration, we found a suite of benefits across all categories in the indigenous Hawaiian cultural ecosystem services framework of Pascua et al. [17], as well as some extensions. This included benefits classified as 'ike (knowledge): families particularly valued the chance for place-based experiential learning of how to grow kalo and provide opportunities for their children to be in and learn in these environments. There was also substantial reference to themes related to mana (spiritual landscapes) as families described the work as a cultural practice linked to ancestral practices and the Kumulipo (legend of Hawaiian origin), and to pilina kānaka (social connections) as the program was discussed as strengthening connections between and among families. Participants also pointed to physical and mental health benefits (ola mau), with many references to the lo'i as a place of individual and collective mental renewal as well as physical health benefits. This connection created a deeper understanding of kuleana (responsibility) to the land and of cultural identity as kānaka (indigenous Hawaiians). The site became a place of healing, love, and self-reflection, providing a higher purpose for the people working to restore the area. Participants described a sense of pride and accomplishment, a sense of belonging (safe space), and valued the opportunity to progress as kānaka (indigenous Hawaiians).

While attention to process is key, an important cultural as well as economic benefit of the restoration is also clearly the production of traditional crops, an outcome highly valued by Kāko'o 'Ōiwi and aligned with the State's goal of doubling local food production. Even in the lowest return scenario, there are still 339 tons of bananas (mai'a), 151 tons of breadfruit (ulu), and 1628 tons of taro (kalo) produced. The food produced would represent an important step towards the State's goal of doubling local food production and would increase the total area dedicated to taro production across the State by 50% [31].

The focus on organic production methods and renewable energy as part of the broader biocultural approach, also produced important energy savings benefits that align with the State's renewable energy goals. In comparison to energy use that would have occurred with the use of industrial fertilizer and oil for electricity, organic fertilizer and solar power also saved 0.87–1.41 million MJ. Adding in potential savings through avoided food imports, this saving is substantial and clearly demonstrates the potential of biocultural restoration of traditional agricultural systems to produce food in a way that also provides synergies for energy sustainability goals.

Our finding that the most financially beneficial model included diversified crop production utilizing a "hybrid" production model that includes the family program also points to important potential synergies between local community and cultural benefits and the program's financial sustainability. Our economic analysis suggests that if the family program were to go to scale, it could roughly increase returns by $2 million USD, which would help to sustain a project that provides many benefits locally and more broadly. However, to go to scale the number of families must increase substantially. While Kāko'o 'Ōiwi is committed to providing these opportunities and "growing farmers," families interviewed also expressed concern that some of the greatest components of the project (close social connections and a sense of sanctuary) could change as more families get involved. Managing scaling in the context of the cultural significance of the quality of the process of restoration is critical to the long-term success of the project.

Finally, our results suggest that biocultural restoration of lo'i kalo in He'eia can provide benefits in terms of the broader social-ecological system including downstream fishpond (managed by Paepae of He'eia) and the Kāne'ohe Bay coral reef ecosystem. In elevated concentrations, both nitrogen and sediment can have adverse impacts on coral reefs and fisheries and likely make these systems less resilient to climate change [65,69,70]. When compared to a hypothetical urban scenario (which was once the fate of He'eia wetland and could be in the future if sustainable models are not developed), we found important benefits of lo'i kalo restoration in terms of reduced nitrogen and sediment loads to these nearshore environments. This aligns with previous research in the Pacific showing that taro fields retain sediment [71], as well as studies of sediment retention in similar rice paddy systems [72]. Lo'i kalo and similar systems have also been shown to have a high capacity to store water compared with invasive wetland grasses [73] and urban environments [74], suggesting that these systems retain and slow more water because of their construction as basins and, therefore, more sediment than alternative land uses addressed in our study. The approach presented in this article accounted for baseflow and small storm conditions; further research is needed to understand how the conversion back to lo'i would be affected by larger storm events.

Within the restored agriculture scenario, however, we found an important tradeoff between enhanced sediment retention (and reduced sediment export to the bay) and increased nitrogen export (due to fertilizer inputs). While the nutrient loads are much lower than they would be with urban expansion or with other forms of conventional agriculture, they are higher than current land cover, and could have impacts on important downstream systems. However, the model used did not directly consider nutrient uptake of drainage channels or by complex microbial and wetland plant communities, and thus can be considered conservative [75]. Careful design of lo'i and the wetland system could mitigate some of the nutrient tradeoffs while also increasing sediment retention of the system. Field design insights from natural and constructed wetlands used to treat waste discharge sites could also be incorporated into lo'i kalo design to further reduce nutrient export [76].

While addressing this potential nitrogen tradeoff is paramount, leaving the area as invasive grasses provides no economic and little to no direct cultural or community benefit and would likely leave the area more susceptible to urban development pressures. The current plant community also provides little to no ecological habitat value, whereas the restored system (including lo'i kalo as well as native wetland restoration) is expected to increase habitat for native fish (e.g., 'o'opu akupa;

Eleotris sandwicensis), insects (e.g., *Pantala flavescens*), plants (e.g., neke; *Cyclosorus interruptus*, 'ahu'awa; *Cyperus javanicus*) and birds, including the endangered Hawaiian stilt (ae'o; *Himantopus mexicanus knudseni*). Overall, our research suggests that the restored system would substantially contribute to prioritized local cultural, economic, and ecological goals while also helping to meet the State of Hawai'i's sustainability goals around food, water, and energy. Thus, the project thus represents a locally viable and beneficial opportunity to meet broad societal environmental objectives, which provides broad lessons for the worldwide challenge of local implementation of the SDGs, in an equitable and effective way.

5. Conclusions

Biocultural approaches to conservation and restoration explicitly recognize the interconnection between biological and cultural diversity and between social and ecological systems that have often been obscured in Western-based conservation efforts. While not a new concept, theories of biocultural restoration that emphasize cultural and place-based perspectives, knowledge, and values are emerging in a context of contemporary conservation and restoration efforts [2,29,43]. Here we demonstrated how existing frameworks of evaluation of synergies and tradeoffs in land management from the food-energy-water and ecosystem services frameworks can be adapted to illuminate potential synergies and tradeoffs among multiple cultural, environmental, and economic goals associated with biocultural restoration projects. An important contribution of such an integrated assessment is that it highlights the potential of biocultural restoration to both achieve locally-relevant cultural, economic, and ecological goals while also contributing meaningfully to broader sustainability goals defined by formal policies.

Kāko'o 'Ōiwi is at the center stage of biocultural restoration of social-ecological systems in Hawai'i. Our collaborative case study from He'eia, Hawai'i suggests that biocultural restoration of traditional agriculture has the potential to simultaneously meet multiple community and statewide sustainability goals, including increasing local food production, reducing energy consumption, increasing cultural connection to place, and decreasing sediment delivery to downstream coastal systems. Yet, there are important tradeoffs to consider in the form of nutrient export, which will be much less than alternate land uses (like urban and conventional agriculture), but still likely an increase from the current fallow, degraded system. By understanding and adapting in light of potential tradeoffs, it is clear that the process of (re)connecting to place inherent in a biocultural approach provides a suite of community and cultural benefits that are essential to the long-term social and financial sustainability of this multi-benefit system.

Supplementary Materials: The following are available online at http://www.mdpi.com/2071-1050/10/12/4502/s1.

Author Contributions: Conceptualization, L.L.B., K.F., C.W., K.B., K.K.-S., N.R., G.C. and T.T.; Data curation, K.F.; Formal analysis, K.F. and C.F.; Funding acquisition, K.F., K.B., K.K.-S. and T.T.; Investigation, L.L.B., K.F., C.F., C.W., K.B. and K.L.L.O.; Methodology, K.F.; Project administration, L.L.B., K.F. and K.K.-S.; Visualization, K.F.; Writing—original draft, L.L.B., K.F., C.F., C.W. and T.T.; Writing—review & editing, L.L.B., K.F., C.F., C.W., K.B., K.K.-S., N.R., G.C., K.L.L.O. and T.T.

Funding: The APC was funded by Hawai'i Community Foundation.

Acknowledgments: This research would not have been possible without the work and dedication of Kāko'o 'Ōiwi staff and volunteers who carry out this restoration and who graciously collaborated with us in this research. We particularly are grateful to the families in the 'ulana program who participated in interviews with Casey Ching and welcomed us into the lo'i with their families. The knowledge and insight gained was inspiring in this work and beyond. We thank Pua'ala Pascua for input and guidance in the beginning of this project in our use of her cultural ecosystem services framework and Kawika Winter, Kevin Chang, and 3 anonymous reviewers for comments on the manuscript. Thanks also to Victoria Ward who assisted with figures and to Sarah Medoff who assisted in the economic analysis and manuscript compilation. Funding for this project was through National Science Foundation Coastal SEES Program (#SES-1325874), National Science Foundation EPSCoR Program (#1557349), DLNR Commission on Water Resource Management Water Security Grant (#66092), NOAA Coastal Resilience Grant (NA17NMF4630007), and the Ha'oli Mau Loa Foundation graduate research fellowship with the Department of Natural Resources and Environmental Management, University of Hawai'i at Mānoa. This is contributed paper WRRC-CP-2019-12 of the Water Resources Research Center, University of Hawai'i at Mānoa, Honolulu, Hawai'i.

Conflicts of Interest: The authors declare no conflict of interest.

References

1. Hong, S.K. Biocultural diversity conservation for island and islanders: Necessity, goal and activity. *J. Mar. Isl. Cult.* **2013**, *2*, 102–106. [CrossRef]
2. Kimmerer, R.W. *Braiding Sweetgrass: Indigenous Wisdom, Scientific Knowledge and the Teachings of Plants*; Milkweed Editions: Minneapolis, MN, USA, 2013.
3. Kurashima, N.; Jeremiah, J.; Ticktin, T. I Ka Wā Ma Mua: The Value of a Historical Ecology Approach to Ecological Restoration in Hawai'i. *Pac. Sci.* **2017**, *71*, 437–456. [CrossRef]
4. Wehi, P.M.; Lord, J.M. Importance of including cultural practices in ecological restoration. *Conserv. Biol.* **2017**, *31*, 1109–1118. [CrossRef] [PubMed]
5. Lyver, P.O.B.; Akins, A.; Phipps, H.; Kahui, V.; Towns, D.R.; Moller, H. Key biocultural values to guide restoration action and planning in New Zealand. *Restor. Ecol.* **2016**, *24*, 314–323. [CrossRef]
6. Kuzivanova, V.; Davidson-Hunt, I.J. Biocultural design: Harvesting manomin with wabaseemoong independent nations. *Ethnobiol. Lett.* **2017**, *8*, 23–30. [CrossRef]
7. Kealiikanakaoleohaililani, K.; Kurashima, N.; Francisco, K.; Giardina, C.; Louis, R.; McMillen, H.; Asing, C.; Asing, K.; Block, T.; Browning, M.; et al. Ritual + Sustainability Science? A Portal into the Science of Aloha. *Sustainability* **2018**, *10*, 3478. [CrossRef]
8. Fernández-Manjarrés, J.F.; Roturier, S.; Bilhaut, A.-G. The emergence of the social-ecological restoration concept. *Restor. Ecol.* **2018**, *26*, 404–410. [CrossRef]
9. Gavin, M.C.; Mccarter, J.; Mead, A.; Berkes, F.; Stepp, J.R.; Peterson, D.; Tang, R. Defining biocultural approaches to conservation. *Trends Ecol. Evol.* **2015**, *30*, 1–6. [CrossRef] [PubMed]
10. Ticktin, T.; Quazi, S.; Dacks, R.; Tora, M.; McGuigan, A.; Hastings, Z.; Naikatini, A. Linkages between measures of biodiversity and community resilience in Pacific Island agroforests. *Conserv. Biol.* **2018**, *32*, 1085–1095. [CrossRef] [PubMed]
11. Gon, S.M., III; Tom, S.L.; Woodside, U.; Gon, S.M.; Tom, S.L.; Woodside, U. 'Āina Momona, Honua Au Loli—Productive Lands, Changing World: Using the Hawaiian Footprint to Inform Biocultural Restoration and Future Sustainability in Hawai'i. *Sustainability* **2018**, *10*, 3420. [CrossRef]
12. Dellasala, D.A.; Martin, A.; Spivak, R.; Schulke, T.; Bird, B.; Criley, M.; Van, C.; Kreilick, J.; Brown, R.; Aplet, G.; et al. A Citizen's Call for Ecological Forest Restoration: Forest Restoration Principles and Criteria. *Ecol. Restor.* **2003**, *21*, 14–23. [CrossRef]
13. Higgs, E. *Nature by Design: People, Natural Process, and Ecological Restoration*; The MIT Press: Cambridge, MA, USA, 2003.
14. Suding, K.; Higgs, E.; Palmer, M.; Callicott, J.B.; Anderson, C.B.; Baker, M.; Gutrich, J.J.; Hondula, K.L.; LaFevor, M.C.; Larson, B.M.H.; et al. Committing to ecological restoration. *Science (80-)* **2015**, *348*, 638–640. [CrossRef] [PubMed]
15. Winter, K.; Lincoln, N.; Berkes, F. The Social-Ecological Keystone Concept: A Quantifiable Metaphor for Understanding the Structure, Function, and Resilience of a Biocultural System. *Sustainability* **2018**, *10*, 3294. [CrossRef]
16. Burnett, K.M.; Ticktin, T.; Bremer, L.L.; Quazi, S.A.; Geslani, C.; Wada, C.A.; Kurashima, N.; Mandle, L.; Pascua, P.; Depraetere, T.; et al. Restoring to the future: Environmental, cultural, and management trade-offs in historical versus hybrid restoration of a highly modified ecosystem. *Conserv. Lett.* **2018**, e12606. [CrossRef]
17. Pascua, P.; McMillen, H.; Ticktin, T.; Vaughan, M.; Winter, K.B. Beyond services: A process and framework to incorporate cultural, genealogical, place-based, and indigenous relationships in ecosystem service assessments. *Ecosyst. Serv.* **2017**, *26*, 465–475. [CrossRef]
18. Folke, C. Resilience: The emergence of a perspective for social–ecological systems analyses. *Glob. Environ. Chang.* **2006**, *16*, 253–267. [CrossRef]
19. Berkes, F.; Ross, H. Community resilience: toward an integrated approach. *Soc. Nat. Resour.* **2013**, *26*, 1–16. [CrossRef]
20. Morishige, K.; Andrade, P.; Pascua, P.; Steward, K.; Cadiz, E.; Kapono, L.; Chong, U. Nā Kilo 'Āina: Visions of Biocultural Restoration through Indigenous Relationships between People and Place. *Sustainability* **2018**, *10*, 3368. [CrossRef]

21. Foley, J.A.; DeFries, R.; Asner, G.P.; Barford, C.; Bonan, G.; Carpenter, S.R.; Chapin, F.S.; Coe, M.T.; Daily, G.C.; Gibbs, H.K.; et al. Global Consequences of Land Use. *Science* **2005**, *309*, 570–574. [CrossRef] [PubMed]

22. Altieri, M.A. Linking ecologists and traditional farmers in the search for sustainable agriculture. *Front. Ecol. Environ.* **2004**, *2*, 35–42. [CrossRef]

23. Bennett, E.M. Changing the agriculture and environment conversation. *Nat. Ecol. Evol.* **2017**, *1*, 1–2. [CrossRef] [PubMed]

24. Lincoln, N.K.; Ardoin, N.M. Cultivating values: Environmental values and sense of place as correlates of sustainable agricultural practices. *Agric. Hum. Values* **2016**, *33*, 389–401. [CrossRef]

25. Endo, A.; Tsurita, I.; Burnett, K.; Orencio, P.M. A review of the current state of research on the water, energy, and food nexus. *J. Hydrol. Reg. Stud.* **2015**. [CrossRef]

26. Bremer, L.L.; Mandle, L.; Trauernicht, C.; Pascua, P.; McMillen, H.L.; Burnett, K.; Wada, C.A.; Kurashima, N.; Quazi, S.A.; Giambelluca, T.; et al. Bringing multiple values to the table: Assessing future land-use and climate change in North Kona, Hawai'i. *Ecol. Soc.* **2018**, *23*, art33. [CrossRef]

27. Díaz, S.; Pascual, U.; Stenseke, M.; Martín-López, B.; Watson, R.T.; Molnár, Z.; Hill, R.; Chan, K.M.A.; Baste, I.A.; Brauman, K.A.; et al. Assessing nature's contributions to people_Supl Mat. *Science (80-)* **2018**, *359*, 270–272. [CrossRef] [PubMed]

28. Delevaux, J.M.S.; Jupiter, S.D.; Stamoulis, K.A.; Bremer, L.L.; Wenger, A.S.; Dacks, R.; Garrod, P.; Falinski, K.A.; Ticktin, T. Scenario planning with linked land-sea models inform where forest conservation actions will promote coral reef resilience. *Sci. Rep.* **2018**, *8*, 12465. [CrossRef] [PubMed]

29. Delevaux, J.; Winter, K.; Jupiter, S.; Blaich-Vaughan, M.; Stamoulis, K.; Bremer, L.; Burnett, K.; Garrod, P.; Troller, J.; Ticktin, T.; et al. Linking Land and Sea through Collaborative Research to Inform Contemporary applications of Traditional Resource Management in Hawai'i. *Sustainability* **2018**, *10*, 3147. [CrossRef]

30. Ruttenberg, K.; Kawelo, H. He'eia Fishpond: Encouraging Volunteerism and Cultural Legacy during the ASLO 2017 Aquatic Sciences Meeting. *Bull. Limnol. Oceanogr.* **2017**, *25*, 131–133. [CrossRef]

31. Hawai'i Green Growth. 2018. Aloha Plus Challenge. Available online: http://aloha-challenge.hawaiigreengrowth.org (accessed on 28 November 2018).

32. Loke, M.K.; Leung, P.S. Hawai'i's food consumption and supply sources: Benchmark estimates and measurement issues. *Agric. Food Econ.* **2013**, *1*, 1–18. [CrossRef]

33. Melrose, J.; Perroy, R.; Cares, S. *Statewide Agricultural Land Use Baseline 2015*; Hawaii Department of Agriculture: Honolulu, HI, USA, 2016.

34. Sharp, R.; Tallis, H.T.; Ricketts, T.; Guerry, A.D.; Wood, S.A.; Chaplin-Kramer, R.; Nelson, E.; Ennaanay, D.; Wolny, S.; Olwero, N.; et al. *InVEST 3.5.0. User's Guide*; Stanford University: Stanford, CA, USA, 2018.

35. Hamel, P.; Falinski, K.; Sharp, R.; Auerbach, D.A.; Sánchez-Canales, M.; Dennedy-Frank, P.J. Sediment delivery modeling in practice: Comparing the effects of watershed characteristics and data resolution across hydroclimatic regions. *Sci. Total Environ.* **2017**, *580*, 1381–1388. [CrossRef] [PubMed]

36. Allan, T.; Keulertz, M.; Woertz, E. The water–food–energy nexus: An introduction to nexus concepts and some conceptual and operational problems. *Int. J. Water Resour. Dev.* **2015**, *31*, 301–311. [CrossRef]

37. Bazilian, M.; Rogner, H.; Howells, M.; Hermann, S.; Arent, D.; Gielen, D.; Steduto, P.; Mueller, A.; Komor, P.; Tol, R.S.J.; et al. Considering the energy, water and food nexus: Towards an integrated modelling approach. *Energy Policy* **2011**, *39*, 7896–7906. [CrossRef]

38. Endo, A.; Burnett, K.; Orencio, P.M.; Kumazawa, T.; Wada, C.A.; Ishii, A.; Tsurita, I.; Taniguchi, M. Methods of the water-energy-food nexus. *Water* **2015**, *7*, 5806–5830. [CrossRef]

39. Ringler, C.; Bhaduri, A.; Lawford, R. The nexus across water, energy, land and food (WELF): Potential for improved resource use efficiency? *Curr. Opin. Environ. Sustain.* **2013**, *5*, 617–624. [CrossRef]

40. Hawai'i Office of Planning. *He'eia National Estuarine Research Reserve Management Plan*; The National Oceanic and Atmospheric Administration: Honolulu, HI, USA, 2016.

41. Giambelluca, T.W.; Chen, Q.; Frazier, A.G.; Price, J.P.; Chen, Y.-L.; Chu, P.-S.; Eischeid, J.K.; Delparte, D.M. Online Rainfall Atlas of Hawai'i. *Bull. Am. Meteorol. Soc.* **2013**, *94*, 313–316. [CrossRef]

42. *Townscape Kāko'o 'Ōiwi Conservation Plan. Windward O'ahu Soil and Water Conservation District, 'Aiea, Hawai'i*; The Hawai'i Community Development Authority: Kāne'ohe, HI, USA, 2011.

43. Sterling, E.J.; Filardi, C.; Toomey, A.; Sigouin, A.; Betley, E.; Gazit, N.; Newell, J.; Albert, S.; Alvira, D.; Bergamini, N.; et al. Biocultural approaches to well-being and sustainability indicators across scales. *Nat. Ecol. Evol.* **2017**, *1*, 1798–1806. [CrossRef] [PubMed]

44. MEA (Millenium Ecosystem Assessment). *Ecosystems and Human Well Being: Biodiversity Synthesis*; World Resources Institute: Washingtong, DC, USA, 2005.
45. USDA-NASS. Hawai'i Bananas Annual Summary. 2008. Available online: https://www.nass.usda.gov/Statistics_by_State/Hawaii/Publications/Archive/Bananas/xban07.pdf (accessed on 28 November 2018).
46. USDA-NASS. Hawai'i Farm Facts—Bananas. 2010. Available online: https://www.nass.usda.gov/Statistics_by_State/Hawaii/Publications/Archive/Bananas/xban09.pdf (accessed on 28 November 2018).
47. USDA-NASS. Hawai'i Tropical Fruit and Crops Report. 2017. Available online: https://www.nass.usda.gov/Statistics_by_State/Hawaii/Publications/Sugarcane_and_Specialty_Crops/Sugarcane/2017/201709tropicalspecialtiesHI.pdf (accessed on 28 November 2018).
48. USDA-NASS. Quick Stats. 2018. Available online: https://quickstats.nass.usda.gov (accessed on 28 November 2018).
49. Fleming, K. *The Economics of Commercial Banana Production in Hawai'i*; Agribusiness, No. 8; Cooperative Extension Service, University of Hawai'i at Mānoa: Honolulu, HI, USA, 1994.
50. Ragone, D. Farm and Forestry Production and Marketing Profile for Breadfruit (Artocarpus altilis). In *Specialty Crops for Pacific Island Agroforestry*; Elevitch, C.R., Ed.; Permament Agriculture Resources (PAR): Holualoa, HI, USA, 2011.
51. Liu, Y.; Jones, A.M.P.; Murch, S.J.; Ragone, D. Crop productivity, yield and seasonality of breadfruit (*Artocarpus* spp., Moraceae). *Fruits* **2014**, *69*, 345–361. [CrossRef]
52. Meilleur, B.A.; Jones, R.R.; Titchenal, C.A.; Huang, A.S. *Hawaiian Breadfruit—Ethnobotany, Nutrition, and Human Ecology*; College of Tropical Agriculture and Human Resources, University of Hawai'i at Mānoa: Honolulu, HI, USA, 2004.
53. Love, K.; Paull, R.E. *Jackfruit*; Fruits and Nuts No. 19; College of Tropical Agriculture and Human Resources, University of Hawai'i at Mānoa: Honolulu, HI, USA, 2011.
54. NREL. Solar Maps—Direct Normal Solar Resource of Hawaii. 2017. Available online: https://www.nrel.gov/gis/solar.html (accessed on 28 November 2018).
55. Galitsky, C.; Worrell, E.; Ruth, M. Energy Efficiency Improvement and Cost Saving Opportunities for the Corn Wet Milling Industry. An ENERGY STAR Guide for Energy and Plant Managers. 2003. Available online: https://www.energystar.gov/ia/business/industry/LBNL-52307.pdf (accessed on 28 November 2018).
56. Fu, R.; Chung, D.; Lowder, T.; Feldman, D.; Ardani, K.; Fu, R.; Chung, D.; Lowder, T.; Feldman, D.; Ardani, K.U.S. Solar Photovoltaic System Cost Benchmark: Q1 2017 U.S. *Nrel* **2017**, 1–66. [CrossRef]
57. Fess, T.L.; Benedito, V.A. Organic versus conventional cropping sustainability: A comparative system analysis. *Sustainability* **2018**, *10*. [CrossRef]
58. Izuka, S.; Hill, B.; Shade, P.; Tribble, G. *Geohydrology and Possible Transport Routes of Polychlorinated Biphyenyls in Haiku Valley, Oahu, Hawaii*; U.S. Geological Survey: Reston, VA, USA, 1991.
59. Hawai'i Department of Health. Clean Water Branch, Chemistry Data. 2018. Available online: http://cwb.doh.hawaii.gov/CleanWaterBranch/WaterQualityData/Chemistry.aspx (accessed on 15 January 2018).
60. U.S. Geological Survey (USGS). National Water Information System data available on the World Wide Web (USGS Water Data for the Nation). 2016. Available online: http://waterdata.usgs.gov/nwis/ (accessed on 10 June 2018).
61. Slaets, J.I.F.; Schmitter, P.; Hilger, T.; Vien, T.D.; Cadisch, G. Sediment trap efficiency of paddy fields at the watershed scale in a mountainous catchment in Northwest Vietnam. *Biogeosci. Discuss.* **2015**, *12*, 20437–20473. [CrossRef]
62. Falinski, K. Predicting sediment export into tropical coastal ecosystems to support ridge to reef managmeent. Ph.D. Dissertation, Tropical Plant and Soil Science, University of Hawaii at Manoa, Honolulu, HI, USA, 2016.
63. Hartemink, A.E.; Poloma, S.; Maino, M.; Powell, K.S.; Egenae, J.; O'Sullivan, J.N. Yield decline of sweet potato in the humid lowlands of Papua New Guinea. *Agric. Ecosyst. Environ.* **2000**, *79*, 259–269. [CrossRef]
64. Whittier, R.; El-Kadi, A.I. Human and Environmental Risk Ranking of Onsite Sewage Disposal Systems. Prepared for the State of Hawai'i, Department of Health, Safe Water Drinking Branch. 2009. Available online: https://health.hawaii.gov/wastewater/files/2015/09/OSDS_OAHU.pdf (accessed on 28 November 2018).
65. Hoover, D.J.; MacKenzie, F.T. Fluvial fluxes of water, suspended particulate matter, and nutrients and potential impacts on tropical coastal water Biogeochemistry: Oahu, Hawai'i. *Aquat. Geochem.* **2009**, *15*, 547–570. [CrossRef]

66. Winter, K.; Beamer, K.; Vaughan, M.; Friedlander, A.; Kido, M.; Whitehead, A.; Akutagawa, M.; Kurashima, N.; Lucas, M.; Nyberg, B. The Moku System: Managing Biocultural Resources for Abundance within Social-Ecological Regions in Hawai'i. *Sustainability* **2018**, *10*, 3554. [CrossRef]

67. Aikau, H.K.; Ann, D. Cultural Traditions and Food: Kānaka Maoli and the Production of Poi in the He'eia Wetland. *Food Cult. Soc.* **2017**, *8014*, 1–22. [CrossRef]

68. Chan, K.M.A.; Balvanera, P.; Benessaiah, K.; Chapman, M.; Díaz, S.; Gómez-Baggethun, E.; Gould, R.; Hannahs, N.; Jax, K.; Klain, S.; et al. Opinion: Why protect nature? Rethinking values and the environment. *Proc. Natl. Acad. Sci. USA* **2016**, *113*, 1462–1465. [CrossRef] [PubMed]

69. Drupp, P.; de Carlo, E.H.; Mackenzie, F.T.; Bienfang, P.; Sabine, C.L. Nutrient Inputs, Phytoplankton Response, and CO_2 Variations in a Semi-Enclosed Subtropical Embayment, Kaneohe Bay, Hawaii. *Aquat. Geochem.* **2011**, *17*, 473–498. [CrossRef]

70. Ringuet, S.; Mackenzie, F.T. Controls on nutrient and phytoplankton dynamics during normal flow and storm runoff conditions, southern Kaneohe Bay, Hawaii. *Estuaries* **2005**, *28*, 327–337. [CrossRef]

71. Koshiba, S.; Besebes, M.; Soaladaob, K.; Isechal, A.L.; Victor, S.; Golbuu, Y. Palau's taro fields and mangroves protect the coral reefs by trapping eroded fine sediment. *Wetl. Ecol. Manag.* **2013**, *21*, 157–164. [CrossRef]

72. Yoon, C.G. Wise use of paddy rice fields to partially compensate for the loss of natural wetlands. *Paddy Water Environ.* **2009**, *7*, 357. [CrossRef]

73. Penn, D.C. Water Needs for Sustainable Taro Culture in Hawai'i. In Proceedings of the Sustainable Taro Culture for the Pacific Conference, Honolulu, HI, USA, 24–25 September 1992; pp. 132–134.

74. Hao, L.; Sun, G.; Liu, Y.; Wan, J.; Qin, M.; Qian, H.; Liu, C.; Zheng, J.; John, R.; Fan, P.; et al. Urbanization dramatically altered the water balances of a paddy field-dominated basin in southern China. *Hydrol. Earth Syst. Sci.* **2015**, *19*, 3319–3331. [CrossRef]

75. Penton, C.R.; Deenik, J.L.; Popp, B.N.; Bruland, G.L.; Engstrom, P.; St. Louis, D.; Tiedje, J. Importance of sub-surface rhizosphere-mediated coupled nitrification–denitrification in a flooded agroecosystem in Hawaii. *Soil Biol. Biochem.* **2013**, *57*, 362–373. [CrossRef]

76. Ockenden, M.C.; Deasy, C.; Quinton, J.N.; Bailey, A.P.; Surridge, B.; Stoate, C. Evaluation of field wetlands for mitigation of diffuse pollution from agriculture: Sediment retention, cost and effectiveness. *Environ. Sci. Policy* **2012**, *24*, 110–119. [CrossRef]

 *sustainability*

Article

The Role of Breadfruit in Biocultural Restoration and Sustainability in Hawai'i

Blaire J. Langston [1,*] and Noa Kekuewa Lincoln [2,*]

[1] Department of Natural Resources and Environmental Management,
College of Tropical Agriculture and Human Resources, University of Hawai'i at Mānoa,
1910 East-West Road, Sherman Laboratory 101, Honolulu, HI 96822, USA

[2] Department of Tropical Plants and Soil Sciences, College of Tropical Agriculture and Human Resources,
University of Hawai'i at Mānoa, 3190 Maile Way, St. John 102, Honolulu, HI 96822, USA

* Correspondence: blairej@hawaii.edu (B.J.L.); nlincoln@hawaii.edu (N.K.L.)

Received: 31 July 2018; Accepted: 18 October 2018; Published: 31 October 2018

Abstract: The Hawaiian Islands today are faced with a complex mix of sustainability challenges regarding food systems. After European arrival, there was a change of dietary customs and decline in traditional Hawaiian agriculture along with the cultural mechanisms which sustained them. Recently, there has been a resurgence for local food and culture alongside an enthusiasm for breadfruit (*Artocarpus altilis*)—a Polynesian staple crop. To investigate the role of breadfruit and biocultural restoration in Hawai'i, we conducted surveys and interviews with local breadfruit producers. Overall, we found that breadfruit has the potential to provide holistic, practical and appropriate solutions to key issues in Hawai'i, including food security, environmental degradation and public health, while simultaneously lending to the revival of cultural norms and social relationships. As breadfruit cultivation expands rapidly in Hawai'i, the opportunities for increased social and environmental benefits can be realized if appropriately encouraged.

Keywords: breadfruit; biocultural restoration; sustainability; food systems; Hawai'i; *Artocarpus altilis*

---

1. Introduction

Sustainability can be defined as improving the quality of human life while living within the carrying capacity of supporting ecosystems [1]. Common frameworks to sustainability may include dimensions of social, economic, environmental, cultural and political needs [2]. Some take the perspective that humans and environment are separate and therefore the major role humans play in the environment is destructive and extractive [3,4]. This sits in opposition to indigenous epistemology that views people as an integral part of the environment—a concept is that is captured in biocultural frameworks, in which we take into consideration relationships between biological and cultural systems in order to conceptualize the link between people and environment [5,6]. Biocultural refers to how human cultures are shaped by their surrounding ecosystems, which in turn, shapes culture itself [6]. This approach is not a new concept by any means but has implications for use today in terms of sustainable management practices. There is an "inextricable link" between cultural diversity and biodiversity and there has been insight on correlation between the two phenomena that illustrates their importance for the resilience of social-ecological systems that sustain life [3,6]. The preservation of cultures and the restoration or protection of the environment are therefore dependent on the other, not only in natural ecosystems but in human managed systems such as agriculture.

Including cultural aspects into environmental approaches is known to lead to sustainable management practices and more resilient systems [6]. Customary values and community capital in Hawai'i facilitate reciprocity and food sharing and the passing of knowledge which can help sustain resources over time and provide effective natural resource management approaches [7]. Community involvement and cultural perpetuation of values is crucial to long term food security and can lead to greater chance of long term success in restoration [7,8]. These cultural perpetuations are likely tied, at least in part, to the key biological species through which biocultural relationships are formed. It terms of losing biocultural diversity, the globalization of food systems and unsustainable agriculture development are substantial contributors [6,9]. Hawai'i has been no exception, with plantation agriculture arguably being the largest driver to cultural displacement and loss over the past 200 years. However, within the last forty years, revitalization of traditional agriculture has increased, paralleling an increase in cultural revival. While there are several crops that hold importance in terms of Hawaiian culture, we focus on the role of breadfruit (*Artocarpus altilis*, (Parkinson, Fosberg)) within contemporary commercial agriculture to explore its broader applications for sustainability in Hawai'i. We argue that in Hawai'i, traditional agriculture plays a pivotal role in the connection between the socioeconomic and cultural aspects of sustainability and follow Barthel et al. [5] in arguing that supporting crops with biocultural importance is critical for long-term success of agriculture. Furthermore, the use and maintenance of such crops is essential for maintaining cultural values and practices, particularly within the modern socioeconomic landscape, and for the health of the environment.

1.1. Traditional Breadfruit Cultivation in Hawai'i

Breadfruit, or 'ulu in Hawaiian, is a tropical tree in the fig family [10] that produces a large (typically 1–3 kg) starchy fruit that tastes much like a potato, banana or plantain depending on the state of maturity (Figure 1). Breadfruit has been consumed as a staple crop throughout Oceania for millennia, appears in many aspects of traditional knowledge, has influenced sociopolitical environments and has multiple resource applications. A sterile, seedless variety of breadfruit was transported to Hawai'i at least 800 years before present [11,12].

Hawai'i, when compared to smaller and more ancient islands of the Pacific, expresses greater opportunities for agricultural development [13] and consequently, Native Hawaiian populations were heavily reliant on intensive cultivation of land [14]. This afforded the emergence of unique agricultural practices, such as mahi'ai—massive, intensive rainfed field systems Lincoln et al. [13] *in this issue*. Breadfruit grows well in marginal habitats and, throughout most of the Pacific, was an important staple crop, often seen as a "gift from the gods" and grown in semi-wild "food forests" with minimal active management [15]. However, in Hawai'i, where extensive cultivation of annual starches was possible, breadfruit assumed a complimentary role and was used to expand cultivable areas [16], increase resilience [17] and enhance place-adapted cropping systems [18]. While breadfruit trees maintained near households were managed much like elsewhere in the Pacific, large-scale breadfruit arboriculture took on a different form in Hawai'i. In addition to the semi-wild and largely unmanaged "food forests," Hawai'i developed breadfruit arboriculture that was well-spaced, highly managed and incorporated the intensive cultivation of multiple annual and perennial crops [11,16].

In Hawai'i, breadfruit likely started as part of individual garden plots near settlements, expanded into semi-wild food forests cultivated on colluvial valley slopes and culminated into the development of sizeable, systematic arboriculture [11,13,16]. Such arboricultural developments cultivated breadfruit within intercropped agroforestry systems planted with kukui (candlenut, *Aleurites moluccanus*), 'ōhi'a 'ai (mountain apple, *Syzygium malaccense*), 'uala (sweet potato, *Ipomoea batatas* L.), maia (banana, *Musa spp*), uhi (yam, *Dioscorea alata*), kalo (taro, *Colocasia esculenta*), kō (sugar cane, *Saccharum officinarum*), wauke (paper mulberry, *Broussonetia papyrifera*), 'olena (turmeric, *Curcuma longa*), pia (arrowroot, *Tacca leontopetaloides*), 'awa (kava, *Piper methysticum*), 'awapuhi (shampoo ginger, *Zingiber zerumbet*) and kī (tī, *Cordyline fruticosa*) [19–21]. These rainfed systems comprised a significant portion of the total agriculture [18,22], created habitat that allowed the optimal cultivation of some crops such as wauke [20] and produced a highly diverse set of food, resource and medicine crops [16].

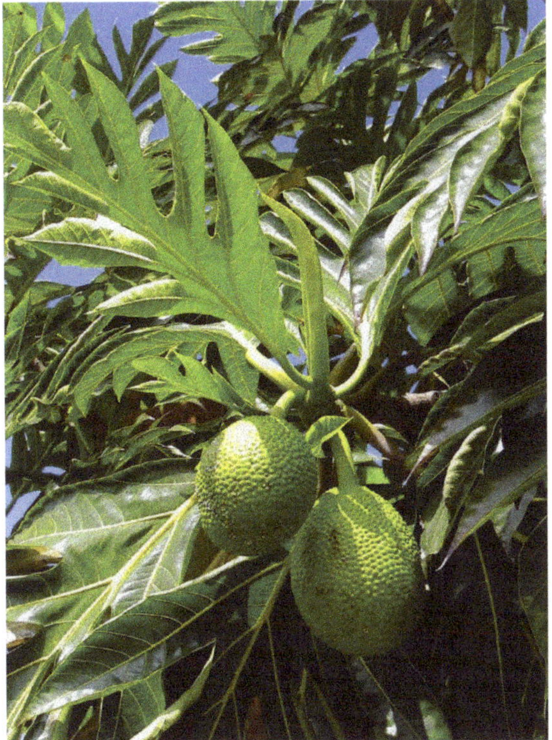

Figure 1. The large fruit and leaves of the Hawaiian breadfruit tree.

1.2. Traditional Importance of Breadfruit in Hawai'i

Breadfruit has made substantial contributions to food production, health and nutrition, environmental quality and culture in the Hawai'i for centuries [11,15,23]. 'Ulu was a staple food source in addition to kalo, 'uala, uhi and mai'a, ranking third in importance for complex carbohydrates and serving as an essential crop in times of human and environmental disturbances [11,22,24]. The fruit is nutritious, especially when compared to contemporary starches such as rice, corn and wheat [25]. 'Ulu trees are relatively low maintenance but provide substantial yields, resulting in high surplus production after establishment. Breadfruit was seasonally abundant but unlike elsewhere in the Pacific, storage of breadfruit in underground pits appears to have been rare, while application as an animal feed appears to have been more extensive [15,26].

Breadfruit was also a valuable resource and medicine. A long list of uses for wood includes; housing and construction, canoes, surfboards, drums, cloth and poi (fermented taro with water) boards, all of which could be polished with its leaves that were used as an abrasive similar to sandpaper [11,15,27]. Virtually all parts of the plant are used medicinally–the leaves, bark, fruit, flowers and latex–to treat ailments including skin conditions, high blood pressure, cardiovascular disease; the flowers are commonly used to repel mosquitos; and the high-latex sap also has an array of applications, including caulking and bird snaring [11,15,27,28].

The development of tree resources, such as breadfruit, can be seen as investing into the land to develop long-term resources for future prosperity [29]. The development of large-scale arboriculture, such as the famous breadfruit groves of Kona and Lāhainā, took place during the "golden age" of Hawai'i in the 16th century [13,30], with the resulting surplus of production further empowering the development of sociopolitical complexity and hierarchy [31–33]. Due to the nature of political and labor organization, extensive systems of agriculture were able to form over time, ensuring the ability for ancient Hawaiians to spend more time into shaping political dynamics. The groves of breadfruit acted as a significant resource that was driven by the desire to increase local economies, which in turn developed stronger political hierarchies (through taxes, military service, etc.) that could further invest into the landscape [16,17].

The seasonal surplus of breadfruit may have powered social dynamics and rituals, such as the emergence of the Makahiki—an extended period where work and certain religious ceremonies were suspended and corresponding to a time of increased recreation and tax collection [34,35]. During Makahiki, breadfruit served many ritualistic purposes including the shaking of the Maoloha net at the end of Makahiki; if the food dropped from the net, there would be no famine during next growing season [34]. The occurrence of the festival in October to January corresponds well with the surplus breadfruit season. The Makahiki developed out of Kona, which was famous for the "breadfruit belt" (expansive stretch of agroforestry largely consisting of breadfruit that crossed over several regional districts throughout the Kona Field System of Hawai'i Island) [14]. One could argue that such an extended period of ceremony could only be made possible by an extensive surplus. This would be similar to other Pacific Islands, where annual time periods that corresponded to religious and ceremonial occurrences aligns with the productive periods of breadfruit and other crops [15,36]. While speculative, that the most famous ceremony of abundance in ancient Hawai'i developed in a region famous for its breadfruit and coincided largely with the season of breadfruit productivity suggests a linkage between 'ulu and abundance.

The abundance from 'ulu also has an important place in Hawaiian cosmology. Stories include that of the god Kū who turned himself into an 'ulu tree to feed his starving wife and children [14]. Breadfruit was depicted in stories of embodiment, in which the gods would enter a breadfruit tree in supernatural form, signifying a close relationship and admiration of the natural world [11]. Many concepts of Hawaiian traditions and values are captured in a range of 'ōlelo nō'eau (Hawaiian proverbs, sayings or stories) that utilize 'ulu as a metaphor for wealth, success and planning. The reoccurring lessons utilize breadfruit to teach values on sharing and hospitality, warnings to be kind to travelers and making sacrifices for the prosperity and success of others [37,38].

1.3. Loss of Traditional Food Systems

The arrival of Europeans in the 18th century marked a decrease in the native Hawaiian population and greatly altered the path of Hawaiian agriculture. This shift resulted in the decline of traditional systems, the establishment of plantation agriculture, the marginalization of native peoples and reshaped the future of Hawai'i [1]. As with all traditional crops, the abundance of breadfruit in Hawai'i declined precipitously (see Kagawa-Viviani et al. *in this issue*).

Upon arrival, early European explorers reported seeing considerable breadfruit groves around all the islands, especially near villages and settlements, with vast arboricultural "belts" in several regions [11,13]. The Kona breadfruit belt was modeled to have consisted of more than 100,000 trees; it is now, however, reduced to a few hundred trees, often neglected and unwanted [16]. These traditional, intercropped systems once produced enough food to sustain the largest population in Polynesia and indeed supported a larger population on each island (excluding O'ahu) than exists today [39,40].

Local economies, cuisines, cultural practices and the state of the environment shifted with the decline in traditional agriculture. Today, 80–90% of food is imported, making Hawai'i one of the most dependent states in the United States [41]. Recently, Hawai'i was assessed as the 48th worst state in terms of farming outlook [42]. Statewide dietary shifts to processed, imported foods have raised public health concerns. Hawaiians and other Pacific Islanders are especially vulnerable to dietary-based health issues, including obesity and diabetes, which have risen alarmingly fast, holding some of the highest rates in the world [25]. Furthermore, we see exceptionally negative environmental impacts that threaten soil, water, ecosystems and human health [43]; and the declining involvement in local food production greatly reduces the overall food security [1] of Hawai'i [44].

With the reduction of breadfruit and other crops, there are fewer opportunities for communities to engage with and learn about them. Furthermore, with decreased number of farms and the connection to cultural foods, there are fewer opportunities to connect with the land, learn cultural practices involving Hawaiian agriculture and to gather around Hawaiian food. It takes time for food to become cuisines and acquire symbolic meanings and once lost the efforts needed to restore those relationships are significant [24].

1.4. Resurgence for Local Food and Breadfruit in Hawai'i

In the 1970s, public awareness increased demands for local, fresh and healthy foods [1]. Coinciding with a growing interest for food self-sufficiency, a resurgence of Hawaiian culture, pride and practices also occurred in what is affectionately known as the "Hawaiian renaissance." These parallel movements coupled with research efforts regarding breadfruit, led to increased demand for fresh breadfruit in Hawai'i. Grassroots and institutional efforts partnered to conduct multiple campaigns to promote the cultivation, application and consumption of breadfruit in Hawai'i [45]. Such efforts included public education, festivals and tree-giveaways. Part of the challenge to popularize breadfruit was due to a social stigma that positioned breadfruit as a second class food in Hawai'i [46]. While outreach programs, marketing, availability, introduction of value-added products and presence in restaurant dishes has educated new users about breadfruit, the stigma was not as easy to overcome. Finding an appropriate role for breadfruit in cultural and agricultural contexts remains an ongoing challenge.

More recently, ecosystem-based management strategies are being promoted for agriculture and conservation [47]. The state government has recognized efforts to strengthen community-based conservation and set goals to have more sustainable and secure food systems in Hawai'i [44,47], while large landowners have similarly promoted improved practices to agricultural leaseholders. However, overall there has been a significant disconnect between stated goals and outcomes [48–50]. Breadfruit cultivation has the ability to contribute to these efforts and more.

2. Materials and Methods

We used a combination of surveys (Table A1), semi-structured interviews (Table A2) and onsite farm observations with 43 individuals and organizations engaged in breadfruit production (Figure 2). Of the 43 participants, 36 were bonafide agricultural producers, while the remaining were non-profit organizations growing breadfruit for preservation, cultural access and community food systems. Participants were initially recruited through open advertisement and targeted approach of well-known growers, with subsequent requests made using a "snowball" approach; all participants that expressed interests were included in the study. Participants represented a range of social and natural environments and included new and long-time farmers. Participants were selected by word of mouth, community connections, internet searches and qualified to participate if they cultivated at least ten breadfruit trees. The breadfruit farmers who engaged in the study are referred to as "participants". The participants varied in their experience of farming breadfruit but were all successful in growing trees that produced fruit and were therefore an expert subset of the general population. The study was conducted on various locations on the islands of Hawai'i (n = 19), O'ahu (n = 15), Kaua'i (n = 5) and Maui (n = 4). Interviews and surveys were conducted onsite beginning in March 2017 and finalized in May 2018, during the spring season (March-May) of each year.

Ethnographic data was collected in two forms; interviews and surveys. Informed consent was received for all participants and survey and interview scripts were granted IRB exemption for work with human subjects. The survey questions focused on the history of the farm and the trees, cultivation techniques, farming practices and tree care methods (Table A1). The interviews were semi-structured and probed the participants on their beliefs, concerns, lessons and ideas concerning breadfruit, contemporary culture and agriculture (Table A2). The interviews, in general, focused on the role breadfruit plays in the connection between culture and place and how those connections provide for the well-being of people and the environment of Hawai'i today. Interview duration was variable, lasting between one-half hour and three hours depending on the individual's willingness and enthusiasm to share information. Quotations and activities reported from the interviews stem solely from the bona-fide agricultural producers and does not pull from the non-profit participants. Survey data was managed and organized in Excel 16.13 (Microsoft Corporation, Redman, WA, USA) and analyzed in SPSS 20.0 (IBM, Armock, NY, USA). Written notes from the interviews and on-site observations were pinned to survey results and coded for thematic descriptions by the authors. Often repeated themes were selected for emphasis and specific farm examples were pulled to demonstrate each theme.

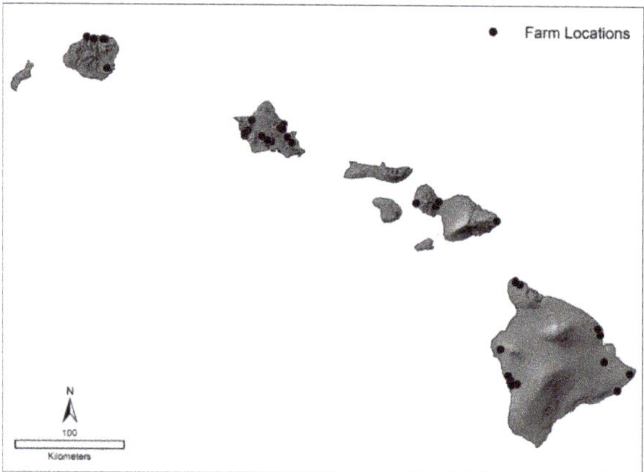

Figure 2. A map of the breadfruit producers that participated in the survey and interview portions of the study.

Participation included on-site sampling to assess soil and tree health at each site. Soil samples were retrieved from each site; composite soil samples were collected by mixing three soil cores taken 2 m from the trunk of the trees at a depth of 30 cm. Soils were assessed for pH and soil moisture in house, then dried, sieved, ground and packaged for analysis at Brookside Laboratory Inc. (New Bremen, OH, USA). Tree health was assessed through chlorophyll and photosynthetic measurements. Chlorophyll counts were measured using a SPAD 502 Plus Chlorophyll Meter (Spectrum Technologies, Aurora, IL, USA) and photosynthetic rates were measured using a miniPPM-100/150 Plant Photosynthesis Meter (EARS, Delft, The Neatherlands). Ten random, consecutive measurements were taken per leaf, with three leaves per tree and three trees per site analyzed (90 total measurements per site); the third leaf from the tip was used for measurement (Lincoln et al. in review). Measurements data was organized in Excel 16.13 and analyzed in JMP Pro 13 (SAS Institute, Cary, NC, USA).

3. Results

Participants were not surveyed for traditional demographic indicators (e.g., age, income, education) but were represented by statistics such as farm size and reliance on farming (Table 1). In these terms, participants overall agreed with statewide statistics [51,52] and previous surveys of farmers in Hawai'i [48,49]. Farmers represented a range of practices, goals, values and demographics. On one end of the spectrum, interviewees included the largest productive orchard in the state, growing over 500 trees in a monocropped section of a highly conventional agricultural operation. In contrast, we also saw smaller farms including Native Hawaiian homesteaders growing diversified agriculture mainly for subsistence purposes. Results included experiences of individuals, non-profit organizations, cooperative organizations and for-profit companies that operated on fee simple, lease and partnership lands (fee simple refers to leases privately owned land, partnership lands are less formal agreements that give access to land without ownership). The individuals we talked to represented Native

Hawaiians, plantation-era multi-generational families and new immigrants to Hawai'i. Despite these vast differences in motivations, backgrounds and structures, there were clear commonalities in the discussions with each as they related to and talked about breadfruit.

Table 1. Farm attributes of survey participants.

Attribute	Agricultural Producers			Non-Profits		
	n	Average	Std Err	n	Average	Std Err
Size of farm (acres)	36	27	10	7	409	339
Area in breadfruit	33	3.4	1.2	7	0.6	0.2
Number of trees	33	69	24	7	38	14
Age of trees	28	14	2.8	7	12	4.6
Production per tree (lbs.)	15	101	39	4	152	68

Analysis of interview notes resulted in the identification of five key themes: customary values, traditional agriculture, food security, ecological health and community (social) capital. Further organization indicated that each theme often overlapped and interacted with other themes, which lead to the development of a conceptual diagram of the themes in the framework of biocultural restoration and sustainability (Figure 3). Building upon this conceptual structure, results are presented within the five identified themes, utilizing data from the participant surveys and farm sampling, along with analysis and specific examples from the interviews. Results are further discussed within the frameworks of biocultural restoration and sustainability.

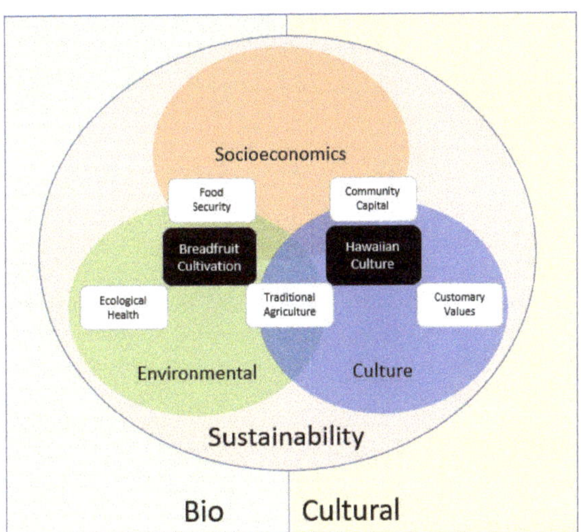

Figure 3. Conceptual diagram illustrating the relationship between the major themes that arose in the research and how they situate within the concepts of biocultural restoration and sustainability. The white boxes show the five themes identified from the interview data. The black boxes represent the basis of the biocultural relationship and the colored circles show common dimensions of sustainability.

3.1. Customary Values

A key theme that emerged from participant interaction centered on customary Hawaiian values, regardless of participant ethnicity. Hawaiian values such as aloha (love or compassion), kuleana (reciprocal responsibility), ha'aha'a (humbleness), pono (righteousness) and ahonui (patience) were prominent in Hawaiian life [53]. One description of aloha from the early 19th century translates, "It was often said...Hawaiians...were full of love when they received visitors wherever they lived...[and]

people...were greatly taken by the hospitality (aloha) of the Hawaiians..." [53]. Expression of Hawaiian values was seen in direct statements by participants, as well as in descriptions of their activities. The importance of showing aloha to guests and between one another, to facilitate bonding and with it a sense of identity was expressed. This was further expressed through generosity, which may be thought of as an extension of aloha and living pono by ensuring comfort and sense of belonging between people.

Offering generous hospitality and food sharing were common themes during interaction with breadfruit farmers. The act of exchanging goods traditionally reaffirmed familial bonds and 'ohana (kinship, family) in ancient Hawai'i [54]. In contemporary times, food sharing is less relevant for food security but still remains a significant cultural tradition that strengthens social ties [55]. For example, a visit to "Farmer A's" farm included a sit-down dinner, food, wine and storytelling extending beyond the purpose of the visit. Another 'ulu farmer, "Farmer C," insists his guests will not leave without being well fed, "Farmer B" shared fruits and homemade products, while "Farmer D" sends visitors on the road with boxes of produce. Like many 'ulu farmers, "Farmer E," an agricultural producer and "Farmer F," a non-profit farmer, will both begin or end work visits with food potluck family-style. Of the 36 agricultural producers, 20 (56%) included food sharing with researchers. More instances of hospitality were demonstrated to the broader community. Of the producers, eight (22%) expressed that they regularly host community gatherings for the farming population where food and food exchange are a critical expression of community commitment and reciprocity. Foods shared among farmers at gatherings are locally grown and wholesome, emphasizing well-being and environmental health. "Farmer E" and "Farmer G" offer a place to stay for travelers and visitors and extend invitations to their community to gather, share stories and spend time together. We have found this to be very different than our previous experience with farmer surveys, in which less than 20% of farmers conducted food-sharing and gifts with the researchers [48,49]. Reciprocity extended beyond physical gifts, often aiding in neighboring farms or even providing direct financial support. Several farmers expressed the importance of supporting each other, with "Farmer E" voicing, "[It's] just the Hawaiian interest and value in me, if someone can do it, they should."

Breadfruit farmers also negotiate with their "customers," who are in reality our communities, which demonstrates a respect and appreciation for the well-being of all. Many 'ulu farmers meet customers halfway on pricing to be able to cover their needs and others provide affordable, local produce. This practice, surprisingly, was seen in both subsistence and commercially-driven producers. Within the interviews this practice was commonly expressed in terms of a "traditional" or "alternative" economy. These instances demonstrate the strong presence of Hawaiian values instilled and alive within the surrounding 'ulu farming communities and importantly, increase resiliency through reciprocity, increased self-reliance, social ties and community building [44,55].

While these values may be prevalent among farmers in general, breadfruit, which carries with it the traditional perspective of a "gift from the gods," seems exceptionally conducive to the customary values of giving. Compared to previous farmer surveys and interviews conducted by the author [48,49] expression of Hawaiian values was much higher among breadfruit cultivators. As "Farmer G" states, "the [breadfruit] trees give us so much for free. It is only right to share it with others."

3.2. Traditional Agricultural Practices

Multiple participating farms shared traditional agricultural practices that were passed down by family. Several farmers including "Farmer B," "Farmer C," "Farmer H," "Farmer I" and "Farmer J" cared for old, remnant breadfruit patches; they call the one large, eldest tree the "mother tree." New trees were then propagated using the keiki (child) root shoots or root cuttings. Similarly, we observed many farmers still using lou—the traditional harvesting tool, although the materials have changed [54]. Many producers (37%) use traditional planting and ground preparation methods, applying green mulch prior to planting. Three (8%) producers used the practice of placing a fish or octopus at the bottom of the planting hole and eight (22%) producers engaged in the ancient practice of planting a placenta under a new tree. More than half (56%) the producers felt they engaged in some sort of cultural protocol when it came to their trees and expressed that these protocols enhanced the health and well-being of the land and plants. Several farmers utilized the traditional practice of planting in accordance with moon phases.

Importantly, participants primarily engaged in the traditional application of breadfruit cultivation within agroforestry systems. Of the farmers, only six (17%) had mono-cropped stands of breadfruit and of those four (11%) grew other crop species in adjacent plots. For the most part, breadfruit cultivation was integrated into diverse agricultural systems, somewhat sporadically interspersed within a variety of different crops or had breadfruit cultivation areas more systematically interplanted with other crops; intercropping included mai'a, kalo, niu, 'ōlena, 'awapuhi, coffee, papaya, ornamentals, citrus trees and leafy green vegetables. "Farmer H" was actively restoring the traditional Hawaiian breadfruit agroforestry system, while "Farmer G" has created a modern agroforestry system with a broad range of tree, shrub and annual crops. Agricultural producers and non-profits, such as "Farmer F," "Farmer K," "Farmer L," "Farmer M," incorporated breadfruit cultivation adjacent to lo'i (flooded terraced) cultivation of kalo. Growing breadfruit in diverse agricultural systems can be tied to cultural values and beliefs regarding the tree. At least three separate growers used the exact same phrase that "'ulu does not want to be alone" and although there is no agronomic evidence to support it, several farmers expressed that they felt breadfruit grew better in mixed agricultural settings rather than in isolation. Although we can likely expect and in fact already see, the development of more mono-cropped breadfruit as the industry continues to develop in Hawai'i (N.K. Lincoln unpublished data), current breadfruit cultivation is employed in mixed agroforestry agriculture that indicates its role in the revival of traditional agricultural practices and usage.

In addition to the physical practices, value-based practices were commonly described by participants discussing their tree management. The most commonly mentioned practice was of kilo (observation of one's surrounding). Farmers often planned their schedules around cycles, weather patterns and observations. For example, "Farmer C" will only water trees according to observing his trees for signs of water deficiency. Similarly, "Farmer F" and "Farmer G" utilize observations of their tree health to determine the best co-crops and "Farmer B" selected individual planting sites based on nuances of microclimates on the farm. Utilization of intuitive management practices guides input adjustment resulting in conservative resource use. This method is reliable and profitable, demonstrated by "Farmer C" having one of the highest yields per tree. Types of local knowledge such as this are thought to have broader implications for restoration projects [8]. It is also important to note that knowledge gained from biocultural relationships typically is formed on small farms [5].

3.3. Community (Social) Capital

Many of the 'ulu farmers leverage their farm to enhance relationships between people, land and plants within their communities by serving as educational and cultural centers. "Farmer F" has regular work days for volunteers and students to participate in traditional farming methods, while "Farmer H" opens her farm to agroecology tours and farmer training to teach about agroforestry, traditional agriculture and dryland agriculture. "Farmer L," one of the largest agricultural producers that participated, has conducted several workshops on breadfruit production, including the promotion of diversified agroforestry despite the fact that he largely employs industrial, monocropping practices. As was often expressed, the education is not only about the food and the plant but the place and the people. "Farmer K" stated, "We really want to get the kids in the area out here to learn about the place where they live. A lot of them do not know the place names of where they live. They do not know what ahupua'a (traditional land division) they are in and I want to teach them that." Linking agriculture, cultural learning and land preservation is not a new concept but is currently being reinforced by 'ulu farmers. In such ways, knowledge and professional development skills are passed on through local and culturally appropriate methods. It is argued that cultural education increases resiliency of food systems because memories and lessons that teach people to be stewards of their land allow them to adapt and reorganize [5].

Growers also commented on how they themselves were educated through engagement with 'ulu. Rebuilding his farm from scratch, "Farmer F" recounted how he learned much of his farming through trial and error; while there were elders who taught him kalo farming, knowledge was not passed to him about growing 'ulu. He was now excited to share what he learned with others and discussed the importance of knowledge exchange between farmers and the younger generation. This was a common theme amongst the breadfruit farmers, who dominantly indicated that working with the crop has been a rewarding learning experience. Their learning experiences are critical in particular because, unlike other crops, breadfruit has not been well studied from an agronomic perspective and especially regarding the vastly different growing conditions that Hawai'i represents. As such, the accumulated knowledge from growers represents a significant knowledge source to share amongst themselves and others.

3.4. Food Security

Breadfruit makes significant contributions to food security and improved health in Hawai'i. Some farmers share breadfruit in the customary sense between familial ties but in general, the practice is broader. One farmer, "Farmer N," does not sell his fruit through any conventional economic pathway but only to friends, family, church members and the broader community, often trading or even giving it away. Every participant, even the most commercial agricultural producers, indicated that they give away fruit or reserve part of their harvest for friends, family, or community. This regard for others before profits is unique when compared to other farmer surveys in Hawai'i (e.g., [48,49]). When "Farmer G" was pressed as to why breadfruit was given away, she indicated that breadfruit was going to be essential for Hawai'i's food security in the future and so that "people need to get ma'a (accustomed to) it now." Similarly, "Farmer H" indicated that breadfruit was traditionally a food that signified wealth and abundance and that "there is no greater wealth than to be able to feed your community." "Farmer H," "Farmer F," "Farmer C," "Farmer M" and others all regularly donated fruit to local schools, food baskets and other non-profit organizations simply to engage in the practice of feeding their communities.

Due to the short shelf-life of the fruit and lack of processing facilities for breadfruit in Hawai'i, a negligible amount of breadfruit is currently exported from the state. Despite supplying a purely local market, breadfruit has seen a dramatic rise in production (Figure 4). Based on our surveys the growth of breadfruit plantings statewide has grown exponentially, from fewer than 500 commercial tree plantings 20 years ago, to over 3000 trees today, with more than a doubling of plantings expected within the next five years. Considering that 'ulu farms are located on all the main Hawaiian Islands (Figure 1) in multiple locations, it has potential to feed multiple communities around the state.

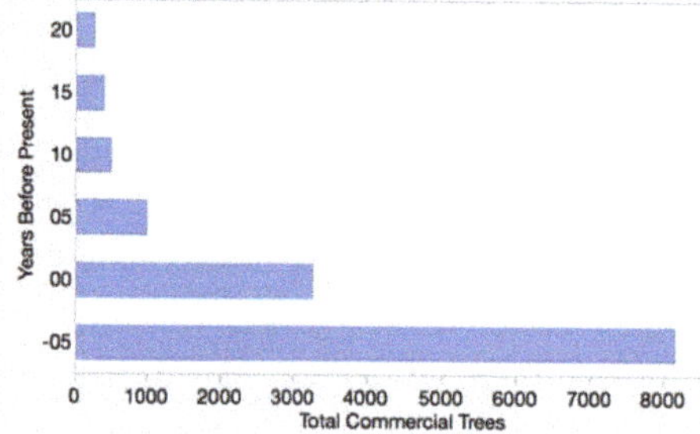

Figure 4. Cumulative number of commercial breadfruit tree plantings identified in Hawai'i in 2017. The negative five-year category represents trees producers have indicated they plan on planting within the next five years.

Some organizations, such as that of "Farmer O" and "Farmer P" have evolved into major community hubs for community food security. Children, community members and farmers participate and learn about local and traditional foods that support good public health. Meetings held at "Farmer P's" farm afford individuals to learn about these foods and make learned decisions about diet. Teaching, for example, that traditional starches, such as breadfruit with its moderate glycemic index, is predicted to greatly reduce diabetes if re-adopted [25]. Similarly, workshops and tree sales by "Farmer H" educate the community on growing backyard trees and the impact that it can make on a family's health, economics and food security.

3.5. Ecological Health

Contemporary 'ulu farmers demonstrate the ability to cultivate breadfruit in a way that protects the environment while remaining profitable. The majority of breadfruit farmers do not use fertilizers, herbicides, or irrigation, with a mere 7% of growers using pesticides (Table 2). Among the users of chemical and water inputs, only two had regular schedules, while the majority of growers applied fertilizers rarely with application times based on tree observations.

Table 2. Rates of various management practices associated with breadfruit cultivation among survey participants.

Practice (n = 38)	% Apply
Fertilizer	25%
Compost	44%
Mulch	54%
Cover Crops	5%
Intercropped	67%
Pesticides	7%
Herbicides	21%
Irrigation	28%

As mentioned previously, most 'ulu farmers surveyed intercropped their breadfruit trees. The diversification of crops in agroforestry systems increases the value of production, enhances profitability of traditional farms and reduces production costs due to a decreased need for inputs such as water, chemicals and energy [56]. Despite low inputs, farmers reported a 97% average success rate for all sapling establishment on average. When farmers added care to their trees, they largely chose environmentally friendly inputs such as recycled or green waste materials for mulch (e.g., clippings, woodchips, leaves, cardboard), organic or homemade fertilizers (e.g., fish emulsion, chicken manure) and natural compost (food scraps, mixed garden waste, human waste from composting toilets). "Farmer C" farmed without chemical inputs and only organic material as fertilizer yet reported exceptionally high yields. "Farmer H," utilized no inputs, also reported high yields and her tree's measurements showed the highest rates of photosynthesis of any producer measured. That breadfruit is highly productive without excessive inputs is captured by agricultural producer, "Farmer L," who said of his trees,

> "Is it a superfood?! It's definitely a tree of abundance. I always joke around it's like a tree of life. Hundreds of pounds. With really minimal [inputs], I fertilize them once in a while but other than that they just go. It's definitely a tree that gives plenty. It's suited for here. I learned to not quite force anything. There's so many trees I've tried to make grow on a farm. It's the ones from here that can endure."

While we did not investigate the role of biodiversity, several farmers made mention of their breadfruit trees being home to the 'io (the Hawaiian hawk, *Buteo solitarius*) and the 'ōpe'ape'a (the Hawaiian hoary bat, *Lasiurus cinereus semotus*), an iconic, endemic and federally endangered species. While this is not an adequate proxy for biodiversity, the presence of these two rare animals nesting within breadfruit agroforestry clearly establishes the potential of increased habitat. Preserving biodiversity is an important concept for biocultural restoration, especially in agricultural development where the loss of biodiversity and environmental destruction are attributed to conventional agriculture methods such as monocropping [4].

Ultimately, breadfruit cultivation provides multiple environmental benefits under current production practices. Tree crops in general have shown to have multiple benefits and those impacts are enhanced in agroforestry settings. Forests are valuable systems needed to carry out ecosystem services essential to sustain the natural resource needs of society [5]. Agroforestry has the ability to replenish litter layer to the soil, increase and maintains soil quality, retain soil nutrients and prevent soil erosion [56,57]. For example, growers had, on average, very high levels of soil organic matter (mean 15.25%, standard error 1.38). These same functions improve water filtration and retention rates, increase water availability, reduce runoff and evapotranspiration, collectively improving water quality [56,57]. The shade and presence of multiple tree species also provide habitat, wind protection and essential microclimates that assure cooler summers and warmer winters [57]. The incorporation of increased biodiversity into farming systems at different trophic levels increases crop yields, wood production, yield stability, presence of natural pollinators and encourages weed and pest suppression [58]. The combination of low inputs, low labor, high productivity and ecosystem services benefits showcase the importance of breadfruit as a sustainable crop.

4. Conclusions

Breadfruit holds a unique position in the concept of biocultural restoration in Hawai'i by contributing to the five key themes described above. As a productive, low maintenance crop, 'ulu has the ability to be grown sustainably, which offers the dual role of food production and ecosystem restoration. Increasing availability of healthy foods in the market and through food sharing contributes to greater food security. The continued cultivation of breadfruit is reliant on the continuation of biocultural relationships, or passing knowledge on its growing needs and its importance as a food. Through these cycles, the cultivation of breadfruit in Hawai'i is supported by Hawaiian culture and Hawaiian culture is supported through breadfruit cultivation, where the restoration of one can affect the other (Figure 2).

Although we presented some of the key findings organized within reoccurring themes, these themes clearly overlap and interact with each other. The revival of customary values, for instance, brings forward practices that contribute to the development of community capital, which in turn provides direct benefits to food security through multiple direct and indirect effects. These interactive effects demonstrate, among other things, the intertwined nature of biocultural systems, in which the growth of one necessitates and requires the growth of the other. In multiple examples, we see both the positive and negative reinforcement that occurs within these systems. In a destructive loop, "Farmer F" was not able to receive cultural knowledge regarding breadfruit because the tree itself was not present in his mentors' lives. Conversely, through growing the tree he has recognized how it has contributed to the growth of his culture and identity within himself. As breadfruit cultivation continues to expand its inseverable connection to the Hawaiian culture will continue to fuel the growth of associated value systems. Scaling up, there is evidence that relatively small farms that practice diversified agriculture could increase food production globally, while simultaneously addressing sustainability issues regionally. Referring to Figure 2, traditional agriculture in Hawai'i is a major mediator between biological and cultural components and at the intersection of the three concepts of sustainability. This is where agriculture is significant in regard to sustainable socioecological systems. Breadfruit cultivation through traditional agriculture methods supports the needs of sustainable agriculture and can provide valuable ecosystem services while simultaneously producing food. From the social and cultural side, breadfruit is present in traditional practices that are necessary to sustain its cultivation long term.

Our findings suggest the support of breadfruit production has the potential to be a pivotal solution to sustainability issues in Hawai'i. Evaluating the experiences and lessons learned from discussions with farmers, we saw that breadfruit is making significant contributions to reconnecting people with place, revitalizing traditional agriculture and Hawaiian culture and that this phenomenon has practical implications for the future. While we cannot say if breadfruit itself promotes these activities or if individuals who engage in these practices are more prone to grow breadfruit, the biocultural restoration of breadfruit can be strongly linked to the furthering of a sustainable socioecological system regardless.

As stated by "Farmer H,"

"We have started calling breadfruit a 'solutionary' food. There are so many problems with our food system. I mean there is the health and nutrition, there are the environmental issues, there is food justice and food security. And there is just the loss of identity and enjoyment of food. So many issues that need to be fixed. We need a revolution of the way we deal with food. What is amazing about breadfruit is that it hits all of these. I mean, more than any other food I can think of. That it is a tree. That it is embedded in the culture. That it is a nutritious staple. It really has the potential to be a solution to many problems in our food system. So, it is the revolutionary solution that we need...'solutionary.' Get it?"

Author Contributions: Conceptualization, N.L.; Methodology, N.L.; Software, N.L.; Validation, N.L., B.L.; Formal Analysis, B.L., N.L.; Investigation, B.L., N.L.; Resources, N.L.; Data Curation, B.L., N.L.; Writing-Original Draft Preparation, B.L.; Writing-Review & Editing, N.L., B.L.; Visualization, B.L., N.L.; Supervision, N.L., B.L.; Project Administration, N.L., B.L.; Funding Acquisition, N.L.

Funding: This research was funded by the USDA Western SARE project, grant number SW17-050 and the USDA National Institute of Food and Agriculture, Hatch project HAW08035 administered by the College of Tropical Agriculture and Human Resources. The APC was funded by Hawai'i Community Foundation.

Acknowledgments: We would like to acknowledge all who offered their time, support and knowledge. This includes the breadfruit farmers who participated in the interviews, staff, interns and volunteers who helped with data collection, input, travel, sample collection, storage and processing including; Amber Au, Kat Hiu, Kimber Troumbley and Maxwell Bendes. We would like to thank the Hawai'i 'Ulu Producers Cooperative for connecting us with local 'ulu farmers. Thank you to Matthew Lucas for technical Geographic Information Systems Assistance. Thank you to friends and colleagues who edited and revised; Maxwell Bendes, Kalisi Mausio, Amber Au and Kawika Winter.

Conflicts of Interest: The authors declare no conflict of interest.

Appendix A

Table A1. List of survey question reported on in this paper as administered to 43 successful breadfruit cultivators in Hawai'i.

1.	What is the size of your farm in acres?
2.	What is the number of acres in breadfruit?
3.	What is the number of trees on your property?
4.	What immediately surrounds your breadfruit trees?
5.	What surrounds your farm?
6.	What type of breadfruit varieties do you grow?
7.	Where are your trees from?
8.	How were your trees propagated?
9.	What form did you get your trees?
10.	Describe the planting method used for your trees.
11.	Do you use other methods to ensure the success of a young plant?
12.	Did you use wind and/or sun protection when establishing your orchard?
13.	What was the survival rate of your plantings?
14.	What is the planting pattern of your trees?
15.	Can you describe the spacing?
16.	How many trees are you growing of each variety?

Table A1. *Cont.*

17.	Do you prune your trees?
18.	At what height do you prune your trees?
19.	What time of year do you prune your trees?
20.	Anything else you'd like to share about how you prune your trees?
21.	How much sunlight did your trees get at establishment?
22.	How much sunlight did your trees get now?
23.	Can you estimate production as pounds per month in pounds or percent annual production?
24.	What was your most productive month?
25.	Can you describe your production seasonally?
26.	What was your production (in pounds) per tree annually in: 2014, 2015, 2016?
27.	Anything else you'd like to share about your production?
28.	Do you use any of the following to assess tree health: Immature Fruit Drop, Leaf Color, Leaf Amounts, Other?
29.	How do you know when you need to do the following or do you follow a schedule for Irrigation, Fertilization, Pruning?
30.	Do you fertilize?
31.	What type/brand?
32.	How often and what time(s) of year?
33.	How much?
34.	Do you use compost?
35.	How often?
36.	How much?
37.	What time(s) of year?
38.	Do you mulch?
39.	What type?
40.	How often and what time of year?
41.	How much?
42.	Do you use soil amendments?
43.	Have you had soil tests done for fertility?
44.	Have you had tissue testing done for nutrients?
45.	Do you use additional nutrient management practices? Please describe.
46.	Do you irrigate?
47.	If yes, why?
48.	How much?
48.	How often?
50.	What time of day?
51.	If no, why not?
52.	Do you use additional water management strategies? Please explain.
53.	Have you noticed any pests?
54.	If yes, what kind?
55.	Do you use insecticides?
56.	What type?
57.	How often?
58.	How much?
59.	Do you use other methods to control pests?
60.	Have you notice any of the following diseases?
61.	Do you use herbicides?
62.	What type?
63.	How often?
64.	How much?
65.	Do you use string trimmers, mowing or ground cover?
66.	Do you use cover crops?
67.	If yes, what types?
68.	Do you regularly remove fallen branches, fallen fruit and fallen leaves?
69.	Do you use additional weed or pest management practices?

Table A2. List of general interview topics that were explored with willing participants.

Interview Questions

1. What were your motivations to grow breadfruit?
2. What do you think about the production of breadfruit throughout the state?
3. How do you cook and eat breadfruit?
4. Who do you sell to/share your breadfruit with?
5. Have you seen more or less breadfruit production and consumption throughout Hawai'i in your life?
6. Would you recommend other farmers to grow breadfruit? If yes, why, if no, why not?
7. What are the benefits and challenges of growing breadfruit?
8. What, if any, additional practices did you or do you use to grow and care for your breadfruit trees?
9. What support would be most beneficial as a farmer to receive from

    a. research in agricultural development for breadfruit?
    b. from government bodies?
    c. your community?

10. What obstacles, if any, do you face producing breadfruit from a

    a. Growers standpoint and
    b. Market stand point?

11. What cultural significance, if any, does breadfruit hold for you?
12. Are you involved in community activities outside of farming that involve breadfruit such as;

    a. Education
    b. Farmers unions
    c. Farmers markets
    d. Community potlucks
    e. Food donations
    f. Agricultural workshops/collaborations with researchers

References

1. Chirico, J.; Farley, G.S. *Thinking Like an Island: Navigating a Sustainable Future in Hawaii*; University of Hawaii Press: Honolulu, HI, USA, 2015.
2. Böhringer, C.; Jochem, P.E. Measuring the Immeasurable—A Survey of Sustainability Indices. *Ecol. Econ.* **2007**, *63*, 1–8. [CrossRef]
3. Maffi, L.; Pretty, J.; Ball, A.S.; Benton, T.S.; Lee, G.J.; Orrm, D.R.; Pfeffer, D.; Ward, M.J. Biocultural Diversity and Sustainability. In *The Sage Handbook of Environment and Society*; Sage Publications: Thousand Oaks, CA, USA, 2007.
4. Pretty, J.; Adams, B.; Berkes, F.; De Athayde, S.F.; Dudley, N.; Hunn, E.; Sterling, E. The Intersection of Biological Diversity and Cultural Diversity: Towards Integration. *Conserv. Soc.* **2009**, *7*, 100–112.
5. Barthel, S.; Crumley, C.; Svedin, U. Bio-Cultural Refugia—Safeguarding Diversity of Practices for Food Security and Biodiversity. *Glob. Environ. Chang.* **2013**, *23*, 1142–1152. [CrossRef]
6. Pretty, J. Interdisciplinary Progress in Approaches to Address Social-Ecological and Ecocultural Systems. *Environ. Conserv.* **2011**, *38*, 127–139. [CrossRef]
7. Vaughan, M.B.; Ayers, A.L. Customary Access: Sustaining Local Control of Fishing and Food on Kauai's North Shore. *Food Cult. Soc.* **2016**, *19*, 517–538. [CrossRef]
8. Wehi, P.M.; Lord, J.M. Importance of Including Cultural Practices in Ecological Restoration. *Conserv. Biol.* **2017**, *31*, 1109–1118. [CrossRef] [PubMed]
9. Godfrey, H.C.; Beddington, J.R.; Crute, I.R.; Haddad, L.; Muir, J.F.; Pretty, J.; Robinson, S.; Thomas, S.M.; Toulmin, C. Food Security: The Challenge Of Feeding 9 Billion People. *Science* **2010**, *327*, 812–819. [CrossRef] [PubMed]
10. Zerega, N.; Wiesner-hanks, T.; Ragone, D. Diversity in the Breadfruit Complex (*Artocarpus*, Moraceae): Genetic Characterization of Critical Germplasm. *Tree Genet. Genomes* **2015**, *11*, 1–26. [CrossRef]

11. Meilleur, B.A.; Jones, R.R.; Tichenal, C.A.; Huang, A.S. *Hawaiian Breadfruit: Ethnobotany, Nutrition, and Human Ecology*; College of Tropical Agriculture and Human Resources, University of Hawaii at Manoa: Honolulu, HI, USA, 2004.

12. McCoy, M.D.; Graves, M.W.; Murakami, G. Introduction of Breadfruit (*Artocarpus altilis*) to the Hawaiian Islands. *Econ. Bot.* **2010**, *64*, 374–381. [CrossRef]

13. Lincoln, N.K.; Vitousek, P.M. *Indigenous Polynesian Agriculture in Hawaii*; Oxford University Press: Evans Road Cary, NC, USA, 2017.

14. Handy, E.S.; Handy, E.G.; Pukui, M.K. *Native Planters in Old Hawai'i: Their Life, Lore, and Environment*; Bishop Museum Press: Honolulu, HI, USA, 1972.

15. Lincoln, N.K.; Ragone, D.; Zerega, N.; Roberts-Nkrumah, L.B.; Merlin, M.; Jones, A.M. Grow Us Our Daily Bread: A Review of Breadfruit Cultivation in Traditional and Contemporary Systems. *Hortic. Rev.* **2018**, *46*, 299–384.

16. Lincoln, N.; Ladefoged, T. Agroecology of Pre-Contact Hawaiian Dryland Farming: The Spatial Extent, Yield and Social Impact of Hawaiian Breadfruit Groves in Kona, Hawai'i. *J. Archaeol. Sci.* **2014**, *49*, 192–202. [CrossRef]

17. Allen, M.S. Bet-Hedging Strategies, Agricultural Change, and Unpredictable Environments: Historical Development of Dryland Agriculture in Kona, Hawaii. *J. Anthropol. Archaeol.* **2004**, *23*, 196–224. [CrossRef]

18. Kurashima, N.; Kirch, P.V. Geospatial Modeling of Pre-Contact Hawaiian Production Systems on Molokai Island, Hawaiian Islands. *J. Archaeol. Sci.* **2011**, *38*, 3662–3674. [CrossRef]

19. Newman, T.S. Hawaiian Agricultural Zones Circa A.D. 1832: An Ethnohistory Study. *Ethnohistory* **1971**, *18*, 335–351. [CrossRef]

20. Kelly, M. *Na Mala o Kona: Gardens of Kona*; Bernice P. Bishop Museum: Honolulu, HI, USA, 1983.

21. Whistler, W.A. *Plants of the Canoe People*; National Tropical Botanical Gardens: Kalaheo, HI, USA, 2009.

22. Winter, K.B.; Lincoln, N.K.; Berkes, F. The Social-Ecological Keystone Concept: A Quantifiable Metaphor for Understanding the Structure, Function, and Resilience of a Biocultural System. *Sustainability* **2018**, *10*, 3294. [CrossRef]

23. Jones, A.M.P.; Murch, S.J.; Ragone, D. Diversity of Breadfruit (*Artocarpus altilis*, Moraceae) Seasonality: A Resource for Year-Round Nutrition. *Econ. Bot.* **2010**, *64*, 340–351. [CrossRef]

24. O'Connor, K. The Hawaiian Luau. *Food Cult. Soc.* **2015**, *11*, 149–172. [CrossRef]

25. Turi, C.E.; Liu, Y.; Ragone, D.; Murch, S.J. Trends in Food Science & Technology Breadfruit (*Artocarpus altilis* and Hybrids): A Traditional Crop with the Potential to Prevent Hunger and Mitigate Diabetes in Oceania. *Trends Food Sci. Technol.* **2015**, *45*, 264–272.

26. MacCaughey, V. The Genus *Artocarpus* in Hawaii. *Torreya* **1917**, *17*, 33–49.

27. Krauss, B.H. *Ethnobotany of Hawaii*; Department of Botany, University of Hawaii: Honolulu, HI, USA, 1986.

28. Jones, A.M.P.; Klun, J.A.; Cantrell, C.L.; Ragone, D.; Chauhan, K.R.; Brown, P.N.; Murch, S.J. Isolation and Identification of Mosquito (*Aedes aegypti*) Biting Deterrent Fatty Acids from Male Inflorescences of Breadfruit (*Artocarpus altilis* (Parkinson) Fosberg). *Agric. Food Chem.* **2012**, *60*, 3867–3873. [CrossRef] [PubMed]

29. Brookfield, H. Intensification Intensified: Prehistoric Intensive Agriculture in the Tropics. *Archaeol. Ocean.* **1986**, *21*, 177–180. [CrossRef]

30. Cachola-Abad, C.K. *An Analysis of Hawaiian Oral Traditions: Descriptions and Explanations of the Evolution of Hawaiian Socio-Political Complexity*; University of Hawaii at Manoa: Honolulu, HI, USA, 2000.

31. Kirch, P.V. Hawaii as a Model System for Human Ecodynamics. *Am. Anthropol.* **2007**, *109*, 8–26. [CrossRef]

32. Kirch, P.V. The Evolution of Sociopolitical Complexity in Prehistoric Hawaii: An Assessment of the Archaeological Evidence. *J. World Prehistory* **1990**, *4*, 311–345. [CrossRef]

33. Kirch, P.V.; Zimmerman, K.S. *Roots of Conflict*; School for Advanced Research Press: Santa Fe, NM, USA, 2011.

34. Malo, D. Concerning the Makahiki. In *Hawaiian Antiquities*; Malo, D., Ed.; Bishop Museum Press: Honolulu, HI, USA, 1951.

35. Handy, E.S.C.; Emory, K.P.; Bryan, E.; Buck, P.H.; Wise, J. Feasts and Holidays. In *Ancient Hawaiian Civilization Revised Edition*; Charles, E., Ed.; Turtle Company: Ibaraki, Japan, 1965; pp. 62–68.

36. Kirch, P.V. Cultural Adaptation and Ecology in Western Polynesia: An Ethnoarchaeological Study. In *Indigenous Crops and Cropping Systems Seminar*; Tropical Plant and Soil Sciences, University of Hawaii: Honolulu, HI, USA, 2017.

37. Beckwith, M.W. *Hawaiian Mythology*; University of Hawaii Press: Honolulu, HI, USA, 1940.

38. Pukui, M.K. *'Ōlelo Nō'eau: Hawaiian Proverbs and Poetical Sayings*; Bishop Museum Press: Honolulu, HI, USA, 1983; Volume 71.

39. Stannard, D.E. *Before the Horror: The Population of Hawai'i on the Eve of Western Contact*; University of Hawaii Press: Honolulu, HI, USA, 1994.

40. Swanson, D.A. The Number of Native Hawaiians and Part Hawaiians. In *Hawai'i, 1778 to 1900: Demographic Estimates by Age, with Discussion*; University of California: Riverside, CA, USA, 2015.

41. Loke, M.; Leung, P. Hawaii's Food Consumption and Supply Sources: Benchmark Estimates and Measurement Issues. *Agric. Food Econ.* **2013**, *1*, 1–18. [CrossRef]

42. Union of Concerned Scientists 50-State Food System Scorecard. Available online: https://www.ucsusa.org/food-agriculture/food-system-scorecard#.W6hw1GhKjIU (accessed on 1 June 2018).

43. Brower, A. Hawaii: GMO Ground Zero—Seeds of Occupation, Seeds of Possibility. Ph.D. Thesis, University of Auckland, Auckland, New Zealand, 2016.

44. Mostafanezhad, M.S. Is Farming Sexy? Agro-Food Initiatives and the Contested Value of Agriculture in Post Plantation Hawaii. *Geoforum* **2018**, *97*, 227–234. [CrossRef]

45. Ragone, D.; Elevitch, C.; Dean, A. Revitalizing Breadfruit in Hawaii—A Model for Encouraging the Cultivation and Use of Breadfruit in the Tropics. *Trop. Agric.* **2016**, *93*, 213–224.

46. Ragone, D.; Cavaletto, C. Sensory Evaluation of Fruit Quality and Nutritional Composition of 20 Breadfruit (*Artocarpus*; Moraceae) Cultivars. *Econ. Bot.* **2006**, *60*, 335–346. [CrossRef]

47. Bremmer, L.; Delevaux, J.M.; Leary, J.J.K.; Cox, L.; Oleson, K.L. Opportunities and Strategies to Incorporate Ecosystem Services Knowledge and Decision Support Tools into Planning and Decision Making in Hawaii. *Environ. Manag.* **2015**, *55*, 884–889. [CrossRef] [PubMed]

48. Lincoln, N.K.; Ardoin, N.M. Cultivating Values: Environmental Values and Sense of Place as Correlates of Sustainable Agricultural Practices. *Agric. Hum. Values* **2016**, *33*, 389–401. [CrossRef]

49. Lincoln, N.K.; Ardoin, N. Farmer Typology in South Kona, Hawaii: Who's Farming, How, and Why? *Food Cult. Soc.* **2016**, *19*. [CrossRef]

50. Perroy, R.L.; Melrose, J.; Cares, S. The Evolving Agricultural Landscape of Post-Plantation Hawaii. *Appl. Geogr.* **2016**, *76*, 154–162. [CrossRef]

51. 2012 Census of Agriculture, State Profile, Hawaii. Available online: https://www.agcensus.usda.gov/Publications/2012/Online_Resources/County_Profiles/Hawaii/cp99015.pdf (accessed on 22 August 2018).

52. 2017 State Agriculture Overview: Hawaii. Available online: https://www.nass.usda.gov/Quick_Stats/Ag_Overview/stateOverview.php?state=HAWAII (accessed on 22 August 2018).

53. Chun, M.C. Welina: Traditional and Contemporary Ways of Welcome and Hospitality. In *No Na Māmo: Traditional and Contemporary Hawaiian Beliefs and Practices*; Curriculum Research & Development Group, College of Education, University of Hawaii: Honolulu, HI, USA, 2011; pp. 14–45.

54. Abbott, I.A. *Lā'au Hawai'i: Traditional Hawaiian Uses of Plants*; Bishop Museum Press: Honolulu, HI, USA, 1992.

55. Vaughan, M.B.; Vitousek, P.M. Mahele: Sustaining Communities through Small Scale Inshore Fishery Catch and Sharing Networks. *Pac. Sci.* **2013**, *67*, 329–344. [CrossRef]

56. Gao, J.; Barbieri, C.; Valdivia, C. A Socio-Demographic Examination of the Perceived Benefits of Agroforestry. *Agrofor. Syst.* **2014**, *88*, 301–309. [CrossRef]

57. Garret, H.E.; Rietvald, W.J.; Fisher, R.F. *North American Agroforestry: An Integrated Science and Practice*; American Society of Agronomy, Inc.: Madison, WI, USA, 2000.

58. Isabell, F.; Adler, P.R.; Eisenhauer, N.; Fornara, D.; Kimmel, K.; Kremen, C.; Letourneau, D.K.; Leibman, M.; Polley, H.W.; Quijas, S.; et al. Benefits of Increasing Plant Diversity in Sustainable Agroecosystems. *J. Ecol.* **2017**, *105*, 871–879. [CrossRef]

Review

I Ke Ēwe 'Āina o Ke Kupuna: Hawaiian Ancestral Crops in Perspective

Aurora Kagawa-Viviani [1], Penny Levin [2,*], Edward Johnston [3,4], Jeri Ooka [4], Jonathan Baker [4,5], Michael Kantar [6] and Noa Kekuewa Lincoln [6,*]

[1] Department of Geography and Environment, University of Hawai'i at Mānoa, Honolulu, HI 96822, USA; kagawa@hawaii.edu
[2] E kūpaku ka 'āina, Wailuku, HI 96793, USA
[3] 'Alia Point 'Awa Nursery, Pepe'ekeo, HI 96783, USA; aliapoint@gmail.com
[4] Association for Hawaiian 'Awa, Pepe'ekeo, HI 96783, USA; jeri@hawaii.edu (J.O.); jonathan.baker@chaminade.edu (J.B.)
[5] Division of Natural Sciences and Mathematics, Chaminade University of Honolulu, Honolulu, HI 96822, USA
[6] Department of Tropical Plant and Soil Sciences, University of Hawai'i at Mānoa, Honolulu, HI 96822, USA; mbkantar@hawaii.edu
* Correspondence: pennysfh@hawaii.rr.com (P.L.); nlincoln@hawaii.edu (N.K.L.)

Received: 19 August 2018; Accepted: 28 November 2018; Published: 5 December 2018

Abstract: Indigenous crops, tremendously valuable both for food security and cultural survival, are experiencing a resurgence in Hawai'i. These crops have been historically valued by agricultural researchers as genetic resources for breeding, while cultural knowledge, names, stories and practices persisted outside of formal educational and governmental institutions. In recent years, and following conflicts ignited over university research on and patenting of kalo (Hāloa, *Colocasia esculenta*), a wave of restoration activities around indigenous crop diversity, cultivation, and use has occurred through largely grassroots efforts. We situate four crops in Hawaiian cosmologies, review and compare the loss and recovery of names and cultivars, and describe present efforts to restore traditional crop biodiversity focusing on kalo, 'uala (*Ipomoea batatas*), kō (*Saccharum officinarum*), and 'awa (*Piper methysticum*). The cases together and particularly the challenges of kalo and 'awa suggest that explicitly recognizing the sacred role such plants hold in indigenous worldviews, centering the crops' biocultural significance, provides a foundation for better collaboration across multiple communities and institutions who work with these species. Furthermore, a research agenda that pursues a decolonizing approach and draws from more participatory methods can provide a path forward towards mutually beneficial exchange among research, indigenous, and farmer communities. We outline individual and institutional responsibilities relevant to work with indigenous crops and communities and offer this as a step towards reconciliation, understanding, and reciprocity that can ultimately work to create abundance through the restoration of ancestral crop cultivar diversity.

Keywords: cultural revitalization; indigenous knowledge; taro; sweet potato; kava; sugarcane; research ethics; restoration

---

1. Introduction

Agriculture throughout the world emerged as communities managed and selected wild crop relatives to produce dependable staple food resources. Over time, thousands of crop cultivars were developed, unique to the complex geographies of place and culture, evolving multi-layered, inseparable biocultural relationships between plants and people, woven together through cosmologies and genealogies that name core crop plants as ancestors and gods (e.g., [1–3]). Within this ethno-biological context, intricate agricultural systems emerged, hefted and slipped into ecological flows and topographical and geological boundaries [4–6]. As people migrated across the world, their accompanying plants and cultural practices both diversified and narrowed dietary choices [7–9]. Within the Pacific, the journeys of core crops such as kalo (taro; *Colocasia esculenta*), 'uala (sweet potato; *Ipomoea batatas*), mai'a (banana, *Musa spp.*), 'awa (kava, *Piper methysticum*), kō (sugarcane, *Saccharum officinarum*), uhi (yam, *Dioscorea alata*), 'ulu (breadfruit, *Artocarpus altilis*) and niu (coconut, *Cocos nucifera*) illuminate connections among culturally related groups.

Commencing in the 15th century, colonial institutions shaped by European epistemologies launched fleets of "discovery" into indigenous territories in part to better understand the world through inventory and inquiry and to identify natural resources for markets at home [10]. Prominent among extracted resources were wild and cultivated plants valued for foods, pharmaceuticals, and spices and resources; indigenous cultivars provided useful new crops or traits to improve existing crop production, which continues to the present day. The collection of indigenous crop cultivars has long played a vital role in their preservation in Hawai'i and elsewhere, where they are maintained in ex situ (off-farm) collections, often for the purposes of breeding (e.g., Secretariat of the Pacific Community Centre for Pacific Crops and Trees in Fiji, University of Hawai'i College of Tropical Agriculture and Human Resources, UH CTAHR). Yet introductions of these crops to new localities and their use in breeding and "crop improvement" programs has had, in some cases, unforeseen consequences, including contributing to the decline of local indigenous crop cultivar diversity [11] or the rise of traits conferring invasiveness [12].

Furthermore, germplasm collection for breeding and genetic manipulation can be experienced by indigenous communities as an ongoing expression of historical injustice, specifically as theft and assault on community-stewarded resources, giving rise to conflicts among researchers, farmers, and indigenous groups [13–15]. In recent decades, explicit concerns have been raised over bioprospecting and biopiracy of traditional plants and their associated knowledge systems, both in Hawai'i and elsewhere. While corporate and institutional appropriation of germplasm and benefits continues, it is also being challenged in several arenas through global treaties and local declarations such as the 1995 Treaty and Related Protocols for a Lifeforms Patent Free Pacific [16], the 2001 International Treaty on Plant Genetic Resources for Food and Agriculture (ITPGRFA) [17], and the 2003 Paoakalani Declaration [18]. Specifically, Article 9 of the ITPGRFA (which was an outgrowth of the Convention on Biological Diversity (CBD), and the Nagoya Protocol on Access and Benefit Sharing) outlines farmers' rights, explaining and recognizing the unique contribution of farmers to the conservation and creation of plant types that contribute to local and global food security. While such global treaties present a useful framework, these documents are non-binding, and institutions that outline best practices, such as the United Nations (UN), the UN Educational, Scientific and Cultural Organization (UNESCO), International Union for Conservation of Nature (IUCN), and the CBD lack the power to enforce them. Complicating this issue is that public funds and institutions continue to support patenting/trademarking and research approaches considered inappropriate by many members of indigenous and farming communities; along with a range of perspectives on crops research among scholars, including indigenous scholars.

The history of indigenous knowledge and crop biodiversity exploitation provides a backdrop for this paper which reviews the process of restoring kalo, 'uala, kō, and 'awa, four Hawaiian crop plants, to contemporary landscapes and communities across Hawai'i. In our restoration efforts and writing, we find the cases fall within a biocultural approach defined by Gavin et al. (2015) as "actions made in the service of sustaining the biophysical and sociocultural components of dynamic, interacting and interdependent social–ecological systems." [19] In describing the cultural significance of the crops, we base our analyses on the kupuna crops themselves (kupuna refers to ancestor or starting point). Furthermore, we draw on Linda Tuhiwai Smith's articulation of 25 "indigenous projects" [20] to illustrate how restoration efforts may align with decolonizing methodologies. Smith [20] proposes such projects as a way to re-indigenize research focused on cultural knowledge and respect indigenous boundaries. Twenty years after its publication, it also provides a measure for assessing what, if any, changes have occurred between research institutions, indigenous crops, farmers and Hawaiians over time. Of the 25 projects, we focus on *remembering, returning, revitalizing, restoring, protecting, celebrating, sharing, connecting* and *envisioning*. We add *resilience* and *abundance* to the framework to capture changing perspectives and approaches. Thus, we operationalize the notion of indigenous projects by reviewing the history of ancestral crop development in Hawai'i, exploring the cultural significance and recent resurgence of four kupuna crops with respect to the 'projects,' and analyzing how the cultural significance of these plants can guide ongoing restoration work both ex situ and in situ (on the land).

We begin by first *remembering*. Our first section provides a brief review of Hawaiian agricultural history and context for the current resurgence. We then share examples of four Hawaiian crop plants: kalo, 'uala, kō, and 'awa by first situating each plant in its Hawaiian context of kinship, relationship, and use. We *remember* specific histories of growth and decline, describe the *return* and *revitalization/regeneration* of names and cultivars themselves through collaborations and the *restoration* of the webs of relationships that accompany these plants, and *celebrate* this restoration (Table 1). We move forward towards resilience and abundance through this re-storying and new planters on the land. In a synthesis of our four cases, we *connect* and compare the cases to each other and lessons from other communities working to restore their biocultural heritage. Finally, we *share* lessons learned as we *envision* future research and agricultural approaches that center biocultural perspectives and employ participatory methods to enable effective collaboration between farmers/indigenous communities and researchers in the work to restore kupuna crops.

In focusing this paper on the need to respect and restore these crops in alignment with cultural values, we pay specific attention to the role of public research institutions in conducting pono (ethical, balanced) decision-making and research and extension initiatives as they intersect with biocultural resources and indigenous and local communities. Our examples bring forward an alternative paradigm of partnership. Thus, we suggest new, specific, and focused frames are needed to guide work on indigenous crops, informed by better understanding where indigenous, farmer, and researcher objectives collide or overlap constructively at the contact zone of indigenous crops research.

2. Background: Ancestral Crops in the Hawaiian Landscape

2.1. Early Hawaiian Agriculture and Crop Trajectories

Polynesians settled in the Hawaiian Islands an estimated 1000 years ago likely from within the Society Islands [21,22] arriving with a group of ~25 plants for food and other uses [23,24] including some that remained essential staples in the Hawaiian diet—kalo, mai'a, niu, 'awa and kō. Evidence suggests 'ulu and 'uala arrived in the islands with later oceanic migrations perhaps around the time of Pa'ao (circa 1300 CE) [25–29] while less is known about the arrival and role of uhi (yam) in early Hawaiian food security. From a limited number of Polynesian founder cultivars, Hawaiians developed an estimated 300-400 varieties of kalo, ~250 'uala, 40 mai'a, 35 'awa and 50 kō, numbers based on recovered germplasm and collected names and accounting for various sources of duplication [30–38]. Just nine cultivar names for uhi and two for niu are recorded by Handy and Pukui [33,39], while 'ulu is represented by only a single Hawaiian cultivar [28]. The primarily vegetative propagation of most these plants implies that Hawaiian cultivars persisting today carry much of the same genetic material as the plants that first arrived in the islands [40].

Archaeological fieldwork and geographic information system (GIS)-based models of the extent and trajectory of traditional agriculture reveal a rich landscape of highly productive farming covering at least 250,000 acres, including immense dryland and wetland agricultural terrace systems at the likely peak of the Hawaiian population in the 1700s [41–43]. These were sustained by diverse Hawaiian soil fertility management strategies [39,44,45] and cultivated according to celestial/seasonal and lunar cycles [46–48]. Water and nutrient flows and topography connected irrigated inland agricultural fields and nearshore fishponds within a cohesive traditional ahupua'a and land tenure system [6,39].

By some estimates, these landscapes supported between 300,000 and 800,000 people [49,50] prior to foreign contact.

Between 1778 and 1900, hundreds of thousands of Hawaiians died from foreign diseases taking with them many of the masterful agriculturalists of their time [50,51]. Wetland taro production declined from ~50,000 acres to ~30,000 acres by the end of the 19th century [41], while expansive dryland field systems were abandoned or became cattle lands [52]. Sugar plantations consumed agricultural lands on each island, securing crop growth through vast amounts of surface and ground water diverted from streams, taro fields and fishpond systems [53–55]. The 1893 overthrow shifted de facto governance of the Hawaiian Kingdom and its lands and people into the hands of American business interests and eventually the United States government [56]. The ensuing changes accelerated the erosion of the traditional Hawaiian land tenure system, making the conversion of agricultural lands and waters seemingly permanent [57–60]. The fight to return water to streams and taro systems has persisted for 150 years [61].

Agricultural systems not supplanted by commercial sugarcane shifted from kalo to rice, from 'uala to vegetable crops, and from local to distant markets. By 1939, barely 1200 acres remained in wetland taro [62]; commercial acreage has hovered around 400 acres since 1965 [63] (HASS/NASS data 1946–2016). Handy [33,39] reported widespread but declining subsistence 'uala cultivation during his surveys in the 1930s, and today, the 500 acres of commercial sweet potato production include few if any Hawaiian cultivars [64,65]. Hawaiian bananas' role as a staple starch was almost completely lost, while uhi, niu, and 'awa remained in the shadows, and the extensive 'ulu agroforests that once banded areas of Maunaloa on Hawai'i Island and covered large swaths of the southern coasts of West Maui were reduced to bare remnants [54,66]. Despite 200,000 acres at the peak of sugar [67], native kō cultivars survived almost exclusively in germplasm collections and botanical gardens. Further influencing this once abundant landscape was a demographic shift from predominantly Hawaiian farmers to other ethnic groups descended from plantation laborers. Through the end of WWII, Hawai'i maintained its ability to feed itself. Today, 90 percent of food consumed in the Hawaiian Islands is imported [68].

The complex influences that altered Hawai'i's endemic ecosystems, Hawaiian society, and its fine-tuned agricultural systems over the last 200 years are well-described [43,69]. Yet through the memories of kupuna (elders) raised in the language and traditions of Hawaiian agriculture, the land that holds onto the remnants of terraces and crop cultivars, and through acts of intentional restoration, ancestral gifts and knowledge endure [70,71].

2.2. A Renaissance of Hawaiian Agricultural Crops

The Hawaiian Renaissance of the 1970s precipitated a movement to reclaim identity, language, navigation skills [72], cultural practices, health, education [73], and resource governance. Struggles over sovereignty, burial grounds [74], military occupation [70], and land and water rights [75] fueled a return to traditional foods through restoration of lo'i (wetland taro fields) [76], loko i'a (fishponds) [77], and the practice of making poi (the staple Hawaiian food made from cooked kalo), including at the University of Hawai'i at Mānoa (UH Mānoa). The era birthed a new generation of taro farmers and students who returned to the land [78–80].

Several decades later, within the context of growing Hawaiian and farmer empowerment, clashes over kalo patenting and genetic manipulation by UH Mānoa researchers opened a wide gap between the institution, Native Hawaiians and farmers while simultaneously fueling a community-based renaissance in traditional Hawaiian agriculture [81], thereby forcing the University to consider the significance of native crops in Hawai'i. Thus, kalo has been at the center of conflict and the stimulus of new growth, especially among small-scale commercial and subsistence taro farmers.

We explore this renaissance through four indigenous crop examples that center a biocultural perspective and review key events in each crop's recovery and restoration that provide lessons for all indigenous crop recovery efforts. We begin with kalo given its tremendous significance in Hawai'i; it also provides a perspective on the revival of ʻuala, kō and ʻawa. In each case, the cultural value of these crops extends far beyond their monetized value as market commodities. Each example demonstrates the proliferation of work to restore crop biodiversity, both in collections and on farm, and suggests new directions for inquiry.

3. Nā Kūpuna: Kalo, ʻUala, Kō, and ʻAwa—Four Ancestral Hawaiian Crops

3.1. Kalo–A Return to the Source and the Societal Role of a Staple Food

3.1.1. Guardians on the Land

Kalo (taro, *Colocasia esculenta*) is grown throughout much of the world as a staple starch root crop and a vegetable green and considered one of the oldest cultivated food crops (ca 10,000 cal BP, [82,83]). Molecular evidence indicates *Colocasia esculenta* was domesticated separately in both Asia and the Pacific [84,85] with the highest genetic diversity (DNA) found in India and the highest clonal diversity (cultivars) in Southeast Asia and the Pacific [86]. Predominantly diploid cultivars from this region provide the basis for Hawaiian kalo cultivar diversity [87,88]. This botanical lineage is recognized in some of the most important genealogical chants of the Hawaiian Islands, together with the plant's cultural significance which begins side by side with the birth of man.

There are several cosmological stories, such as the Kumuhonua, Mele a Pākui and the Kumulipo, a more than 2000 line chant, [3,25] that parallels scientific theory in the primordial birthing of the organisms, the creation of the world, the islands, and the birth of Hāloa. In the very first wā (era) of the *Kumulipo*, the distant red Kalo Manauea, a reference to a South Pacific taro, appears as a kiaʻi, a guardian on the land, to the manauea of the sea (a smoky red seaweed) [3,25]. The "dusky black ʻApe" appears in the fourth wā, another important food crop member of the Aroid family to which *Colocasia esculenta* belongs. Here, ʻApe is the kapu (taboo) chiefly progenitor, setting the stage for the birth of Hāloa (the kalo) into Hawaiʻi and a symbolic link to Kahiki (the ancient past; Tahiti) [89].

Hāloa-naka-lau-kapalili, the plant, is born in the 13th wā of the *Kumulipo*, emerging from the burial place of the stillborn child of the gods Wākea and Hoʻohōkūkalani [25]. Hāloa, the man, is born, and the chiefly ancestors from which Hawaiians descend follow; the two (kalo and man) are bound together as brothers through parentage and a profound reciprocal kiaʻi relationship. Each feeds the other; both survive. Out of this cosmology emerges the sacred trust that is the foundation of Hawaiian agriculture and cultural identity. Hawaiian oral traditions and written literature link kalo to the principal gods Kāne and Lono, associated with fresh water and agriculture, and the demi-god, Kānepuaʻa [25,33,90–92]. The plant is food, medicine, dye, feed for nearshore fish, tax payments to the chiefs, and offerings for the gods [25,46,62,93,94]. Kalo is also replete with symbolism. From the structure of the parent corm and side shoots (nā ʻohā) comes the word, ʻohana (family) [95]; from its lineage, hardiness and its waving leaves have come a flag of resistance, especially among taro farmers for the return of water to their steams [96]; the sap of the kalo is referred to as koko (blood) reminding us of the sibling connection [89].

3.1.2. Remembering: Dissonant Histories from Collection and Conservation to Hybrids and Genetic Engineering

In the 1920s, driven by the realization that Hawaiian crop biodiversity and agricultural practices were in rapid decline, Gerritt P. Wilder and E.S. Craighill Handy, Bishop Museum researchers, initiated the first formal taro germplasm collections in Hawaiʻi, including cultivars gathered from the Pacific [62]. The fledgling UH Agricultural Experiment Station received its original kalo collection in the mid-1930s from these two sources and individual taro farmers. The authors of *Bulletin 84: Taro Varieties in Hawaii* (1939) recognized that cross-breeding and new cultivar selections were proceeding without a systematic nomenclature for existing Hawaiian taro varieties [62] and developed a taxonomic key that became the seminal work on Hawaiian taro varietal identification.

Since the 1940s, the University of Hawai'i's experiment stations have maintained kalo collections, primarily for research and breeding programs. An important exception to this is the Moloka'i Research and Demonstration Farm which maintains a primary focus on cultivar conservation and production for taro, serving for the last 40 years as the largest and most important true-type Hawaiian kalo plant material distribution center in the islands for local farmers. In contrast, the University's Harold Lyon Arboretum kalo collection focuses on germplasm storage, conservation, and botanical study.

During the 1950–70s, additional cultivars were brought from the Pacific to Hawai'i (botanical garden accession records), some of which were distributed to the agricultural experiment stations. A common pattern throughout this collection history was that names of origin, an important source of information, were frequently replaced or left unrecorded. Later, breeder-assigned codes further distanced cultivars from their identities (e.g., P1–P20 for 20 cultivars from Palau).

At this junction, two trends appear in the University's taro research. The first is an increased focus on a variety of commercial targets [97,98] by selecting and promoting a limited number of Hawaiian kalo varieties felt to be best-suited for these purposes. This, along with demand for product uniformity (i.e., purple poi), contributed towards further decline in taro diversity in farmers' fields. By 1990, a single variety, Maui Lehua (a farmer-developed cross between two Hawaiian varieties, Lehua maoli and Moi, [29]) was the preferred choice in commercial markets and hence farmers' fields, developing over time the usual pest and disease challenges associated with monocropping [99–101].

Second, researchers responded to declining taro yields and wetland soil health with fertilizer and breeding trials. Initially, these were among Hawaiian varieties, followed by Pacific x Hawaiian and later Southeast Asian x Hawaiian crosses [102–104]. An outbreak of taro leaf blight (*Phytophthora*) in Samoa cemented the UH College of Tropical Agriculture and Human Resources (CTAHR) emphasis on developing disease-resistant taro, first by conventional cross-breeding. Three such taro hybrids were patented [102] causing an uproar in the Hawaiian community over indigenous rights to germplasm. In 2006, after student and community standoffs at the University President's office and Board of Regents meetings, the patents were torn up and released. Hawaiians rejected initial offers to hand the patents over, clarifying that kalo could not be owned, even by them [105]. Work on development of Hawaiian kalo hybrids continues, along with patenting of ornamental taro crosses [106,107].

In Hawai'i, numerous undescribed and poorly-tracked introductions and crosses present a significant challenge to maintaining indigenous crop biodiversity and knowledge of kalo varieties [61]. Similarities between these introductions and Hawaiian cultivars complicate efforts to retain true-type Hawaiian and Pacific kalo collections and to help farmers differentiate among them. A tendency for invasiveness (aggressive runners) in crop cultivar introductions from outside Hawai'i and among new hybrids has become a serious issue for some kalo farmers (Figure 1).

Figure 1. An almost cormless invasive hybrid taro with Lehua maoli-like characteristics (bronzing in youngest leaves; lilac purple base) removed from a wetland taro patch, Hawai'i (**left**) and runners expressed in field trials in the hybrid, Pa'lehua, one of the patented taros (**right**). Photo by P. Levin 2006/2011.

At the same time the patent issue arose, individual researchers at UH Mānoa were already investigating the development of disease resistant taro through genetic engineering. Genes for rice chitinase, wheat oxalate oxidase, and grapevine stilbene synthase were inserted into taro [108–110], a process considered a violation of the sacred relationship between kalo and kanaka. The very thought of changing the being and genetic structure of an ancestor was unacceptable to many Native Hawaiian farmers and cultural practitioners [111,112]. Moreover, taro farmers and Hawaiians only learned of the project by chance through a brief report [113] rather than pre-research consultation. Clearly expressing that genetic modification was an undesirable solution that carried high socio-economic risk, including the potential to jeopardize their livelihoods and way of life, taro farmers requested that the research be ended and the plants destroyed. Proactive recommendations were made for more relevant and useful agroecological research directions [61]. Taro growers also reached out to department chairs and chancellors at the university, and the Taro Task Force submitted a formal letter to UH's President offering to work with the university to develop guidance for future kalo investigations without response.

In the years-long effort to persuade the University administration, which carried into the Hawai'i State Legislature, Native Hawaiians, taro farmers, and their supporters passionately demonstrated and lobbied for the cultural integrity of kalo, as a plant sacred to the Hawaiian people, to be honored and maintained. University responses tended to polarize the issue, characterizing such advocacy as a threat to the intellectual freedom of researchers. Inherent in the conflict was the absence of a more holistic approach to the complex challenges of taro and taro field health. The overt focus on "fixing the plant," a concept embedded in commercial agriculture, ignored farm conditions, kinships between Hawai'i and the Pacific, potential for economic injustice, and evidence from the Pacific of cross-pollination risks in diploid taros [84,87]. The outcome was a legislated 5-year moratorium on genetic engineering, specifically protecting Hawaiian kalo varieties but keeping the door open for all other taros [114]; it did not end the conflict. The physical, financial, emotional and spiritual costs of the research and subsequent University response for many taro growers and Hawaiians has never been fully quantified, and many remain deeply wounded and distrustful of the University of Hawai'i. The University's mission as a land grant institution was called into question for the disenfranchisement of communities it was meant to serve. This history has also been described and analyzed in several publications from outsider [111,115,116] and insider [112] perspectives. The framing of farmer/Hawaiian science, knowledge and rights in opposition to- and of lesser value than- the expertise and intellectual freedom of researchers during such conflicts stands out both as a wall and an opening where transformation in research and a higher standard of research ethics might occur.

3.1.3. Connecting Voices—Turning the Canoe

At the height of these conflicts, the Hawai'i Department of Agriculture was called on to convene stakeholders and develop a research program addressing "taro security and purity" without genetic engineering [114]. The resulting Taro Security and Purity Task Force was formalized by the Hawai'i State Legislature in 2008 to address the many other issues affecting taro farming in the islands [117]. The task force was unique in that, by statute, it was required to consist of no less than 50 percent taro farmers along with representatives from 'Onipa'a Nā Hui Kalo, a respected statewide taro farmers' organization, three state agencies, and the University of Hawai'i.

The task force spent a year researching and meeting with farmers on each island and developed 87 recommendations which were submitted in a report to the 2010 legislature.. The web-accessible Taro Security and Purity Task Force Report provided a template to help the University better align research initiatives with community-identified needs for kalo [61]. Notably, the development of disease-resistant taro cultivars is absent from the recommendations. Instead, the Task Force asked all stakeholders to consider that "the development of new hybrid taros does not resolve the underlying responsibilities we have to take care of the soil," and, it "encourages a focus on improving taro yields, and pest and disease resistance and reduction through the improvement of soil and water conditions and the study of the preferred conditions of traditional Hawaiian kalo cultivars ... in the search for the most robust matches between taro varieties, farm practices and locations" (p58) [61]. A major recommendation was to "establish policy to guide and encourage taro research that supports taro farmer needs and concerns." While the report represents taro farmer expertise and experiences and is often used by growers to remind agencies of priority issues and established recommendations, informal exchanges suggest most university researchers are unfamiliar with, or even unaware of, these recommendations.

Ku'i Ka Pōhaku, 'Anapa Ke Ahi o Ka Lewa: From the Right to Make Traditional Foods, a Flame

In 2011, the grassroots Legalize Pa'i 'ai Movement fought for and re-established the legal right to pound kalo into pa'i 'ai (the almost waterless stage before poi) and poi using a traditional board and stone and to sell the product directly to consumers (HRS 321-4.7, [118]). This fed an increased demand for Hawaiian kalo varieties in local exchanges, markets and at taro events and raised the revenues to taro growers 200 percent. It also gained the attention of several of Hawaii's rising chefs [119–121], an important factor in expanding consumer demand and improving farmer incomes for lesser known indigenous crops [122,123]. To meet family demands, board and stone workshops and ku'i clubs (poi-making clubs) have proliferated in the islands. Young commercial growers are now planting a more diverse set of Hawaiian taro varieties for these specialty markets rather than monocropping for larger poi mills (Levin, field obs). At least one ku'i club is growing their own kalo supply in the taro patch with established farmers, potentially creating new farmers in the process. As a reminder of the agency and authority of Hāloa, over the last 10 years, an annual poi pounding event at the state capitol (Ku'i at the Capitol) has drawn thousands of taro farmers, cultural practitioners, and students from across the islands in the simple, yet political, act of making food, in recent years preparing 1,000 pound of poi in a single day.

The strength and breadth of the community-based network that enabled the Pa'i 'ai Movement, cultivar protection, the contesting of water allocation decisions that affected taro farmers, and many other efforts can be traced in part to 'Onipa'a Nā Hui Kalo (and its precursors, the Mauka Lo'i and Kalopa'a O Waiāhole), an organization and an idea that providing hands-on experiences in reclaiming and restoring ancient kalo-growing sites and communing around food together also taught the collective power of community. [124,125] Named in honor of Queen Lili'uokalani whose message to Hawaiians during the era of the Overthrow was to be steadfast ('onipa'a), ONHK became a support network for isolated taro farmers; an information exchange ahupua'a to ahupua'a on planting practices, pest and disease management, markets, water and land rights challenges, food security and family wellbeing. Members and participants record that it "awakened laulima" (cooperation) for thousands of youth and community members. [126] Huli (planting material) exchanges, 'ai pono movements (health centered around traditional Hawaiian foods), education from in the lo'i (wetland taro patches) to school gardens and curriculum in schools, and a significant rise in volunteerism on small-scale taro farms (a modern version of collective agriculture labor traditions) are offshoots of this movement. Today, reflecting on the resilience and abundance that the network embodies, the traditions of Hāloa are alive, and are, as Osorio (2016) writes, increasingly "meaningful and instructive" the more that people are involved in the restorative work of 'āina (land) and growing kalo [127].

3.1.4. *Ola Ka Inoa*: The Importance of Identity and the Power of Restoring Names

Early kalo survey work [30,31,39] suggests Hawaiian agriculturalists developed both highly site-specific varieties and more broadly adapted cultivars. The longest-standing and most recent work in Hawaiian kalo biodiversity has documented almost 2000 kalo names from Hawaiian and English language source materials [29]. Even with Whitney, Bowers and Takahashi's (1939) conservative rule of 50 percent duplication in naming, this may more than double Handy's estimates in kalo cultivar diversity (roughly 300–400) to 800 or more unique varieties; a spectacular example of natural cultivar mutation nurtured by human selection and invention. Kalo names reflect keen observational skills, farmer connections to place, often a delicious sense of humor and affirm the poetic beauty of Hawaiian language and the world of the kalo farmer. Kalo names based on the color and patterns of fish were frequent, such as the Mana ʻōpelu kalo with its dark flecking and pinkish cast that mimics the iʻa ʻōpelu (mackerel scad; *Decapterus pinnulatus*; Figure 2).

Figure 2. Kalo Mana ʻōpelu and iʻa ʻōpelu (*Decapterus pinnulatus*). P. Levin 2014 (taro) and J. Konanui 2014 (fish).

Accurately identifying Hawaiian kalo cultivars (approx. 60 remain) relies on recognizing morphological characteristics, still the most practical method for varietal recognition. Thirteen Hawaii-based botanical gardens, university research stations and individuals now maintain kalo assemblages in the islands. Over the last two decades, regular collaboration between taro experts and plant managers has improved identification and shifted some collections from solely conservation to interactive education, active propagation and distribution. With that, a larger public is being reintroduced to the cultivars and their names.

Identification guides have been essential to improving botanical collections. They have also been unexpected catalysts in community-based biocultural restoration efforts. The 1939 publication of *Bulletin 84: Taro Varieties in Hawaii* by the Hawaiʻi Agricultural Experiment Station [62] is an example of how such information can ignite a tenacious pursuit of rediscovery, even for backyard growers. This humble book has remained a beacon for novices of Hawaiian crop cultivar diversity for almost 80 years and has stood as a silent challenge to learn more and do better. Building on the original work of *Bulletin 84*, cultural practitioner and kalo and ʻawa expert Jerry Konanui and others worked tirelessly over the last 40 years to expand, refine, and reverify cultivar descriptions. A lack of original names and descriptions for Pacific island varieties introduced to Hawaiʻi, as well as parentage and descriptions for hybrid cultivars persists.

Kalo and ʻawa varieties' identification work has gone hand in hand in Hawaiʻi under the guidance of Konanui. The frequency of site visits to collections and workshops increased exponentially as a response to the challenges to kalo that arose within and outside the University. Between 2005 and 2015, Konanui shared his knowledge with enthusiasm and passion in more than 100 kalo varieties workshops; average attendance rose from 15 to over 100 people. Demand for kalo varieties and *Bulletin 84* has grown in the process. Farmers and poi-making practitioners, with a focus on local markets, are adding value to their product by sharing what they learn with consumers. Growers are also beginning to experiment and make discerning choices about cultivars for planting based on what the names tell them and where they farm. Thus, the power of kupuna crop names is in the libraries of knowledge they carry. Relearning which cultivars belong where in the landscape expands our food security capacity.

3.1.5. Envisioning: Appropriate Research Directions and Applications of Technology

Refocusing our attention on molecular technologies, we note that phylogenetic and population genetic studies of Hawaiian kalo varieties have lagged behind research on taro in Asia and the Pacific, and current tools are unable to clearly distinguish between individual Hawaiian cultivars within 'ohana groupings or beyond regional relationships in some cases [40,128,129]. Analysis based on ISSR-PCR methods suggest strong connections within the Mana and Manini kalo 'ohana of Hawai'i to those elsewhere in the Pacific [129]. The group names appear in the Tahitian, Samoan, and Hawaiian languages and they bear common characteristics, strengthening the argument. Another recent study appears to identify the Hawaiian cultivars as a unique set of plant material relative to the rest of the Asia and the Pacific [130]. It is critical to note, however, that the validity of all such studies requires that source material is properly identified at the start of research and living material is maintained for reverification; this involves the expertise of indigenous crop cultivar specialists and the careful maintenance of cultivars and points to some of the limitations of genomic technologies without such protocols and relationships in place.

An effective and high-demand application of proven technologies is the tissue culture work being done by the Hawai'i Rare Plant Micropropagation Lab at UH Harold L. Lyon Arboretum (HRPM Lab). The Lab has collaborated for many years with local kalo experts to provide carefully curated, long-term germplasm storage and backup for living collections of Hawaiian kalo cultivars [40]. Helping to safeguard the future of kupuna crops through tissue culture is an example of culturally acceptable technological applications, especially as this contributes to development and maintenance of disease-free material. Accompanied by good land-management practices, it provides tools for strengthening kalo disease resistance from the ground up without necessitating loss of cultivar integrity.

The current network of collaboration between the HRPM Lab, kalo experts, key botanical gardens, agricultural stations, non-governmental organizations (NGOs) and taro growers demonstrates how listening to farmer needs and kupuna crop voices, along with education, has maintained bridges and made invitations possible from the community that are restoring Hawaiian kalo diversity, improving kalo vigor and farmer economics from multiple directions. Shared among them is a respect for kalo origins, a sense of 'ohana, and clear track records of thoughtful engagement in questions of indigenous crops restoration, local food security and kuleana.

Better understanding of the strengths and limitations of various technologies in the rapidly changing field of molecular biology, appropriate application to study questions, and ethical implications will improve research rigor and outcomes. Research attentive to questions of interest to farmers and indigenous communities may improve engagement and acceptance further. Which taro varieties came first to Hawai'i and how they are related to each other and to the rest of the Pacific remains a tantalizing question linked to the oceanic migrations of Hawai'i's past. Such inquiry might also be part of a journey towards reconciliation, improved credibility and research vigor for the University provided the canoe is properly prepared for, built, provisioned and manned.

As kin (elder brother) and akua (god) in the Hawaiian cosmology, kalo is as powerfully symbolic and present today for some as it was a millennium ago. As we plant more kalo and align agriculture research with kuleana, so, too, will greater attention to indigenous science and thought, and relationship with Hāloa grow.

3.2. 'Uala—Reclaiming History to Restore Diversity

3.2.1. Celestial and Oceanic Connections

While Hāloa (kalo) plays a central role in Hawaiian cosmology and present-day understanding of food systems and family system [131], 'uala (sweet potato) plays a central role in traditional food systems for Māori in Aotearoa for whom its celestial whakapapa (genealogy) is common knowledge [132,133]. Although 'uala holds a less revered position in Hawaiian origin stories, the stories of 'uala that emerge provide us a view into connections of Hawai'i and the rest of Oceania. Through the names of deities and ancestors emerge common associations of 'uala with Hina and 'Aikanaka, ancestor of Kaha'i, a famous navigator. In Hāna, Maui, hualani (a reference to an 'uala) is associated with Hinahanaiakamalama (also known as Hinaaimalama or Lonomuku), whose foot is cut off by her husband 'Aikanaka as she leaps toward the moon [1,134]. Veiled references to 'uala are found in the *Kumulipo* as the seeds of Makali'i, Hina, and Lonomuku, and the food of Hinahanaiakamalama (wā 14) [3,25], and later, Hinaaiakamalama is associated with 'Aikanaka, Hema, Kaha'i, and Wahieloa (wā 16) [3,25], navigators referenced in other traditions across Oceania [1,48,135]. These names allude to 'uala's association with celestial bodies, lunar cycles, seasonal planting, and navigation. In yet another epic, following a destructive tsunami of Kahina'aimalama and Kahinali'i, the priest Kanalu opens his malo and finds kalo, 'uala, kō, mai'a and 'uhi which he plants and prays to grow, spread, and repopulate the land [136].

The presence of sweet potato across Oceania has fueled academic discussion among anthropologists for decades (see [137,138]). Archaeologists now contend sweet potato arrived in Hawai'i with a second wave of Polynesian arrivals following the plant's introduction to central Polynesia from the South American continent [29,137,139]. R.C. Green suggests the timing of this as 1000–1100 CE [29], and at present, the earliest 'uala radiocarbon date in Hawai'i is 1290-1430 CE for a charcoal fragment from the leeward Kohala field system [140]. Regardless of the exact timing of arrival, what is known is that 'uala enabled expansion of cultivation and populations into areas generally too dry for kalo cultivation [140,141].

Cultivation in these drier regions was dependent on the skill of farmers and the will of the gods. Documented prayers associated with 'uala and dryland farming appeal to natural forms associated with the deities of Kū, Lono, and Kāne, imploring clouds to shade and bring rain and calling Kānepua'a to root in the fields (for some examples, see [142]). Thus, the cosmological connections of 'uala with atmospheric phenomena and celestial bodies (as calendars) reflect not only its dispersal but also serve a practical purpose in seasonally dry environments where it was a major staple. 'Uala planting here was determined by the onset of rains and planters were acutely attentive to the skies. "*Hā'ule ka ua i Kahoaea/Papakōlea, i hea 'oe?*" (When it rained at Kahoaea or Papakōlea, both 'uala growing areas, where were you? [143,144]). Across Hawai'i, 'uala was planted primarily with vegetative slips, and planters would have needed access to living material to establish new fields during windows of opportunity. At the same time, 'uala yields tubers as early as three months, making it the fastest-maturing of the Hawaiian planter's staple crops. "*He 'uala ka 'ai ho'ōla koke i ka wī,*" sweet potato is the food that quickly ends the famine [143].

3.2.2. Naming and Reclaiming: Ongoing Efforts to Deepen and Expand Our Understanding of 'Uala Diversity

Approximately 300 names have been documented for various 'uala in the Hawaiian Islands [33,36]. These include an unknown number of synonyms and Hawaiian cultivars that may be extinct or known today only as accession numbers within botanical collections [36]. The most comprehensive published descriptions of identifying traits (English language) are found in The Hawaiian Planter [33], informed by E.S. Craighill Handy's visits by to several communities across the Hawaiian archipelago in the 1930s. Although the 300 names and Handy's survey include a handful of introduced varieties (e.g., 'Okinawa,' 'Pukiki'), the large number suggests high cultivar diversity and local specificity (Figure 3). Regardless, the disconnect between documented names from Hawaiian language records, varieties grown by Handy's informants (some preserved as pressed specimens in the Bishop Museum Herbarium), and current day living collections presents significant challenges. Handy's herbarium specimens did not retain colors and do not include tuber descriptions important for identification.

Figure 3. Four of an unknown number of extant Hawaiian 'uala cultivars (leaves and tubers): top row (L-R) 'Pala'ai,' "Ele'ele,' and bottom row (L-R) 'Kalia,' 'Piko.' Photos by A. Kagawa-Viviani

Another daunting challenge is understanding the nature of the cultivars that persist today in collections and gardens. Cultivars of living sweet potato collections are easily mixed during bed rotations, requiring collection managers to be particularly attentive to distinguishing traits and labeling. Yet another challenge is understanding how much kahuli (mutation, sporting) has shaped the plants we know today; sports are a natural phenomenon and are well-known and described for both Māori kumara [145] and to a lesser extent for Hawaiian 'uala [33]. In conservation-oriented botanical settings where active farmer-informed selection is not occurring, mutants might be perpetuated over multiple planting cycles, leading to altered traits in key identifiers such as leaf shapes and pigments as well as differences in tuber pigments. On the flip side, breeding efforts through the 1900s [146–149] not only produced numerous hybrids, but may have also contributed to disproportionate preservation of flowering cultivars over non-flowering lineages. The extent and nature of hybrids—both natural (open-pollinated) and intentional (hand-pollinated)—in current 'uala collections remains poorly understood. Finally, the natural plasticity of 'uala traits (variation under different environmental conditions) also presents challenges in comparing 'uala observations made at different locations and times. All of the above- mix-ups, sporting, hybrids, trait plasticity, contribute to the challenge of matching today's living cultivars with older written descriptions, names, and stories.

Molecular tools, utilized for some decades to understand the movements and connectivity of sweet potato across the Pacific [150,151], may aid in the effort to reconnect living plants to herbarium collections, and older documented names and cultural knowledge. Current research efforts focus on the genetic relatedness of preserved 'uala specimens in herbaria and Hawaiian varieties maintained in botanical collections [152]. Similar research in Aotearoa has found that Māori kumara have genetics reflecting hybridization with introduced varieties [133]. Despite this, these kumara are still considered a native food. It remains to be seen whether "traditional" Hawaiian varieties that persist today have distinct genetics or reflect hybridization with introduced varieties, and to what extent this may influence cultural valuation of these crops as spiritual and physical sustenance.

3.2.3. Returning ʻUala Diversity to Diets and Landscapes

Understanding how we moved from high crop diversity to the current situation of only a handful of commercial sweet potato varieties is a complicated tale involving pests, climatic variability, population collapse and migration, and changing land tenure, demographics, dietary preference, and market forces [43]. A quick dive into records from Hawaiʻi's Kingdom and Territorial periods implicates a shift toward commercial production favoring the cultivation of introduced varieties. The 1850 formation of the Royal Hawaiian Agricultural Society was catalyzed by the prospect of exporting produce to the recently acquired American coast of California [153]. In 1857, Kalaupapa was exporting ʻuala 'likolehua,' 'apo,' and 'halonaipu' along with introduced varieties 'Iapana' and 'Kaleponi' [154]. By 1916, the US Department of Agriculture (USDA) Hawaii Agricultural Experiment Station system was promoting the distribution of "improved" sweet potato varieties such as 'New Era', 'Kauai/Madera', 'Merced/Jersey Sweet' [155,156]. On Kauaʻi, G.W. Sahr advertised, "*Uniform products demand the highest prices. The county agent can help you market a product that can be depended upon. One of the greatest obstacles in marketing sweet potatoes is the existence of so many different varieties, many of which are of inferior quality. The pink Kauai Madera sweet potato is one of the best varieties to plant. Let the county agent aid you in getting cuttings.*" [157] During this Territorial period, USDA agricultural stations actively hosted the import of and breeding of new varieties including crosses with native varieties which continued at the University of Hawaiʻi [146,147], although activities beyond these thesis projects are not well documented. Although introduced sweet potato varieties were already in cultivation at the turn of the 19th century, distribution of new varieties and hybrids oriented toward commercial production and increasing reliance on markets over home gardens may have accelerated the move away from locally-adapted native cultivars.

Changing the trajectory of over 100 years of history is no small task. Yet distributed efforts are underway to reclaim ʻuala as a Hawaiian crop. Restoration efforts in dryland field systems (see Lincoln et al., this issue) have stimulated interest in rain-fed cultivation techniques and crop varieties with varied tolerances to drought and introduced pests (see also [158,159]). Efforts are also underway to increase capacity for citizen and student observers to document their cultivars including those persisting in home gardens that may not be included in botanical collections. Ethnobotanical collections across the Hawaiian Islands have safeguarded several dozen ʻuala varieties for decades. Field trials of documented ʻuala are being conducted now at University research stations to better characterize them and encourage farmers to plant heirloom varieties with uniquely Hawaiian histories. Many in Hawaiʻi are still only familiar with the purple-fleshed, white-skinned 'Okinawan' varieties or imported moist orange-fleshed 'yams.' It will take effort and education to restore appreciation for the older varieties and their place on dinner tables across Hawaiʻi.

As interest in Hawaiian ʻuala grows, it is natural that research interests may focus on creating "better" hybrids to feed the masses. Yet the history of ʻuala highlights its value in connecting people and providing medicine [160] and sustenance over long voyages, dry regions, and in the most difficult times. Recognizing this context, we are now challenged to move beyond existing paradigms of crop improvement to consider research agendas centered on ʻuala's broad accessibility and local specificity, that foster ʻuala as a catalyst for restoring indigenous foods to families across Hawaiʻi. *He ʻuala ka ʻai hoʻōla koke i ka wī.*

3.3. Kō—A Knowledge Revival and Changes in Crop Use over Time

3.3.1. Kō in Ancient Hawaiʻi

Prior to European arrival, kō played a significant role in Hawaiian agricultural traditions. Kō is associated with the god Kāne, the first of four primary dieties to come to Hawaiʻi; a variety of kō, "Ele', appears second to the kalo 'Manauea' in the *Kumulipo* creation chant [3,25], and plays a central role in the *Kumuhonua* genealogy that provides the lineage of traditional medical experts [161,162]. This importance of kō in these essential religious sagas is manifested in its extensive use in ceremonial offerings and medicinal practices. Indigenous horticulturalists distinguished upwards of 50 sugarcane cultivars that differed in their appearance, usage, and environmental tolerance (Figure 4) [163]. Agriculturally, kō could be found growing in most arable habitats: along the banks of lo'i (flooded terraces) with mai'a and other moisture-loving crops, stabilizing the banks and shading the water; in backyard gardens where the canes were meticulously manicured; in extensive rainfed systems forming thick hedges that extended for miles, acting as windbreaks; in young lava flows growing in excavated pits, heavily mulched; in boggy lowlands persisting even on brackish waters; and in other conditions. The cultivation of kō, in many cases, appears to have played a critical role in the cropping system, and the almost excessive cultivation of kō may have been to manage landscape level patterns of moisture and nutrients to better cultivate staple crops such as kalo and 'uala [33,39,164,165].

3.3.2. Remembering: Plantations and Germplasms

The earliest development of the sugarcane industry in Hawai'i beginning in 1836 utilized native kō cultivars. Although the Hawaiian cultivars were developed and selected for use in diversified cropping systems, many still exhibited superior production in monoculture cultivation and were exported to plantations around the world. Starting in 1854 with a Tahitian cultivar, an influx of introduced varieties replaced Hawaiian kō in cultivation and, ultimately, even in backyard and other small-scale plantings. Well-known varieties were given Hawaiian names and adopted into cultural norms.

In the early 1900s, a world-renowned breeding program at the Hawaii Sugar Planters' Association (HSPA) began which produced tens of thousands of new hybrid varieties that displaced both Hawaiian and imported sugarcanes. In the late 1800s, as part of the establishment of its breeding program, HSPA conducted statewide reconnaissance of kō, collecting several hundred accessions along with minimal local ethnographic information (typically only a name). From this effort, about 65 unique cultivars were identified, including both Hawaiian and introduced heirloom varieties. In 1932, a bulletin publication, *The Native Hawaiian Canes*, was released describing, with very little detail, the Hawaiian cultivars [166]. Although HSPA has maintained the collection, it has been neglected, resulting in mislabeling and losses of cultivars. Starting in the 1970s and into the 1990s, multiple ethnobotanical gardens took interest in indigenous Hawaiian crops, leading to the dissemination of the HSPA collection. The gardens each added new canes, often presumed to be Hawaiian varieties, from various private sources.

3.3.3. Revitalizing Knowledge of Kō

In the early 2000s, a volunteer effort began to restore the knowledge systems of Hawaiian kō. With a certain irony, the sugarcane industry simultaneously preserved the physical germplasm while eroding the cultural knowledge regarding it. After initial efforts, it was quickly realized that multiple shortcomings in the germplasms existed: the names attached to the modern collection were often incorrect, there was poor documentation or differentiation between native Hawaiian, introduced heirloom, and early hybrid varieties, and there was no identification process for verifying or identifying a cultivar.

Mahi'ai (native farmers) and *kūpuna* (elders) confirmed that contemporary knowledge of kō was almost non-existent; when talking about native varieties, most described only a "red" and "white" type, with no names or more detailed descriptions. A handful of practitioners with more substantial knowledge were identified, but it became clear that the ethnobotanical gardens were considered the primary source of knowledge. Unfortunately, that knowledge was primarily handed down from HSPA, which was minimal to begin with and severely affected by errors and a lack of concern for the accuracy of the ethnographic information.

The effort to rebuild knowledge of kō over the past 15 years has been an ongoing series of partnerships between research and practice. The first major effort involved detailed morphological documentation. After all, if a cane is not identifiable, it cannot be accurately reconnected to its history. Over a ten-year period, this work served to spark increased interest in the specific sugarcane varieties through discussions with germplasm managers and farmers. The efforts resulted in a botanical key that is now freely available online along with the cultivar descriptions and pictures [167]. Importantly, this work provides clarity for germplasm management by distinguishing Hawaiian kō from other heirloom varieties and providing a clear identification guide to varieties. Efforts are now underway to verify these findings through genetic assessment of the Hawaiian sugarcane germplasms.

Concurrent efforts to rebuild cultural knowledge of kō included interviews and discussions with practitioners that had knowledge of names, uses, and descriptions, tracing germplasms accession back to their roots, examining herbarium specimens and collection notes, and translating primary sources such as past writings by Hawaiian scholars, recordings of ethnographic interviews conducted in the early 1900s, and the Hawaiian newspapers that published over 58,000 pages of articles between 1834 and 1948. From these sources, thorough historical descriptions of the cane varieties, new names and descriptions of lost varieties, unique uses, growing conditions, and agricultural practices were identified.

Simultaneously, efforts were made to share cultivars to have more material available for propagation, education and use, and to connect growers interested in kō to facilitate knowledge exchange. The entire collection was provided to over 20 private growers, with hundreds of growers receiving multiple or individual cultivars. The ethnographic and botanical research provided a stronger sense of the cultivars, and their usage and significance, which enabled outreach and education, still ongoing. Since 2010, 23 workshops on growing, identifying, and cultural uses of kō have been conducted that have reached some 600 famers and practitioners.

Figure 4. A subset of kō varieties illustrating some of the diversity showing, from left to right, 'Nānahu,' 'Moano,' "Ula,' 'Laukona,' 'Hinahina,' 'Hālāli'i,' 'Pua'ole,' and 'Wai'ōhi'a.' Photo by N. Lincoln.

3.3.4. Kō in the Contemporary Period

The efforts to revive knowledge and use of kō have resulted in enhanced interest and awareness, such as the inclusion of five Hawaiian kō cultivars in Slow Food's Ark of Taste, multiple feature articles chronicling kō diversity [168–170], and vibrant exchanges in social media groups. Significant growth in the cultivation of kō has occurred, with an unquantified, but substantial, increase in backyard farmers and the incorporation of kō into small-scale diversified production. One issue remains in that, unlike most other crops, sugarcane is not easily prepared at the home-scale as it requires, at minimum, a press to extract the juice for use.

The enhanced availability of the ethnographic and agronomic aspects of kō have facilitated emerging industries. Processors and restaurants have incorporated cane juice into products, and the establishment of multiple rhum agricole (alcohol directly distilled from sugarcane juice rather than molasses or sugar as is the case for most rums) producers has led to increased markets for sugarcane production in Hawai'i. The growth of these industries has occurred as the last sugarcane plantation in the state closed its doors at the end of 2016. This has led to significant cultivation of the heirloom varieties, and it is safe to say that more Hawaiian kō varieties are being grown today than at any time in the last 100 years.

The current challenge is to ensure the continued use of kō in the growing markets, and to ensure that it is done in a proper way. Some producers have latched on to kō cultivars because of the rich history and stories that make for good marketing of products, while others exclusively grow kō because of cultural connections. Despite this, few care about the identity of the source cultivar. Research into the nutritional qualities of pure cane juice, which is high in antioxidants, micronutrients, and has half the glycemic index of processed sugar, clearly indicates that heirloom sugarcane cultivars in general offer greater health benefits, although specific studies examining kō have not yet occurred. Steps have been discussed to form an appropriate group to represent the kō, and, in particular, to safeguard the cultivars from excessive commercial exploitation and ensure appropriate giveback from the economic benefits derived from the cultivars and cultural knowledge regarding kō.

3.4. 'Awa—A Cultural Treasure

3.4.1. Dominance and Fate

'Awa (*Piper methysticum*) is a multi-functional plant in the broad categories of ceremony, medicine, and recreation. Cultivars of 'awa may be restricted for use in a narrow sub-category or may be used in all categories. Prepared as a beverage, nothing else was more commonly drunk other than water and coconuts. Names, traditional uses and practices associated with 'awa in ceremony and medicine are embedded in legends and chants and were recorded by Hawaiian and foreign scholars [34]. Upon first arrival to Hawai'i, the ancient gods of Kāne and Kanaloa set about searching for springs so that they could indulge in 'awa; although the land was dry, Kāne struck the earth with his staff and water gushed forth [171]. 'Awa played many roles, but was dominant in Hawaiian ceremony, being considered the favorite of the gods [172,173]. The act of drinking 'awa was important in maintaining the link with one's 'aumākua (deified ancestors) and 'unihipili (family spirits) [1]. For warriors, farmers and fishermen, it was a tonic to relax the mind and body in everyday use [172,174] and it is this factor that may have contributed to its survival into the 20th century.

Yet Western contact challenged and eroded the strong biocultural relationships developed during Hawai'i's centuries of isolation. By the mid-1820s, Western ideologies and Christianity adopted by Hawaiian ali'i eventually resulted in de facto banning of 'awa production and consumption and churches did much to discourage the use of 'awa [172]. In the early 1900s, Hawaiian 'awa still could be found in great groves within the forested places of the islands abandoned by the collapse of the Hawaiian population in the 1800s. Entrepreneurs were given leases to harvest 'awa on lands under the jurisdiction of the new Territorial Forestry Department [175] contributing to the gradual mining of historic 'awa groves as availability and usage declined. Today alcoholic beverages have largely replaced 'awa as the traditional drink among many Hawaiians [172,176].

3.4.2. Revitalizing 'Awa in Hawai'i: A Grassroots Effort

Due to the decades of decline in production and consumption, Hawaiian 'awa cultivars, and the knowledge associated with them, were in danger of being lost. After seven years of searching remote areas to gather planting material and document heritage 'awa plantings, often with the Hawai'i Biodiversity and Mapping Program, a backyard nursery was started in 1992 which has grown into the Ālia Point 'Awa Nursery (APAN). In 1994, APAN's donation of 'awa plants to a Canoe Club fundraiser in Hilo provided an introduction to Jerry Konanui. A friendship was established and a partnership developed to restore and enhance the understanding of Hawaiian 'awa here in our islands and the world.

Our first project was with the Center for International Research in Agricultural Developments, Vanuatu and France, to research and publish a comprehensive study of all known Hawaiian 'awa cultivars—DNA, kavalactones, and morphology. The data clearly demonstrated that Hawaiian 'awa cultivars are exceptional [176]. The publication of these data coincided with the desire of the Rural Economic Transitional Assistance-Hawai'i (RETA-H) program to inspire 'awa growing in Hawai'i.

The Association for Hawaiian 'Awa (AHA) was created in 1997 for research, education and preservation of the cultural and medicinal values associated with the 'awa plant, and as a vehicle through which funding to support conservation and education efforts related to Hawaiian 'awa could be sought. RETA-H funded a program of AHA-led educational workshops on Hawai'i, Maui, Moloka'i, O'ahu and Kaua'i. AHA also published a quarterly 'awa newsletter. These years of re-building the historic significance of 'awa involved the efforts of many individuals from a wide range of ethnic, social, economic and educational backgrounds.

Figure 5. Hawaiian 'Awa Cultivars: 8 of 13 known cultivars of Hawaiian 'awa: Top (L to R): Pana'ewa; Papakea, Kumakua, Nene. Bottom (L to R): Hiwa; Uliuli a opulepule no ho'i (also known as Hanakapi'ai); Opihikao; Mahakea. Photos by Ed Johnston.

3.4.3. 'Awa Growing: Boom and Bust

In 1992, the authors of *Kava, the Pacific Elixir* [177] (p. 96) declared "'Awa is now a relic of traditional Hawaiian culture—an uncommon but attractive plant that can be found for sale as an ornamental in a few commercial nurseries." Six years later this was no longer true. By 1998, the unique Hawai'i cultivars were well characterized and being planted throughout the state (Figure 5). By the early 2000s, an estimated 200 acres of 'awa were in the ground, with the largest acreage on Hawai'i Island. Larger farms were Wainani Farms, Ho'owaiwai Farm, Hawaiian Pacific Kava Company, Pu'u'ala Farm, and Kulana Ki'i. Most of the smaller farms were in Puna. Pu'u O Hoku Farm/Ranch was the major grower on Moloka'i. Most Hawai'i 'awa farmers were keenly aware of cultivar names and often devoted sections of their plantings to specific cultivars, a practice rarely done in other Pacific Kava growing regions.

During the decade prior to 2004, kava was becoming very popular worldwide as a dietary supplement. Pharmaceutical companies were buying metric ton quantities of dried kava for extraction from the South Pacific and Hawai'i growers at prices much higher than the traditional drink market, encouraging early harvesting, sloppy processing and adulteration to meet market demand and grasp the price windfall. During this period, kava dietary supplements became linked with liver injury in some individuals. Health advisories from national agencies such as the US Food and Drug Administration about potential hepatotoxicity associated with kava did not distinguish between supplement and beverage preparations as the source of the problem. The traditional water-based beverage market was thus not only adversely affected by higher cost for product, but it also was tainted by the broad-brush definitions used to warn of possible liver damage [178]. The market crashed worldwide and nearly all Hawai'i farms closed. Three exceptions, at that time, were smaller farms who primarily produced fresh frozen 'awa pulp using modern chopping and straining machinery for making the traditional beverage (originally by chewing the root to a fine pulp, mixing with water and straining through the sedge, ahu'awa).

Despite more than a decade of research into the possible causes of liver injury associated with kava, there is still no clear understanding of mechanisms by which hepatotoxicity could occur. During the time since the first cases were reported, kava has continued to be consumed both as a supplement and as a beverage, but new cases of hepatotoxicity have not been reported. The Food and Agriculture Organization of the United Nations (UN FAO) and World Health Organization (WHO), in a review of the safety of traditional and recreational kava beverage consumption, concluded that there are significant gaps in the knowledge about kava's effects on health, but that "There is little documented evidence of adverse health effects associated with traditional moderate levels of consumption of kava beverage, with only anecdotal reports of general symptoms of lethargy and headaches" [179]. An editorial in the *Journal of the American Medical Association* further noted that: "Historic use shows that kava is safe under the strict control of the rituals of Pacific cultures. The traditional beverage is consumed on a daily basis without apparent adverse effects, and kava cultivars considered as noble ones have a long tradition of safe use" [180]. Noble cultivars are cultivars with history of safe, traditional use. All Hawaiian 'awa cultivars are considered noble. This is an example of traditional knowledge being right all along. Given that the majority of evidence suggests traditionally prepared 'awa beverage is safe, AHA has been working with the UN FAO and WHO's *Codex Alimentarius* to add the Hawaiian cultivars to their safe foods and beverage list. This is an on-going effort, not yet completed.

3.4.4. Envisioning How to Sustain and Increase 'Awa Growing and Use

Currently in Hawai'i there is a growing market for quality fresh frozen and dry 'awa, and the revival of traditional knowledge has supported a market for single strain products using traditional varieties. The release of quality ethnobotanical resources has been essential for grower knowledge. Until recently, the authoritative volume on Hawaiian 'awa was a 1948 publication *Kava in Hawai'i* [172]. This collection of information was built upon in 2004 in *Hawaiian 'Awa: A study in ethnobotany* [174]; and most recently the Association for Hawaiian 'Awa released its original studies and 'official' varietal documentation in the free publication *Hawaiian 'Awa, Views of an Ethnobotanical Treasure* [34] (online at www.awadevelopment.org).

Hawai'i must grow more of its own 'awa cultivars if preservation of these plants is to be successful. Education is critical in understanding the role of 'awa in Hawaiian history and what important research is currently occurring with 'awa beverage in a healthful lifestyle. In the past, much of 'awa research focused solely on kavalactones—often extracted with solvents producing a fragmented product far from the traditional aqueous beverage so revered across Oceania. The 'awa beverage is much more than just the concentrated kavalactones found in extracts and pills. Dr. Amanda Martin published a study [181] which revealed high levels of variation in chemical content and cytotoxicity of such commercially available kava products. These findings support the traditional knowledge of 'awa as a beverage, not a pill. As Martin [182] has said: "It's not necessarily that more is better. What you're looking for is the right amount. And with kava an aqueous solution produces the right amount."

4. Discussion

4.1. Connecting: from Restoring Cultivar Diversity to Restoring Abundance

Several themes emerge from the cases examined. Each plant, regardless of function as staple food, medicine, ceremonial or agronomic resource, embodies a sacred relationship with a key role in Hawaiian society. As Hawaiian populations declined and underwent major societal change in the 1800s, each crop experienced a decline in cultivated area and inventory of native varieties, and those that became market commodities were displaced by introduced varieties. Internationally, the loss of local varieties is among the four factors identified by the FAO as threatening traditional agricultural systems, livelihoods and food security (along with unsustainable models of agriculture, land access pressures, and climate change) [183]. The historical threats to Hawai'i's indigenous crops and sustainability mirror challenges faced by communities around the world.

Despite the decline, a resurgence in interest and revitalization of food and food production systems extends across all four crops, albeit at different stages. Traditional crop cultivar diversity was conserved for decades in ex situ collections for largely research purposes while in situ and ex situ revival and restoration of crop diversity was driven by significant individual and collective community-based efforts external to these institutions. The renewal of crop cultivar uses and values was fed by and feeds cultural revival and involved intensive community and farmer engagement, local entrepreneurship, and strong advocacy at the grassroots level. Each case reminds us that the recent recovery of these ancestral crops emerged out of peoples' desires to restore their relationships with plants, foods, and places.

Table 1 summarizes actions across the four cases as five strategies relevant to the contemporary restoration of Hawai'i's indigenous crops: revitalize and regenerate knowledge of cultivars, return them to the landscape (in situ), protect and steward, restore biocultural relationships and heritage, and resilience and abundance. Each case illustrates how various types of research and collaboration have played a role in the resurgence, from the curation and recovery of living cultivars and archival records to the shift from ex situ conservation settings back into the hands of cultural practitioners and farmers so they can be cultivated in situ. Community stewards, educators, collections managers, agricultural extension agents, families and farmers have all participated in the return of kupuna crops to the landscape through varied pathways. Although not discussed explicitly, in all four cases, the voices of kūpuna through mo'olelo (stories), the sensory experience of taste and smell, and the familial experience of preparing and sharing foods are all important elements of remembering and rekindling interest in such crops.

The expansion and diversification of cultivation evident in the examples of kalo and 'awa commercial plantings also illustrate how shifting perspectives from global back to local, and reconnection with kupuna crop traditional knowledge, can result in products better aligned with traditional values, practice, and use (e.g., pa'i 'ai; fresh 'awa root). Such products not only resonate with consumers, but also diversify direct-to-consumer and value-added markets for farmers to choose from, raise farm-gate prices, and restore honor and dignity to the livelihoods of small growers, a strategy that is gaining ground world-wide [122,123,184,185].

Table 1. Strategies supporting biocultural restoration of indigenous crops in Hawai'i and creation of resilience and abundance. Values in the four columns reflect the current state of progress for each crop (5: advanced, 1: in infancy, x: occurring but not assessed).

Strategies [1]	Kalo	'Uala	Kō	'Awa
Revitalize **and** *regenerate* **the knowledge base**				
Maintain one or more verified (ex situ) crop collections; cultivate kupuna crop specialists; kuleana	5	2	4	4
Research: archival, oral history/ethnographic; storytelling; naming (participatory; community-driven; academic)	4	2	4	4
Cultivar identification resources	Update in prep		In press, online	Completed
Return **to the landscape (in situ)**				
Restore traditional agricultural sites and practices (educational, non-governmental organization (NGO), community, 'ohana); awakening laulima	3	1	2	2
Diversify, restore, envision novel markets and products (farmers' cooperatives; chef connections)	3	x	3	4
Subsistence/sustenance plantings: expand area and diversify cultivars (food security and surplus)	3	x	2	3
Commercial plantings: expand area and diversify cultivars	2	x	3	3
(Re)Honor farmer livelihoods and lifestyles and practitioner knowledge	3	x	x	x
Farmer to farmer mentoring	4		2	3
Protect and steward				
Awakening to kuleana, onipa'a; envisioning	4		2	3
Legislation and policy	2		1	2
Pono protocols, relationships and research with tools of modern science	1	x	2	4
Restore **biocultural heritage**				
Workshops, education, outreach (collections managers trainings; community; growers, K-12 schools; university)	5	2	4	5
Festivals; honoring and celebrating; (re)connecting	5		1	5
Community networks of information and resource exchange/sharing; (re)connecting; cultivating resilience	4	1	3	4
Resilience and abundance				
Climate change and disaster resilience; collections on each island	5	1	5	3
Production meets and exceeds local needs and markets	2	1	2	2
Regenerative agricultural practices support kupuna crops	2	1	2	2
Maintaining ongoing kuleana and laulima; multiplying practitioners	4	1	2	3

[1] The first four subheadings are borrowed from Tuhiwai Smith's (1999) twenty-five indigenous projects [20].

Workshops, celebrations, and festivals create a multiplier effect that increases outreach and awareness. For kalo and 'awa, annual festivals provide opportunities for community members to gather and reconnect as well as exchange knowledge and cultivars. As gathering places, they embrace youth, local communities, farmers, visitors, and the general public as consumers and supporters. Celebration is recognized as a vital regenerative act in many cultures where festivals and agricultural ceremonies persist, including those that call the rain and mark harvests and planting seasons and which are highly refined systems that coordinate agricultural activities with natural cycles (e.g, the Hawaiian makahiki season, the Ngan Bongfai of Northeast Thailand, the royal rites and duties of the ploughing ceremonies of India and Southeast Asia, or the many agricultural rituals of the Ifugao [186,187]). Celebration is also recognized among many indigenous NGO groups as an important element in rekindling people's interest and passion for conserving indigenous crop biodiversity (e.g., Fiesta del Maíz of Mexico and the Indigenous Crop Biodiversity Festival which paralleled the IUCN World Conservation Congress in 2016). Art and the art of growing and preparing food entice, engage, inspire and stretch the boundaries of our senses and view of indigenous crops as merely agricultural products. Farmer-to-farmer mentoring, working one-on-one with collections managers and individual researchers reclaims expertise back into community and suggests a model for balancing roles and guiding research efforts [188–190].

The enthusiasm surrounding indigenous crops resurgence can, however, generate new problems when steps to avoid old patterns of abuse of indigenous knowledge and resources and people are not made. As the kalo case revealed, the seemingly laudable effort to increase disease resistance and yields through purely technical fixes met with strong opposition (kū'ē; resistance and protest) from members of Hawaiian and farming communities, and conflict escalated through the institutional privileging of intellectual freedom over the multiple concerns voiced by community members. Advancing legislation and proposing policy to protect kalo required exhaustive and committed effort to educate communities and decision-makers. The parallel example of the 'awa boom and bust highlighted how export-driven commodification ultimately left small growers with fewer market options and how, conversely, sustainable solutions can be found in heeding traditional knowledge and use in the modern context. Both cases illustrate the capacity and power of strong networks to overcome local and international judgements.

As interest and acreage in kō and 'uala expands, the kalo and 'awa lessons point to the importance of centering and prioritizing the kinship and stewardship relationships of each plant. These cases remind us that a kuleana to a reciprocal and familial relationship to these plants places us in service to the needs of the plants, i.e., tending to soil health, planting and harvesting at proper times, listening, remembering names and mo'olelo and proper usage, sharing abundance. By giving primacy to this relationship, we suggest ethical, acceptable, and sustainable practices, markets, and research can naturally emerge to restore indigenous crop biodiversity and sustainable food systems. Ultimately, such an approach is respectful and acknowledges not only the generations who developed and diversified these distinctively Hawaiian crops, but also the work of the many farmers and caretakers, storytellers, recorders, and families, Hawaiian or not, who have stewarded and continue to care for these kupuna crops today. Across the five key strategies, awakening to the kuleana of caretaking these crops evolves for some into 'onipa'a, a steadfast protection of a treasured resource embedded in cultural identity and a deep commitment to the work of recovery. Maintaining both kuleana and 'onipa'a in the long term, as in the example of Jerry Konanui, provides leadership, vision and inspiration that guides the larger network of collaborative efforts. Resilience and abundance are both a perception from within community and an overall goal. For Pacific islands, in particular, ensuring cultivar survival in the face of climate change and increasing hurricane frequency involves establishing collections and propagation projects across multiple islands with multiple partners. The cases also tell us that fulfilling local market needs is still a long way off (e.g., while lands in kalo and 'uala production have increased, to put them on the table of every family requires thousands of acres of land and water not yet accessible). Kalo farmers are noted for the abundance they share with family and community (food, time, knowledge) that contributes to 'ohana and community wellbeing. This has also transferred to guiding the next generation of young farmers and researchers together—a reflection of both kuleana and envisioning the future.

While our cases highlight steps toward reinvigorating and restoring traditional agriculture across Hawai'i, we can look to the Parqe de la Papa in the Peruvian Andes for a model of biocultural heritage restoration where the repatriation of traditional potatoes, agrobiodiversity, and community-led solutions are underpinned by cosmologies of kinship (cosmovisión) [191,192]. The FAO's Globally Important Agricultural Heritage Systems Programme [183] provides a model of more intact biocultural heritage landscapes now threatened in the face of economic globalization that might be adapted and leveraged locally to afford better protections for traditional agricultural systems and the knowledge and lifestyles they support. We also learn from challenges and solutions of Ifugao rice-growing communities in the Philippines as traditional rice varieties enter global markets as commodities in an effort to save the heritage landscape [187,193]. Ficiciyan et al (2018) make a case for the complex set of ecosystem services that traditional crop varieties provide [194] which raises important questions about the resilience of such biocultural heritage to climatic change [195] and the role of biocultural innovation in generating new possibilities in the context of modern economic forces [196].

4.2. Reframing Kupuna Crops Research—'Auhea Kō Kuleana?

Kuleana is defined by Pukui and Elbert as a "right, privilege, concern, or responsibility," as well as conveying title and authority (e.g., over land) and the liability attached to that [89]. We define kuleana as to be in service, to practice and exercise accountability, commitment to a reciprocal and familial relationship such as that exemplified by the metaphor of Hāloa [58]. Noted cultural practitioner, Sam Kaai described how when kupuna talked about 'āina it meant family, not land, with that same reciprocal relationship bringing the circle back to Hāloa. Kuleana in the context of this paper would mean putting mana (spirit energy of character) [197] into the care of indigenous crops. By contrast with the largely transactional exchanges of modern life, it necessitates a sense of reverence for living beings and the histories that created them. In more straightforward terms, it implies taking seriously the work attached to these plants, including attending to associated complexities and ethics, and to 'āina.

4.2.1. Reflecting on the Lessons of Hāloa

Given broad interest in Hawai'i to return to greater local food production and sustainability, we suggest researchers and institutions must shift from the dominant perspective that indigenous crops are merely "genetic resources." Of note is Governor Ige's promise on the stage of the IUCN 2016 World Conservation Congress to double Hawai'i's capacity to grow food by 2030. To move in this direction, we examine the kalo conflict to identify pathways towards more pono (ethical) research protocols and investigator practices (pono is a core concept in Hawaiian decision-making and action, [198]). However, the restoration of pono (ho'oponopono), a prerequisite for moving forward, requires willingness to acknowledge previous wrongs and make good faith efforts to correct them.

The conflict over patenting and genetic modification of kalo exposed a key underlying issue with indigenous crops research at the University of Hawai'i. The rush toward biotech solutions for kalo and the ensuing struggle revealed vastly different worldviews held by university researchers and members of the taro farming/Hawaiian community which were not considered prior to the formation of the research proposal. For some, the cultural significance of kalo (and other crops) means that many activities are simply kapu (taboo), and that plants, animals and the environment as a whole cannot be owned and should be treated with the utmost care and respect, just as any other member of the family [112]. That local taro growers' concerns also went unheard suggests that researchers continue to see farmers and stakeholders as largely *recipients of* rather than *partners in* the research process or *knowledge keepers* from whom they should seek direction, and thus their perspectives are also not considered in the development of research priorities.

Thus, while we agree academic freedom is an important value for public educational institutions to protect, we contend that deploying this defense in the kalo controversy was a straw man and ultimately a means of shirking institutional responsibility to consult and engage with a diversity of views. We challenge such institutions instead to provide space to discuss the limits of so-called "academic freedom" in the context of their local communities and to address, on a truly open and proactive basis, the hotspots and complexities where ethics, history, culture, and power dynamics meet research agendas, and to consider them holistically in terms of both individual and institutional kuleana.

More immediately, a starting point for restoring trust and laying the groundwork for future collaboration is acknowledgement that there are many ways of knowing. Within the metaphors and poetry of indigenous languages and lifeways often exist astute and practical observations and decision-making logics relevant to contemporary study of agriculture and sustainability [183,199,200].

4.2.2. Acknowledge Different World Views, Objectives, and Time Frames

We suggest that an openness and willingness to listen, hear, and understand different worldviews is necessary to enable deeper empirical insight and also mend historic rifts and wounds. By listen, we mean to be attentive to, or have heightened perceptiveness towards people and research "subjects," which is enabled by actual care—the kind one provides to family members, hence the indigenous metaphors of kinship. A productive exchange might also acknowledge that the modern researcher is often bound to short timeframes of observation while indigenous/local communities tend to be concerned with both immediate survival and livelihoods as well as much longer (intergenerational) windows of observation and outcomes where empirical insights guide decisions to benefit a broader set of individuals over decades and lifetimes. The cases we describe result from the work and commitments made by various individuals to care for a given crop, reaching far beyond any job description. This extends to other indigenous crops not discussed in this paper ('ulu, mai'a; breadfruit and banana) as well. As the successes of crop recovery efforts in Hawai'i indicate, an effective collaborative relationship among academic researchers and members of indigenous and farmer communities recognizes the value of indigenous kinships and the importance of active stewardship of these. How can meaningful and respectful "hybrid" approaches inform—or transform—agricultural research and practice?

4.2.3. Exercise Individual Accountability

Researchers, whether indigenous or not, come with varying identities such as outsider, insider, ally, and everything in between. They must navigate their roles and, in the case of working with indigenous and local communities, may have minimal experience or guidance on how to do this effectively. In general, the onus to build trust is upon individual university researchers working with indigenous and local communities who must recognize the differential authority they are afforded through titles (PhD, etc.) and affiliation. This often comes with a default assumption that they are "expert" even when community members are the keepers of biocultural knowledge, which raises another important aspect of discussion: the need to recognize and respect different forms and sources of expertise. The kupuna crops in this paper illustrate that tremendous expertise is held by cultural practitioners, farmers and other skilled knowledge holders outside of academic institutions, whose knowledge and experience are all too often marginalized by mainstream cultural norms. These forms of knowing can be fundamentally different than that familiar to the researcher, and the complementarity has the potential to provide mutual benefit if partnerships can be established. It is imperative that researchers, particularly those at large public institutions such as universities, recognize such indigenous or local knowledge and acknowledge its caretakers.

At the same time, members of rural, farmer, and indigenous communities may be hesitant to trust representatives of research enterprises and Western educational systems that have played a role in historic dispossession or marginalization [20,201]. This adds another layer of complexity to exchanges, as researchers are all too often unaware of these histories or are poorly equipped to collaborate effectively and find themselves stepping into existing wounds. Individuals can begin fulfilling their kuleana by learning the history and context of their respective place, and drawing guidance from research protocols and best practices including, for example, the International Society of Ethnobiology International Code of Ethics [202], the University of Otago Pacific Research Protocols [203], the still-developing University of Hawai'i Kūlana Noi'i initiative [204], and the Taro Task Force 2010 Legislative Report [61]. Many resources providing guidance on research ethics exist online (for example, see [205]). Simply investing time in relationships though good faith discussion and scoping early in the process is also critical. Effective collaboration would involve asking and receiving permission, keeping in touch, engaging throughout the process, openly sharing outcomes and returning results in a timely manner, based on core principles of participatory action research [206]. We suggest researchers learn history, read broadly, listen openly (including to silences, an early form of dissent), make time for face-to-face interaction, be humble, and pay attention. Such practices are common in social sciences research; applied here, protocols and kapu accompanying indigenous crops establish kuleana at the beginning of a more informed research scoping process.

In the years since the massive community pushback over the patenting and genetic modification of kalo, few changes have been institutionalized at the University. The advocacy and efforts of Taro

Security and Purity Task Force members did, however, lead to creation of an Indigenous Crops position within CTAHR, but true institutional commitment to change remains to be seen. For example, the publicly articulated CTAHR philosophy on agricultural technologies still states: "We uphold the values of academic freedom and respect the rights of farmers and consumers to decide which technologies are most appropriate." [207] Absent are the specifics of how "respect" is enacted, or explicit acknowledgment of the rights of Hawaiians and other indigenous peoples in maintaining and stewarding ancestral germplasm, and how these translate to University decisions. Depsite the disconnect between UH CTAHR and indigenous perspectives in agriculture, change continues along consumer, farmer, economic and cultural pathways.

4.2.4. Formalize Institutional Support, Policies, and Practices for Improved Accountability and Collaboration

While the development of a taro research policy responsive to farmer needs and concerns remains unfulfilled, the Taro Task Force report [61] provides a template for work better aligned with community-identified needs which can be extended for indigenous crops in general. UH Mānoa is the largest public research institution in Hawai'i and was founded as a land grant college on ceded lands by the Hawai'i Territorial Legislature. Although it historically has had a complex and often contentious relationship with members of the Hawaiian community, in 2002, the University's vision statement included the statement, "Mānoa celebrates its diversity and uniqueness as a Hawaiian place of learning." [208] We interpret this as the institutional recognition that UH is situated within a broader context of Hawaiian history and community. Yet persisting attitudes that traditional knowledge and ethnographic research are less valuable and less rigorous than research utilizing modern technologies continue to privilege academic research over community-held expertise. Our cases illustrate the need for a more egalitarian approach and, indeed, a recent initiative to promote greater researcher responsiveness to community needs and collaboration has begun to advance this discussion on the UH Mānoa campus [204]. We observe such efforts are often initiated outside of the university and championed by individual (and often untenured) faculty, staff, and students sometimes with significant risk of retribution. Recent strategic planning documents including a proposal to improve cultural competency among faculty and staff [209] and the recent appointment of a Native Hawaiian Affairs officer could signal an institutional shift. Historically, however, the operationalization of such recommendations is usually met with institutional resistance and requires pressure [210,211]. Beyond policy, an institutional mission to serve the public good in the context of Hawai'i must allow for discussion and change of course, especially when working with matters concerning indigenous and farming communities and where free, prior, and informed consent [212] has not been obtained before embarking on a project. Helping researchers avoid conflict could involve developing an Institutional Review Board (IRB)—like process for working with indigenous crops that effectively holds researchers accountable for considering implications of their research early in the process. A less top-down approach involves institutional support to train researchers in ethical best practices, such as those described earlier, and increase researcher awareness and understanding of local (see other articles in this issue) and global biocultural diversity and heritage conservation/restoration and biocultural design [183,187,191–193,196]. Such an initiative could facilitate opportunities for collaboration and innovation provided trust is present.

Institutions interested in continuing work with indigenous crops must also explore ways to support staff and researchers committed to longer-term collaborations with indigenous crops or communities since developing relationships necessary for collaboration requires significant time and energy. In addition to the cases presented, we note the work of the Breadfruit Institute (also home to the Global Breadfruit Initiative), whose long-term commitment to protecting and honoring cultivar origins goes beyond conservation to maintaining indigenous knowledge associated with each variety, along with agreements for sharing breadfruit varieties and ensuring benefits from distribution return directly to Pacific nations. The two-year grant, pressure to "publish or perish," and the existing tenure review process prioritizing these can strongly deter researchers from pursuing rewarding but time-intensive relationships and collaborations required for this work.

4.3. Kanu: Envision, Plant, and Cultivate with Intention

A fitting metaphor for productive collaborative work with kupuna crops is Kimmerer's (2013) description of the Three Sisters of corn, beans, and squash [201]. In this, she describes the mutualism made possible when traditional ecological knowledge (TEK) stewarded by indigenous/local communities and scientific knowledge of the academic research community engage in respectful and productive relationships. The elder sister of corn (TEK) is planted first and is the foundation of life, providing food, medicine, materials, and spiritual health. Beans as the metaphor for curiosity-driven scientific ecological knowledge (SEK) whip tendrils as they stretch and climb, reliant on the corn and unable to stand on its own. The plant's nutrient-acquisitive nature provides nitrogen to the corn, and its searching roots can unearth soil resources and bring them to the surface. Left unchecked, however, its excessive growth can take over the garden and choke out all else. Kimmerer asks, "What would knowledge generation look like if we created a mutualism in which the climbing 'beans' of scientific inquiry are guided by the 'maize' of indigenous principles?" The third sister of squash embodies "the climate of mutual respect, intellectual pluralism and critical thinking in which both knowledges, TEK and SEK can grow." The low, spreading layer of leaves, by cooling the soil and suppressing weeds, fosters conditions necessary to the symbiosis, Ermine's "ethical space of engagement" about which he writes "The ethical space, at the field of convergence for disparate systems, can become a refuge of possibility in cross-cultural relations … for the effect of shifting the status quo of an asymmetrical social order to a partnership model between world communities." [213,214]. Yet, Kimmerer observes, "the squash is the slowest to germinate, and when young requires the greatest care." The role of the fourth sister is thus critical. "She's the one who noticed the ways of each species and imagined how they might live together … We are the planters, the ones who clear the land, pull the weeds and pick the bugs; we save the seeds over winter and plant them again next spring … We too are part of the reciprocity." Kimmerer's metaphor provides us a vision for how various worldviews might work together productively to support restoration of kupuna crops.

If those working with kupuna crops are the fourth sister, they/we must recognize the choices made shape the future of our kupuna crops and have important implications for the trajectories of our institutions, our landscapes, and communities in which we live. Below, we suggest general themes to guide pono decision-making:

(1) *Center kinship relationships, biocultural perspectives.* Moving the formerly marginal kinships associated with these plants toward the center also acknowledges the values and work of those that have been largely ignored in formal (funded) crops work; this can also restore balance to previously asymmetrical relationships of power and authority [20,215]. An indigenous crops research agenda for Hawai'i's current multicultural setting will necessarily be informed by and be responsive to local values, ecosystems, history and priorities of representative communities. Such an approach would not only support the ongoing movement to restore kupuna crops and traditional relationships, but also more effectively serve, engage, and learn from Hawai'i and other Pacific communities. As we grow more kupuna crops, can we also grow more attentive to indigenous science and the ethics of responsibility; to how we cultivate mutualisms in our gardens and our social institutions.

(2) *Cultivate with the tools of decolonizing methodologies and participatory action research.* Many of the ethical codes and best practices we suggest align with decolonizing/indigenous methodologies that support communities reclaiming traditional ways of knowing and existing. Related tools can be found in participatory action research [206,216], which crosses multiple disciplines, similarly centers community concerns, and provides tools to researchers interested in enabling change. Work begins with listening, observing and asking to understand community priorities and interests and may yield novel insights, open doors for partnerships, grow more

indigenous/farmer scientists, and garner greater public support. Farmers working from their own fields, and within the context of the natural environment and the production methodologies they use, are often well positioned to define questions for both pure and applied research [189,190]. Such a return to center might give rise to more thoughtful attention to issues of invasiveness, economies of scale, and resistance/resilience to natural disasters and climate change. Interest in longer-term solutions may emerge such as integrated pest management, better soil management, crop cultivar diversification, fallowing regimes or other more ecologically and culturally informed approaches, as well as innovative infrastructure, market and supply chain models. In this approach, community members are partners in the research process, and, we suggest, can and should be empowered to hold the greatest opportunity possible to shape research directions.

(3) *Cultivate spaces of healthy exchange.* Hoʻoponopono is a critical first step in the larger process of forward reaching shifts in policy, research and action. We have observed a need for translators and spaces of safe exchange where hard discussions can be held. Yet these safe spaces must be facilitated to aid in closing historic rifts and wounds between indigenous/farmer communities and academic institutions. Until then, progress will continue to be made through individuals shouldering kuleana to care for indigenous crops, one collaboration at a time.

(4) *Allocate institutional support for culturally important facilities, projects and research.* In Hawaiʻi, the University's network of agricultural stations is a critical interface for research, extension and growers. In exploring ideas of how the University might "decolonize" existing research agendas, one potential avenue is reframing the purpose and operations of the stations to better steward ancestral crops and associated knowledge systems. The four cases clearly demonstrate that mechanisms exist for this, including expanding support for Hawaiian crop collections that prioritize cultivar conservation and maintenance, distribution of clean and verified propagation material, and leveraging existing agricultural infrastructure to better support community needs around developing farmer skills and food security. Kupuna crops have much to teach us and serve as a model for many key principles in sustainable agriculture.

(5) *Foster innovation.* Just as ancestral agriculturalists traveled vast distances with precious kupuna crops in hand, diversified them, and developed complex agroecosystems with great ingenuity, there is need for creativity and innovation in today's work with kupuna crops. In advocating for this, however, we challenge the common and limited perception of "innovation"—a word largely defined by mainstream values and industrial technologies. We return instead to Kimmerer's fourth sister who innovates in cultivation; those who observe, weed, balance, and select, too, are innovators working through an iterative, responsive, reflective process, not unlike the iterative *inspiration, ideation, and implementation* principles of "design thinking." [217] Biocultural design, at the nexus of biodiversity and heritage, decolonizing methodologies, and whole systems thinking offers a process for creative solutions to emerge in a bottom-up, community-led setting [196]. A key consideration in this process is understanding the roles university researchers, cultural practitioners, and farmers play in helping maintain sustainable cultivation and biodiversity in the field and on the landscape. The fourth sister, the planter- farmers, cultural practitioners, researchers, and individuals in positions of authority, must be attentive and plant with intention.

5. Closing

The history of ancestral Hawaiian crops is complex. At the same time, there has been tremendous renewal of interest in restoring their original cultivar diversity and strong community-driven revival of knowledge, germplasm, production, and traditional practices associated with these kupuna crops. Although individual and institutional actions ignited and exacerbated the kalo conflict, other efforts and partnerships to sustain ex situ collections have been instrumental for the return and revitalization of cultivar diversity of the four kupuna crops examined here. Moving forward, we suggest reconciliation is needed along with more informed and thoughtful protocols for researchers and research institutions engaging in work with Hawaiian crops. We envision institutions and collaborations that are based on a foundation of respect for plants, culture, and people, that work proactively to protect and steward kupuna crops, and that thoughtfully engage with communities to collectively determine appropriate pathways and the next critical steps forward. From such an approach, meaningful and impactful programs can emerge to strengthen the ongoing restoration of kupuna crops, agricultural landscapes, and Hawai'i's communities. We end on a quote from Osorio and Osorio (2016) [127] (p. 193) outlining how such a vision might be achieved: "We need to tell the old stories and also new stories about ourselves. We need to allow people to discover themselves in the stories that describe how we live responsibly for the 'āina, our ancestors, and 'aumākua. We need to demonstrate … the depth and richness of lives that acknowledge the spirit of other living beings and how they are a part of us. And here is the point. It really is not possible to think of a mountain summit, a fishery, a taro garden, an ulu tree, or an entire island as both a being that shares your spirit and as property in the same breath … Ultimately, it is about reverence."

Author Contributions: This work was conceived by a team of individuals, each considered a subject matter expert on their respective crop and/or topic. Conceptualization, N.K.L., P.L., M.B.K., and A.K.V.; Methodology, A.K.V. and P.L.; Writing-Original Draft Preparation, N.K.L., E.J., J.O., J.B., M.B.K., A.K.V., and P.L.; Writing-Review and Editing, P.L., A.K.V. and N.K.L.

Funding: The APC was funded by Hawai'i Community Foundation. A.K.V. was supported by the Ford Foundation dissertation fellowship program. N.K.L. was funded, in part, by the USDA National Institute of Food and Agriculture, Hatch project HAW08035 administered by the College of Tropical Agriculture and Human Resources.

Acknowledgments: We dedicate this paper to Anakala Jerry Konanui, mahi'ai lepo popolo (MLP), kalo and 'awa master, gifted storyteller, and inspiration to us all. It is not possible, in these few lines, to describe all that Jerry accomplished. He had an enduring devotion to all these kupuna crops. Puhalu ka ihu, nānā i ke kā'ao (One only notices the many good things a person does when it is too late to show appreciation). With regards to the Association for Hawaiian 'Awa and kalo cultivar recovery efforts, he was the heart and without his devotion we would not have had the success that we did. We are deeply grateful for the steadfast dedication to the Hawaiian kalo collection provided by Alton Arakaki that has made it possible for the varieties to reach so many people. We also dedicate this paper to the many, many stewards, mahi'ai, and scholars who have kept alive both these plants, lands, and 'ike we rely on today—amazing gifts from our kūpuna that will feed us long into the future if we take up the work.

Conflicts of Interest: The authors declare no conflict of interest. No funders influenced the writing of this manuscript.

References and Notes

1. Beckwith, M.W. *Hawaiian Mythology*; University of Hawaii Press: Honolulu, HI, USA, 1970; ISBN 978-0-8248-0514-2.

2. Taube, K.A. *The Major Gods of Ancient Yucatan*; Studies in Pre-Columbian Art and Archaeology, No. 32; Dumbarton Oaks Research Library and Collection: Washington, DC, USA, 1992; ISBN 978-0-88402-204-6.

3. Campbell, J.K. (Ed.) *The Kumulipo: An Hawaiian Creation Myth*; Queen Lili'uokalani (Lydia Lili'u Loloku Walania Kamaka'eha), Translator; Pueo Press: Kentfield, CA, USA, 1997.

4. Nabhan, G. (Ed.) *Ethnobiology for the Future: Linking Cultural and Ecological Diversity*; Southwest Center Series; University of Arizona Press: Tucson, AZ, USA, 2016.

5. Vitousek, P.M.; Ladefoged, T.N.; Kirch, P.V.; Hartshorn, A.S.; Graves, M.W.; Hotchkiss, S.C.; Tuljapurkar, S.; Chadwick, O.A. Soils, Agriculture, and Society in Precontact Hawai'i. *Science* **2004**, *304*, 1665–1669. [CrossRef] [PubMed]

6. Kirch, P.V. *The Wet and the Dry: Irrigation and Agricultural Intensification in Polynesia*; University of Chicago Press: Chicago, IL, USA, 1994; ISBN 978-0-226-43749-1.

7. Khoury, C.K.; Bjorkman, A.D.; Dempewolf, H.; Ramirez-Villegas, J.; Guarino, L.; Jarvis, A.; Rieseberg, L.H.; Struik, P.C. Increasing homogeneity in global food supplies and the implications for food security. *Proc. Natl. Acad. Sci. USA* **2014**, *111*, 4001–4006. [CrossRef] [PubMed]

8. Khoury, C.K.; Achicanoy, H.A.; Bjorkman, A.D.; Navarro-Racines, C.; Guarino, L.; Flores-Palacios, X.; Engels, J.M.M.; Wiersema, J.H.; Dempewolf, H.; Sotelo, S.; et al. Origins of food crops connect countries worldwide. *Proc. R. Soc. B* **2016**, *283*, 20160792. [CrossRef]

9. Castañeda-Álvarez, N.P.; Khoury, C.K.; Achicanoy, H.A.; Bernau, V.; Dempewolf, H.; Eastwood, R.J.; Guarino, L.; Harker, R.H.; Jarvis, A.; Maxted, N.; et al. Global conservation priorities for crop wild relatives. *Nat. Plants* **2016**, *2*, 16022. [CrossRef] [PubMed]

10. Schiebinger, L.; Swan, C. *Colonial Botany: Science, Commerce, and Politics in the Early Modern World*; University of Pennsylvania Press: Philadelphia, PA, USA, 2007; ISBN 978-0-8122-2009-4.

11. Esquinas-Alcázar, J. Protecting crop genetic diversity for food security: Political, ethical and technical challenges. *Nat. Rev. Genet.* **2005**, *6*, 946–953. [CrossRef]

12. Hovick, S.M.; Whitney, K.D. Hybridisation is associated with increased fecundity and size in invasive taxa: Meta-analytic support for the hybridisation-invasion hypothesis. *Ecol. Lett.* **2014**, *17*, 1464–1477. [CrossRef]

13. Elias, P. Who owns taro? *Honolulu Star-Bulletin*, 21 January 2006.

14. Andow, D.; Bauer, T.; Belcourt, M.; Bloom, P.; Child, B.; Doerfler, J.; Eule-Nashoba, A.; Heidel, T.; Kokotovich, A.; Lodge, A.; et al. *Preserving the Integrity of Manoomin in Minnesota*; Wild Rice White Paper; University of Minnesota: St. Paul, MN, USA, 2011; Available online: http://www.cfans.umn.edu/sites/cfans.umn.edu/files/WhitePaperFinalVersion2011.pdf (accessed on 4 August 2018).

15. Mgbeoji, I. *Global Biopiracy: Patents, Plants, and Indigenous Knowledge*; UBC Press: Vancouver, BC, Canada, 2014; ISBN 978-0-7748-5170-1.

16. Mead, A.T.P.; Ratuva, S. (Eds.) *Pacific Genes & Life Patents: Pacific Indigenous Experiences & Analysis of the Commodification & Ownership of Life*; Call of the Earth Llamado de la Tierra and The United Nations University Institute of Advanced Studies: Wellington, New Zealand, 2007; ISBN 0-473-11237-X.

17. Rose, G. International Law of Sustainable Agriculture in the 21st Century: The International Treaty on Plant Genetic Resources for Food and Agriculture. *Georget. Int. Environ. Law Rev.* **2002**, *15*, 583.

18. Lindsey, R.H. Responsibility with Accountability: The Birth of a Strategy to Protect Kanaka Maoli Traditional Knowledge Symposium: Intellectual Property and Social Justice. *How. L.J.* **2004**, *48*, 763–786.

19. Gavin, M.C.; McCarter, J.; Mead, A.; Berkes, F.; Stepp, J.R.; Peterson, D.; Tang, R. Defining biocultural approaches to conservation. *Trends Ecol. Evol.* **2015**, *30*, 140–145. [CrossRef]

20. Smith, L.T. *Decolonizing Methodologies: Research and Indigenous Peoples*; Zed Books: New York, NY, USA, 1999; ISBN 978-1-85649-624-7.

21. Athens, J.S.; Rieth, T.M.; Dye, T.S. A Paleoenvironmental and Archaeological Model-Based Age Estimate for the Colonization of Hawai'i. *Am. Antiq.* **2014**, *79*, 144–155. [CrossRef]

22. Kirch, P.V. When Did the Polynesians Settle Hawaii? A Review of 150 Years of Scholarly Inquiry and a Tentative Answer. *Hawaii. Archaeol.* **2011**, *12*, 3–26.

23. Sohmer, S.H.; Gustafson, R. *Plants and Flowers of Hawai'i*; University of Hawaii Press: Honolulu, HI, USA, 1987; ISBN 978-0-8248-1096-2.

24. Abbott, I.A. *Lā'au Hawai'i: Traditional Hawaiian Uses of Plants*; Bishop Museum Press: Honolulu, HI, USA, 1992; ISBN 978-0-930897-62-8.

25. Beckwith, M.W. (Ed.) *The Kumulipo: A Hawaiian Creation Chant*; University of Hawaii Press: Honolulu, HI, USA, 1972; ISBN 978-0-8248-0771-9.

26. Morton, J.F. Breadfruit. In *Fruits of Warm Climates*; Dowling, C.F., Ed.; J.F. Morton: Miami, FL, USA, 1987; pp. 50–58. ISBN 978-0-9610184-1-2.

27. Ragone, D. Ethnobotany of Breadfruit in Polynesia. In *Islands, Plants, and Polynesians: An Introduction to Polynesian Ethnobotany*; Cox, P.A., Banack, S.A., Eds.; Dioscorides Press: Portland, OR, USA, 1991; pp. 203–220. ISBN 978-0-931146-18-3.

28. Zerega, N.; Ragone, D.; Motley, T.J. Breadfruit Origins, Diversity, and Human-Facilitated Distribution. In *Darwin's Harvest: New Approaches to the Origins, Evolution, and Conservation of Crops*; Motley, T.J., Zerega, N., Cross, H., Eds.; Columbia University Press: New York, NY, USA, 2006; pp. 213–238. ISBN 978-0-231-50809-4.

29. Green, R.C. Sweet potato transfers in Polynesian prehistory. In *The Sweet Potato in Oceania: A Reappraisal*; Ethnology Monographs 19, Oceania Monograph 56; University of Sydney: Sydney, NSW, Australia, 2005; pp. 43–62. ISBN 978-0-945428-13-8.

30. McCaughey, V.; Emerson, J. The Kalo in Hawaii. In *The Hawaiian Forester and Agriculturist; a Quarterly Magazine of Forestry, Entomology, Plant Inspection and Animal Industry*; Advertiser Pub. Co., Ltd.: Honolulu, HI, USA, 1913; pp. 186–193, 225–231, 280–288, 315–323, 349–358, 371–375.

31. McCaughey, V.; Emerson, J. The Kalo in Hawaii. In *The Hawaiian Forester and Agriculturist; a Quarterly Magazine of Forestry, Entomology, Plant Inspection and Animal Industry;* Board of Commissioners of Agriculture and Forestry: Honolulu, HI, USA, 1914; pp. 17–23, 44–51, 111–112, 201–216.

32. Wilder, G. *The Kalo of Hawaii;* Wilder Box 24, Bishop Museum Archives: Honolulu, HI, USA, 1933; Unpublished manuscript.

33. Handy, E.S.C. *The Hawaiian Planter, Volume I: His Plants, Methods, and Areas of Cultivation;* Bernice P Bishop Museum: Honolulu, HI, USA, 1940.

34. Johnston, E.; Rogers, H. (Eds.) *Hawaiian 'Awa: Views of an Ethnobotanical Treasure;* Association for Hawaiian 'Awa: Hilo, HI, USA, 2006.

35. Kepler, A.K.; Rust, F.G. *World of Bananas in Hawai'i: Then and Now: Traditional Pacific & Global Varieties, Cultures, Ornamentals, Health & Recipes;* Pali-o-Waipio Press: Haiku, HI, USA, 2011; ISBN 978-0-9837266-0-9.

36. Kagawa-Viviani, A. *Untangling 'Uala: Toward Re-Diversifying and Re-Placing Sweet Potato in the Hawaiian Landscape;* E kūpaku ka 'āina- The Hawai'i Land Restoration Institute: Wailuku, HI, USA, 2016.

37. Konanui, J.; Konanui, G.; Levin, P. *Bulletin 84: Taro Varieties in Hawaii, Revised Edition;* JAK KAW Press: Mount Horeb, WI, USA, 2019; Unpublished.

38. Lincoln, N.K. *Kō: An Ethnobotanical Guide to Hawaiian Sugarcane Varieties;* University of Hawai'i Press: Honolulu, HI, USA, 2018; in press.

39. Handy, E.S.C.; Handy, E.G.; Pukui, M.K. *Native Planters in Old Hawaii: Their Life, Lore, and Environment;* Revised Edition; Bishop Museum Press: Honolulu, HI, USA, 1991.

40. Levin, P.; Sugii, N. Ho'olanalana—Learning from the Spider: Generational Integrity in Vegetative Seed Storage and Recovery. In Proceedings of the Hawaii Native Seed Conservation Conference, Honolulu, HI, USA, 15–18 May 2018.

41. Ladefoged, T.N.; Kirch, P.V.; Gon, S.M., III; Chadwick, O.A.; Hartshorn, A.S.; Vitousek, P.M. Opportunities and constraints for intensive agriculture in the Hawaiian archipelago prior to European contact. *J. Archaeol. Sci.* **2009**, *36*, 2374–2383. [CrossRef]

42. Kurashima, N.; Kirch, P.V. Geospatial modeling of pre-contact Hawaiian production systems on Moloka'i Island, Hawaiian Islands. *J. Archaeol. Sci.* **2011**, *38*, 3662–3674. [CrossRef]

43. Levin, P. Searching for Sustainable Agriculture in Hawai'i. In *Thinking Like an Island: Navigating a Sustainable Future in Hawaii;* Chirico, J., Farley, G.S., Eds.; University of Hawai'i Press: Honolulu, HI, USA, 2015; pp. 46–78. ISBN 978-0-8248-5416-4.

44. Queen Emma (Emma Kalanikaumakaamano Kaleleonālani Na'ea Rooke). *Observations on Varieties and Culture of Taro;* HEN Collection Box 71 Folio 76-83; Bishop Museum Archives: Honolulu, HI, USA, 1860.

45. Iokepa, J. Taro Culture, Kelsey Collection, Hilo. HEN MS DOC 253, Bishop Museum Archives nd.

46. Kamakau, S.M. *The Works of the People of Old: Na Hana a ka Po'e Kahiko;* Barrère, D.B., Ed.; Pukui, M.K., Translator; Bishop Museum Press: Honolulu, HI, USA, 1976.

47. Tsuha, A.K. Kaulana Mahina: He 'ōnaehana 'alemanaka Hawai'i. Master's Thesis, University of Hawai'i at Mānoa, Honolulu, HI, USA, 2007.

48. Johnson, R.K.; Mahelona, J.K.; Ruggles, C.L.N. *Na Inoa Hoku: Hawaiian and Pacific Star Names, Revised Edition;* Ocarina Books: West Sussex, UK, 2015; ISBN 978-0-9540867-5-6.

49. Stannard, D.E. *Before the Horror: The Population of Hawai'i on the Eve of Western Contact;* Social Science Research Institute, University of Hawaii: Honolulu, HI, USA, 1989; ISBN 978-0-8248-1232-4.

50. Swanson, D. *The Number of Native Hawaiians and Part-Hawaiians in Hawai'i, 1778 to 1900: Demographic Estimates by Age, with Discussion;* Canadian Population Society: Calgary, AB, Canada, 2016.

51. Schmitt, R.C.; Nordyke, E.C. Death in Hawai'i: The epidemics of 1848–1849. *Hawaii. J. Hist.* **2001**, *35*, 1–13.

52. Allen, M.S. *Gardens of Lono: Archaeological Investigations at the Amy B.H. Greenwell Ethnobotanical Garden, Kealakekua, Hawai'i;* Bishop Museum Press: Honolulu, HI, USA, 2001; ISBN 978-1-58178-008-6.

53. Farber, J.M. *Ancient Hawaiian Fishponds: Can Restoration Succeed on Moloka'i?* Neptune House Publications in Association with the East-West Center's Pacific Islands Development Program: Encinitas, CA, USA, 1997; ISBN 0-965978 2-0-6.

54. Sterling, E.P. *Sites of Maui;* Bishop Museum Press: Honolulu, HI, USA, 1998; ISBN 978-0-930897-97-0.

55. Tong, N.W.T. He 'Āina Wai: Remembering Water Narratives of Wai'anae Kai. Master's Thesis, University of Hawai'i at Mānoa, Honolulu, HI, USA, 2014.

56. Osorio, J.K. *Dismembering Lāhui: A History of the Hawaiian Nation to 1887*; University of Hawai'i Press: Honolulu, HI, USA, 2002; ISBN 978-0-8248-2432-7.

57. Cooper, G.; Daws, G. *Land and Power in Hawaii: The Democratic Years*; Benchmark Books: Honolulu, HI, USA, 1985; ISBN 978-0-9615052-0-2.

58. Kame'eleihiwa, L. *Native Land and Foreign Desires: Pehea Lā E Pono Ai?* Bishop Museum Press: Honolulu, HI, USA, 1992.

59. Preza, D.C. The Empirical Writes Back: Re-Examining Hawaiian Dispossession Resulting from the Māhele of 1848. Master's Thesis, University of Hawai'i at Mānoa, Honolulu, HI, USA, 2010.

60. Beamer, B.K. Huli Ka Palena. Master's Thesis, University of Hawai'i at Mānoa, Honolulu, HI, USA, 2005.

61. Taro Security and Purity Task Force (Taro Task Force). *E ola hou ke kalo; ho'i hou ka 'āina lē'ia: The Taro Lives, Abundance Returns to the Land*; Taro Security and Purity Task Force Report to the Hawaii State Legislature; Office of Hawaiian Affairs: Honolulu, HI, USA, 2009; Available online: https://www.oha.org/culture/taro-security-and-purity-task-force/ (accessed on 4 August 2018).

62. Whitney, L.D.; Bowers, F.A.I.; Takahashi, M. *Bulletin 84: Taro Varieties in Hawaii*; Hawaii Agricultural Experiment Station, University of Hawaii: Honolulu, HI, USA, 1939.

63. USDA National Agricultural Statistics Service Quick Stats Database. Available online: https://quickstats.nass.usda.gov/ (accessed on 4 August 2018).

64. USDA NASS. *Hawaii 2016 Vegetable and Melon Crops Report, Revised January 2018*; United States Department of Agriculture in Cooperation with the Hawaii Department of Agriculture: Honolulu, HI, USA, 2018.

65. USDA NASS. *Hawaii 2017 Vegetable and Melon Crops Report*; United States Department of Agriculture in Cooperation with the Hawaii Department of Agriculture: Honolulu, HI, USA, 2018.

66. Lincoln, N.; Ladefoged, T. Agroecology of pre-contact Hawaiian dryland farming: The spatial extent, yield and social impact of Hawaiian breadfruit groves in Kona, Hawai'i. *J. Archaeol. Sci.* **2014**, *49*, 192–202. [CrossRef]

67. Wilcox, C. *Sugar Water: Hawaii's Plantation Ditches*; Reprint Edition; University of Hawaii Press: Honolulu, HI, USA, 1997; ISBN 978-0-8248-2044-2.

68. Kent, G. Food Security in Hawai'i. In *Food and Power in Hawai'i: Visions of Food Democracy*; Kimura, A.H., Suryanata, K., Eds.; University of Hawaii Press: Honolulu, HI, USA, 2016; pp. 36–53. ISBN 978-0-8248-5861-2.

69. Bushnell, O.A. *The Gifts of Civilization: Germs and Genocide in Hawai'i*; University of Hawaii Press: Honolulu, HI, USA, 1993; ISBN 978-0-8248-1457-1.

70. McGregor, D. *Na Kua'aina: Living Hawaiian Culture*; University of Hawaii Press: Honolulu, HI, USA, 2007; ISBN 978-0-8248-2946-9.

71. Peralto, N. Kokolo Mai ka Mole Uaua o 'Ī: The Resilience & Resurgence of Aloha 'Āina in Hāmākua Hikina, Hawai'i. Ph.D. Thesis, Department of Political Science, University of Hawai'i at Mānoa, Honolulu, HI, USA, 10 May 2018.

72. Low, S. *Hawaiki Rising: Hokule'a, Nainoa Thompson and the Hawaiian Renaissance*; University of Hawai'i Press: Honolulu, HI, USA, 2013; ISBN 978-1617102004.

73. Kahakalau, K.H. Kanu o ka 'Āina: Natives of the Land from Generations Back. A Pedagogy of Hawaiian Liberation. Ph.D. Dissertation, Union Institute and University, Cincinnati, OH, USA, 2002.

74. Kawelu, K. *Kuleana and Commitment*; University of Hawai'i Press: Honolulu, HI, USA, 2015; ISBN 978-0-8248-5712-7.

75. Trask, H.-K. Birth of the Modern Hawaiian Movement: Kalama Valley, O'ahu. *Hawaii. J. Hist.* **1987**, *21*, 126–153.

76. Goodyear-Ka'ōpua, N. Rebuilding the 'Auwai: Connecting Ecology, Economy and Education in Hawaiian Schools. *Altern. Int. J. Indig. Peoples* **2009**, *5*, 46–77. [CrossRef]

77. Keala, G. "Buddy" *Loko i'a: A Manual on Hawaiian Fishpond Restoration and Management*; College of Tropical Agriculture and Human Resources, University of Hawai'i at Mānoa: Honolulu, HI, USA, 2007; ISBN 978-1-929325-20-7.

78. Fox, C.T. The Essential Guide to Taro: A Kalo Culture Comeback. *Honolulu Magazine*, 28 December 2017.

79. Kubota, G.T. *Hawaii Stories of Change: Kokua Hawaii Oral History Project*; Kokua Hawaii Oral History Project: Honolulu, HI, USA, 2018; ISBN 978-0-9799467-2-1.

80. Osorio, J.K.K. Hawaiian Issues. In *The Value of Hawai'i: Knowing the Past, Shaping the Future*; Howes, C., Osorio, J.K.K., Eds.; University of Hawai'i Press: Honolulu, HI, USA, 2010; pp. 15–21.

81. Goodyear-Kaopua, N. Kuleana Lahui: Collective Responsibility for Hawaiian Nationhood in Activists' Praxis. *Affin. J. Radic. Theory Cult. Action* **2011**, *5*, 130–163.

82. Harris, D.R. Origins and Spread of Agriculture. In *The Cultural History of Plants*; Prance, S.G., Nesbitt, M., Eds.; Routledge: New York, NY, USA, 2005.

83. Denham, T. Early Agriculture and Plant Domestication in New Guinea and Island Southeast Asia. *Curr. Anthropol.* **2011**, *52*, S379–S395. [CrossRef]

84. Lebot, V. Section II: Sweet Potato. In *Tropical Root and Tuber Crops: Cassava, Sweet Potato, Yams and Aroids*; CABI: Wallingford, UK, 2008; pp. 91–179. ISBN 978-1-84593-621-1.

85. Kreike, C.M.; Eck, H.J.V.; Lebot, V. Genetic diversity of taro, *Colocasia esculenta* (L.) Schott, in Southeast Asia and the Pacific. *Theor. Appl. Genet.* **2004**, *109*, 761–768. [CrossRef]

86. Chaïr, H.; Traore, R.E.; Duval, M.F.; Rivallan, R.; Mukherjee, A.; Aboagye, L.M.; Rensburg, W.J.V.; Andrianavalona, V.; de Carvalho, M.A.A.P.; Saborio, F.; et al. Genetic Diversification and Dispersal of Taro (*Colocasia esculenta* (L.) Schott). *PLoS ONE* **2016**, *11*, e0157712. [CrossRef]

87. Matthews, P.J.; Nguyen, D.V. Taro: Origins and Development. In *Encyclopedia of Global Archaeology*; Springer: New York, NY, USA, 2014; pp. 7237–7240.

88. Lebot, V. *Tropical Root and Tuber Crops: Cassava, Sweet Potato, Yams and Aroids*; CABI: Wallingford, UK, 2008; ISBN 978-1-84593-621-1.

89. Pukui, M.K.; Elbert, S.H. *Hawaiian Dictionary: Revised and Enlarged Edition*; University of Hawaii Press: Honolulu, HI, USA, 1986; ISBN 978-0-8248-0703-0.

90. Pukui, M.K.; Elbert, S.H. Appendix A: Glossary of Hawaiian gods, demigods, family gods, and a few heroes. In *Hawaiian Dictionary: Revised and Enlarged Edition*; University of Hawaii Press: Honolulu, HI, USA, 1971; ISBN 978-0-8248-0703-0.

91. Johnson, R.K. *The Hawaiian Kinolau: Manifestations of Deity in the Natural and Spiritual World*; Workshop of Primal Spirituality; Institute of Culture and Communication, East-West Center: Honolulu, HI, USA, 1991.

92. Dorton, L. A Legendary Tradition of Kamapua'a, The Hawaiian Pig-God. Master's Thesis, University of Hawai'i at Mānoa, Honolulu, HI, USA, 1982.

93. Chun, M.N.; Hawaii Board of Health. *Native Hawaiian Medicines*; A New Revised and Enlarged Translation; First People's Productions: Honolulu, HI, USA, 1994.

94. Maly, K.; Maly, O. *Ka Hana Lawai'a a me Nā Ko'a O Na Kai 'Ewalu: A History of Fishing Practices and Marine Fisheries of the Hawaiian Islands Compiled from Native Hawaiian Traditions, Historical Accounts, Government Communications, Kama'āina Testimony and Ethnography, Vol I and II*; Kumu Pono Associates LLC Prepared for the Nature Conservancy: Honolulu, HI, USA, 2003.

95. Pukui, M.K.; Haertig, E.W.; Lee, C.A. *Nānā i ke kumu (Look to the Source) Volume I*; Hui Hanai, a Queen Lili'uokalani Children's Center Publication: Honolulu, HI, USA, 1972; ISBN 978-0-916630-15-7.

96. E nalohia ana ka oni ame ka luli ana o ka lau kalo (Gone will be the fluttering and waving of the taro leaves). *Kuokoa Home Rula*, 18 August 1911.

97. Payne, J.H.; Ley, G.J.; Akau, G. *Processing and Chemical Investigations of Taro*; Bulletin 86; Hawaii Agricultural Experiment Station, University of Hawaii: Honolulu, HI, USA, 1941.

98. Bowers, F.A.I.; Plucknett, D.L.; Younge, O.R. *Specific Gravity Evaluation of Corm Quality in Taro*; Circular No. 61; Hawaii Agricultural Experiment Station, University of Hawaii: Honolulu, HI, USA, 1964.

99. Lin, B.B. Resilience in Agriculture through Crop Diversification: Adaptive Management for Environmental Change. *BioScience* **2011**, *61*, 183–193. [CrossRef]

100. Jacques, P.; Jacques, J.; Jacques, P.J.; Jacques, J.R. Monocropping Cultures into Ruin: The Loss of Food Varieties and Cultural Diversity. *Sustainability* **2012**, *4*, 2970–2997. [CrossRef]

101. Nelson, S. Poly- and Monocultures: The Good, the Bad, and the Ugly: Trees for Improving Sustainability, Resource Conservation, and Profitability on Farms and Ranches. In Proceedings of the Pacific Island Agroforestry Workshop-Hawai'i, Kona, HI, USA, 16–19 May 2006.

102. Trujillo, E.E.; Menezes, T.D.; Cavaletto, C.G.; Shimabuku, R.; Fukuda, S.K. *Promising New Taro Cultivars with Resistance to Taro Leaf Blight: "Pa'lehua", "Pa'akala", and "Pauakea"*; New Plants for Hawaii, NPH-7; College of Tropical Agriculture and Human Resources Cooperative Extension Service, University of Hawaii: Honolulu, HI, USA, 2002.

103. Cho, J.J. *Breeding Hawaiian Taros for the Future*; Maui Agricultural Research Center, University of Hawaii: Kula, HI, USA, 2003.

104. De la Peña, R.S. Development of New Taro Varieties through Breeding. In *Proceedings of the Taking Taro into the 1990s: A Taro Conference*; Hollyer, J.R., Sato, D.M., Eds.; University of Hawaii: Hilo, HI, USA, 1990; pp. 32–36.

105. TenBruggencate, J. UH expected to abandon controversial taro patents. *The Honolulu Advertiser*, 20 June 2006.

106. USDA Research, Education & Economics Information System. Preservation of Hawaiian Taro Cultivars and the Development of Pest Resistant Commercial Taro Hybrids, Project No. HAW00948-H (2005–2010). Available online: https://portal.nifa.usda.gov/web/crisprojectpages/0205446-preservation-of-hawaiian-taro-cultivars-and-the-development-of-pest-resistant-commercial-taro-hybrids.html (accessed on 6 August 2018).

107. College of Tropical Agriculture and Human Resources. *CTAHR and Taro: Taro Research by the College of Tropical Agriculture and Human Resources*; Background Paper; College of Tropical Agriculture and Human Resources: Honolulu, HI, USA, 2009.

108. USDA Research, Education & Economics Information System. Genomic and Biotechnological Approaches for Evaluating and Improving Tropical Crops, Project No. 5320-21000-010-00D (2003–2006). Available online: https://portal.nifa.usda.gov/web/crisprojectpages/0407217-genomic-and-biotechnological-approaches-for-evaluating-and-improving-tropical-crops.html (accessed on 6 August 2018).

109. Miyasaka, S. *Update on Genetic Engineering of Chinese Taro (Variety Bunlong) for Increased Disease Resistance*; University of Hawaii: Honolulu, HI, USA, 2006.

110. He, X.; Miyasaka, S.C.; Fitch, M.M.M.; Moore, P.H.; Zhu, Y.J. Agrobacterium tumefaciens-mediated transformation of taro (*Colocasia esculenta* (L.) Schott) with a rice chitinase gene for improved tolerance to a fungal pathogen *Sclerotium rolfsii*. *Plant Cell Rep.* **2008**, *27*, 903–909. [CrossRef] [PubMed]

111. Gugganig, M. The Ethics of Patenting and Genetically Engineering the Relative Hāloa. *Ethnos* **2017**, *82*, 44–67. [CrossRef]

112. Ritte, W.; Kanehe, L.M. Kuleana No Hāloa (Responsibility for Taro) Protecting the Sacred Ancestor from Ownership and Genetic Modification. In *Pacific Genes & Life Patents: Pacific Indigenous Experiences & Analysis of the Commodification & Ownership of Life*; Mead, A.T.P., Ratuva, S., Eds.; Call of the Earth Llamado de la Tierra and The United Nations University Institute of Advanced Studies: Wellington, New Zealand, 2007; ISBN 0-473-11237-X.

113. Zhu, J.; Fitch, M.M.M.; Moore, P.H.; He, L.; Miyasaka, S.C.; Tanabe, M.; Cho, J. Development of a Transformation and Regeneration System for Taro. In *2003 Hawaii Agriculture Research Center Annual Report*; Schenck, S., Vance, B., Eds.; HARC: Hawaii, HI, USA, 2003.

114. Senate Concurrent Resolution 206: Protection of Taro. 2007. Available online: http://lrbhawaii.org/legis07/req07.pdf (accessed on 6 August 2018).

115. Conway-Jones, D. Safeguarding Hawaiian Traditional Knowledge and Cultural Heritage: Supporting the Right to Self-Determination and Preventing the Co-Modification of Culture Symposium: Intellectual Property and Social Justice. *How. L.J.* **2004**, *48*, 737–762.

116. Schlais, G.K. The Patenting of Sacred Biological Resources, the Taro Patent Controversy in Hawai'i: A Soft Law Proposal Recent Developments. *U. Haw. L. Rev.* **2006**, *29*, 581–618.

117. Act 211: Relating to Taro; Volume SB2915 SD2 HD1 CD1 (CCR 175-08); 2008.

118. Hawaii Revised Statutes 321-4.7: Producers of Hand-Pounded Poi; Exemption. 2011. Available online: https://www.capitol.hawaii.gov/hrscurrent/Vol06_Ch0321-0344/HRS0321/HRS_0321-0004_0007.htm (accessed on 6 August 2018).

119. Shimabukuro, B. Taro Traditions: Farmers find spirit; chefs find inspiration in humble root. *Honolulu Star-Bulletin*, 12 November 2003.

120. Rummell, N. Pa'i'ai: Hawaii's Link to the Past and Glimpse at Its Future. StarChefs.com, October 2012. Available online: https://www.starchefs.com/cook/savory/product/paiai (accessed on 6 August 2018).

121. Mishan, L. The Chefs Redefining Polynesian Cuisine. *The New York Times*, 6 August 2018.

122. Thiam, P. A Forgotten Ancient Grain That Could Help Africa Prosper (TED 2017). Available online: https://www.ted.com/talks/pierre_thiam_a_forgotten_ancient_grain_that_could_help_africa_prosper/transcript (accessed on 9 October 2018).

123. Rao, S.J. History on your plate. *The Hindu*, 11 January 2018.

124. Gregory, R. USA—Hawaii—Restoring the Life of the Land: Taro Patches in Hawai'i. The EcoTipping Points Project. August 2014. Available online: http://www.ecotippingpoints.org/our-stories/indepth/usa-hawaii-taro-agriculture.html (accessed on 27 October 2018).

125. 'Onipa'a Nā Hui Kalo. *Guidelines for Grassroots Lo'i Kalo Rehabilitation: Pono, Practical Procedures for Lo'i Kalo Restoration*; Queen Lili'uokalani Childrens Center: Honolulu, HI, USA, 2003.

126. 'Onipa'a Nā Hui Kalo. *A Kanaka Brown Paper. Unpublished Report to Queen Liliuokalani Childrens Center*; Queen Lili'uokalani Childrens Center: Honolulu, HI, USA, 2016.

127. Osorio, J.; Osorio, J. Two Perspectives on Political Narrative in One Activist Family. *Hūlili* **2016**, *10*, 185–201.

128. Konanui, J. (cultural practitioner; kalo and 'awa specialist). Personal communication, 2005–2017.

129. James, S.; Bolick, H.; Imada, C. *Genetic Variability within and Identification Markers for Hawaiian Kalo Varieties (Colocasia esculenta (L.) Schott–Araceae) Using ISSR-PCR*; Final Report. Contribution No. 2012-018; Hawai'i Biological Survey and Pacific Center for Molecular Biodiversity, Bishop Museum: Honolulu, HI, USA, 2012.

130. Helmkampf, M.; Wolfgruber, T.K.; Bellinger, M.R.; Paudel, R.; Kantar, M.B.; Miyasaka, S.C.; Kimball, H.L.; Brown, A.; Veillet, A.; Read, A.; et al. Phylogenetic Relationships, Breeding Implications, and Cultivation History of Hawaiian Taro (*Colocasia esculenta*) Through Genome-Wide SNP Genotyping. *J. Hered.* **2018**, *109*, 272–282. [CrossRef] [PubMed]

131. Handy, E.S.C.; Pukui, M.K. *The Polynesian Family System in Ka-'u, Hawai'i*; Later Printing Edition; Charles E. Tuttle Company: Rutland, VT, USA, 1972.

132. Roberts, M. Ways of seeing: Whakapapa. *Sites: J. Soc. Anthropol. Cult. Stud.* **2013**, *10*, 93–120. [CrossRef]

133. Roskruge, N. *Rauwaru, the Proverbial Garden: Ngaweri, Maori Root Vegetables, Their History and Tips on Their Use*; Institute of Agriculture and Environment: Palmerston North, New Zealand, 2014.

134. Kamakau, S.M. Ka Moolelo Hawaii. *Ke Au Okoa*, 21 October 1869.

135. Fornander, A.; Stokes, J.F.G. *An Account of the Polynesian Race: Its Origins and Migrations, and the Ancient History of the Hawaiian People to the Times of Kamehameha I*; Trubner & Company: Heidelberg, Germany, 1878.

136. Namakaokeahi, B.K. *The history of Kanalu: Mo'okū'auhau 'elua*; First Peoples Productions: Honolulu, HI, USA, 2004.

137. Yen, D.E. *The Sweet Potato and Oceania: An Essay in Ethnobotany*; Bernice P. Bishop Museum Bulletin 236; Bishop Museum Press: Honolulu, HI, USA, 1974; ISBN 978-0-910240-17-8.

138. Ballard, C.; Brown, P.; Bourke, R.M.; Harwood, T. *The Sweet Potato in Oceania: A Reappraisal*; Oceania Monographs No. 56; University of Sydney: Sydney, NSW, Australia, 2005; ISBN 978-0-945428-13-8.

139. Kaschko, M.W.; Allen, M.S. The impact of the sweet potato on prehistoric Hawaiian cultural development. In Proceedings of the Second Conference in Natural Sciences, Hawaii Volcanoes National Park, HI, USA, 1–3 June 1978; Cooperative National Park Resources Studies Unit, University of Hawaii at Manoa, Department of Botany: Honolulu, HI, USA, 1978.

140. Ladefoged, T.N.; Graves, M.W.; Coil, J.H. The Introduction of Sweet Potato in Polynesia: Early Remains in Hawai'i. *J. Polyn. Soc.* **2005**, *114*, 359–374.

141. Coil, J.; Kirch, P.V. An Ipomoean landscape: Archaeology and the sweet potato in Kahikinui, Maui, Hawaiian Islands. In *The Sweet Potato in Oceania: A Reappraisal*; Ballard, C., Brown, P., Bourke, R.M., Harwood, T., Eds.; Oceania Monographs No. 56; University of Sydney: Sydney, NSW, Australia, 2005; pp. 71–84. ISBN 978-0-945428-13-8.

142. Solis, R.K. Kekahi Mau Pule Mahi'ai. Master's Thesis, University of Hawai'i at Mānoa, Honolulu, HI, USA, 1999.

143. Pukui, M.K. *'Ōlelo No'eau: Hawaiian Proverbs & Poetical Sayings*; Bishop Museum Press: Honolulu, HI, USA, 1983; ISBN 978-0-910240-93-2.

144. Trapp, K. No ka mahi'ai 'ana, māhele 3. *Ka Ho'oilina* **2003**, *2*, 2–15.

145. Yen, D.E. The New Zealand Kumara or Sweet Potato. *Econ. Bot.* **1963**, *17*, 31–45. [CrossRef]

146. Chung, H.L. Sweet Potato Breeding and Selection. Master's Thesis, College of Hawaii, Honolulu, HI, USA, 1920.

147. Takahashi, M. Self and Cross Fertility and Sterility Studies of the Sweet Potato (Ipomoea batatas (L.) Poir). Master's Thesis, University of Hawai'i at Mānoa, Honolulu, HI, USA, 1937.

148. Poole, C.F. *Sweet Potato Genetic Studies*; Technical Bulletin No. 27; Hawaii Agricultural Experiment Station, College of Tropical Agriculture, University of Hawaii: Honolulu, HI, USA, 1955.

149. Poole, C.F. *Seedling Improvement in Sweet Potato*; Technical Bulletin No. 17; Hawaii Agricultural Experiment Station, College of Tropical Agriculture, University of Hawaii: Honolulu, HI, USA, 1952.

150. Roullier, C.; Benoit, L.; McKey, D.B.; Lebot, V. Historical collections reveal patterns of diffusion of sweet potato in Oceania obscured by modern plant movements and recombination. *Proc. Natl. Acad. Sci. USA* **2013**, *110*, 2205–2210. [CrossRef]

151. Roullier, C.; Rossel, G.; Tay, D.; Mckey, D.; Lebot, V. Combining chloroplast and nuclear microsatellites to investigate origin and dispersal of New World sweet potato landraces. *Mol. Ecol.* **2011**, *20*, 3963–3977. [CrossRef] [PubMed]

152. Winnicki, E.; Perez, K.; Kagawa-Viviani, A.; Radovich, T.; Kantar, M.B. *Genetic Diversity of 'Uala (Sweet Potato) in Hawai'i*; Presented at American Society of Plant Biology: Montreal, QC, Canada, 14–18 July 2018.

153. Reynolds, S.; Newcomb, W.; Marshall, J.F.B.; Wood, R.W.; Lee, W.L. Circular (Issued by the Committee on the first of June). *Trans. Royal Hawaii. Agri. Society* **1850**, *1*, 5–9.

154. Napihelua, M.L. Uala! Uala! *Ka Hae Hawaii*, 4 March 1857.

155. Corn and potatoes for local planting: Experiment stations tells about results obtained on Haiku demonstration farm.—Seed for distribution. *The Maui News*, 4 February 1916; p. 6.

156. Maui agricultural notes: Distribution of choice varieties: Sweet potato cuttings. *The Maui News*, 7 December 1917; p. 7.

157. Sahr, G.W. Items of Interest to Our Homesteaders. *The Garden Island*, 31 December 1918; p. 3.

158. Marshall, K.; Koseff, C.; Roberts, A.; Lindsey, A.; Kagawa-Viviani, A.; Lincoln, N.; Vitousek, P. Restoring people and productivity to Puanui: Challenges and opportunities in the restoration of an intensive rain-fed Hawaiian field system. *Ecol. Soc.* **2017**, *22*. [CrossRef]

159. Kagawa, A.K.; Vitousek, P.M. The Ahupua'a of Puanui: A Resource for Understanding Hawaiian Rain-Fed Agriculture. *Pac. Sci.* **2012**, *66*, 161–172. [CrossRef]

160. Kaaiakamanu, D.M.; Akina, J.K. *Hawaiian Herbs of Medicinal Value, Found among the Mountains and Else Where in the Hawaiian Islands and Known to the Hawaiians to Possess Curative and Palliative Properties Most Effective in Removing Physical Ailments*; Territory of Hawaii Board of Health: Honolulu, HI, USA, 1922.

161. Chun, M.N. *Hawaiian medicine book/He buke laau lapaau*; Bess Press: Honolulu, HI, USA, 1986.

162. Chun, M.N.; Hawaii Board of Health. *Native Hawaiian medicines*; A new revised and enlarged translation; First People's Productions: Honolulu, HI, USA, 1994.

163. Lincoln, N.K. *Kō: An Ethnobotnical Guide to Hawaiian Sugarcane Cultivars*; University of Hawai'i Press: Honolulu, HI, USA, in press.

164. Lincoln, N.K.; Vitousek, P. Nitrogen fixation during decomposition of sugarcane (*Saccharum officinarum*) is an important contribution to nutrient supply in traditional dryland agricultural systems of Hawai'i. *Int. J. Agric. Sustain.* **2016**, *14*, 214–230. [CrossRef]

165. Lincoln, N.; Kagawa-Viviani, A.; Marshall, K.; Vitousek, P.M. Observations of sugarcane and knowledge specificity in traditional Hawaiian cropping systems. In *Sugarcane: Production Systems, Uses and Economic Importance*; Murphy, R., Ed.; Nova Science Publishers, Inc.: Hauppauge, NY, USA, 2017; ISBN 978-1-5361-0898-9.

166. Moir, W.W.G. The Native Hawaiian Canes. In Proceedings of the Fourth International Congress of the Association of Sugar Cane Technologists, San Juan, Puerto Rico, 1–16 March 1932.

167. Lincoln, N. Kō: An Ethnobotanical Guide to Hawaiian Sugarcane Varieties. Available online: https://cms.ctahr.hawaii.edu/cane (accessed on 2 December 2018).

168. Ruppenthal, S. Sweet Disposition. *Kikaha by Island Air*, 6 January 2016.

169. McNarie, A.D. Sugarland. *Hana Hou!* July 2012. Available online: https://hanahou.com/15.3/sugarland (accessed on 4 August 2018).

170. Speere, B. Kō: Hawaiian Sugarcane. *Maui No Ka 'Oi Magazine*. April 2017. Available online: https://mauimagazine.net/ko-hawaiian-sugarcane/ (accessed on 4 August 2018).

171. Green, L.S.; Pukui, M.K.; Beckwith, M.W. *The Legend of Kawelo: And Other Hawaiian Folk Tales*; Territory of Hawai'I: Honolulu, HI, USA, 1936.

172. Titcomb, M. Kava in Hawaii. *J. Polyn. Soc.* **1948**, *57*, 105–171.

173. Fornander, A. *Fornander Collection of Hawaiian Antiquities and Folk-Lore*; Memoirs of the Bernice Pauahi Bishop Museum of Polynesian Ethnology and Natural History v. 4–6; Bishop Museum Press: Honolulu, HI, USA, 1916.

174. Winter, K. Hawaiian 'Awa, *Piper methysticum*: A Study in Ethnobotany. Master's Thesis, University of Hawai'i at Mānoa, Honolulu, HI, USA, 2004.
175. Board of Commissioners of Agriculture and Forestry. The Gathering of 'Awa. Volume 7, p. 203. Available online: https://www.biodiversitylibrary.org/item/181573#page/273/mode/1up (accessed on 4 August 2018).
176. Lebot, V.; McKenna, D.J.; Johnston, E.; Zheng, Q.Y.; McKern, D. Morphological, phytochemical, and genetic variation in Hawaiian cultivars of 'Awa (Kava, *Piper methysticum*, Piperaceae). *Econ. Bot.* **1999**, *53*, 407–418. [CrossRef]
177. Lebot, V.; Merlin, M.; Lindstrom, L. *Kava: The Pacific Elixir: The Definitive Guide to Its Ethnobotany, History, and Chemistry*; Healing Arts Press: Rochester, VT, USA, 1992; ISBN 978-0-89281-726-9.
178. Baker, J.D. Tradition and toxicity: Evidential cultures in the kava safety debate. *Soc. Stud. Sci.* **2011**, *41*, 361–384. [CrossRef] [PubMed]
179. Food and Agriculture Organization of the United Nations and World Health Organization. *Kava: A Review of the Safety of Traditional and Recreational Beverage Consumption*; Technical Report; Food and Agriculture Organization of the United Nations: Rome, Italy; World Health Organization: Geneva, Switzerland, 2016.
180. Teschke, R.; Schulze, J. Risk of Kava Hepatotoxicity and the FDA Consumer Advisory. *JAMA* **2010**, *304*, 2174–2175. [CrossRef] [PubMed]
181. Martin, A.C.; Johnston, E.; Xing, C.; Hegeman, A.D. Measuring the Chemical and Cytotoxic Variability of Commercially Available Kava (*Piper methysticum* G. Forster). *PLoS ONE* **2014**, *9*, e111572. [CrossRef] [PubMed]
182. McNarie, A.D. Root Medicine. *Hana Hou!* November 2012, pp. 91–97. Available online: https://hanahou.com/15.5/root-medicine (accessed on 4 August 2018).
183. Food and Agriculture Organization of the United Nations. *Globally Important Agricultural Heritage Systems: Combining Agricultural Biodiversity, Resilient Ecosystems, Traditional Farming Practices and Cultural Identity*; GIAHS Programme; FAO: Rome, Italy, 2018.
184. Winter, I.S. Invest in ancient fig cultivars in Morocco, invest in the future. *Green Prophet*, 10 November 2015.
185. Nabhan, G.P. *Desert Terroir: Exploring the Unique Flavors and Sundry Places of the Borderlands*; University of Texas Press: Austin, TX, USA, 2012; ISBN 978-0-292-72589-8.
186. Álvarez-Buylla Roces, E.; Carreón García, A.; San Vicente Tello, A. *Haciendo Milpa: La Protección de las Semillas y la Agricultura Campesina [The Protection of the Seeds of Small Farmers]*; Semillas de Vida. Ciudad Universitaria, Delegación Coyoacán: México City, Mexico, 2011; ISBN 978-607-02-2456-0.
187. Acabado, S.; Martin, M. Between pragmatism and cultural context. Continuity and change in Ifugao wet-rice agriculture. In *Water & Heritage: Material, Conceptual and Spiritual Connections*; Willems, W.J.H., van Schaik, H.P.J., Eds.; Sidestone Press: Leiden, The Netherlands, 2015; pp. 273–295. ISBN 978-90-8890-278-9.
188. Jansen, T. *Hidden Taro, Hidden Talents: A Study of on-Farm Conservation of Colocasia esculenta (taro) in Solomon Islands*; Solomon Islands Planting Material Network and Kastom Garden Association: Honiara, Solomon Islands, 2002.
189. Coomes, O.T.; McGuire, S.J.; Garine, E.; Caillon, S.; McKey, D.; Demeulenaere, E.; Jarvis, D.; Aistara, G.; Barnaud, A.; Clouvel, P.; et al. Farmer seed networks make a limited contribution to agriculture? Four common misconceptions. *Food Policy* **2015**, *56*, 41–50. [CrossRef]
190. Pfeiffer, J.M.; Dun, S.; Mulawarman, B.; Rice, K.J. Biocultural diversity in traditional rice-based agroecosystems: Indigenous research and conservation of mavo (*Oryza sativa* L.) upland rice landraces of eastern Indonesia. *Environ. Dev. Sustain.* **2006**, *8*, 609–625. [CrossRef]
191. Graddy, T.G. Regarding biocultural heritage: In situ political ecology of agricultural biodiversity in the Peruvian Andes. *Agric. Hum. Values* **2013**, *30*, 587–604. [CrossRef]
192. Graddy, T.G. Situating in situ: A critical geography of agricultural biodiversity conservation in the Peruvian Andes and beyond. *Antipode* **2014**, *46*, 426–454. [CrossRef]
193. Glover, D.; Stone, G.D. Heirloom rice in Ifugao: An 'anti-commodity' in the process of commodification. *J. Peasant Stud.* **2018**, *45*, 776–804. [CrossRef]
194. Ficiciyan, A.; Loos, J.; Sievers-Glotzbach, S.; Tscharntke, T. More than Yield: Ecosystem Services of Traditional versus Modern Crop Varieties Revisited. *Sustainability* **2018**, *10*, 2834. [CrossRef]
195. Sayre, M.; Stenner, T.; Argumedo, A. You Can't Grow Potatoes in the Sky: Building Resilience in the Face of Climate Change in the Potato Park of Cuzco, Peru. *Cult. Agric. Food Environ.* **2017**, *39*, 100–108. [CrossRef]

196. Davidson-Hunt, I.J.; Turner, K.L.; Mead, A.T.P.; Cabrera-Lopez, J.; Bolton, R.; Idrobo, C.J.; Miretski, I.; Morrison, A.; Robson, J.P. Biocultural Design: A New Conceptual Framework for Sustainable Development in Rural Indigenous and Local Communities. *SAPIENS Surv. Perspect. Integr. Environ. Soc.* **2012**, *5*, 33–45.

197. Andrews, L. *A Dictionary of the Hawaiian Language*; Bilingual Edition; Printed by Henry M. Whitney, Honolulu, HI. 2003 Reprint; Island Heritage Publishing: Honolulu, HI, USA, 1865; ISBN 978-0-89610-374-0.

198. Chun, M.N. *No Nā Mamo: Traditional and Contemporary Hawaiian Beliefs and Practices*; University of Hawai'i Press: Honolulu, HI, USA, 2011; ISBN 978-0-8248-3624-5.

199. Zimmerer, K.S. Understanding agrobiodiversity and the rise of resilience: Analytic category, conceptual boundary object or meta-level transition? *Resilience* **2015**, *3*, 183–198. [CrossRef]

200. Akana, C.L.; Gonzalez, K. *Hānau Ka Ua: Hawaiian Rain Names*; Bilingual Edition; Kamehameha Schools Press: Honolulu, HI, USA, 2015; ISBN 978-0-87336-246-7.

201. Kimmerer, R.W. The Fortress, the River and the Garden. In *Contemporary Studies in Environmental and Indigenous Pedagogies*; Kulnieks, A., Longboat, D.R., Young, K., Eds.; Sense Publishers: Rotterdam, The Netherlands, 2013; pp. 49–76. ISBN 978-94-6209-293-8.

202. International Society of Ethnobiology. International Society of Ethnobiology Code of Ethics (with 2008 additions). Available online: http://ethnobiology.net/code-of-ethics/ (accessed on 9 August 2018).

203. University of Otago. *Pacific Research Protocols*; University of Otago: Dunedin, New Zealand, 2011; Available online: https://www.otago.ac.nz/research/otago028670.html (accessed on 11 October 2018).

204. Kūlana Noi'i Working Group Kūlana Noi'i v. 1 2018. Available online: http://seagrant.soest.hawaii.edu/wp-content/uploads/2018/06/Kulana-Noii-low-res-web.pdf (accessed on 30 September 2018).

205. Research Ethics. Available online: http://www.indigenousgeography.net/ethics.shtm (accessed on 11 October 2018).

206. Fals-Borda, O.; Rahman, M.A. *Action and Knowledge: Breaking the Monopoly with Participatory Action Research*; Rowman & Littlefield Publishers: New York, NY, USA, 1991; ISBN 978-0-945257-57-8.

207. College of Tropical Agriculture and Human Resources CTAHR Philosophy on Agricultural Technology. Available online: https://cms.ctahr.hawaii.edu/CTAHRPhilosophy (accessed on 19 August 2018).

208. UH Mānoa Office of the Chancellor. *Defining Our Destiny: Strategic Plan 2002–2010*; University of Hawai'i at Mānoa: Honolulu, HI, USA, 2002; Available online: https://manoa.hawaii.edu/strategicplan/dod-2002-2010/pdf/DOD_English.pdf (accessed on 8 October 2018).

209. Hawaiian Place of Learning Implementation Task Force. *Ka Ho'okō Kuleana: Fulfilling Our Responsibility to Establish the University of Hawai'i at Mānoa as a Hawaiian Place of Learning, An Implementation Report for the Ke Au Hou Recommendations*; University of Hawai'i at Mānoa: Honolulu, HI, USA, 2016; Available online: https://manoa.hawaii.edu/strategicplan/ (accessed on 8 October 2018).

210. Kame'eleihiwa, L. How Do We Transform the University of Hawai'i at Mānoa into a Hawaiian Place of Learning? Generational Perspectives: Part 1. *Hūlili* **2016**, *10*, 203–225.

211. Lipe, K. How Do We Transform the University of Hawai'i at Mānoa into a Hawaiian Place of Learning? Generational Perspectives: Part 2. *Hūlili* **2016**, *10*, 227–243.

212. United Nations. *United Nations Declaration on the Rights of Indigenous Peoples*; United Nations: New York, NY, USA, 2007.

213. Ermine, W. The Ethical Space of Engagement. *Indigenous L.J.* **2007**, *6*, 193–204.

214. Ermine, W.; Sinclair, R.; Jeffery, B. *The Ethics of Research Involving Indigenous Peoples*; Indigenous Peoples' Health Research Centre: Saskatoon, SC, Canada, 2004.

215. Smith, G.H. Indigenous Struggle for the Transformation of Education and Schooling. In Proceedings of the Alaskan Federation of Natives Convention, Anchorage, AK, USA, 23 October 2003.

216. Bentley, J.W. Facts, fantasies, and failures of farmer participatory research. *Agric. Hum. Values* **1994**, *11*, 140–150. [CrossRef]

217. Brown, T.; Wyatt, J. Design Thinking for Social Innovation. *Stanford Social Innovation Review*, 23 November 2009.

MDPI
St. Alban-Anlage 66
4052 Basel
Switzerland
Tel. +41 61 683 77 34
Fax +41 61 302 89 18
www.mdpi.com

Sustainability Editorial Office
E-mail: sustainability@mdpi.com
www.mdpi.com/journal/sustainability